# Wood Properties and Processing

# Wood Properties and Processing

Special Issue Editor

**Miha Humar**

MDPI • Basel • Beijing • Wuhan • Barcelona • Belgrade • Manchester • Tokyo • Cluj • Tianjin

*Special Issue Editor*
Miha Humar
University of Ljubljana
Slovenia

*Editorial Office*
MDPI
St. Alban-Anlage 66
4052 Basel, Switzerland

This is a reprint of articles from the Special Issue published online in the open access journal *Forests* (ISSN 1999-4907) (available at: https://www.mdpi.com/journal/forests/special_issues/ Wood_properties).

For citation purposes, cite each article independently as indicated on the article page online and as indicated below:

LastName, A.A.; LastName, B.B.; LastName, C.C. Article Title. *Journal Name* **Year**, *Article Number*, Page Range.

ISBN 978-3-03928-821-2 (Hbk)
ISBN 978-3-03928-822-9 (PDF)

Cover image courtesy of Miha Humar.

# Contents

# About the Special Issue Editor

**Miha Humar**, Profesor of Wood Science and Technology at University of Ljubljana.

# Preface to "Wood Properties and Processing"

Wood is one of the most important building materials, and its importance has been increasing in recent decades. This trend is also likely to continue in the future. This book addresses relevant problems in wood science and technology. These problems are not limited to a certain country or territory but are generally experienced all over the world. In the first part, contributions addressing the question of how to assess wood quality in forests. Novel non-destructive techniques offer cost- and time-effective solutions. In the second part, contributions are dedicated to primary wood processing, wood drying, machining, and the development and performance of advanced wood-based composites. The third section is dedicated to the performance of wood in outdoor applications. Special emphasis is given to esthetic performance. The fourth section is devoted to the development of final products (furniture) and market analysis. The last section examines the use of bamboo-based materials in the harsh conditions of cooling towers. Thus, this book brings good insight into the recent developments within wood science and technology.

**Miha Humar**
*Special Issue Editor*

 *forests*

*Article*

# Effect of Rotation Age and Thinning Regime on Visual and Structural Lumber Grades of Douglas-Fir Logs

**Eini C. Lowell [1,*], Eric C. Turnblom [2], Jeff M. Comnick [2] and CL Huang [3]**

[1]   USDA Forest Service Pacific Northwest Research Station, 620 SW Main Street Suite 502, Portland, OR 97205, USA
[2]   School of Environmental and Forest Sciences, College of the Environment, University of Washington, Seattle, WA 98195, USA; ect@uw.edu (E.C.T.); jcomnick@u.washington.edu (J.M.C.)
[3]   School of Environmental and Forest Sciences, College of the Environment, Formerly (While Contributing to This Project) University of Washington, Seattle, WA 98195, USA; clhuang@uw.edu
*   Correspondence: elowell@fs.fed.us; Tel.: +1-503-808-2072

Received: 24 July 2018; Accepted: 8 September 2018; Published: 18 September 2018

**Abstract:** Douglas-fir, the most important timber species in the Pacific Northwest, US (PNW), has high stiffness and strength. Growing it in plantations on short rotations since the 1980s has led to concerns about the impact of juvenile/mature wood proportion on wood properties. Lumber recovered from four sites in a thinning trial in the PNW was analyzed for relationships between thinning regime and lumber grade yield. Linear mixed-effects models were developed for understanding how rotation age and thinning affect the lumber grade yield. Log small-end diameter was overall the most important for describing the presence of an appearance grade, generally exhibiting an indirect relationship with the lower quality grades. Stand Quadratic Mean Diameter (QMD) was found to be the next most uniformly important predictor, its influence (positive or negative) depending on the lumber grade. For quantity within a grade, as log small-end diameter increased, the quantity of the highest grade increased, while decreasing the quantity of the lower grades differentially. Other tree and stand attributes were of varying importance among grades, including stand density, tree height, and stand slope, but logically depicted the tradeoffs or rebalancing among the grades as the tree and stand characteristics change. Structural lumber grade presence was described best by acoustic wave flight time, log position (decreasing presence in upper logs), and an increasing presence with rotation age. A smaller set of variables proved useful for describing quantity within a structural grade. Forest managers can use these results in planning to best capture value in harvesting, allowing them to direct raw materials (logs) to appropriate manufacturing facilities given market demand.

**Keywords:** Douglas-fir; lumber; non-destructive testing; modulus of elasticity (MOE); stiffness; thinning; silviculture

---

## 1. Introduction

Douglas-fir (*Pseudotsuga menziesii* (Mirb.) Franco), the most important commercial timber species in the Pacific Northwest (PNW), is predominantly recognized for its stiffness and strength [1]. About 70% of the harvested Douglas-fir is for lumber products, which includes less than 5% machine graded lumber (machine-stress-rated (MSR) and machine-evaluated-lumber (MEL)). Due to its value, intensively managed stands in the PNW are primarily Douglas-fir [2]. Intensive management and an improved genetic stock have increased the growth and yield amounts and tree size in young Douglas-fir plantations. Geneticists are also studying the heritability of the stiffness trait [3,4]. What has not been addressed fully are the effects of this management choice on wood quality. In the 1980s, an emphasis on volume production and short rotations in plantations led to concerns about the proportion of juvenile wood to mature wood [5,6] and its impact on stiffness and strength. Properties of juvenile wood, such as a lower wood density and a higher microfibril angle [7], can make it unsuitable for higher value, structural products. Megraw [8] reported that there was a broader juvenile wood zone than that found in other species. Increasing the complexity of determining the impact of juvenile wood on wood quality are the findings of Abdel-Gadir and Krahmer [9] who wrote that variation in the age of wood density maturation for Douglas-fir ranged from 15–38 years old. Aubry [10] found that wood density significantly influenced economic value using MSR grading rules. A symposium held in 1985, "Douglas-fir Stand Management for the Future", spoke to these concerns as did a report prepared by Forintek Canada Corp. for the British Columbia (BC) Ministry of Forests Douglas-fir Task Force [11].

Faster grown trees have a higher proportion of juvenile wood in the core. Barrett and Kellogg [12,13] found a decline in strength and stiffness properties of second-growth Douglas-fir 5.1 × 10.2 cm (2 × 4 in) lumber relative to established standards for young-growth Douglas-fir and related it to the proportion of juvenile wood. They examined changes in the Modulus of Elasticity (MOE, or stiffness) and Modulus of Rupture (MOR, or strength) based on visual grade, log position, and percent of juvenile wood and found that MOE and MOR decreased with increasing height in the tree and with an increased overall percentage of juvenile wood.

In examining the product potential of Douglas-fir from young-growth, managed stands, Fahey [14] conducted a lumber recovery study in western Oregon, USA (OR) and Washington, USA (WA). They found that knot size and the amount of juvenile wood had a significant impact on the yields of visually and machine-stress-rated lumber and visually graded veneer. This study demonstrated that there can be a wide range of wood quality within the young-growth resource as a result of the management strategies employed and confirmed the results from other studies [11,13,15] that examined the structural properties of lumber manufactured from juvenile wood.

Knots are another wood quality concern, as noted in several research studies [14,16]. Silvicultural regimes that promote fast grown trees, such as wide initial spacing, also impact crown length, rate of crown recession [17], and branch longevity thus attainable branch size [18]. Weiskettel [19] found the maximum branch size to be very responsive to silvicultural treatment and Brix [20] saw thinning effects predominately in the bottom half of the crown. Predicting branch size has been the focus of several studies including those by Maguire [21,22] and Briggs [23]. The timing of thinning is also influential. Pre-commercial thinning in younger stands will have more of an impact on branch size in the lower bole [24] than a thinning conducted later (e.g., 40 years or more) [25].

Branches translate to knots in products and are considered defects that impact both the visual grades and structural properties. Visual lumber grading rules [26] have criteria for knot size, location (center or edge), number, and condition (sound or unsound) for a given width board in assigning a grade. In a study by Middleton and Munro [27], knots prevented lumber from being assigned to the highest grade Select Structural about 30% of the time. Grain deviation around knots has a strength- and stiffness-reducing effect [5].

Barratt and Kellogg [13] found that it was hard to recognize lumber from second-growth trees with high stiffness and strength by visual grades. A continued reliance on visual grades for Douglas-fir lumber grading may be due, in part, to its intrinsic microfibril angle (MFA) patterns. When compared to other species, the volume of low MFA or low shrinkage wood in a Douglas-fir log is large, which renders a stable lumber product [14]. Therefore, the lumber value of a Douglas-fir tree is mainly driven by the size (volume) of the log. The volume of logs in a tree is principally affected by tree diameter, height, and taper/form, all of which can be impacted by silviculture. These findings have led to additional research on the ability to predict lumber quality from a standing tree or bucked log attributes. Briggs [28] found that measuring the largest branch in the breast height region or calculating the branch index (the average of the largest four branches in each quadrant) could be used to predict product (lumber or veneer) quality in the first 4.9 m (16 ft) log (butt log) and was easy to measure. The use of non-destructive testing (NDT) tools (acoustic velocity) for predicting the potential of a log or a tree to produce stress-rated lumber is also becoming more common [23,29–35]. Branches can influence acoustic readings as Amishev and Murphy [36] found a negative correlation between branches size and acoustic velocity. They also noted that branches accounted for some of the variation noted in the acoustic measurements.

The impacts of the increased mix of juvenile wood and branch size on the performance of lumber products are of concern to the wood products industry. A diverse structural-grade yield among different plantation stands is not uncommon. The range of MOEs found among logs of the same morphology or grade is very large and with the increasing amount of juvenile wood in the log market, it is becoming more challenging to find high stiffness logs for mills producing structural and engineered wood products (EWP) in the PNW [6].

The effects of rotation age and cultural treatments on lumber grade recovery and the effects of MOE on different lumber products is presented first, followed by findings from exploratory modeling efforts to further understand and explain more rigorously how rotation age and other factors (site, silviculture, tree characteristics) act together to determine the presence and amount of lumber in particular grade classes. These results will assist land managers (a) in assessing if stands and stand treatments are within desired specifications and (b) in making improved marketing decisions. The acoustic data related to first-log lumber MOE results were previously summarized [37]. Here we quantify the distribution of lumber grades by silviculture and rotation age for all logs produced.

## 2. Materials and Methods

### 2.1. Study Sites

The Stand Management Cooperative (SMC), based at the University of Washington, established long-term research installations designed to address the effects of forest management regimes and silvicultural treatments on stand and tree growth and development. The SMC Type II installations were designed to provide data representative of plantations reaching a commercial thinning stage of development at the time of study establishment [38]. In 2006, these installations had reached the end of their designed measurement cycle and land owners were free to harvest them. Four of these installations (numbered 803, 805, 807, and 808) representing a wide geographic range and two rotation age levels (third and fifth decades) were selected for this study (Figure 1) in order to assess the relationship between lumber quality and stand/tree/log variables and to assess the effects of thinning on lumber stiffness at rotation (final harvest).

Each installation contained five plots (0.4 ha) representing different thinning regimes (Table 1) all harvested in the same calendar year (2006) to provide the material for this study. Thinnings were triggered (implemented) when the stand attained a particular value of Curtis' [39] stand relative density (RD). Curtis' relative density measures the extent that trees have captured available growing space. In the case of Douglas-fir, stands that have a measured RD lower than ~15 are essentially open-grown.

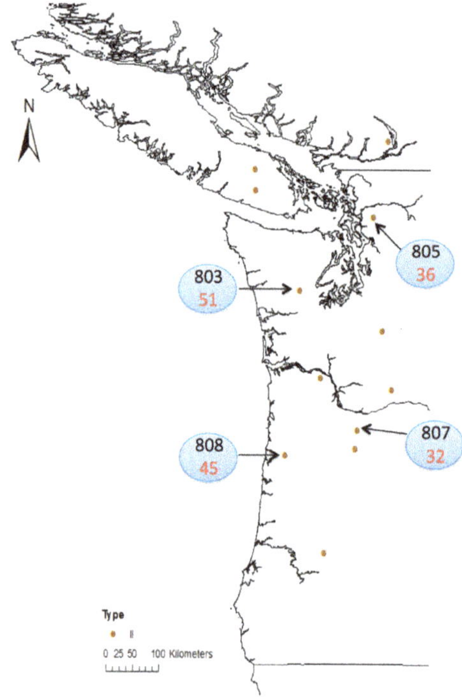


**Figure 1.** Locations of the Stand Management Cooperative (SMC) installations selected for this study. The numbers in red are the rotation ages of the stands when harvested.

**Table 1.** Thinning regime, Relative Density triggers, and thinning dates (with corresponding stand ages) with the last row containing the rotation ages (final harvest ages in years since planting) for each installation. Study trees came from the final harvest.

| Treatment Code | Thinning Regime | RD [a] Trigger Sequence | Installation Thinning Dates (Age at Thinning) | | | |
|---|---|---|---|---|---|---|
| | | | 803 | 805 | 807 | 808 |
| A | No thinning (Control) | | none | none | none | none |
| B | Thin heavy once | RD55-RD30; no further thinning | 1987 (33) | 1990 (21) | 1989 (15) | 1991 (31) |
| C | Delayed thinning | RD65-RD35; no further thinning | none | none | 1993 (19) | 1993 (33) |
| D | Repeated, heavy thinning | RD55-RD30; subsequent thinnings RD50-RD30 | 1987 (33) | 1990 (21) 2004 (34) | 1989 (15) 2001 (30) | 1993 (33) |
| E | Repeated, light thinning | RD55-RD35; RD55-RD40; subsequent thinnings RD60-RD40 | 1987(33) | 1996(27) | 1989(15) | 1991(31) |
| | Rotation age | (Final harvest age) | 51 | 36 | 45 | 32 |

[a] RD = relative density [39].

Stands in which the RD is between ~15 to 30 are growing large individual trees rapidly, but generally at the expense of per hectare production. Stands that have a measured RD between 30 and 55 may have slower individual tree growth rates but are still increasing the per hectare production with increases in RD. Whereas stands beyond an RD of 55 are beginning to lose stems due to competition-induced mortality though the individual tree size continues to increase as does the per hectare yield. Generally speaking, for the objective of producing wood volume, one strategy is to commercially thin either once or several times to earn income during a rotation, which serves at the same time to maintain or enhance tree and stand growth rates after thinning. If multiple thinnings are contemplated, a general strategy is to thin less frequently and more lightly as the stand ages. For example, the thinning regime behind treatment code D allows a stand to just reach the imminent competition-induced mortality boundary and thins back nearly to a condition where the per hectare production may be sacrificed; subsequent thinnings maintain an overall lower density stand condition by keeping the RD no higher than 50.

Site, stand, and average tree characteristics by treated plot at the age when harvested appear in Table 2. The delayed thinning plot (C) of installation 808 was lost due to a windstorm, so no data were available for this treatment from that site.

Table 2. Selected site, stand and tree characteristics for the four installations by plot.

| Inst. (elev, m) | Plot | SI [a] | Density | BA [b] | QMD [c] | Ht | Avg Stem Taper | LCR [d] | LLAD Butt Log [e] |
|---|---|---|---|---|---|---|---|---|---|
| (slope, %) | | m@50y | trees/ha | $m^2$/ha | cm | m | cm/m | % | cm |
| 803 | A | 36 | 791 | 56 | 30.0 | 36.6 | 1.05 | 30 | 2.54 |
| (585) | B | 37 | 306 | 45 | 43.4 | 39.0 | 1.12 | 37 | 2.54 |
| (1) | C | 35 | 899 | 52 | 26.9 | 35.1 | 0.97 | 29 | 1.78 |
| | D | 35 | 336 | 45 | 41.4 | 35.4 | 1.17 | 33 | 3.30 |
| | E | 34 | 459 | 47 | 36.1 | 33.8 | 1.05 | 30 | 1.78 |
| 805 | A | 38 | 860 | 52 | 27.7 | 30.8 | 0.94 | 36 | 4.57 |
| (168) | B | 40 | 454 | 40 | 33.5 | 31.4 | 1.03 | 37 | 3.05 |
| (15) | C | 39 | 366 | 33 | 34.0 | 32.3 | 1.00 | 41 | 4.32 |
| | D | 41 | 420 | 42 | 35.6 | 32.0 | 1.24 | 39 | 2.79 |
| | E | 39 | 405 | 36 | 33.8 | 32.3 | 1.12 | 39 | 5.33 |
| 807 | A | 33 | 1398 | 49 | 21.1 | 24.1 | 1.13 | 27 | 1.27 |
| (152) | B | 33 | 741 | 39 | 25.9 | 24.7 | 1.12 | 34 | 2.03 |
| (1) | C | 30 | 825 | 36 | 23.4 | 23.2 | 1.12 | 37 | 2.54 |
| | D | 35 | 395 | 27 | 29.2 | 25.9 | 1.17 | 43 | 3.05 |
| | E | 37 | 929 | 40 | 23.4 | 27.1 | 1.05 | 33 | 0.76 |
| 808 | A | 34 | 731 | 62 | 32.8 | 31.1 | 1.28 | 37 | 2.03 |
| (762) | B | 33 | 296 | 44 | 43.4 | 30.9 | 1.42 | 47 | 2.79 |
| (5) | C | - | - | - | - | - | - | - | - |
| | D | 31 | 247 | 42 | 46.7 | 28.7 | 1.57 | 51 | 2.29 |
| | E | 33 | 351 | 50 | 42.4 | 30.8 | 1.46 | 39 | 3.05 |

[a] SI = site index; [b] BA = basal area; [c] QMD = quadratic mean diameter; [d] LCR = live crown ratio; [e] LLAD = largest limb average diameter in first 4.9 m (16 ft).

## 2.2. Tree, Log, and Lumber Measurements

As there is a relationship among the NDT values of trees, logs, and lumber [34], TreeSonic velocity (TSV) was measured on the standing tree at breast-height as time-of-flight (nanoseconds) of an acoustic wave over a 1 m distance using the Fakopp TreeSonic instrument on 50 plot-centered trees on each of the plots. Mill trial trees were selected based on the distribution of the TSV. Twelve trees were selected from each plot. Two, four, four, and two trees from each plot were randomly selected from the following four TSV categories: The lowest 10%, medium-low 11–50%, medium-high 51–90%, and the top 10%, respectively for processing into lumber and veneer. Six trees were randomly chosen using the above TSV distribution and allocated to the lumber recovery study. The lumber recovery trees were bucked into 10 m (33 ft) logs in the woods, delivered to the South Union Sawmill in Elma, WA, and cut into 4.9 m (16 ft) logs in the log yard. From each tree, the resonant acoustic velocity of the merchantable bole and the 10 m (33 ft) length logs was measured in the woods, and the 4.9 m (16 ft) logs were measured at the sawmill using the Director HM200 [34]. Logs were processed into predominantly 5.1 × 10.2 cm (2 × 4 in) and 5.1 × 15.2 cm (2 × 6 in) lumber. Most of the MSR and MEL lumber produced were of these sizes. In addition, the location and size of the largest knot in each quadrant of the 4.9 m (16 ft) log segments were measured to calculate the large limb average diameter (LLAD) also known as branch index (BIX). The location and size of any ramicorn branches were also recorded.

Logs were sawn using a Mighty Mite circular saw (7.1 mm or 0.28 in saw kerf) with horizontal edger blades. Each piece of lumber was labeled to identify the tree and log it came from as well as the log position within the tree. The lumber was kiln-dried and surfaced. Finished lumber was visually graded by a certified lumber grader from the Western Wood Products Association and grade-limiting defects were recorded for each piece. All lumber was shipped to the USDA Forest Products Laboratory (FPL), Madison, WI for MOE determination using the MetriGuard e-computer. All data were collected from fall 2006 through spring 2007.

The MOE was adjusted to 15% moisture content [40] to calculate the volume-weighted log MOE. The percentage of lumber that met the MSR/MEL grade requirements was calculated based on the moisture content adjusted MOE.

### 2.3. Data Analysis

First, the branch index, grade-limiting defect, and lumber stiffness were summarized by the stand and silviculture regime. Next, the lumber grade distributions among the silviculture regimes and MOE distributions by lumber grade were examined more rigorously through an explanatory modeling effort. One set of equations was developed to assess the effects of site, stand, tree and log attributes on the proportion of log volume by visual lumber grade and another set of equations generated to assess how the same attributes affect the proportion of lumber volume that meets the structural design specifications for each grade. Each set of equations was developed using a two-step process.

The proportion of log volume in a visual grade was modeled first. In the first step, a model was developed to predict the presence of a grade within a log. The presence was indicated with a one (1), absence with a zero (0). In the second step, the abundance of the grade was estimated given that the grade was present. Although on the surface our data contained what appeared to be a very large number of observed zeroes, which indicate the absence of a grade within a log, methods to account for such a condition [41] showed no improvement to the fit when the model was recast as a fractional regression. Therefore, a generalized linear mixed-effects model was chosen to describe the presence of a grade, linked to a logistic error distribution, which maximized the likelihood of the parameters when the response is Bernoulli. The model appears in Equation (1).

$$p = \frac{1}{\left(1 + e^{-(\beta_0 + b_{0i} + \beta_1 X_1 + \beta_2 X_2 + \ldots)}\right)} + \delta_i \tag{1}$$

where $p$ denotes presence ($p = 1$) or absence ($p = 0$) of a grade within a log, $e$ denotes the base of the natural logarithm, $X$s denote the set of predictor variables examined, $\beta$s are the fixed model coefficients, $b_{0i}$ are random deviations due to the plot from the fixed component of the model coefficient, $\beta_0$, and $\delta_i$ are random error terms describing the residual variation unexplained by the predictor variables and random plot effects.

In the second step, an abundance model to predict the proportion of log volume in a particular visual grade given its presence was developed using a linear mixed-effects regression model. Since the proportion of a particular grade within a log is, by definition, any number between zero (0) and one (1), the log odds-ratio transformation (logit) was applied to the observed response values to normalize their distribution. The model form appears in Equation (2).

$$logit(\theta) = \beta_0 + b_{0i} + \beta_1 X_1 + \beta_2 X_2 + \ldots e_i \tag{2}$$

where $\theta$ denotes the proportion of log volume in one of the grades, logit($\theta$) denotes the natural logarithm of the ratio of the proportion in the grade to the proportion that is not, or "log odds-ratio", $X_i$s denote the set of predictor variables examined, $\beta_i$s are fixed model coefficients, $b_{0i}$ is the random deviation due to the plot from the fixed component of the model coefficient, $\beta_0$, and $e_i$ are the random error terms describing the residual variation unexplained by the predictor variables and random plot effects.

The design values were assigned [42] to meet the engineering requirements of the intended end use of the lumber (structural capability). They differ not only by end use but also by species and are influenced by such features as knots and slope of grain. The size of lumber was also a consideration in assigning design value. Structural lumber (including dimension lumber) can be visually and/or mechanically (MSR) graded for its strength and physical working properties. The set of models, derived to estimate the proportion of lumber volume meeting the structural design values for Douglas-fir, were derived similarly to the visual grade models.

In the first step, a model was derived to predict the presence of lumber meeting the structural value within a grade using the same model as Equation (1), while the second step model estimated the proportion of volume meeting the structural design value given that it was present using the same model as Equation (2). A two-step modeling process, such as used here, has previously been used quite successfully in other contexts where the conceptual framework is analogous (see for example Reference [43]).

Installation, or geographic location, effects were accounted for as fixed effects in the models in the form of site attributes, such as slope, aspect, elevation, among others. The effects of the plot were considered random in all fitted models, to assess and characterize the magnitude of uncontrollable noise, i.e., variation that is unaccounted for by the treatments applied. The only exceptions to this were the Economy visual grade and No. 3 structural grade abundance models, each of which lacked a sufficiently large sample size to assess the plot variation adequately. Thinning methods were expected to express their influences in the form of differing stand density, basal area, and average stand diameter that were attained over the course of time through stand dynamics processes as moderated by silvicultural thinning. Tree variables (Diameter at Breast Height [DBH] total height, taper, height-diameter ratio) were considered to be fixed, measurable effects. Each model set was developed using a forward selection of variables, with a chosen significance level of 0.1 for all models. Given the high level of variation observed among plots and trees, this less conservative significance level was chosen in order to capture all important variables influencing the presence and abundance of the grades. Twenty candidate predictor variables were evaluated for each model including treatment (silviculture regime and harvest age), site (latitude, longitude, slope, aspect, and elevation), plot (Trees Per Hectare [TPH], Quadratic Mean Diameter [QMD], basal area, relative density, site index), tree (DBH, height, height-DBH ratio, taper, and acoustic velocity), and log (small-end diameter, position along the stem, and LLAD) attributes. Only main effects were considered. The lme4 R package was used for fitting the models [44].

For binary presence/absence models, methods to assess the fit in a meaningful way are not well defined, which also holds true for logit models, or models where proportions represent the response. We chose the following method for evaluating the overall combined fit for the two-model sets. After fitting the models, the mean predicted abundance for each grade was determined using a Monte Carlo simulation. In this process, a random number between zero and one was generated 500 times per grade for each log in the dataset and compared to the modeled probability of presence. The abundance was then tallied as either the predicted abundance when the random number did not exceed the modeled presence probability or zero otherwise. The results were averaged to calculate the mean response per grade for each log. Finally, the predicted abundance for all grades within a log was scaled proportionally to sum to one. The set of equations were evaluated as a system by comparing these predictions to the observed values by calculating the following set of fit statistics: The adjusted R-squared, root mean squared error, mean absolute deviation, mean bias, and mean percent error.

## 3. Results

### 3.1. Initial Data Summary

A total of 1758 pieces of lumber were sawn from 317 logs out of 97 trees. The reduced number of trees from the 112 trees originally selected was primarily due to weather conditions on installation 807 (13 trees were not transported to the mill) and the inability to saw the remaining four trees due to defects (sweep) and size (log small-end scaling diameter). About 25% of the lumber produced was 5.1 × 10.2 cm (2 × 4 in) and the remaining 75% was 5.1 × 15.2 cm (2 × 6 in). A very small amount of 2.5 cm (1 in) lumber was sawn. Restricting the lumber sizes allowed for a better assessment of the impact of knots on the lumber grade. The yield of No. 2 and better (a grade grouping often found in marketing Douglas-fir) lumber ranged from 91 to 95%. Table 3 shows the lumber data and grade yield from the study.

**Table 3.** Sample data and percent volume yield by lumber grade of the sampled installations.

| Inst. | Age | Tree | DBH | Height | Logs Processed | Lumber Pieces | Lumber Grade | | | | |
|-------|-----|------|-----|--------|----------------|---------------|--------------|------|------|------|------|
| | yr | n | cm | m | n | n | | | pct | | |
| | | | | | | | Sel Str [a] | No. 1 | No. 2 | No. 3 | Econ [b] |
| 803 | 51 | 29 | 39.11 | 35.97 | 119 | 718 | 29 | 40 | 26 | 4 | 1 |
| 805 | 36 | 27 | 35.05 | 31.70 | 77 | 368 | 23 | 44 | 29 | 3 | 2 |
| 807 | 32 | 17 | 28.70 | 24.69 | 33 | 119 | 7 | 35 | 49 | 6 | 3 |
| 808 | 45 | 24 | 43.43 | 30.18 | 88 | 553 | 23 | 37 | 31 | 7 | 2 |
| Total | | 97 | | | 317 | 1758 | | | | | |

[a] Sel Str = select structural lumber grade; [b] Econ = economy lumber grade.

### 3.1.1. Branch Index

Another of the main factors directly affecting the lumber grade is the size of knots, which start as branches on the tree. The branch index (LLAD) on the bottom 4.9 m (16 ft) log ranged from a low of 1.8 cm (0.7 in) to a high of 5.3 cm (2.1 in). Douglas-fir is not prone to self-pruning dead branches below the live crown [22]. Thus, thinning that impacts the crown structure can impact knot type and size. The butt log typically contains the highest value lumber.

### 3.1.2. Grade-Limiting Defect

Wane (85%) and knots (6%) were the dominant reasons for trimming lumber (about 190 pieces or 11%) to increase the grade (Figure 2a). These were also the two factors that were the grade-limiting defects, causing the lumber to be downgraded (Figure 2b). Intensive management can increase taper in a tree, especially in the upper stem, that leads to the presence of wane in lumber. Forty-two percent of the lumber was downgraded for wane. Face knots (the primary grade-limiting knot type) accounted for 37% of the downgrade. Spike and edge knots accounted for an additional 16% of the downgrade.

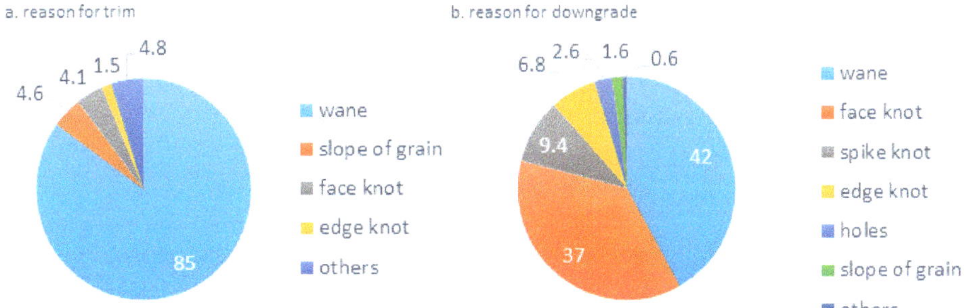



**Figure 2.** Percent of end-trim of lumber (**a**) and visual grade downgrade factors (**b**) for all lumber combined.

### 3.1.3. Lumber Stiffness

The volume-weighted MOE by log position of the tested installations is shown in Figure 3. The MOE exhibits gradients from the tree base to the tree top, being more variable in the younger stands (805 and 808). The log lumber MOE is related to other variables besides log diameter and log position, including juvenile wood proportion and wood density. Upper segments near the top of the tree (e.g., segment 4) generally have a higher proportion of juvenile wood and a lower wood density.

The percentage of Douglas-fir visual grade lumber that meets the design value is related to the amount of juvenile wood [13]. In this study, about 50% of the visually graded lumber met or exceeded the design value [26] (Figure 4). Of note is the small sample size of No. 3 grade lumber and the number of 5.1 × 10.2 cm (2 × 4 in) lumber that met the Select Structural grade.

Except for installation 803, the higher grades of the tested plantation lumber had average MOE values that fell below the published MOE design values (Table 4).

**Table 4.** Average density (weighted by lumber volume) and lumber MOE by site and visual grade (number in parentheses is MOE design value of the grade).

| Site | Density | MOE | Sel Str (13,100) | No. 1 (11,721) | No. 2 (11,032) | No. 3 (9653) | Econ |
|---|---|---|---|---|---|---|---|
| | (kg/m$^3$) | | | MPa | | | |
| 803 | 569 | 13,334 | 14,162 ^ | 12,473 * | 12,638 * | 12,555 ** | 12,052 |
| 805 | 551 | 11,625 | 12,114 | 11,438 | 11,052 ^ | 10,587 ** | 11,101 |
| 807 | 521 | 10,004 | 11,749 | 9694 | 9894 | 11,018 * | 10,949 |
| 808 | 580 | 11,521 | 13,017 | 11,321 | 10,839 | 9218 | 10,018 |

^ MOE average meets the specification but its distribution does not; * MOE average meets MEL; ** MSR meets MOE specifications (the amount of below-grade MOE pieces is more lenient for MEL grading rules).

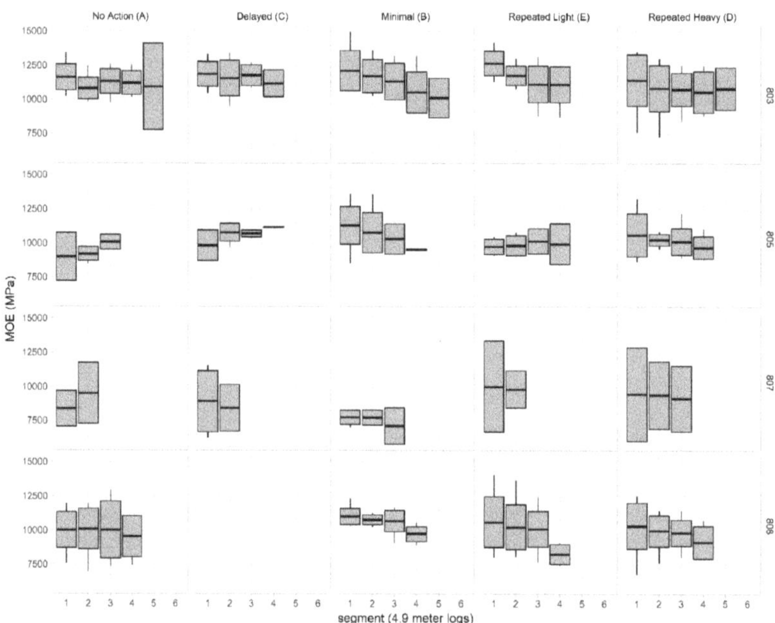

**Figure 3.** Log lumber MOE by segment (log position within the tree with the butt log being segment 1) and thinning type by installation. Note: Segments are 4.9 m logs (16 ft logs).

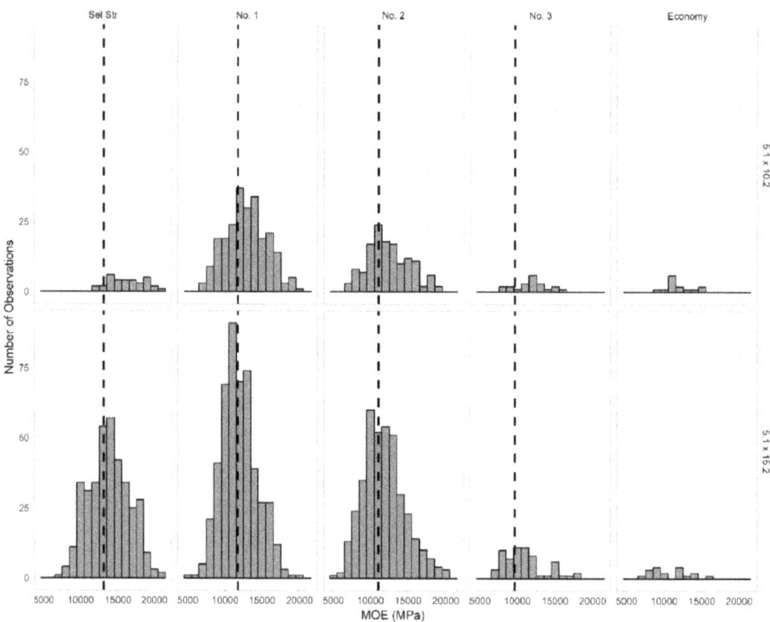

**Figure 4.** Distribution of lumber MOE by lumber size and visual grades. The dotted vertical line is the published MOE value for the specific grade. Economy lumber does not have a structural design value assigned to it.

The calculations used to derive the proportions meeting the design values relied only on the MOE limit, so if additional design limits were incorporated in those calculations, the amounts of below-grade lumber will probably be larger. One interesting observation of the test results is that when compared to pine species, the MOE of the lower visual grade Douglas-fir lumber was relatively high, which may be explained by the intrinsic stiffness property of Douglas-fir. There were many high MOE pieces within the visual grade plantation-grown lumber that keep the average MOE of the grade on par, but the amount of low MOE lumber in the distribution can fail the grading rules. The cause of the additional low MOE pieces could most likely be due to the increased proportion of juvenile wood within the log.

### 3.2. Modeling Summary

### 3.2.1. Lumber Grade Distribution

In the first step, the presence/absence of each grade within a log was modeled. This resulted in the selected predictor variables listed in Table 5 showing the coefficient estimates and significance. Intercepts were kept in all models, even if not significant, so that models remain unbiased and would at minimum be capable of predicting a mean presence value, in cases where there may be no significant tree or stand variables.

**Table 5.** Selected parameters and level of significance for lumber visual grade presence/absence models.

| Variable | Sel Str | No. 1 | No. 2 | No. 3 | Econ |
|---|---|---|---|---|---|
| Intercept | $-3.4227$ ** | 0.1283 | 1.4725 | $-4.6907$ *** | $-5.9568$ *** |
| Plot QMD | 0.0926 * | | $-0.0539$ * | | |
| Tree DBH | | | | 0.0943 *** | |
| Tree height | | | $-0.1158$ ** | | |
| Tree taper | $-4.3120$ *** | | | | |
| Log SED | 0.2006 *** | 0.0707 * | 0.1554 *** | | 0.1007 ** |
| Log LLAD | | | 0.1194 ** | | |
| Log position | | | | $-1.6142$ * | |
| $\sigma^2_{b_0}$ | 0.4847 | 0.0789 | 0.0097 | 0.0958 | 0.9674 |

Significance level symbols ***, **, and * indicate $p$-value ranges of $p < 0.001$, $0.001 < p < 0.01$, and $0.01 < p < 0.05$, respectively. Intercept terms were always included regardless of significance to maintain unbiasedness.

In the second step, the abundance of each grade within a log, given that it was present, was modeled. This resulted in the selected predictor variables listed in Table 6, showing the coefficient estimates and level of significance. Here as well, intercepts were kept in all models, even if not significant, so that models remain unbiased and will at minimum be capable of predicting a mean abundance value, in cases where there may be no significant tree or stand variables.

**Table 6.** Selected parameters and level of significance for the lumber visual grade abundance models.

| Variable | Sel Str | No. 1 | No. 2 | No. 3 | Econ |
|---|---|---|---|---|---|
| Intercept | 5.7471 ** | 1.9021 | −1.1781 + | −0.1635 | 1.620 |
| Installation slope | | 0.0985 + | | | |
| Plot TPA | | | 0.0009 * | | |
| Plot QMD | 0.0491 + | | | | |
| Plot site index | −0.1135 * | −0.1440 ** | | | |
| Tree DBH | −0.0481 * | | | | |
| Tree height | | 0.0994 *** | | | |
| Tree taper | −1.6227 + | | 1.3858 ** | | |
| Tree velocity | | 7.7888 * | | | |
| Log SED | | −0.1264 *** | −0.0593 *** | −0.0384 * | −0.2373 + |
| Log LLAD | | | 0.0773 * | | |
| $\sigma^2_{b_0}$ | 0.0661 | 0.0097 | 0.0132 | 0.0058 | NA |
| $\sigma^2_e$ | 1.4872 | 1.6036 | 1.3930 | 0.4116 | 1.8578 |

Significance level symbols ***, **, *, and + indicate *p*-value ranges of $p < 0.001$, $0.001 < p < 0.01$, $0.01 < p < 0.05$, and $0.05 < p < 0.1$, respectively. Intercept terms were always included regardless of significance to maintain unbiasedness.

As stated in the methods section, each set of equations was evaluated as a system by comparing the Monte Carlo predicted values to the observed values. Summary statistics were calculated based on 1575 data points (315 logs × 5 lumber grades) and 34 parameters (Table 7).

**Table 7.** Summary of the fit statistics for the final lumber visual grade model system.

| Statistic | Value |
|---|---|
| Adjusted R-squared | 0.4279 |
| Root Mean Squared Error | 0.2015 |
| Mean Absolute Deviation | 0.1409 |
| Mean Bias | $2.4939 \times 10^{-18}$ |
| Mean Percent Error | 0.7045 |

The behaviors of the visual grade models as a system were explored by comparing the effects of log small-end diameter (SED), harvest age, and treatment regime on the predicted grade proportions. To accomplish this, linear regression models were first developed to predict model input parameters, including TPH, QMD, tree height, and LLAD, from harvest age and treatment regime. Harvest ages were chosen to range from 30 to 55 years by 5-year steps. The log SED values were chosen to range from 10.2 to 45.7 cm (6 to 18 in) by 7.6 cm (3 in) steps. The log position parameter was chosen to be 0.25, representing the butt log of the tree. The remaining parameters were set to median values for the data set. The results are illustrated in Figure 5.

Log SED and harvest age have the largest effects on the model. As the small-end diameter of a log increases, the proportion of the Select Structural grade increases at the expense of the No. 1 grade, while No. 2 and the remaining grades stay relatively flat. For a given harvest age, grade No. 2 has a positive relationship with the log SED at its low end but turns negative for larger SEDs. Increasing the harvest age results in proportionally larger amounts of Select Structural in the lower SED range, but proportionally smaller amounts in the larger SEDs. Proportionally more No. 1 grade was produced over all SEDs as the harvest age increased. The No. 2 grade decreases proportionally over the range of the SED with harvest age, while No. 3 and Economy (E) were predicted in very small proportions in all scenarios.

The thinning regime appears to have a much smaller effect on the distribution of grade, for reasons stated previously. Grades Select Structural and No. 1 occurred in slightly larger proportions in a smaller diameter, denser stands for a given harvest age and small end diameter. These results may differ for absolute abundance, as logs with larger small-end diameters would be expected to occur more frequently in stands with a lower density (fewer trees per hectare that are larger in diameter for a given harvest age). This system of models provides a methodology to better understand the influences of silvicultural thinning on the tree and stand attributes that can be used directly to predict the proportion of lumber grades to expect under the different regimes. This will lead to greater accuracy and precision when appraising/valuing the resultant products produced.

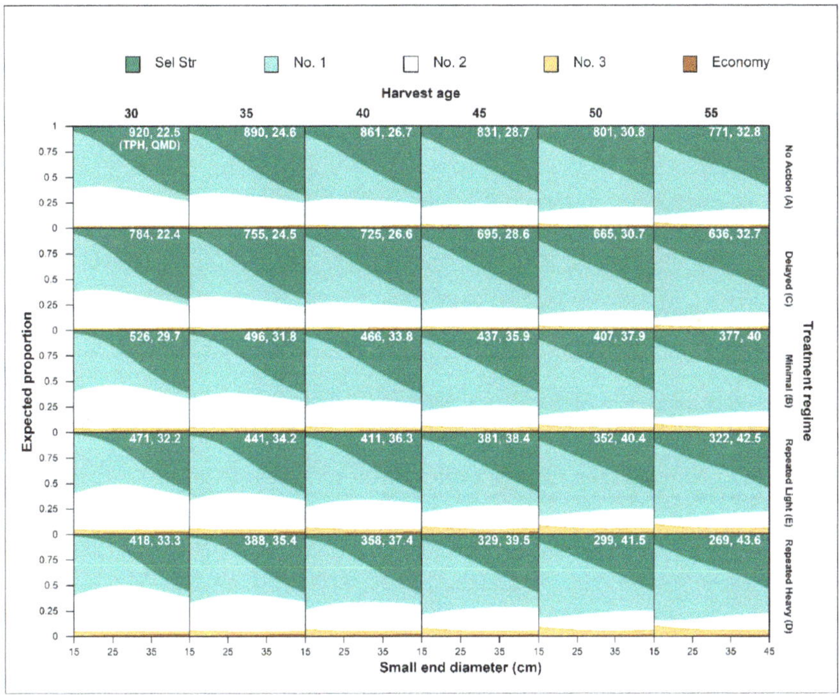


**Figure 5.** Expected proportion of log volume by lumber visual grade by harvest age, treatment regime, and log small end diameter. Trees per hectare (TPH) and quadratic mean diameter (QMD) for each rotation age by treatment panel are shown in the upper right corner.

### 3.2.2. Lumber Structural Grade

The proportion of lumber volume for a visual grade that met the MOE standard was calculated for each log. For presence/absence modeling purposes, presence was defined as any proportion greater than zero that met the MOE standard for a particular grade. Economy grade does not have a structural design standard. The selected variables and estimates of the coefficients are reported in Tables 8 and 9 for the presence/absence and abundance models, respectively.

**Table 8.** Selected parameters and level of significance for lumber structural grade presence/absence models.

| Variable | Sel Str | No. 1 | No. 2 | No. 3 |
|---|---|---|---|---|
| Intercept | −10.2863 ** | −3.5808 *** | −4.1380 ** | −29.5200 * |
| Harvest age | 0.0997 ** | 0.0642 ** | | |
| Tree height | | | 0.2028 *** | |
| Tree velocity | 21.5060 * | | | 83.0200 * |
| Log position | −3.0699 ** | | −2.1948 ** | |
| Log SED | | 0.0642 ** | | |
| $\sigma^2_{b_0}$ | 0.0629 | 0.0770 | 0.1012 | 3.5360 |

Significance level symbols ***, **, *, and + indicate *p*-value ranges of $p < 0.001$, $0.001 < p < 0.01$, $0.01 < p < 0.05$, and $0.05 < p < 0.1$, respectively. Intercept terms were always included regardless of significance to maintain unbiasedness.

**Table 9.** Selected parameters and level of significance for lumber structural grade abundance models.

| Variable | Sel Str | No. 1 | No. 2 | No. 3 |
|---|---|---|---|---|
| Intercept | −6.4760 + | −4.9053 + | 7.7580 *** | 9.495 *** |
| Tree velocity | 22.7280 ** | 16.6633 * | | |
| Tree taper | | | −3.6568 *** | −4.2934 *** |
| Log LLAD | | | −0.1663 * | |
| $\sigma^2_{b_0}$ | 0.2970 | $9.142 \times 10^{-16}$ | $4.113 \times 10^{-16}$ | NA |
| $\sigma^2_{e}$ | 4.4090 | 4.7690 | 4.3180 | 1.9061 |

Significance level symbols ***, **, *, and + indicate *p*-value ranges of $p < 0.001$, $0.001 < p < 0.01$, $0.01 < p < 0.05$, and $0.05 < p < 0.1$, respectively. Intercept terms were always included regardless of significance to maintain unbiasedness.

Overall, the structural grade models explained a very low amount of variation in the response variables. Summary statistics are reported in Table 10.

**Table 10.** Summary of the fit statistics for the two-equation structural grade model sets by visual grade.

| Statistic | Sel Str | No. 1 | No. 2 | No. 3 |
|---|---|---|---|---|
| Adjusted R-squared | 0.1533 | 0.0407 | 0.1405 | 0.3649 |
| Root Mean Squared Error | 0.3714 | 0.3804 | 0.3821 | 0.3092 |
| Mean Absolute Deviation | 0.3269 | 0.3320 | 0.3403 | 0.2153 |
| Mean Bias | $-1.601 \times 10^{-17}$ | $2.0497 \times 10^{-17}$ | $-5.7301 \times 10^{-19}$ | $-4.8720 \times 10^{-18}$ |
| Mean Percent Error | 0.4631 | 0.5559 | 0.4952 | 0.2539 |

## 4. Discussion

The data themselves are highly variable, leading to somewhat low R-squared values in all models, but the main objective, again, was assessing the significance of factor effects to explain the responses, not necessarily creating a model with a high precision for predictive use; though that remains an outcome to be desired and eventually achieved. It should be noted that although treatment regime variables were actually tested throughout in all the models, they were always supplanted by the actual stand and tree attributes at final harvest. This does not mean that treatments were ineffective in producing differences in log and lumber grades, only that treatment affects themselves appear indirectly through their accumulated impact over the rotation on the responses by way of their influence on stand dynamics processes [19,45].

Considering first the lumber visual grade presence model (Table 5), it is seen that the greatest single impact on a single grade is tree taper. Logically, as taper increases, the presence of the Sel Str grade is less likely. The single predictor that had influence over most of the grades was the small-end diameter of the log (log SED) variable. This is completely expected since this is the main driving variable in the visual log grading system that is used in the Pacific Northwest [46]. Log small-end diameter was most influential in the select structural lumber grade (largest coefficient), but it will be seen that a large diameter log benefits the presence of all grades (positive signs). The average diameter of the stand (plot QMD) positively influenced the presence of the Select Structural grade, as expected, further enhancing the positive effect of the log SED. For the Select Structural grade, a greater degree of taper negatively influenced its presence, likely due to less solid central wood in the log that is capable of producing higher grade lumber, i.e., a larger proportion of wood in jacket boards (lumber sawn from the outer portion of a log), slabs and edgings. The presence of the No. 2 grade was negatively impacted by stands with an overall larger plot QMD, and especially so if the tree was among the taller component. Though a large Tree DBH will help the probability that the No. 3 grade is present, it will be less likely in the upper logs. The Economy grade presence seems insensitive to all other stand and tree attributes, perhaps because it captures all the lumber not in the other grades.

When considering the abundance of visual grades, given their presence, we see again that log small-end diameter was the most important variable overall (Table 6), because it remained significant in four of the five grade models, which no other predictor variable did. Though the coefficients were all negative, we can interpret the magnitudes of the coefficients as indicating tradeoffs in lumber volume between grades. For example, a log with a large SED may produce all grades of lumber, but will produce the most Select Structural lumber (coefficient is zero, i.e., no negative impact from the SED), followed by No. 3 (smallest magnitude negative coefficient), then, No. 2 (next larger magnitude coefficient), No. 1 (even larger negative coefficient), and finally Economy (largest negative coefficient), respectively. The next overall most important variable might be considered to be either the plot site index or tree taper. A higher site index decreases the Select Structural and No. 1 grades, likely due to fewer rings per inch in the logs produced since the site index has been shown to be reflected visibly in rings per inch, the two variables being essentially interchangeable [47]. A greater tree taper negatively impacts the Select Structural grade, which seems to be offset by more No. 2 grade, another tradeoff. The slope of the ground at the site (installation slope) positively impacted the abundance of No. 1 grade lumber. This result was somewhat unexpected, and while further exploration of why this might be important is beyond the scope of this study, it is interesting to speculate how this and other environmental attributes or climatic variables may influence tree growth, wood production and subsequent lumber grade turnout; currently under investigation elsewhere [48]. Overall, stands with a larger average diameter produced relatively more Select Structural lumber, though it was tempered by individual tree DBH; its abundance was decreased to a greater degree if the tree had a DBH larger than QMD. As expected, grade No. 2 tolerates larger knots (LLAD) [26]. The net effect of high-density stands is to produce trees with less taper and seems to positively influence the abundance of No. 2 grade lumber. The magnitude of between-plot variation for the visual grade abundance models was quite small compared to residual error.

For the structural grade models (Tables 8 and 9), log position (inversely related to log diameter) or log SED were chosen for the Select Structural, No. 1, and No. 2 grades. Log diameter has been correlated to lumber grade recovery and thus value [27]. Harvest age and TSV (Tree Velocity) were each selected for multiple models. The presence of the Select Structural and No. 1 grades showed positive relationships with harvest age. This might be expected since as trees age, annual ring widths tend to become narrower, even if the growth rate doesn't slow because the annual wood layer would be laid down on an ever increasing diameter. This, in turn, would lead to an increased density in the outer rings (higher proportion of LW), leading to a greater stiffness. The tree velocity (TSV) showed a positive relationship with the response variables, also as expected, because the speed of an acoustic wave through wood is directly and positively correlated with wood stiffness; an important structural attribute [34]. The Select Structural and No. 1 grades of lumber were more likely to be present in larger diameter logs (occurring lower in the tree) harvested from older stands, as expected. The presence of No. 2 grade was more likely in logs from taller trees located lower in the bole. No. 3 grade presence was predicted by only the Tree velocity. The Economy grade is known not to yield any structural lumber, so there is no design value assigned to it.

The structural abundance models (Table 9), largely exhibited variables with signs that were easily interpreted. As expected, the greater the TSV, the greater the abundance of the Select Structural and No. 1 grades, given that they were present. Given the presence of the No. 2 and 3 grades, a greater tree taper reduced abundance. The abundance of the No. 2 grade was further negatively impacted when the LLAD was large, likely due to grain distortion around the knots. Fahey [14] also found that the LLAD influenced lumber grade recovery in Douglas-fir. The magnitude of plot-to-plot variation for the structural grade abundance models was relatively small compared to residual error.

Both sets of models for both the visual grade presence and abundance and presence and abundance of structural lumber within a grade clearly demonstrated that visual lumber grade alone is insufficient for predicting the actual quantity of lumber produced that meets the structural design values for each grade. The incorporation of other tree and stand variables, resulting from stand treatment, into the models helped the prediction of visual lumber grades more so than for structural lumber, as judged by the fit statistics evaluated (Tables 7 and 10).

## 5. Conclusions

Decision support tools need to be integrated at every step in the value chain, from stand management to log marketing. For a lumber mill, the amount of high MOE material is enough to satisfy the current small MSR/MEL market and the visual grade lumber is the main product. Therefore, the effects of low MOE wood on Douglas-fir lumber mills are relatively small as long as the majority of the lumber meets the visual specifications. On the other hand, the MOE is directly related to the value of engineered wood products (EWP), so for manufacturers producing EWP, the additional low MOE materials directly reduce mill profit. Not only does having surplus low MOE material cause waste in an EWP facility, but additional high MOE materials need to be purchased on the open market to fill customer orders. Unlike visually graded lumber, the internal strength and stiffness are the key value factors of EWP. Balancing the MOE in a log mix for EWP mills is getting more complicated with the increasing amount of low MOE juvenile wood in the wood basket.

The use of non-destructive, in-woods testing equipment to measure acoustic velocity was found to be the most important variable for predicting the presence of structural grades in the lumber produced.

The MOE is but one factor among other considerations in making various types of timberland investment and forest management decisions. Plantation forests are a long-term investment and knowledge gained from operational research, such as a mill trial, enables tree growers to tailor their prescriptions to meet customer needs and allocate logs for maximum profit. A clear understanding of the internal quality of standing timber provides flexibility for landowners to capture established and emerging markets and for manufacturers to meet product specifications, adapt to changing grading rules, and develop new products. Such knowledge is necessary to gain a market share and price advantage. Internal wood quality sorting technologies are necessary for log suppliers to deliver the right log to the right mill; however, the vendors may not have a sufficient understanding in operation constraints for developing cost-effective tools for the timberlands and the mills. Logs account for 50–70% of the operational cost of a mill, and a consistent and reliable supply of log mix is a necessity for mill managers. For some reason, communication barriers are frequently found between mills and log suppliers. The disappearance of vertically integrated forest companies makes the information sharing even more difficult.

**Author Contributions:** E.C.L. and E.C.T. conceived, designed, and performed the experiments. J.M.C. and E.C.T. conducted the analysis. E.C.L, E.C.T., C.H., and J.M.C. all contributed to the writing of the manuscript.

**Funding:** This project was funded through the Sustainable Forestry component of Agenda 2020, a joint effort of the USDA Forest Service Research & Development and the American Forest & Paper Association.

**Acknowledgments:** Research partners include the Stand Management and Precision Forestry Cooperatives and Rural Technology Initiative Program at the University of Washington, the School of Environmental and Forest Sciences, USDA Forest Service Pacific Northwest Research Station, CHH FibreGen, and the USDA Forest Service Forest Products Laboratory. The authors also wish to acknowledge the anonymous journal reviewers who provided input and insight thus improving this manuscript.

**Conflicts of Interest:** The authors declare no conflicts of interest.

## References


1. Barbour, R.J.; Kellogg, R.M. Forest management and end-product quality: A Canadian perspective. *Can. J. For. Res.* **1990**, *20*, 405–414. [CrossRef]
2. Vance, E.D.; Maguire, D.A.; Zalesney, R.S., Jr. Research Strategies for Increasing Productivity of Intensively Managed Forest Plantations. *J. For.* **2010**, *108*, 183–192.
3. Cherry, M.; Vikas, V.; Briggs, D.; Cress, D.W.; Howe, G.T. Genetic variation in direct and indirect measures of wood stiffness in coastal Douglas-fir. *Can. J. For. Res.* **2008**, *38*, 2476–2486. [CrossRef]
4. Vikram, V.; Cherry, M.L.; Briggs, D.; Cress, D.W.; Evans, R.; Howe, G.T. Stiffness of Douglas-fir lumber: Effects of wood properties and genetics. *Can. J. For. Res.* **2011**, *41*, 1160–1173. [CrossRef]
5. Jozsa, L.A.; Middleton, G.R. *A Discussion of Wood Quality Attributes and Their Practical Implications, SP-34*; Forintek Canada Corp.: Vancouver, BC, Canada, 1994.
6. Kennedy, R.W. Coniferous wood quality in the future: Concerns and strategies. *Wood Sci. Technol.* **1995**, *29*, 321–338. [CrossRef]
7. Gartner, B.L. Assessing wood characteristics and wood quality in intensively managed plantations. *J. For.* **2005**, *103*, 75–77.
8. Megraw, R. Douglas-fir Wood Properties. In *Douglas-Fir: Stand Management for the Future*; Oliver, C.D., Hanley, D.P., Johnson, J.A., Eds.; College of Forest Resources, University of Washington: Seattle, WA, USA, 1986; pp. 81–96.
9. Abdel-Gadir, A.Y.; Krahmer, R.L. Estimating the age of demarcation of juvenile and mature wood in Douglas-fir. *Wood Fiber Sci.* **2007**, *25*, 242–249.
10. Aubry, C.A.; Adams, W.T.; Fahey, T.D. Determination of relative economic weights for multitrait selection in coastal Douglas-fir. *Can. J. For. Res.* **1998**, *28*, 1164–1170. [CrossRef]
11. Kellogg, R.M. *Second Growth Douglas-Fir: Its Management and Conversion for Value, SP-32*; Forintek Canada Corp.: Vancouver, BC, Canada, 1989.
12. Barratt, J.D.; Kellogg, R.M. Lumber Quality from second growth managed forests. In *A Technical Workshop: Juvenile Wood–What Does It Mean to Forest Management and Forest Products?* Forest Products Research Society: Madison, WI, USA, 1986; pp. 57–71.
13. Barrett, J.D.; Kellogg, R.M. Strength and Stiffness of Dimension Lumber. In *Second Growth Douglas-Fir; Its Management and Conversion for Value, Special Publ. SP-32*; Kellogg, R.M., Ed.; Forintek Canada Corp.: Vancouver, BC, Canada, 1989; pp. 50–58.
14. Fahey, T.D.; Cahill, J.M.; Snellgrove, T.A.; Heath, L.S. *Lumber and Veneer Recovery from Intensively Managed Young-Growth Douglas-Fir*; US Department of Agriculture, Forest Service: Portland, OR, USA, 1991.
15. Bendtsen, B.A.; Plantinga, P.L.; Snellgrove, T.A. The influence of juvenile wood on the mechanical properties of 2 by 4's cut from Douglas-fir plantations. In Proceedings of the International Conference on Timber Engineering, Pullman, WA, USA, 19–22 September 1988; Washington State Univ.: Pullman, WA, USA, 1988; pp. 226–240.
16. Lowell, E.C.; Maguire, D.A.; Briggs, D.G.; Turnblom, E.C.; Jayawickrama, K.J.; Bryce, J. Effects of silviculture and genetics on branch/knot attributes of coastal Pacific Northwest Douglas-fir and implications for wood quality—A Synthesis. *Forests* **2014**, *5*, 1717–1736. [CrossRef]
17. Curtis, R.O.; Reukema, D.L. Crown development and site estimates in a Douglas-fir plantation spacing test. *For. Sci.* **1970**, *16*, 287–301.
18. Grah, R.F. Relationship between tree spacing, knot size, and log quality in young Douglas-fir stands. *J. For.* **1961**, *59*, 270–272.
19. Weiskittel, A.R.; Maguire, D.A.; Monserud, R.A.; Rose, R.; Turnblom, E.C. Intensive management influence on Douglas fir stem form, branch characteristics, and simulated product recovery. *N. Z. J. For. Sci.* **2006**, *36*, 293–312.
20. Brix, H. Effects of thinning and nitrogen fertilization on branch and foliage production in Douglas-fir. *Can. J. For. Res.* **1981**, *11*, 502–511. [CrossRef]
21. Maguire, D.A.; Kershaw, J.A., Jr.; Hann, D.W. Predicting the effects of silvicultural regime on branch size and crown wood core in Douglas-fir. *For. Sci.* **1991**, *37*, 1409–1428.

22. Maguire, D.A.; Johnston, S.R.; Cahill, J. Predicting branch diameters on second-growth Douglas-fir from tree-level descriptors. *Can. J. For. Res.* **1999**, *29*, 1829–1840. [CrossRef]

23. Briggs, D.G.; Kantavichai, R.; Turnblom, E.C. Predicting the diameter of the largest breast-height region branch of Douglas-fir trees in thinned and fertilized plantations. *For. Prod. J.* **2010**, *60*, 322–330. [CrossRef]

24. Reukema, D.L. Crown expansion and stem radial growth of Douglas-fir as influenced by release. *For. Sci.* **1964**, *10*, 192–199.

25. Barbour, R.J.; Parry, D.L. *Log and Lumber Grades as Indicators of Wood Quality in 20- to 100-Year Old Douglas-Fir Trees from Thinned and Unthinned Stands*; US Department of Agriculture, Forest Service: Portland, OR, USA, 2001.

26. Western Wood Products Association (WWPA). *Western Lumber Grading Rules*; Western Wood Products Association: Portland, OR, USA, 2017.

27. Middleton, G.R.; Munro, B.D. Log and Lumber Yields. In *Second Growth Douglas-Fir: Its Management and Conversion for Value*; Kellogg, R.M., Ed.; Forintek Canada Corp.: Vancouver, BC, Canada, 1989; Chapter 7; pp. 66–74.

28. Briggs, D.G.; Ingaramo, L.; Turnblom, E.C. Number and Diameter of Breast-height Region Branches in a Douglas-fir Spacing Trial and Linkage to Log Quality. *For. Prod. J.* **2007**, *57*, 28–34.

29. Ross, R.J.; McDonald, K.A.; Green, D.W.; Schad, K. Relationship between log and lumber modulus of elasticity. *For. Prod. J.* **1997**, *47*, 89–92.

30. Ross, R.J.; Willits, S.A.; VonSegen, W.; Black, T.; Brashaw, B.K.; Pellerin, R.F. A stress wave based approach to NDE of logs for assessing potential veneer quality. Part I. Small diameter ponderosa pine. *For. Prod. J.* **1999**, *49*, 60–62.

31. Ridoutt, B.G.; Wealleans, K.R.; Booker, R.E.; McConchie, D.L.; Ball, R.D. Comparison of log segregation methods for structural lumber yield improvement. *For. Prod. J.* **1999**, *49*, 63–66.

32. Carter, P.; Briggs, D.; Ross, R.J.; Wang, X. Acoustic testing to enhance western forest values and meet customer wood quality needs. In *Productivity of Western Forests: A Forest Products Focus*; Harrington, C.A., Schoenholz, S.H., Eds.; US Department of Agriculture Forest Service: Portland, OR, USA, 2005; pp. 121–129.

33. Wang, X.; Ross, R.J.; McClellan, M.; Barbour, R.J.; Erickson, J.R.; Forsman, J.W.; McGinnis, G.D. Nondestructive evaluation of standing trees with a stress wave method. *Wood Fiber Sci.* **2001**, *33*, 522–533.

34. Wang, X.; Carter, P.; Ross, R.J.; Brashaw, B.K. Acoustic assessment of wood quality of raw forest materials—A path to increased profitability. *For. Prod. J.* **2007**, *57*, 6–14.

35. Wang, X.; Ross, R.J.; Carter, P. Acoustic evaluation of wood in standing trees. Part I. Acoustic wave behavior. *Wood Fiber Sci.* **2007**, *39*, 28–38.

36. Amishev, D.; Murphy, G.E. In-forest assessment of veneer grade Douglas-fir logs based on acoustic measurement of wood stiffness. *For. Prod. J.* **2008**, *58*, 42–47.

37. Briggs, D.G.; Thienel, G.; Turnblom, E.C.; Lowell, E.; Dykstra, D.; Ross, R.J.; Wang, X.; Carter, P. Influence of thinning on acoustic velocity of Douglas-fir trees in western Washington and western Oregon. In Proceedings of the 15th International Symposium on Nondestructive Testing of Wood, Duluth, MN, USA, 10–12 September 2007; pp. 113–123.

38. Maguire, D.A.; Bennett, W.S.; Kershaw, J.A., Jr.; Gonyea, R.; Chappell, H.N. *Establishment Report Stand Management Cooperative Project Field Installations, Institute of Forest Resources Contrib. 72*; College of Forest Resources, University of Washington: Seattle, WA, USA, 1991.

39. Curtis, R.O. A simple index of stand density for Douglas-fir. *For. Sci.* **1982**, *28*, 92–94.

40. Evans, J.W.; Kretschmann, D.E.; Herian, V.L.; Green, D.W. *Procedures for Developing Allowable Properties for a Single Species Under ASTM1900 and Computer Programs Useful for the Calculation*; US Department of Agriculture, Forest Service: Madison, WI, USA, 2001.

41. Ramalho, E.A.; Ramalho, J.J.; Murteira, J.M. Alternative estimating and testing empirical strategies for fractional regression models. *J. Econ. Surv.* **2011**, *25*, 19–68. [CrossRef]

42. American Forest and Paper Association/American Wood Council (AF & PA/AWC). *Wood Structural Design Data*; American Forest and Paper Association: Washington, DC, USA, 2004.

43. Zhao, D.; Borders, B.; Wang, M. Survival model for fusiform rust infected loblolly pine plantations with and without mid-rotation understory vegetation control. *For. Ecol. Manag.* **2006**, *235*, 232–239. [CrossRef]

44. Bates, D.; Maechler, M.; Bolker, B.; Walker, S.; Christensen, R.H.B.; Singmann, H.; Dai, B.; Scheipl, F.; Grothendieck, G.; Green, P. Linear Mixed-Effects Models Using 'Eigen' and S4: Package 'lme4'. 2018. Available online: https://cran.r-project.org/web/packages/lme4/lme4.pdf (accessed on 19 February 2018).

45.   Oliver, C.D.; Larson, B.C. *Forest Stand Dynamics*; Update Edition; Wiley: New York, NY, USA, 1996; 520p.

46.   Northwest Log Rules Advisory Group. *Official Rules for the Following Log Scaling and Grading Bureaus: Columbia River, Grays Harbor, Northern California, Puget Sound, Southern Oregon, Yamhill*; Northwest Log Rules Advisory Group: Eugene, OR, USA, 1998; 48p.

47.   Hoibo, O.A.; Turnblom, E.C. Models of knot characteristics in young coastal U.S. Douglas-fir: Are the effects of tree and site data visibly rendered in the annual ring width pattern at breast height? *For. Prod. J.* **2017**, *67*, 29–38.

48.   Todoroki, C.L.; Lowell, E.C.; Kantavichai, R. Growth and mortality in response to climatic extremes and competition in thinning trials of Douglas-fir. manuscript in preparation.

*Article*

# Acoustic Velocity—Wood Fiber Attribute Relationships for Jack Pine Logs and Their Potential Utility

Peter F. Newton

Canadian Wood Fiber Centre, Canadian Forest Service, Natural Resources Canada, Sault Ste. Marie, ON P6A 2E5, Canada; peter.newton@canada.ca; Tel.: +1-705-541-5615

Received: 18 October 2018; Accepted: 23 November 2018; Published: 30 November 2018

**Abstract:** This study presents an acoustic-based predictive modeling framework for estimating a suite of wood fiber attributes within jack pine (*Pinus banksiana* Lamb.) logs for informing in-forest log-segregation decision-making. Specifically, the relationships between acoustic velocity (longitudinal stress wave velocity; $v_l$) and the dynamic modulus of elasticity ($m_e$), wood density ($w_d$), microfibril angle ($m_a$), tracheid wall thickness ($w_t$), tracheid radial and tangential diameters ($d_r$ and $d_t$, respectively), fiber coarseness ($c_o$), and specific surface area ($s_a$), were parameterized deploying hierarchical mixed-effects model specifications and subsequently evaluated on their resultant goodness-of-fit, lack-of-fit, and predictive precision. Procedurally, the data acquisition phase involved: (1) randomly selecting 61 semi-mature sample trees within ten variable-sized plots established in unthinned and thinned compartments of four natural-origin stands situated in the central portion the Canadian Boreal Forest Region; (2) felling and sectioning each sample tree into four equal-length logs and obtaining twice-replicate $v_l$ measurements at the bottom and top cross-sectional faces of each log ($n = 4$) from which a log-specific mean $v_l$ value was calculated; and (3) sectioning each log at its midpoint and obtaining a cross-sectional sample disk from which a $2 \times 2$ cm bark-to-pith radial xylem sample was extracted and subsequently processed via SilviScan-3 to derive annual-ring-specific attribute values. The analytical phase involved: (1) stratifying the resultant attribute—acoustic velocity observational pairs for the 243 sample logs into approximately equal-sized calibration and validation data subsets; (2) parameterizing the attribute—acoustic relationships employing mixed-effects hierarchical linear regression specifications using the calibration data subset; and (3) evaluating the resultant models using the validation data subset via the deployment of suite of statistical-based metrics pertinent to the evaluation of the underlying assumptions and predictive performance. The results indicated that apart from tracheid diameters ($d_r$ and $d_t$), the regression models were significant ($p \leq 0.05$) and unbiased predictors which adhered to the underlying parameterization assumptions. However, the relationships varied widely in terms of explanatory power (index-of-fit ranking: $w_t$ (0.53) > $m_e$ > $s_a$ > $c_o$ > $w_d$ >> $m_a$ (0.08)) and predictive ability ($s_a$ > $w_t$ > $w_d$ > $c_o$ >> $m_e$ >>> $m_a$). Likewise, based on simulations where an acoustic-based $w_d$ estimate is used as a surrogate measure for a Silviscan-equivalent value for a newly sampled log, predictive ability also varied by attribute: 95% of all future predictions for $s_a$, $w_t$, $c_o$, $m_e$, and $m_a$ would be within $\pm 12\%$, $\pm 14\%$, $\pm 15\%$, $\pm 27\%$, and $\pm 55\%$ and of the true values, respectively. Both the limitations and potential utility of these predictive relationships for use in log-segregation decision-making, are discussed. Future research initiatives, consisting of identifying and controlling extraneous sources of variation on acoustic velocity and establishing attribute-specific end-product-based design specifications, would be conducive to advancing the acoustic approach in boreal forest management.

**Keywords:** end-product-based fiber attribute determinates; longitudinal stress wave velocity; mixed-effects hierarchical linear models; predictive performance

## 1. Introduction

The Canadian forest sector has been increasingly embracing an aspirational trivariate management proposition for which the goal is to maximize end-product value while simultaneously realizing volumetric yield and ecological sustainability objectives (sensu [1,2]). Tangential to this transformative shift is the increased apprehension regarding the quality and associated end-product potential of the increasing wood volumes harvested from more intensely managed second-growth forests [3]. Among the many prerequisites required for realizing this goal and addressing knowledge gaps in second-growth fiber quality, is the provision of enhanced in-forest intelligence in regards to the forecasting of end-product potential at the time of harvest or immediately afterwards (n., in-forest generically refers to all forest operations conducted or decisions made before mill-processing commences inclusive of operational felling areas, landings and log sorting yards).

End-product forecasts can be used to inform and optimize log-segregation, allocation and merchandizing decision-making thus contributing to the potential realization of the value maximization objective [4–6]. More precisely, the end-product potential of the logs extracted from the main stem of individual coniferous trees is a function of (1) external morphological stem features (e.g., diameter, height, taper, sweep, crook, and eccentricity), and (2) internal anatomical characteristics (e.g., microfibril angle, tracheid wall thickness, tracheid radial and tangential diameters) and associated physical properties (e.g., modulus of elasticity, wood density, fiber coarseness and specific surface area) of the xylem tissue. Various performance measures which are used to define and classify the overall type and grade of derived products have been shown to be explicitly linked to these internal fiber-based attributes. Although only stiffness and wood density measures are currently used in machine stress grading systems for solid wood products (e.g., dimensional lumber [7]), microfibril angle, tracheid cell wall thickness, radial and tangential tracheid diameters, fiber coarseness and specific surface area, are also important determinates of end-product quantity and quality [8] (sensu Table 1). Thus, as the fiber supply increases in diversity with the arrival of wood from density manipulated forests (e.g., pre-commercially and/or commercially thinned natural-origin stands, and plantations) combined with increasing pressures to find economic efficiencies within the upper portion of the forest products supply chain, provision of explicit information on these additional attributes may become consequential.

**Table 1.** Product-based performance measures and their relationship with fiber attributes for boreal conifers.

| Product Category | Performance Measure | Relationship with Fiber Attribute [a] |
|---|---|---|
| Biomass (e.g., pellets) | Calorific value | $\propto$ xylem density |
| Pulp and paper (e.g., paperboards, newsprint, facial tissues, and specialized coated papers) | Tensile strength<br>Tear strength<br>Stretch<br>Bulk<br>Light scattering<br>Collapsibility<br>Yield | $\propto$ (tracheid wall thickness)$^{-1}$, specific surface area<br>$\propto$ fiber coarseness<br>microfibril angle<br>$\propto$ tracheid wall thickness, (tracheid diameter)$^{-1}$<br>$\propto$ (tracheid wall thickness)$^{-1}$<br>$\propto$ tracheid wall thickness<br>$\propto$ xylem density |
| Solid wood and composites (e.g., dimensional lumber; glulam-based beams) | Strength<br>Stiffness | $\propto$ xylem density, (microfibril angle)$^{-1}$<br>$\propto$ xylem density, modulus of elasticity, (microfibril angle)$^{-1}$ |
| Poles and squared timbers (e.g., utility poles and solid wood beams) | Strength<br>Stiffness | $\propto$ xylem density, (microfibril angle)$^{-1}$<br>$\propto$ xylem density, modulus of elasticity, (microfibril angle)$^{-1}$ |

[a] Generic-based inferences derived from generalized end-product-specific performance metric—attribute relationships for boreal conifers (sensu [8]).

Attribute determination for use in forecasting end-product potential deploying destructive sampling approaches can be logistically challenging, time consuming and expensive (e.g., in-forest extracting of 12 mm diameter xylem cores followed by Silviscan-based laboratory processing and testing). Conversely, non-destructive sampling methods in which attributes can be estimated indirectly through their relationship with acoustic velocity has been shown to be viable alternative [9–12]. More specifically, in-forest acoustic-based sampling tools provide forest practitioners with an ability to estimate the dynamic modulus of elasticity within harvested logs and standing trees. The dynamic modulus of elasticity is a wood stiffness measure that is used as a surrogate measure of its static counterpart (static modulus of elasticity). This latter static measure is commonly employed along with maximum knot size, sweep and physical dimensions, to categorize solid wood products into specific grade classes [7,9,13]. Thus, the ability to forecast one of the key underlying determinates of end-product potential within the forest at the time of harvest or soon after, has important utility in terms of informing segregation, allocation and merchandizing decision-making [6].

Mathematically, as derived from engineering principles, the dynamic modulus of elasticity (MOE$_{\mathrm{dyn}}$ denoted $m_e$ in this study; GPa) of a log can be expressed as a function of the density-weighted velocity of a longitudinal stress wave ($v_l$; km/s) that propagates through the xylem tissue upon its generation arising from a mechanically induced impact on one of the open cross-sectional log faces (Equation (1) and denoted the primary relationship herewith; [12]).

$$m_e = f\left(w'_d v_l^2\right) \tag{1}$$

where $w'_d$ is the green (fresh) wood density (kg/m$^3$) of the log's xylem tissue. The dynamic modulus of elasticity has also been shown to be correlated with other attributes underlying end-product potential [14–16].

More specifically, based on extracted clear xylem samples obtained from black spruce (*Picea mariana* (Mill) B.S.P.), red pine (*Pinus resinosa* Ait.) and jack pine (*Pinus banksiana* Lamb.) trees growing in the central North America, Silviscan-derived estimates of wood density (oven-dried wood density denoted $w_d$ in this study; kg/m$^3$), microfibril angle (MFA denoted $m_a$ in this study; $^\circ$), tracheid cell wall thickness ($w_t$; μm), radial ($d_r$; μm) and tangential ($d_t$; μm) tracheid diameters, fiber coarseness ($c_o$; μg/m) and specific surface area ($s_a$; m$^2$/kg), have been shown to be correlated with the dynamic modulus of elasticity (Table 2). Statistically significant ($p \leq 0.05$) (1) directly proportional relationships between $m_e$ and $w_d$, $w_t$ and $c_o$ ($m_e \propto w_d, w_t, c_o$), and (2) indirectly proportional relationships between $m_e$ and $m_a$, $d_r$, $d_t$ and $s_a$ $\left(m_e \propto m_a^{-1}, d_r^{-1}, d_t^{-1}, s_a^{-1}\right)$, have been established for these species (Table 2). Supporting correlative evidence has also been reported for non-boreal conifers. For example, a directly proportional correlative relationship between $m_e$ and wood density ($r = 0.75$) and an inversely proportional relationship between $m_e$ and $m_a$ ($r = -0.83$) has been reported for Silviscan-derived attribute estimates obtained from clear wood samples extracted from 76 radiata pine (*Pinus radiata* D. Don.) boards [13]. Based on this collective correlative evidence, an empirical-based expansion of the acoustic-based inferential framework was proposed for boreal conifer logs [15], as summarized in Table 2. Since wood density is also one of the principal end-product determinates (sensu Table 1) and analytically required for acoustic-based attribute predictions (e.g., Equation (1); Table 2), a simplified directly proportional relationship between acoustic velocity and wood density was included within this framework (i.e., $w_d \propto v_l^2$). Collectively, this expanded acoustic inferential framework represents a more encompassing analytical platform for forecasting commercially relevant wood quality attributes than previous univariate approaches solely based on the primary relationship (Equation (1)).

**Table 2.** Linear associations between the dynamic modulus of elasticity and other Silviscan-derived wood quality attributes for 3 boreal conifers and the resultant acoustic-based relationship inferred.

| Attribute | Association with Dynamic Modulus of Elasticity as Measured by the Pearson Product Moment Correlation Coefficient [a] | | | Statistical Inference | Empirical Linkage to Acoustic Velocity [b] |
| --- | --- | --- | --- | --- | --- |
| | Black Spruce | Red Pine | Jack Pine | | |
| Wood density | 0.7765 * | 0.7585 * | 0.6717 * | $m_e \propto w_d$ | $w_d \propto v_l^2$ |
| Microfibril angle | −0.8981 * | −0.8075 * | −0.7101 * | $m_e \propto m_a^{-1}$ | $m_a \propto \left(w_d v_l^2\right)^{-1}$ |
| Tracheid wall thickness | 0.6765 * | 0.6871 * | 0.6648 * | $m_e \propto w_t$ | $w_t \propto w_d v_l^2$ |
| Radial tracheid diameter | −0.4477 * | −0.3833 * | −0.3169 * | $m_e \propto d_r^{-1}$ | $d_r \propto \left(w_d v_l^2\right)^{-1}$ |
| Tangential tracheid diameter | −0.3037 * | −0.0886 | −0.0814 | $m_e \propto d_t^{-1}$ | $d_t \propto \left(w_d v_l^2\right)^{-1}$ |
| Fiber coarseness | 0.2044 | 0.4385 * | 0.5412 * | $m_e \propto c_o$ | $c_o \propto w_d v_l^2$ |
| Specific surface area | −0.5812 * | −0.6355 * | −0.6368 * | $m_e \propto s_a^{-1}$ | $s_a \propto \left(w_d v_l^2\right)^{-1}$ |

[a] Species-specific bivariate linear associations between Silviscan-determined accumulative area-weighted mean fiber attributes as previously presented in Table 2 of Newton [16]: Pearson product moment correlation coefficient for the relationship between the dynamic modulus of elasticity and each attribute at breast-height (1.3 m) for 50 semi-mature black spruce trees [14], 54 mature red pine trees [16], and 61 semi-mature jack pine trees (this study) where * denotes a significant correlation at the 0.05 probability level. [b] Based on the primary $m_e \propto w_d v_l^2$ relationship for all attributes except wood density.

Although these statistical-based secondary relationships lack the causal underpinnings of the primary relationship (sensu [17]), the empirical results to date have been supportive. Specifically, applying this analytical framework to red pine logs, revealed that statistically viable acoustic-based relationships could be established for five of the eight attributes examined ($m_e$, $w_d$, $w_t$, $c_0$ and $s_a$) [15]. However, the applicability of this expanded framework of acoustic-based wood attribute relationships to other intensely managed softwood species is largely unknown. Furthermore, even the most fundamental acoustic velocity—attribute relationship (Equation (1)) has yet to be parameterized for most of the commercially important Canadian boreal conifers (e.g., an "acoustic velocity" keyword search for articles published within Canada's primary applied journal (Forestry Chronicle) was unsuccessful in identifying previous research in this field). Conversely, other forest regions, such as New Zealand, United States of America, Australia, Europe, and Asia, have adapted this acoustic-based innovation pathway in order improve segregation decision-making (e.g., [18,19]). Hence, the evaluation of the acoustic approach and the provision of prediction equations for the estimating a suite of key attributes underlying end-product potential for jack pine, could contribute to addressing this innovation void. Principally, by providing the forest practitioner with the prerequisite prediction equations, in-forest segregation decision-making could be improved and potentially yield overall efficiency gains within the upper portion of the jack pine forest products supply chain. Consequently, the objective of this study was to investigate the empirical applicability of the expanded acoustic-based inferential framework by examining the nature, strength and predictability of the relationships between acoustic velocity and the dynamic modulus of elasticity, wood density, microfibril angle, tracheid cell wall thickness, radial and tangential tracheid diameters, fiber coarseness and specific surface area, specifically for jack pine.

## 2. Materials and Methods

### 2.1. Sample Stands, Plot Establishment and Sample Tree Selection

Two geographically separated jack pine thinning experiments that were previously established in northeastern (denoted the Sewell site) and northcentral (denoted the Tyrol site) regions of Ontario, were selected for sampling. The experimental treatments were representative of the crop planning strategy historically followed in Ontario for this species. Specifically, allowing jack pine to naturally regenerate following a stand-replacing disturbance and if sufficiently successful in terms of stocking levels and free-to-grow status, (1) implementing precommercial thinning in order to moderate overall site occupancy and encourage individual tree growth thereby reducing the time-to-operability status in order to mitigate the effects of projected wood supply deficits, or (2) implementing a temporal sequence of thinning treatments (precommercial and commercial thinning) in order to optimize site occupancy and maximize end-product potentials throughout the rotation [20].

At the Sewell site, sample trees were selected within 6 sample plots that were established in 3 fire-origin jack pine stands (2 plots/stand). These approximately 53-year-old stands were situated on sites of medium-to-good quality (mean site index of 18 m at a breast-height age of 50 [21]) and geographically located within Forest Section B.7—Missinaibi-Cabonga of the Canadian Boreal Forest Region [22]. The soils and topography were characterized as deep (>1 m) coarse-to-medium textured sandy types situated on gently undulating terrain. Silviculturally, the stands were subjected to 1 of 3 density manipulation treatments yielding 3 unique crop plans: (1) natural stand development with no density management (unthinned); (2) precommercial thinned (PCT) at age 11 (1971); and (3) PCT at age 11 followed by a light pseudo-commercial thinning (CT) at age 43 (2003). Procedurally, for the first stand, sample trees were selected within 2 previously established 0.06 ha circular monitoring plots. For these 2 plots, a stratified random sample protocol was used to select sample trees: the observed diameter frequency distribution at the time of sampling was stratified into 3 size classes from which 1 or 2 trees per class were selected. However, to preserve the original sample design within the remaining stands, temporary prism-based variable-size plots where established adjacent to 2 of the existing long-term monitoring plots within the second and third stands. Deploying the observed diameter frequency distribution derived from the prism sweeps, 3 size classes were similarly formulated from which 1 or 2 trees per class were selected. In total, 31 jack pine sample trees were selected at the Sewell site: 3 stands × 2 plots/stand × 3 size classes/plot × 1–2 trees/size class.

At the Tyrol site, sample trees were selected within 4 0.07 ha circular sample plots that were established in 2 fire-origin jack pine stands. These stands were situated on sites of good-to-excellent quality (mean site index of 21 m at a breast-height age of 50 [21]), geographically located within Forest Section B9 (Superior) of the Canadian Boreal Forest Region [22], and approximately 73 years of age when sampled. The soils and topography were characterized as deep (>1 m) finely textured sandy types on gently undulating terrain. Silviculturally, the stands were subjected to 2 density manipulation treatments: PCT at an age of approximately 20 (1962) followed by a light pseudo-CT treatment at an age of approximately 56 (1998). A size-based stratified random sample protocol was used to select trees resulting in approximately 2–3 trees being chosen from each of the 3 size (diameter) classes. In total 30 jack pine sample trees were selected from the Tyrol site: 2 stands × 2 plots/stand × 3 size classes/plot × 2–3 sample trees/size class. In summary, a total of 61 sample trees were selected from the 2 experimental areas which were grown under a range of density management regimes reflective of past and current forest management practices for this species and region (e.g., unthinned, PCT, PCT+CT).

*2.2. Sample Tree Measurements, Stem Analysis Procedures, Log-Based Acoustic Measurements and Disk Sampling*

Prior to felling and sectioning each sample tree using destructive stem analysis, diameter at breast-height (1.3 m) outside-bark diameter ($D_b$; cm), total height ($H_t$; m) and height-to-live crown ($H_c$; m) measurements were obtained. Note, all measurements and destructive sampling were completed on the trees at Sewell and Tyrol sites at the conclusion of the 2013 and 2015 growing seasons, respectively. The destructive stem analysis protocol consisted of felling each sample tree at stump height (0.3 m), delimbing the stem and then topping the stem at an 80% relative height position. The stem was then sectioned into 0%–20%, 20%–40%, 40%–60% and 60%–80% percentile-based log-length intervals employing a percent height sampling protocol. Immediately thereafter, the velocity of a mechanically induced longitudinal stress wave propagating throughout each log was measured using a Director HM200 acoustic velocity resonance tool (Fibre-gen Inc., Christchurch, New Zealand; www.fibre-gen. com). Specifically, a twice-replicated set of $v_l$ (km/s) measurements were obtained at the bottom and top of each log from which an arithmetic mean value was calculated. Log lengths, ambient air, and xylem temperatures were also taken at the time of each acoustic velocity measurement.

For each tree, cross-sectional samples were then extracted from the center of each log, corresponding to the relative height positions of 10%, 30%, 50% and 70%. The samples were placed in temporary cold storage within 8 h of sectioning and then transported to a permanent cold storage (<0 °C) facility until further processing. Overall, the deployed sampling design yielded 124 cross-sectional disks from the Sewell site (1 disk/log × 4 logs/tree × 31 trees) and 120 cross-sectional disks from the Tyrol site (1 disk/log × 4 logs/tree × 30 trees). Due to logistical challenges during field sampling at the Sewell site, however, one of the cross-sectional samples had to be disregarded. Hence a final total of 243 cross-sectional disk samples obtained from the midpoint position on each of the 243 sample logs were available for analysis. Table 3 summarizes the mensuration characteristics of the standing sample trees prior to felling. The trees from both sites exhibited a similar degree of variation for each measured characteristic as quantified by the coefficient of variation. Diameter and live crown ratio (($H_t - H_c$)/$H_t$) exhibited the greatest amount of variation given that a size-based stratified random sampling protocol was used to select the sample trees (e.g., trees from all 3 diameter size-groups (small, medium, and large) where included). The minimal degree of age variation is largely due to the even- aged nature of the sampled stands. Table 4 provides a descriptive summary of the derived logs in terms of their dimensions and associated acoustic velocity measurements.

**Table 3.** Descriptive statistical summary of the mensuration characteristics of the 61 sample trees by site (*n* = 31 and 30 for Sewell and Tyrol, respectively).

| Variable | Site | Mean | Standard Error | Minimum | Maximum | CV [a] (%) |
|---|---|---|---|---|---|---|
| Diameter at | Sewell | 18.8 | 2.11 | 14.7 | 22.6 | 11.2 |
| breast-height (cm) | Tyrol | 24.4 | 2.17 | 19.8 | 29.1 | 8.9 |
| Breast-height age (year) | Sewell | 50 | 0.96 | 47 | 51 | 1.9 |
| | Tyrol | 69 | 1.27 | 66 | 71 | 1.9 |
| Total height (m) | Sewell | 21.1 | 1.26 | 18.3 | 22.9 | 6.0 |
| | Tyrol | 22.2 | 1.57 | 19.5 | 24.6 | 7.1 |
| Live crown ratio (%) | Sewell | 26.1 | 4.50 | 15.0 | 35.3 | 17.3 |
| | Tyrol | 28.2 | 7.20 | 14.1 | 41.5 | 25.5 |

[a] Coefficient of variation.

**Table 4.** Descriptive statistical summary of the dimensions of the 243 sample logs and associated acoustic velocity measurements by log position (*n* = 243 of which 61, 61, 61 and 60 were first, second, third and fourth positioned logs, respectively).

| Variable | Log [a] | Mean | Median | Standard Error | Minimum | Maximum | CV [b] (%) |
|---|---|---|---|---|---|---|---|
| Log length (m) | 1st | 4.30 | 4.36 | 0.04 | 3.48 | 5.00 | 7.7 |
| | 2nd | 4.28 | 4.35 | 0.04 | 3.47 | 4.91 | 7.8 |
| | 3rd | 4.28 | 4.37 | 0.04 | 3.18 | 4.91 | 7.9 |
| | 4th | 4.25 | 4.37 | 0.06 | 2.27 | 4.89 | 10.8 |
| Mean log diameter (inside-bark; cm) | 1st | 19.49 | 19.39 | 0.35 | 13.70 | 26.92 | 14.1 |
| | 2nd | 16.97 | 17.03 | 0.31 | 10.89 | 21.76 | 14.4 |
| | 3rd | 14.38 | 14.56 | 0.28 | 9.41 | 19.25 | 15.4 |
| | 4th | 11.30 | 10.99 | 0.26 | 7.54 | 15.71 | 17.5 |
| Longitudinal stress wave velocity ($v_l$; km/s) | 1st | 3.59 | 3.58 | 0.02 | 3.20 | 4.29 | 5.2 |
| | 2nd | 3.59 | 3.62 | 0.03 | 3.15 | 4.32 | 5.5 |
| | 3rd | 3.39 | 3.42 | 0.02 | 2.86 | 4.12 | 6.0 |
| | 4th | 3.08 | 3.08 | 0.02 | 2.69 | 3.72 | 6.0 |

[a] Proceeding upwards from the stump to the stem apex, 1st, 2nd, 3rd and 4th denotes the ordinal position of the first, second, third and fourth log extracted from the main stem of the 61 jack pine sample trees. [b] Coefficient of variation.

### 2.3. Silviscan-3 Estimation of Fiber Attributes and Preliminary Computations

A transverse 2 cm × 2 cm bark-to-pith-to-bark sample was extracted along the geometric mean diameter of each of the frozen cross-sectional disks. Annual area-weighted ring measures of $m_e$, $w_d$, $m_a$, $w_t$, $d_r$, $d_t$, $c_o$ and $s_a$ were determined along one of the transverse pith- to-bark radii using the SilviScan-3 system. SilviScan-3 analyses yields a set of commercially relevant fiber attribute estimates through the deployment of an integrated hardware-software processing system involving automatic image acquisition and analysis (cell scanner), X-ray densitometry, and X-ray diffractometry technologies. In this study, the following attribute estimates were used from the Silviscan-3 analysis: (1) wood density as measured at a 25 μm resolution and 8% moisture content (dry basis) and determined following the removal of resins using X-ray densitometry (i.e., $w_d = f$ (intensity of incident and transmitted X-ray beams; sample thickness; and X-ray travel distance); [23,24]); (2) microfibril angle as determined via X-ray diffraction patterns (i.e., $m_a = f$ (variance of the cellulose-I 002 azimuthal X-ray diffraction pattern; and dispersion of the microfibril orientation distribution); [25]); (3) dynamic modulus of elasticity as determined from a combination of X-ray densitometry and diffraction measurements (i.e., $m_e = f$ (X-ray densitometry density estimate; and coefficient of variation of the normalized intensity profile); [26]); and (4) fiber dimensions and derived metrics consisting of tracheid wall thickness, radial and tangential tracheid diameters, fiber coarseness and specific surface area as determined using the results from the image analyses (radial and tangential diameters) in combination with the wood density estimate (sensu [27]).

Computationally, deploying the annual area-weighted ring estimates for $m_e$, $w_d$, $m_a$, $w_t$, $d_r$, $d_t$, $c_o$ and $s_a$ and proceeding in a pith-to-bark direction, an attribute-specific cumulative area-weighted moving average value was calculated for each of the 243 radial sequences (Equation (2)).

$$F = \frac{\sum\limits_{i=1}^{I} f_i a_i}{\sum\limits_{i=1}^{I} a_i} \tag{2}$$

where $F$ is the cumulative moving area-weighted fiber attribute average for each sequence, $a_i$ is the Silviscan-3-determined area of the $i$th annual ring (mm$^2$; $i = 1, \ldots ,I$; $I$ = total cambial age), and $f_i$ is the mean Silviscan-3-determined fiber attribute value within the $i$th annual ring. Table 5 provides an attribute-specific descriptive statistical summary by log position, inclusive of measures of central tendency (arithmetic means and medians), ranges (minimums and maximums) and relative variation measures (coefficients of variation).

**Table 5.** Descriptive statistical summary of the fiber attributes of by log position ($n$ = 243 of which 61, 61, 61 and 60 were first, second, third and fourth positioned logs, respectively).

| Variable | Log [a] | Mean | Median | Minimum | Maximum | CV [b] (%) |
|---|---|---|---|---|---|---|
| Modulus of elasticity ($m_e$; GPa) | 1st | 12.72 | 12.68 | 8.59 | 16.73 | 15.0 |
| | 2nd | 12.81 | 12.95 | 8.24 | 16.16 | 14.2 |
| | 3rd | 12.38 | 12.82 | 7.93 | 15.89 | 13.7 |
| | 4th | 11.33 | 11.57 | 7.14 | 14.57 | 12.7 |
| Wood density ($w_d$; kg/m$^3$) | 1st | 430.38 | 421.66 | 372.88 | 489.67 | 6.5 |
| | 2nd | 416.48 | 418.68 | 337.88 | 482.02 | 6.6 |
| | 3rd | 407.47 | 405.99 | 359.76 | 467.79 | 6.0 |
| | 4th | 394.23 | 391.66 | 356.95 | 442.25 | 5.1 |
| Microfibril angle ($m_a$; °) | 1st | 12.98 | 12.79 | 7.49 | 19.71 | 21.1 |
| | 2nd | 11.47 | 10.95 | 6.33 | 19.23 | 22.7 |
| | 3rd | 11.25 | 10.92 | 6.24 | 17.84 | 22.6 |
| | 4th | 12.52 | 12.21 | 6.84 | 20.23 | 22.8 |
| Tracheid wall thickness ($w_t$; μm) | 1st | 2.70 | 2.67 | 2.35 | 3.16 | 7.6 |
| | 2nd | 2.60 | 2.60 | 2.08 | 3.01 | 7.8 |
| | 3rd | 2.52 | 2.49 | 2.22 | 2.92 | 7.3 |
| | 4th | 2.39 | 2.37 | 2.17 | 2.77 | 5.9 |
| Tracheid radial diameter ($d_r$; μm) | 1st | 30.80 | 30.90 | 28.51 | 33.00 | 3.7 |
| | 2nd | 30.75 | 30.95 | 28.69 | 32.74 | 3.5 |
| | 3rd | 30.52 | 30.68 | 27.99 | 32.79 | 3.7 |
| | 4th | 30.03 | 30.18 | 26.44 | 31.97 | 4.1 |
| Tracheid tangential diameter ($d_t$; μm) | 1st | 27.87 | 27.93 | 26.41 | 29.88 | 2.4 |
| | 2nd | 28.13 | 28.13 | 26.89 | 30.25 | 2.5 |
| | 3rd | 28.10 | 28.16 | 26.53 | 29.72 | 2.5 |
| | 4th | 27.97 | 27.99 | 26.38 | 29.63 | 2.6 |
| Fiber coarseness ($c_o$; μg/m) | 1st | 406.02 | 406.67 | 360.44 | 471.20 | 6.8 |
| | 2nd | 395.23 | 394.97 | 339.39 | 447.85 | 7.1 |
| | 3rd | 383.33 | 383.76 | 328.30 | 455.34 | 7.2 |
| | 4th | 364.51 | 365.10 | 311.38 | 420.68 | 6.3 |
| Specific surface area ($s_a$; m$^2$/kg) | 1st | 314.32 | 315.76 | 275.59 | 351.46 | 6.4 |
| | 2nd | 322.71 | 321.31 | 284.37 | 372.46 | 6.4 |
| | 3rd | 328.16 | 327.71 | 287.81 | 366.30 | 6.4 |
| | 4th | 338.15 | 341.99 | 297.17 | 369.33 | 5.0 |

[a] Proceeding upwards from the stump to the stem apex, 1st, 2nd, 3rd and 4th denotes the ordinal position of the first, second, third and fourth log extracted from the main stem of the 61 jack pine sample trees. [b] Coefficient of variation.

### 2.4. Preliminary Data Stratification for Cross-Validation Assessment

To minimize the potential effects of spatial correlation arising from the selection of the multiple logs from the same sample tree while at the same time providing a validation data subset, the full data set consisting of 243 $m_e$, $w_d$, $m_a$, $w_t$, $d_r$, $d_t$, $c_o$, $s_a$—$v_l$ observational pairs was systematically subdivided into calibration and validation data subsets of approximately equal size. This involved randomly selecting from each sample tree a sequential set of measurements consisting of either those associated with the 1st and 3th positioned logs or those associated with the 2nd and 4th positioned logs. The resultant pairs were then allocated to either the calibration or validation subsets. At the conclusion of this stratification process, the calibration and validation data subsets were comprised of 122 and 121 $m_e$, $w_d$, $m_a$, $w_t$, $d_r$, $d_t$, $c_o$, $s_a$—$v_l$ observational pairs, respectively. As presented in Table 6, the acoustic velocity measurements and attribute values within the subsets were very similar in terms of measures of central tendency, magnitudes, and relative variation. A further similarly structured comparative assessment of the data subsets by log position and location (not shown) also confirmed the equivalency between the data subsets.

**Table 6.** Descriptive statistical summary of the calibration ($n$ = 122 logs) and validation ($n$ = 121 logs) data subsets.

| Variable | Data Subset | Mean | Minimum | Maximum | CV [a] (%) |
|---|---|---|---|---|---|
| Longitudinal stress wave velocity | Calibration | 3.42 | 2.74 | 4.32 | 7.8 |
| ($v_l$; km/s) | Validation | 3.41 | 2.69 | 4.29 | 8.8 |
| Modulus of elasticity ($m_e$; GPa) | Calibration | 12.28 | 7.14 | 16.57 | 14.8 |
| | Validation | 12.34 | 7.93 | 16.73 | 14.7 |
| Wood density ($w_d$; kg/m$^3$) | Calibration | 412.41 | 343.02 | 489.67 | 6.9 |
| | Validation | 412.02 | 337.88 | 482.02 | 6.8 |
| Microfibril angle ($m_a$; °) | Calibration | 12.10 | 6.84 | 19.23 | 21.7 |
| | Validation | 12.01 | 6.24 | 20.23 | 24.3 |
| Tracheid wall thickness ($w_t$; µm) | Calibration | 2.56 | 2.14 | 3.13 | 8.6 |
| | Validation | 2.55 | 2.08 | 3.16 | 8.3 |
| Tracheid radial diameter ($d_r$; µm) | Calibration | 30.59 | 27.99 | 33.00 | 3.8 |
| | Validation | 30.47 | 26.44 | 32.74 | 3.9 |
| Tracheid tangential diameter | Calibration | 28.01 | 26.41 | 29.88 | 2.5 |
| ($d_t$; µm) | Validation | 28.02 | 26.38 | 30.25 | 2.5 |
| Coarseness ($c_o$; µg/m) | Calibration | 388.35 | 318.32 | 459.90 | 8.1 |
| | Validation | 386.38 | 311.38 | 471.20 | 7.7 |
| Specific surface area ($s_a$; m$^2$/kg) | Calibration | 325.22 | 278.70 | 369.90 | 6.8 |
| | Validation | 326.36 | 275.59 | 372.46 | 6.4 |

[a] Coefficient of variation.

### 2.5. Specifying Functional Forms and Parameterization Methods Utilized

Deploying the expanded acoustic-based inferential framework, the Silviscan-derived (1) $m_e$, $m_a$, $w_t$, $d_r$, $d_t$, $c_o$ and $s_a$ estimates were expressed as a function of the Silviscan-derived wood density estimates and HM200-based acoustic velocity measures ($m_e$, $m_a$, $w_t$, $d_r$, $d_t$, $c_o$, $s_a = f\left(w_d v_l^2\right)$), and (2) $w_d$ estimates were expressed solely as a function of HM200-based acoustic velocity measures ($w_d = f\left(v_l^2\right)$). Statistically, the model specification procedure consisted of assessing the results from extensive graphical and correlation analyses to determine the most appropriate functional form for each of these 8 relationships. Employing the full data set, bivariate scatterplots were used to determine the degree of linearity between the each dependent and independent variable (i.e., $m_e$, $m_a$, $w_t$, $d_r$, $d_t$, $c_o$, and $s_a$ versus $w_d v_l^2$, and $w_d$ versus $v_l^2$). A visual interpretation of the graphics for $m_e$, $w_d$, $m_a$, $w_t$, $c_o$, and $s_a$ indicated mostly linear trends whereas there were no detectable trends for $d_r$ and $d_t$. Consequently, log-linear, log-log and power-based transformed relationships were also examined for $d_r$ and $d_t$ using both scatterplots and statistical measures of association. However, the graphical trends and related measures of linear association (Pearson moment correlation coefficient) from this supplementary analysis did not reveal the presence of viable linear relationships.

Similar to the approach used for quantifying acoustic—attribute relationships for red pine logs (i.e., [15]), a two-level hierarchical mixed-effects linear model specification consisting of fixed and random effects was used (sensu [28]). Specifically, at the first level, log-specific mean attribute values were expressed as a linear function of the density-weighted or density-unweighted acoustic velocity (Equation (3)).

$$F_{(k')_l} = \beta_{0_l(k')} + \beta_{1(k')}\left(w_{d_l} v_{l_i}^2\right) + \varepsilon_{(k')_l} \tag{3a}$$

$$F_{(w_d)_l} = \beta_{0_l(w_d)} + \beta_{1(w_d)}\left(v_{l_i}^2\right) + \varepsilon_{(w_d)l} \tag{3b}$$

where $F_{(k')_l}$ is the cumulative area-weighted moving average value for the $k'$th attribute where $k' = m_e, m_a, w_t, d_r, d_t, c_o$ or $s_a$ for the $l$th log, $F_{(w_d)l}$ is the cumulative area-weighted moving average wood density value for the $l$th log, $v_{l_i}$ is the mean acoustic velocity value for the $l$th log, $\beta_{0_l(k)}$ is a first-level random effects intercept parameter specific to the $k$th attribute ($k = k'$ attributes and $w_d$ inclusive) that is allowed to vary across the $L$ logs, $\beta_{1(k)}$ are first-level fixed effects slope parameter specific to the $k$th attribute, and $\varepsilon_{(k)_l}$ is a random error term specific to the $k$th attribute for the $l$th log. The random errors ($\varepsilon_{(k)_l}$) are assumed to be independent, uncorrelated and have constant variance. The second level expressed the first-level intercept parameter as a function of a grand mean plus a random error term whereas the slope parameter was considered a fixed effect (Equations (4a) and (4b), respectively).

$$\beta_{0(k)_l} = \gamma_{0(k)} + u_{0(k)_l} \tag{4a}$$

$$\beta_{1(k)} = \gamma_{1(k)} \tag{4b}$$

where $\gamma_{0(k)}$ and $\gamma_{1(k)}$ are attribute-specific grand mean values and $u_{0(k)_l}$ is an attribute-specific random error term specific to the $l$th log. The final mixed-effects model specifications were derived by combining the level one- and two-level expressions into a single model (i.e., Equation (5)).

$$F_{(k')_l} = \gamma_{0(k')} + \gamma_{1(k')}\left(w_{d_l} v_{l_l}^2\right) + u_{0(k')_l} + \varepsilon_{(k')_l} \tag{5a}$$

$$F_{(w_d)l} = \gamma_{0(w_d)} + \gamma_{1(w_d)}\left(v_{l_l}^2\right) + u_{0(w_d)_l} + \varepsilon_{(w_d)l} \tag{5b}$$

Given the complex error structures, the parameter estimates and model statistics were obtained via the deployment of the hierarchical linear and nonlinear modeling software program HLM7 [29]. Statistically, the program provides empirical-Bayes first-level parameter estimates for each randomly varying coefficient (intercept term), generalized least squares-based estimates for second-level terms, and maximum likelihood estimates for the variance and covariance components. Procedurally, employing the observational pairs from the calibration data subset, the models were parameterized and subsequently assessed on their compliance with the underlying statistical assumptions. Based on the protocol developed by Raudenbush and Bryk [28], this evaluation included testing the (1) constant variance assumption among first stage residuals, and (2) presence of significant random effects as determined via testing the null hypothesis ($u_0 \neq 0$) versus the alternative hypothesis ($u_0 = 0$). Park's homogeneity test was used to evaluate the constant variance assumption for each of the significant regression relationships (sensu [30]). This involved regressing the first stage Bayes residual values (logarithmic square values; dependent variable) against the predictive variable values (logarithmic values; independent variable) using a simple linear regression model specification, and then testing the null hypothesis that the resultant slope parameter estimate was not significantly different from zero at the 0.01 probability level. Resultant slope values not significantly different from zero were supportive of the homoscedasticity assumption. Conversely, slope values significantly different from zero are suggestive of the presence of heteroscedasticity (non-constant variance).

Additionally, given the potential effect of spatial correlation on statistical inference in terms of producing inefficient estimators without the minimum variance best linear unbiased estimator property when present, the spatial correlation among the empirical-Bayes residuals derived from the first-level models was assessed. The approach consisted of employing residual regressions in order to detect the presence of a first-order spatial correlation scheme among adjacent residuals for logs sampled from the same tree (i.e., either the 1st and 3rd log pair or the 2nd and 4th log pair): i.e., fitting the relationship $e_{l+2} = \hat{\rho} e_l + v_a$ where $e_l$ and $e_{l+2}$ are the residual values for the $l$th and $l$th + 2 ordinal positioned log, respectively (1st and 3rd ordinal log, or 2nd and 4th ordinal log, respectively), $\hat{\rho}$ is the first-order spatial correlation coefficient estimate ($-1 \leq \hat{\rho} \leq +1$), and $v_a$ is a random error term [30]. For significant ($p \leq 0.01$) regression relationships, the resultant $\hat{\rho}$ estimate reflected the magnitude of the first-order spatial correlation. Otherwise, for regressions that were not significant, spatial correlation was assumed to be absent and the hence the independence assumption was not rejected. The presence of potential outliers or influential observations, systematic lack-of-fits, and non-constant variance was also assessed through the examination of predictor—residual error bivariate scatterplots.

These selected specifications acknowledge the potential log-to-log variation in relationships that may be present and hence the intercept term was allowed to vary randomly. However, the slope term was treated as fixed. This latter constraint was established during the preliminary model specification phase by initially treating both the intercept and slope terms as random. In cases were convergence could not be achieved, the model specification was changed by defining the intercept as fixed and the slope as random and vice versa. Overall, the results indicated that (1) convergence could not be achieved when both terms were treated as random irrespective of the relationship, (2) convergence could not be achieved when the intercept term was treated as fixed and the slope was treated as random for all relationships but that for wood density, and (3) convergence for all relationships was only achieved when the slope was treated as fixed and the intercept term was treated as random. These statistical results were used to inform the selection of the final model specifications (i.e., Equations (5a) and (5b)) and hence were considered the most applicable for the sample population used.

### 2.6. Goodness-of-fit, Lack-of-fit, and Predictive Ability of Fitted Models

The parameterized models were evaluated based on goodness-of-fit, lack-of-fit, and predictive ability employing the validation data subset. More specifically, the proportion of variability in the dependent variable ($k$th attribute) explained by the parameterized model as measured by the index-of-fit squared statistic ($I^2_{(k)}$), was used as a goodness-of-fit metric (Equation (6)). Parameterized relationships which had mean absolute bias ($\overline{B}_{a(k)}$; Equation (7)) or mean relative bias ($\overline{B}_{r(k)}$; Equation (8)) that were significantly ($p \leq 0.05$) different from zero as determined through the use of the 95% confidence intervals (Equation (9)), where considered as demonstrating a consequential lack-of-fit. The linear regression relationship between the observed and predicted values was employed as a secondary lack-of-fit measure in terms of detecting systematic bias (sensu [31]). Based on Reynolds [32] statistical framework in combination with Gribko and Waint [33] SAS macro program extension, prediction and tolerance intervals for absolute and relative error where used to quantify the predictive accuracy of the parameterized relationships (Equations (10) and (11), respectively).

$$I^2_{(k)} = 1 - \frac{\sum\limits_{l=1}^{n_{(k)}} \left( V_{(k)_l} - \hat{V}_{(k)_l} \right)^2}{\sum\limits_{l=1}^{n_{(k)}} \left( V_{(k)_l} - \overline{V}_{(k).} \right)^2} \tag{6}$$

$$\overline{B}_{a(k)} = \frac{\sum\limits_{l=1}^{n_{(k)}} \left( V_{(k)_l} - \hat{V}_{(k)_l} \right)}{\sum\limits_{l=1}^{n_{(k)}} l} \tag{7}$$

$$\overline{B}_{r(k)} = \frac{\sum\limits_{l=1}^{n_{(k)}} 100 \frac{\left( V_{(k)_l} - \hat{V}_{(k)_l} \right)}{V_{(k)_l}}}{\sum\limits_{l=1}^{n_{(k)}} l} \tag{8}$$

$$\overline{B}_{a,r(k)} \pm \frac{S_{a,r(k)} \cdot t_{(n_{(k)}-1,0.975)}}{\sqrt{n_{(k)}}} \tag{9}$$

$$\overline{B}_{a,r(k)} \pm \sqrt{1/n_{(k)} + 1/n_p} \cdot S_{a,r(k)} \cdot t_{(n_{(k)}-1,0.975)} \tag{10}$$

$$\overline{B}_{a,r(k)} \pm g(\lambda, n_{(k)}, P') \cdot S_{a,r(k)} \tag{11}$$

where $V_{(k)_l}$ and $\hat{V}_{(k)_l}$ are the observed and predicted values, respectively, for the $k$th attribute within the $l$th sample log belonging to the validation data subset, $n_{(k)}$ is the number of predicted-observed pairs specific to the $k$th attribute within the validation data subset, $S_{a,r(k)}$ is the standard deviation of the absolute ($S_{a(k)}$) or relative ($S_{r(k)}$) biases specific to the $k$th attribute, respectively, $t_{(n_{(k)}-1,0.975)}$ is the 0.975 quantile of the $t$-distribution with $n_{(k)} - 1$ degrees of freedom specific to the $k$th attribute, $n_p$ is the number of future predictions under evaluation ($n_p = 1$; individual log level), and $g$ is a normal distribution tolerance factor specifying the probability ($\lambda$; i.e., 0.05) that a specified minimum proportion (i.e., 95%) of the distribution of errors ($P'$) will be contained within the stated tolerance interval.

### 2.7. Predictive Performance when Deploying Acoustic-Derived Wood Density Estimates

Given the practical reality of not having access to Silviscan-equivalent wood density estimates when acoustic sampling, necessitated the evaluation of the density-weighted acoustic relationships when a surrogate acoustic-derived $w_d$ estimate is used. Procedurally, this involved (1) generating a density estimate for each log within the validation data subset utilizing the parameterized wood density prediction model in combination with the acoustic velocity measurement, (2) given (1), deploying the resultant density estimate along with its acoustic velocity measurement to generate attribute predictions for each of the successfully parameterized models, and (3) using the observed and predicted values to calculate (i) mean absolute and relative biases along with associated 95% confidence intervals, and (ii) prediction and tolerance error intervals. The computations used to generate the mean absolute and relative biases and associated confidence intervals, and the prediction and tolerance error intervals, are similar to those described above (i.e., Section 2.6; Equations (7)–(11)).

## 3. Results

### 3.1. Attribute—Acoustic Velocity Relationships: Parameter Estimates, Regression Statistics and Tenability of Associated Assumptions

The resultant parameter estimates, regression statistics and validation metrics for assessing compliance with underlying assumptions for the attribute-specific acoustic velocity relationships, as described by the mixed-effects model specifications (Equation (5)), are given in Table 7. The successful parameterization of 6 out of the 8 relationships assessed ($m_e$, $w_t$, $m_a$, $c_o$, $s_a = f(\hat{\gamma}_{0,1}, w_d v_l^2)$ and $w_d = f(\hat{\gamma}_{0,1}, v_l^2)$) in terms of the (1) statistical significance of the parameter estimates and random effect term, (2) proportion of variability in the dependent variables explained ($I^2$), and (3) compliance with the independence and constant variance assumptions underlying the parameterization approach used, as assessed by the degree of spatial correlation among adjacent level one residuals employing regression analysis and Park's homogeneity of variance test [30], respectively, confirmed the applicability of the chosen mixed-effects regression specification in quantifying the attribute—acoustic velocity relationships for jack pine.

**Table 7.** Attribute-specific regression results for the mixed-effects model specification (Equation (5)): parameter estimates, regression statistics and compliance indicators.

| Relationship | Parameter Estimates [a] | | Regression Statistics and Compliance Indices | | | | |
| --- | --- | --- | --- | --- | --- | --- | --- |
| | $\hat{\gamma}_0$ | $\hat{\gamma}_1$ | Degrees of Freedom [b] $(n_{reg}, n_{res})$ | $I^2$ [c] | Random Effects [d] | Homogeneity of Variance [e] | Spatial Correlation [f] |
| $m_e - w_d v_l^2$ | 6.5439 | 0.001194 | 1, 120 | 0.466 | * | $H_0$ | $H_0$ |
| $w_d - v_l^2$ | 321.6483 | 7.7201 | 1, 120 | 0.315 | * | $H_0$ | $H_0$ |
| $m_a - w_d v_l^2$ | 15.0699 | −0.00063 | 1, 120 | 0.079 | * | $H_0$ | $H_0$ |
| $w_t - w_d v_l^2$ | 1.8190 | 0.000150 | 1, 120 | 0.603 | * | $H_0$ | $H_0$ |
| $d_r - w_d v_l^2$ | 30.6445 | $-0.000037^{ns}$ | 1, 120 | - | - | - | - |
| $d_t - w_d v_l^2$ | 28.2502 | $-0.000048^{ns}$ | 1, 120 | - | - | - | - |
| $c_o - w_d v_l^2$ | 300.0200 | 0.017790 | 1, 120 | 0.456 | * | $H_0$ | $H_0$ |
| $s_a - w_d v_l^2$ | 388.1280 | −0.012724 | 1, 120 | 0.497 | * | $H_0$ | $H_0$ |

[a] $\hat{\gamma}_0$ is a random effect intercept parameter estimate specific to the $k$th attribute and $\hat{\gamma}_1$ is fixed effect slope parameter estimate specific to the $k$th attribute. Note, parameter estimates not significantly ($p > 0.05$) different from zero are superscripted by ns. [b] Degrees of freedom for regression ($n_{reg}$) and residual error ($n_{res}$). [c] $I^2$ is the index-of-fit squared (Equation (6)). [d] Non-significant ($p > 0.05$) and significant ($p \leq 0.05$) random effects are denoted ns and *, respectively (i.e., testing the null hypothesis that $u_0 = 0$ versus the alternative hypothesis $u_0 \neq 0$ in Equation (5)). [e] Non-rejection and rejection of the constant variance assumption at the 0.01 probability level are denoted $H_0$ and $H_1$, respectively, as determined from Park's test for homoscedasticity (see text for details). [f] Non-rejection and rejection of the independence assumption (spatially uncorrelated residuals) at the 0.01 probability level are denoted $H_0$ and $H_1$, respectively, as determined from the residual regression approach ([30]; see text for details).

Specifically, for these fitted relationships, the intercept and slope parameter estimates and the random effect terms were significantly ($p \leq 0.05$) different from zero, and the percentage of variation explained varied from a relatively low minimum value of 8% for $m_a$ to a relatively moderate maximum value of 60% for $w_t$. The values for the remaining relationships ordered by magnitude of $I^2$, were as follows: 32% for $w_d$, 46% for $c_o$, 47% for $m_e$ and 50% for $s_a$. The lack of significant ($p \leq 0.01$) regression relationships among adjacent residuals for the log pairs derived from the same sample trees were indicative of the absence of spatial correlation effects (sensu [29]). Furthermore, the constant variance assumption was not rejected given the results of Park's test for homoscedasticity along with the interpretation of the Bayes residual error—predictor bivariate scatterplots. Note, during the initial parameterizations, examination of the circular-shape data point clusters within the residual error—predictor scatterplots, did reveal the presence of 2 potential outliers or influential observations (i.e., these observational pairs were visually apart from the concentrated residual cloud within the scatterplots for attributes $m_e$, $w_d$, $m_a$, $w_t$, $c_o$, and $s_a$). Thus a review of the field records and laboratory procedures in terms of identifying possible data recording errors, compiling errors or incorrect processing procedures for these suspect data pairs was implemented. Additionally, the attribute values for the other logs sampled from the suspect trees were also examined. Although this inquiry revealed the absence of recording or processing errors, the measurements for the suspect log within the calibration data subset and another one from the same sample tree but within the validation data subset, were substantially different from the measurements for the other logs within their respective data subsets. Based on the acknowledgment that these attributes may be occasionally influenced by uncontrollable and largely undetectable sources of variation such as complex internal knot distributions within logs, it was concluded that these suspect observations should be removed. Consequently, the calibration and validation data subsets were reduced by one observational pair each yielding a final total of 121 and 120 attribute-specific—acoustic velocity observational pairs within calibration and validation data subsets, respectively.

The attribute—acoustic velocity observational pairs within the calibration data subset for each of the 8 relationships evaluated are presented in Figure 1A,B. The successfully parameterized models are also superimposed on the observational pairs where applicable. Examination of the subgraphs for the attributes that were not successfully parameterized by the mixed-effects model specification (i.e., $d_r$ and $d_t$; Table 7), reinforced the lack of a functional relationship between each of these attributes and density-weighted acoustic velocity: i.e., random cloud of the $d_r$–$w_d v_l^2$ and $d_t$–$w_d v_l^2$ observational pairs that were devoid of any obvious linear or nonlinear relationship. Conversely, for the attributes that were successfully parameterized by the mixed-effects model specification (i.e., $m_e$, $m_a$, $w_d$, $w_t$, $c_o$, and $s_a$; Table 7) which are superimposed on their respective subgraphs confirmed the statistical findings (i.e., relationships were unbiased and in accord with the $m_e$–$w_d v_l^2$, $w_d$–$v_l^2$, $m_a$–$w_d v_l^2$, $w_t$–$w_d v_l^2$, $c_o$–$w_d v_l^2$ and $s_a$–$w_d v_l^2$ linear trends).

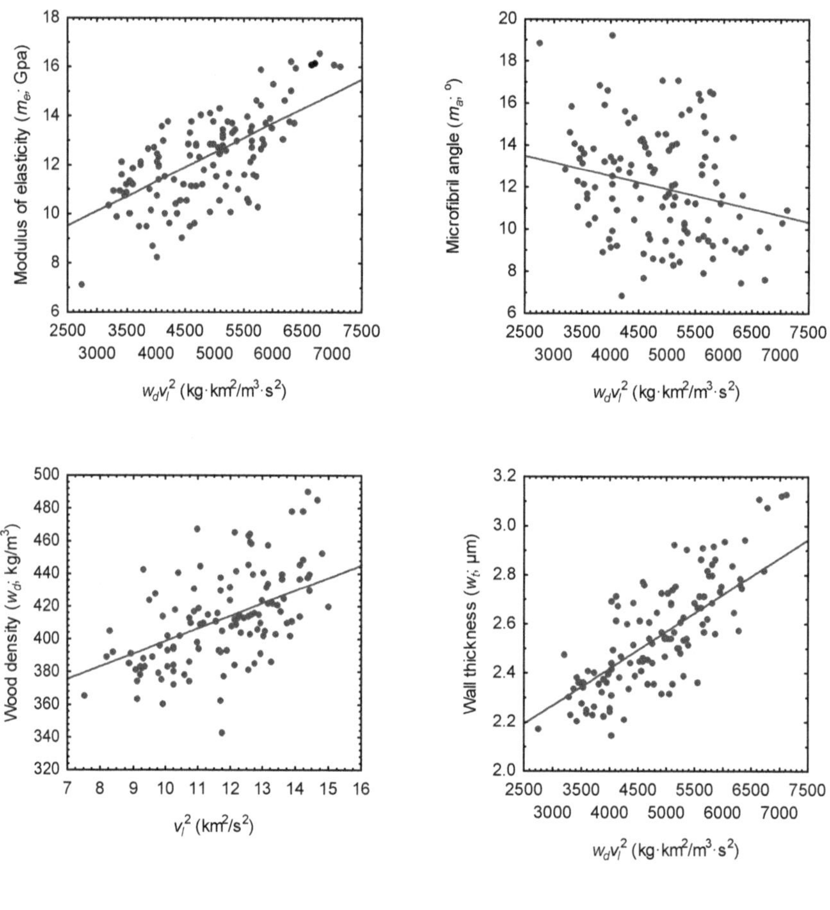

(**A**)

**Figure 1.** *Cont.*

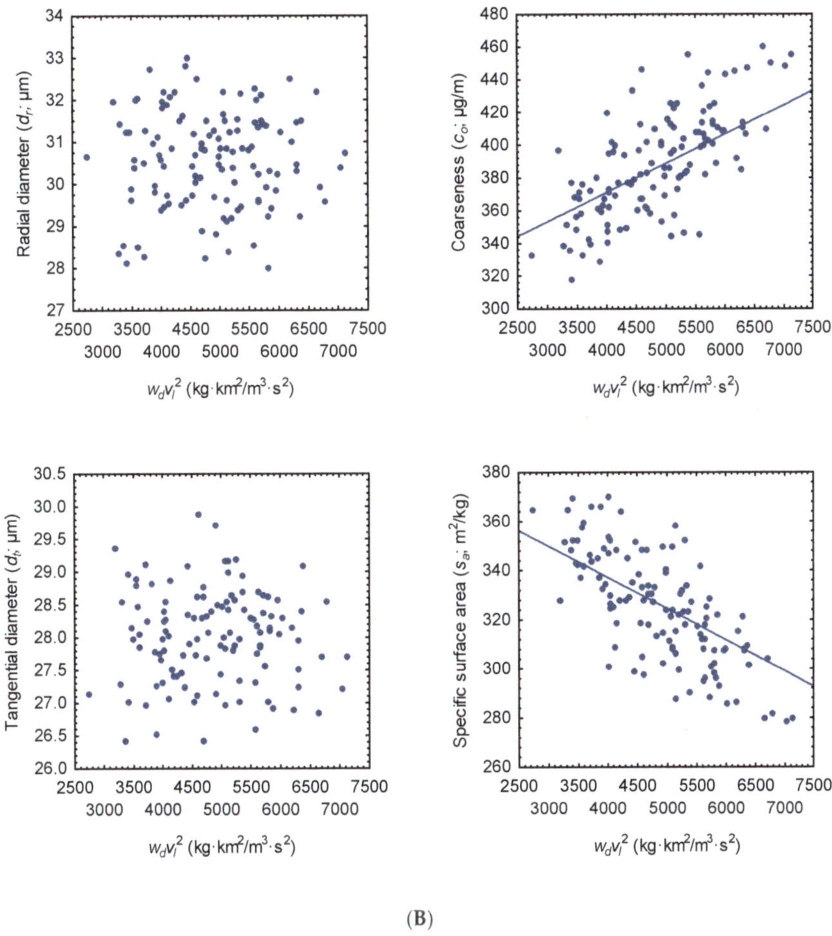

(**B**)

**Figure 1.** Scatterplots illustrating the relationship between each attribute and density-weighted and density-unweighted acoustic velocity ($v_l$) for the jack pine logs within the calibration data subset: (**A**) $m_e = f(w_d v_l^2)$, $w_d = f(v_l^2)$, $m_a = f(w_d v_l^2)$, and $w_t = f(w_d v_l^2)$ relationships; and (**B**) $d_r = f(w_d v_l^2)$, $d_t = f(w_d v_l^2)$, and $c_o = f(w_d v_l^2)$ and $s_a = f(w_d v_l^2)$ relationships. The parameterized model is denoted by solid line (Table 7).

### 3.2. Goodness-of-fit and Lack-of-fit Assessment

The goodness-of-fit evaluation consisted of assessing the magnitude of the variability explained by the parameterized models when applied to the validation data subset. Specifically, the $I^2$ statistic was calculated for the acoustic-based $m_e$, $w_d$, $m_a$, $w_t$, $c_o$ and $s_a$ mixed-effects models (Table 8). Results indicated that the models explained a relatively low to moderate level variation (ordered by magnitude): 8% for $m_a$, 30% for $w_d$, 43% for $c_o$, 44% for $s_a$, 53% for $m_e$ and 58% for $w_t$. In agreement with expectation, the values for 5 of 6 attributes were slightly less than those observed for the relationships parameterized using the calibration data subset (cf. Table 7 versus Table 8). The exception being that of the result for $m_e$ which was slightly greater than that derived from the calibration data subset. Based on the regression results for the linear relationship between the observed and predicted values, there was insufficient evidence to indicate systematic lack-of-fit for any of the 6 parameterized relationships: i.e., intercept and slope values were not significant ($p > 0.01$) different from zero and unity, respectively (sensu [31]). Similarly, based on the mean absolute biases and their associated 95% confidence intervals, there was no evidence of lack-of-fit for any of the 6 relationships (Table 8): mean absolute biases were not significantly ($p \leq 0.05$) different from zero as inferred from the 95% confidence intervals. Although approximately equivalent results were obtained when assessing lack-of-fit using mean relative biases, the mean relative bias generated for the $m_a$ relationship was significantly ($p \leq 0.05$) different from zero (Table 8).

The density-weighed and density-unweighted acoustic velocity—attribute observational pairs within the validation data subset for each of the relationships including those that did not exhibit a significant relationship with acoustic velocity, are graphically presented in Figure 2A,B. The parameterized mixed-effects regression relationships derived using the calibration data subset were also superimposed on the $m_e$, $w_d$, $m_a$, $w_t$, $c_o$ and $s_a$ subgraphs. It was visually evident that the parameterized relationships were representative of the linear trends between the $m_e$–$w_d v_l^2$, $w_d$–$v_l^2$, $m_a$–$w_d v_l^2$, $w_t$–$w_d v_l^2$, $c_o$–$w_d v_l^2$ and $s_a$–$w_d v_l^2$ observational pairs. The general concordance between the linear trends exhibited by the observational pairs and that established by the regressions, were also suggestive of the absence of systematic bias. For the attributes that were not successfully described by the mixed-effects model specification (i.e., $d_r$ and $d_t$; Table 7), an examination of the subgraphs reconfirmed the statistical result, that is, there was a random scatter of $d_r$–$w_d v_l^2$ and $d_t$–$w_d v_l^2$ observational pairs devoid of discernable linear or nonlinear trends.

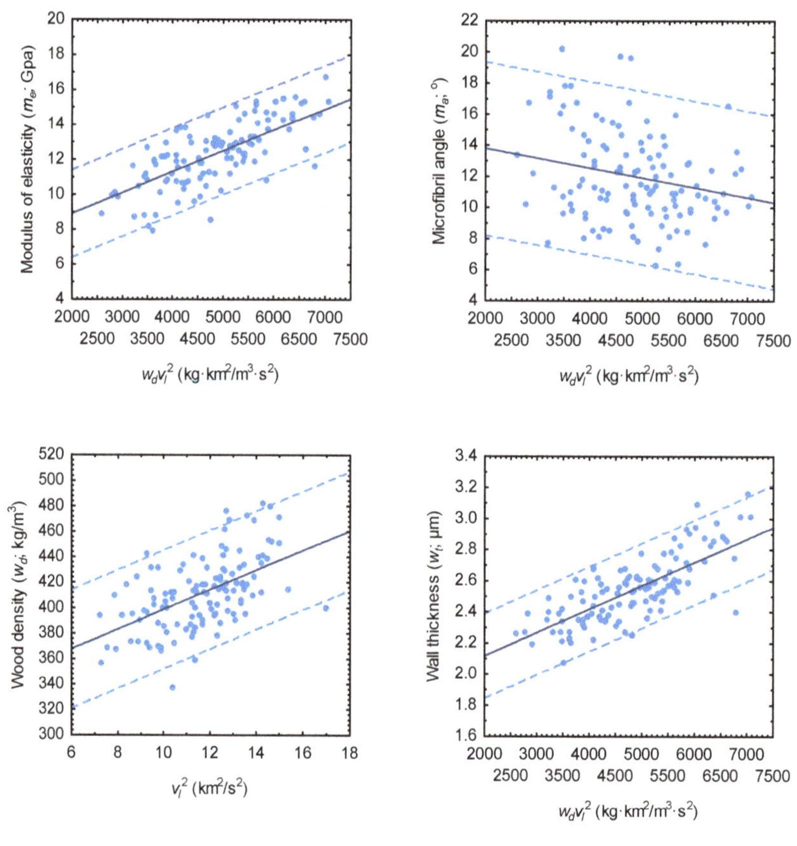

(A)

**Figure 2.** *Cont.*

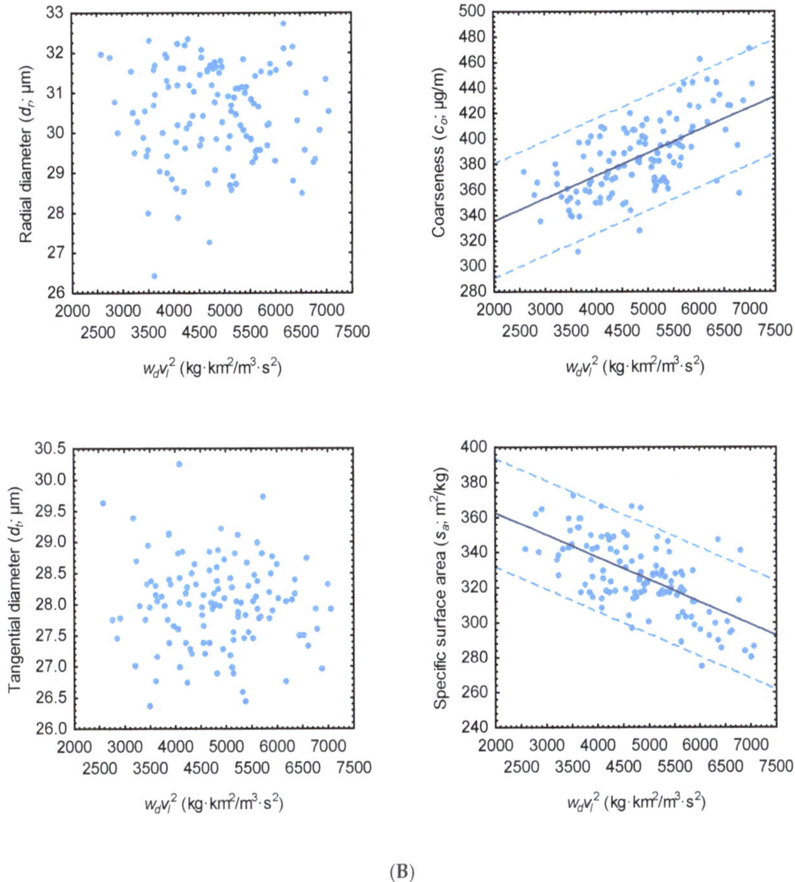

(**B**)

**Figure 2.** Scatterplots illustrating the relationship between each attribute and density-weighted and density-unweighted acoustic velocity ($v_l$) for the jack pine logs within the validation data subset: (**A**) $m_e = f(w_d v_l^2)$, $w_d = f(v_l^2)$, $m_a = f(w_d v_l^2)$, and $w_t = f(w_d v_l^2)$ relationships; and (**B**) $d_r = f(w_d v_l^2)$, $d_t = f(w_d v_l^2)$, and $c_o = f(w_d v_l^2)$ and $s_a = f(w_d v_l^2)$ relationships. Note, where applicable, the superimposed parameterized model is denoted by solid line (Table 7) and superimposed contextual 95% prediction error limits are denoted by the parallel dashed lines.

**Table 8.** Evaluation metrics for the parameterized models: goodness-of-fit, lack-of-fit statistics and predictive ability metrics.

| Relationship | Goodness-of-fit Statistic | Hypotheses [b] | | Lack-of-fit Measures | | | | Predictive Ability: 95% Error Intervals [d] | | | |
| | | | | Absolute [c] | | Relative [c] (%) | | Prediction | | Tolerance | |
| | $I^2$ [a] | $\alpha_0=0$ | $\alpha_1=1$ | Mean Bias | 95% CL | Mean Bias | 95% CL | Absolute 95% CL | Relative (%) 95% CL | Absolute 95% CL | Relative (%) 95% CL |
|---|---|---|---|---|---|---|---|---|---|---|---|
| $m_e - w_d v_l^2$ | 0.526 | $H_0$ | $H_0$ | $-0.031$ | $\pm0.226$ | 0.986 | $\pm2.049$ | $\pm2.489$ | $\pm\underline{22.543}$ | $\pm2.759$ | $\pm\underline{24.993}$ |
| $w_d - v_l^2$ | 0.304 | $H_0$ | $H_0$ | $-0.125$ | $\pm4.232$ | 0.292 | $\pm1.028$ | $\pm46.550$ | $\pm11.306$ | $\pm51.610$ | $\pm12.535$ |
| $m_a - w_d v_l^2$ | 0.075 | $H_0$ | $H_0$ | 0.050 | $\pm0.506$ | 5.998 * | $\pm\underline{4.553}$ | $\pm5.565$ | $\pm\underline{50.087}$ | $\pm6.170$ | $\pm\underline{55.531}$ |
| $w_t - w_d v_l^2$ | 0.583 | $H_0$ | $H_0$ | 0.004 | $\pm0.025$ | 0.151 | $\pm0.971$ | $\pm0.272$ | $\pm10.679$ | $\pm0.301$ | $\pm11.840$ |
| $c_o - w_d v_l^2$ | 0.426 | $H_0$ | $H_0$ | $-0.399$ | $\pm4.082$ | 0.250 | $\pm1.070$ | $\pm44.903$ | $\pm11.764$ | $\pm49.784$ | $\pm13.043$ |
| $s_a - w_d v_l^2$ | 0.438 | $H_0$ | $H_0$ | 0.224 | $\pm2.812$ | 0.305 | $\pm0.861$ | $\pm30.937$ | $\pm9.468$ | $\pm34.299$ | $\pm10.497$ |

[a] $I^2$ is the index-of-fit squared (Equation (6)). [b] Testing the (1) null hypothesis that $\alpha_0 = 0$ versus the alternative hypothesis $\alpha_0 \neq 0$ where $H_0$ and $H_1$ denote the non-rejection and rejection of the null hypothesis at the 0.05 probability level, and (2) null hypothesis that $\alpha_1 = 1$ versus the alternative hypothesis $\alpha_1 \neq 1$ where $H_0$ and $H_1$ denote the non-rejection and rejection of the null hypothesis at the 0.05 probability level, where $\alpha_0$ and $\alpha_1$ are derived from the simple linear ordinary least squares regression relationship between observed ($y$) and predicted ($x$) values ($y = \alpha_0 + \alpha_1 x + \varepsilon$ where $\varepsilon$ is an error term [31]). [c] Mean absolute (Equation (7)) and relative (Equation (8)) bias and the limits (CL) of the associated 95% confidence interval (Equation (9)) where mean values not significantly ($p > 0.05$) different from zero were indicative of an unbiased relationship (i.e., mean bias $\pm$ 95% confidence limits (CL)); note, underlined CL values denote approximations given the presence of significant ($p \leq 0.05$) non-normality within the underlying error distribution; absolute error units are attribute-specific: GPa, kg/m$^3$, $\mu$m, $\mu$g/m and m$^2$/kg for $m_e$, $w_d$, $m_a$, $w_t$, $c_o$ and $s_a$, respectively. [d] CL for the 95% prediction and tolerance error intervals for absolute and relative errors (Equations (10) and (11), respectively): mean bias $\pm$ 95% CL; specifically, there is a 95% probability that a future error will be within the stated prediction interval and that there is a 95% probability that 95% of all future errors will be within the stated tolerance interval [32,33]; note, underlined CL values denote approximations given the presence of significant ($p \leq 0.05$) non-normality within the underlying error distribution.

### 3.3. Predictive Ability

The predictive ability of the successfully parameterized mixed-effects models was quantified via the use of prediction and tolerance error intervals (Equations (9) and (10), respectively): mean bias $\pm$ 95% confidence limit. The intervals were generated from the absolute and relative bias values generated from the validation data subset. Statistically, these error intervals attempt to quantify the performance of the models when they are actually deployed [32]. Specifically, the prediction intervals quantify the boundaries of the expected range of absolute and relative errors when applying the models for a newly sampled jack pine log (e.g., there is 95% probability that the future error will be within the stated prediction interval). The tolerance intervals quantify the boundaries of the range of the expected population of absolute and relative errors which would be generated from the models when repeatedly deploying them to newly sampled jack pine log populations (e.g., there is 95% probability that 95% of all future errors will fall within the stated tolerance interval).

Accordingly, there is a 95% probability that a future error associated with a $m_e$, $w_d$, $m_a$, $w_t$, $c_o$ or $s_a$ prediction for a newly sampled jack pine log would be expected to fall within the following absolute and relative intervals (Table 8): (1) $-2.5 \leq m_e$ error (GPa) $\leq 2.5$; $-46.7 \leq w_d$ error (kg/m$^3$) $\leq 46.4$; $-5.5 \leq m_a$ error ($^\circ$) $\leq 5.6$; $-0.3 \leq w_t$ error ($\mu$m) $\leq 0.3$; $-45.3 \leq c_o$ error ($\mu$g/m) $\leq 44.5$; and $-30.7 \leq s_a$ error (m$^2$/kg) $\leq 31.2$; and (2) $-21.6 \leq m_e$ error (%) $\leq 23.5$; $-11.0 \leq w_d$ error (%) $\leq 11.6$; $-44.1 \leq m_a$ error (%) $\leq 56.1$; $-10.5 \leq w_t$ error (%) $\leq 10.8$; $-11.5 \leq c_o$ error (%) $\leq 12.0$; and $-9.2 \leq s_a$ error (%) $\leq 9.8$. Similarly, based on the tolerance error intervals as presented in Table 8, there is a 95% probability that 95% of all future errors generated from the $m_e$, $w_d$, $m_a$, $w_t$, $c_o$ and $s_a$ models would fall within the following absolute and relative intervals: (1) $-2.8 \leq m_e$ error (GPa) $\leq 2.7$; $-51.7 \leq w_d$ error (kg/m$^3$) $\leq 51.5$; $-6.1 \leq m_a$ error ($^\circ$) $\leq 6.2$; $-0.3 \leq w_t$ error ($\mu$m) $\leq 0.3$; $-50.2 \leq c_o$ error ($\mu$g/m) $\leq 49.4$; and $-34.1 \leq s_a$ error (m$^2$/kg) $\leq 34.5$; and (2) $-24.0 \leq m_e$ error (%) $\leq 26.0$; $-12.2 \leq w_d$ error (%) $\leq 12.8$; $-49.5 \leq m_a$ error (%) $\leq 61.5$; $11.7 \leq w_t$ error (%) $\leq 12.0$; $-12.8 \leq c_o$ error (%) $\leq 13.3$; and $-10.2 \leq s_a$ error (%) $\leq 10.8$. Although these intervals also reveal that the parameterized equations would generate unbiased predictions, the level of predictive accuracy attained would vary widely among the attributes (e.g., $s_a > w_t > w_d > c_o >> m_e >>> m_a$ based on the width of relative error tolerance intervals).

### 3.4. Predictive Performance of Parameterized Models When Deploying Acoustic Generated Wood Density Estimates

Employing the validation data subset, the empirical performance of the density-weighted models in cases where an acoustic-based wood density estimate was used as a surrogate for a Silviscan-based density estimate, was also evaluated. Specifically, based on the functional expressions, $m_e$, $m_a$, $w_t$, $c_o$, $s_a = f\left(\hat{\gamma}_{0,1}, \hat{w}_d v_l^2\right)$ where $\hat{w}_d$ is derived from the $w_d = f\left(\hat{\gamma}_{0,1}, v_l^2\right)$ relationship, the following set of computations were carried out: (1) inputting the acoustic velocity measurement value for each log in the validation data subset into the parameterized $w_d$ model (Table 7) and generating associated estimates for each attribute; and (2) given (1), treating these resultant attribute estimates as predicted values and the corresponding Silviscan-determined estimates as observed values, absolute and relative biases and associated 95% confidence, and 95% prediction and tolerance intervals, were calculated via the deployment of the computational structure given by Equations (7)–(11).

Upon review of the resulting predictive performance metrics as provided in Table 9, lack-of-fit issues were non-concerning for either of the error measures: 95% confidence intervals indicated that the mean absolute and relative biases were not significantly ($p \leq 0.05$) different from zero. Numerically, the magnitude of error generated when estimating $m_e$, $m_a$, $w_t$, $c_o$ and $s_a$ for a newly sampled jack pine log deploying the acoustic-based density estimate, would be captured within the following absolute and relative prediction error intervals: (1) $-2.8 \leq m_e$ error (GPa) $\leq 2.7$; $-5.5 \leq m_a$ error ($^\circ$) $\leq 5.6$; $-0.3 \leq w_t$ error ($\mu$m) $\leq 0.3$; $-51.8 \leq c_o$ error ($\mu$g/m) $\leq 50.9$; and $-36.5 \leq s_a$ error (m$^2$/kg) $\leq 37.0$; and (2) $-23.5 \leq m_e$ error (%) $\leq 25.8$; $-43.7 \leq m_a$ error (%) $\leq 55.7$; $-10.5 \leq w_t$ error (%) $\leq 10.8$; $-13.1 \leq c_o$ error (%) $\leq 13.7$; and $-10.9 \leq s_a$ error (%) $\leq 11.7$. Similarly, the corresponding 95%

tolerance intervals for which 95% of all future errors would be expected to be captured are delineated as follows: (1) $-3.1 \leq m_e$ error (GPa) $\leq 3.0$; $-6.1 \leq m_a$ error (°) $\leq 6.2$; $-0.3 \leq w_t$ error (µm) $\leq 0.3$; $-57.4 \leq c_o$ error (µg/m) $\leq 56.5$; and $-40.5 \leq s_a$ error (m$^2$/kg) $\leq 41.0$; and (2) $-26.2 \leq m_e$ error (%) $\leq 28.4$; $-49.1 \leq m_a$ error (%) $\leq 61.1$; $-11.7 \leq w_t$ error (%) $\leq 12.0$; $-14.6 \leq c_o$ error (%) $\leq 15.2$; and $-12.2 \leq s_a$ error (%) $\leq 12.9$. In summary, these results suggest that the attribute prediction models in which an acoustic-based wood density estimate is deployed, would yield unbiased predictions which would be slightly less precise than that attained when using a Silviscan-derived or equivalent density estimate, and would exhibit a similar among-attribute performance ranking (cf. Table 8 versus Table 9).

## 4. Discussion

### 4.1. Hierarchical Mixed-Effects Acoustic-Based Attribute Prediction Models for Jack Pine

The mixed-effects linear model for the primary relationship between the dynamic modulus of elasticity and density-weighted acoustic velocity ($m_e = f(\hat{\gamma}_{0,1}, w_d v_l^2)$) explained 47% of the variation in the $m_e$ for the logs within the calibration data subset (Table 7) and 53% of the variation in the $m_e$ for the logs within the validation data subset (Table 8). Based on the logs within the validation data subset, the model generated unbiased predictions when assessed using both absolute and relative error measures. The tolerance intervals indicated that 95% of the absolute and relative errors generated from repeatedly applying the parameterized model to a new sample population of jack pine logs would be expected (95% probability level) not to exceed $\pm 2.8$ GPa and $\pm 25.0$% respectively, when using Silviscan-equivalent $w_d$ estimate (Table 8). For the deployment scenario in which an acoustic-based wood density estimate is employed as a surrogate measure for its Silviscan-based counterpart, the model's predictions would be similarly unbiased (Table 9). However, the predictions would be less precise as evident from the wider tolerance intervals for absolute and relative error that were generated when using the acoustic-based density estimate (i.e., 95% tolerance limits for error of $\pm 3.1$ GPa and $\pm 27.3$%, respectively; Table 9). Newton [15] reported slightly better results in a similarly structured analysis of red pine logs in terms of explanatory power (e.g., explaining 71% of variation in $m_e$ for red pine logs versus 50% of variation in $m_e$ for jack pine logs) and precision of predictions when deploying either a Silviscan-equivalent estimate of wood density (e.g., relative tolerance error intervals of $\pm 19$% for red pine versus $\pm$ 25% for jack pine) or acoustic-based estimate (e.g., relative tolerance error intervals of $\pm 19$% for red pine versus $\pm$ 27% for jack pine).

The results for the mixed-effects linear regression model for the relationship between wood density and acoustic velocity ($w_d = f(\hat{\gamma}_{0,1}, v_l^2)$), revealed that the parameterized model generated unbiased predictions (Tables 7 and 8). Overall, however, the model performed moderately in terms of (1) explanatory power given that only 32% and 31% of the $w_d$ variation for the logs within the calibration and validation data subsets, respectively, was explained (Tables 7 and 8), and (2) predictive performance given that 95% of absolute and relative errors generated from repeatedly applying the equation to a new sample population of logs would be expected (95% probability level) not to exceed $\pm 51.6$ kg/m$^3$ or $\pm 12.5$%, respectively (Table 8). Considerably better results were reported previously for red pine logs [15] in terms of explanatory power (e.g., explaining 80% of variation in $w_d$ for red pine logs versus 31% of variation in $w_d$ for jack pine logs). However, differences in the precision of the predictions were only marginally better (e.g., relative tolerance error intervals of $\pm 10$% for red pine versus $\pm$ 13% for jack pine).

**Table 9.** Predictive performance of parameterized models when deploying acoustic-based wood density estimates.

| Relationship | Lack-of-fit Measures [a] | | | | Predictive Ability: 95% Error Intervals [b] | | | |
| | Absolute | | Relative (%) | | Prediction (Stand-Level) | | Tolerance | |
| | Mean bias | 95% CL [e] | Mean bias | 95% CL | Absolute 95% CL | Relative (%) 95% CL | Absolute [b] 95% CL | Relative (%) 95% CL |
|---|---|---|---|---|---|---|---|---|
| $m_e - w_d v_l^2$ | −0.033 | ±0.251 | 1.118 | ±2.240 | ±2.756 (±0.562) | ±24.640 (±5.028) | ±3.055 | ±27.318 |
| $m_a - w_d v_l^2$ | 0.051 | ±0.505 | 6.000 * | ±4.522 | ±5.554 (±1.129) | ±49.738 (±10.111) | ±6.158 | ±55.144 |
| $w_t - w_d v_l^2$ | −0.004 | ±0.025 | 0.151 | ±0.971 | ±0.272 (±0.055) | ±10.679 (±2.171) | ±0.301 | ±11.840 |
| $c_o - w_d v_l^2$ | −0.440 | ±4.671 | 0.300 | ±1.219 | ±51.383 (±10.445) | ±13.406 (±2.725) | ±56.968 | ±14.863 |
| $s_a - w_d v_l^2$ | 0.253 | ±3.340 | 0.369 | ±1.027 | ±36.734 (±7.467) | ±11.291 (±2.295) | ±40.727 | ±12.519 |

[a] Mean absolute (Equation (7)) and relative (Equation (8)) bias and the limits (CL) of the associated 95% confidence interval (Equation (9)) where mean values not significantly ($p > 0.05$) different from zero were indicative of an unbiased relationship; note, underlined CL values denote approximations given the presence of significant ($p \leq 0.05$) non-normality within the underlying error distribution; absolute error units are attribute-specific: GPa, °, µm, µg/m and m²/kg for $m_e$, $m_a$, $w_t$, $c_o$ and $s_a$, respectively; and * denotes significant bias. [b] Confidence limits for the 95% prediction and tolerance error intervals for absolute and relative errors (Equations (10) and (11), respectively): mean bias ± 95% CL; specifically, there is a 95% probability that a future error will be within the stated prediction interval and that there is a 95% probability that 95% of all future errors will be within the stated tolerance interval (sensu [32]); notes, (1) absolute error units are attribute-specific as stated above, (2) underlined CL values denote approximations given the presence of significant ($p \leq 0.05$) non-normality within the underlying error distribution, and (3) stand-level CL intervals denote the 95% prediction limits for the mean error generated when group sampling (e.g., assigning a mean attribute estimate to 30-log sample (pile) derived from a single stand).

Although the results for the relationship where microfibril angle is expressed as a function of density-weighted acoustic velocity ($m_a = f(\hat{\gamma}_{0,1}, w_d v_l^2)$) generated unbiased predictions (Tables 7 and 8), the parameterized model exhibited a low level of explanatory power and a relatively high degree of imprecision. Specifically, the model only explained 8% of the variation in $m_a$ for the logs within both the calibration and validation data subsets, respectively (Tables 7 and 8). The predictive intervals indicated that 95% of absolute and relative errors generated from repeatedly applying the equation to a new sample population of logs would be expected (95% probability level) not to exceed (1) ±6.2 ° and ±55.5%, respectively, when using Silviscan-equivalent $w_d$ estimates (Table 8), and (2) ±6.2 ° and ±55.1%, respectively, when deploying acoustic-based wood density estimates. Although these weak results are superior to those obtained previously for red pine logs in which no viable acoustic-based relationship was found for microfibril angle [15], the results are inferior to the level of variation described by the acoustic-based relationship reported in other studies (e.g., 86% for radiata pine logs [6]).

The relationship for cell wall thickness ($w_t = f(\hat{\gamma}_{0,1}, w_d v_l^2)$) exhibited the highest levels of explanatory power as evident from the level of variation explained: 60% and 58% for the calibration and validation data subsets, respectively (Tables 7 and 8). Furthermore, the parameterized model exhibited no lack-of-fit issues and produced unbiased predictions at a relatively high level of precision. Specifically, 95% of absolute and relative errors generated from repeatedly applying the equation to a new sample population of jack pine logs would be expected (95% probability level) to not exceed ±0.3 µm and ±11.8%, respectively, when using Silviscan-equivalent or acoustic generated $w_d$ estimates (Tables 8 and 9). Newton [15] reported a superior result for red pine logs in terms of explanatory power (e.g., explaining 90% of variation in $w_t$ for red pine logs versus 59% of variation in $w_t$ for jack pine logs), but much less so in terms of predictive precision. For example, relative tolerance error intervals of ±9% for red pine versus ±12% for jack pine when deploying a Silviscan-equivalent estimate of wood density, and ±12% for red pine versus ±12% for jack pine when using an acoustic-based estimate. The graphical and statistical results for the relationships in which radial and tangential diameters were expressed as functions of density-weighted acoustic velocity, did not support the existence of viable relationships for these attributes. This result is similar to that obtained for red pine logs (i.e., [15]). Combining these regression results with the observed lack of linear associations exhibited in the scatterplots (Figures 1 and 2 for jack pine logs; and Figures 1 and 2 for red pine logs in [15]), suggest that these attributes may be intrinsically unrelated to acoustic velocity for these species.

The regression relationships for fiber coarseness and specific surface area ($c_o$, $s_a = f\left(\hat{\gamma}_{0,1}, w_d v_l^2\right)$) exhibited moderate levels of explanatory power (Tables 7 and 8): (1) 46% and 43% of the variation in fiber coarseness was explained for the logs within the calibration and validation data subsets, respectively; and (2) 50% and 44% of the variation in specific surface area was explained for the logs within the calibration and validation data subsets, respectively. These parameterized models produced unbiased and relative precise predictions when compared with the other attribute models. For the fiber coarseness model, 95% of absolute and relative errors generated from repeatedly applying the parameterized equation to a new sample population of logs would be expected (95% probability level) not to exceed (1) ±49.8 µg/m and ±13.0%, respectively, when using Silviscan-equivalent wood density estimates (Table 8), and (2) ±57.0 µg/m and ±14.9%, respectively, when deploying acoustic-based wood density estimates (Table 9). Similarly, for the model developed for specific surface area, the corresponding precision limits for absolute and relative error were (1) ±34.3 m$^2$/kg and ±10.5%, respectively, when using Silviscan-equivalent wood density estimates (Table 8), and (2) ±40.7 m$^2$/kg and ±12.5%, respectively, when deploying acoustic-derived wood density estimates (Table 9).

In relation to acoustic-based prediction of fiber coarseness, Newton [15] reported much improved results for red pine logs in terms of explanatory power (e.g., explaining 81% of variation in $c_o$ for red pine logs versus 44% of variation in $c_o$ for jack pine logs) but again much less so in terms of predictive precision (e.g., relative tolerance error intervals of ±12% for red pine versus ±13% for jack pine when deploying a Silviscan-equivalent estimate of wood density, and ±14% for red pine versus ±15% for jack pine when using an acoustic-based estimate). In comparison of the model developed for predicting specific surface area for red pine logs, the result was considerably superior in terms of explanatory power (e.g., 84% of the variation in $s_a$ explained for red pine logs versus 47% of the variation in $s_a$ explained for jack pine logs). However, similar to the interspecies comparison of the coarseness model, the difference in predictive ability was minimal between the species: (1) relative tolerance error intervals of ±8% for red pine versus ±11% for jack pine when deploying a Silviscan-equivalent estimate of wood density, and (2) ±10% for red pine versus ±13% for jack pine when using an acoustic-based estimate.

*4.2. Potential Utility of the Expanded Acoustic-Based Inferential Framework for Jack Pine*

Estimation of the modulus of elasticity is of primary importance in log-segregation operations where the objective is to forecast the potential type and grade of extracted solid wood products (e.g., dimension lumber). However, other wood attributes (e.g., wood density, microfibril angle, tracheid dimensions) and derived composite metrics (e.g., fiber coarseness and specific surface area), are also prime determinates of end-product potential (sensu Table 1). Consequently, the expanded acoustic inferential framework presented in this study along with its empirical validation for six of the eight jack pine attributes examined (dynamic modulus of elasticity, wood density, microfibril angle, cell wall thickness, fiber coarseness and specific surface area), yields a more comprehensive system for non-destructive forecasting of end-product potential than systems based solely on wood stiffness (modulus of elasticity). Operationally, however, the in-forest deployment of the parameterized relationships will be largely dependent on the objectives and precision requirements of the end-user. For example, the relative error intervals were quite large for $m_e$ and $m_a$ as exemplified by the resultant tolerance errors of ±25% and ±56%, respectively, when using a Silviscan-equivalent wood density estimate (Table 8), and ±27% and ±55%, respectively, when using an acoustic-based wood density estimate (Table 9). Thus, the ability to stratify individual jack pine logs into narrow-grade class-based acoustic estimates of $m_e$ and $m_a$ would be difficult if not impossible. However, in situations for which the objective is to segregate groups of logs into end-product categories according to their average attribute values, then the magnitude of estimation error associated with the acoustic-based population mean estimates may be more operationally acceptable. For example, the mean absolute and relative error to be expected when collectively sampling 30 new jack pine logs deploying an acoustic-based wood density estimate would be, respectively, 0.6 GPa and 5% for $m_e$, and 1° and 10% for $m_a$ (Table 9).

Comparably, the relative tolerance error intervals for $w_d$, $w_t$, $c_o$ and $s_a$ were considerably less (i.e., errors in the order of $\pm 10$ to $\pm 15\%$ (Table 8)). Thus, depending on the specified width of the $w_d$, $w_t$, $c_o$ and $s_a$ intervals the end-user employs to designate logs into specific end-product categories or a binary grade class, differentiating logs according to these attribute predictions may be feasible. Similarly, at the stand-level where the objective is to classify a population of logs in a specific end-product category based on their mean $w_d$, $w_t$, $c_o$ or $s_a$ values, the precision of the estimates would be improved considerably. For example, the mean absolute and relative error to be expected when collectively sampling 30 new jack pine logs deploying an acoustic-based wood density estimate would be respectively, 0.1 μm and 2% for $w_t$, 10.4 μg/m and 3% for $c_o$, and 7.5 m$^2$/kg and 2% for $s_a$ (Table 9).

More generally however, based on the explanatory power and precision of acoustic-based attribute prediction models developed to date, increasing the precision of the point-estimates derived from acoustic-based models will be required if they are to be used to stratify individual logs into grade categories given the relatively narrow range of static modulus of elasticity values that currently delineate product grade classes. For example, the mean difference between the 14 machine stress-rated (MSR) lumber grade categories established for Canadian softwood species [34] is approximately 0.7 GPa and hence the explicit stratification of individual logs into such narrow-width grade classes would not be possible given the empirical results to date. Even if it is assumed that the dynamic estimate is equivalent to its static counterpart which in turn is reflective of the actual end-product-based value, the precision of most acoustic-based estimates would be inadequate for stratifying logs into MSR-based classes (e.g., $\pm 3$ GPa expected error for individual jack pine logs; Table 9). Thus reducing the number of grade categories from 14 to a smaller number of discrete quality classes based on the expected prediction error range(s) through clustering, and (or) using multiple individual log estimates to generate a mean site, landing or log pile value, may represent a viable alternative when using the acoustic approach in log-segregation operations for jack pine.

Further research into analytical advancements that increase the amount of variability explained and decrease the standard error of estimate, possibly by including covariates within the acoustic-attribute model specification as suggested by Bérubé-Deschênes et al. [35] and Butler et al. [36], may be warranted. Acoustic velocity is known to be affected by a multitude of internal and external factors which includes knot distributions, embedded voids, environmental conditions during acoustic sampling (temperature and moisture conditions), tree size, local competition, and overall site conditions (e.g., [35,37,38]). Provision of operational solutions for minimizing these effects could improve the predictive accuracy of acoustic-based attribute estimation. Recent results examining the effect of xylem temperature and moisture on acoustic velocity within standing semi-mature jack pine trees during the vegetative growing season indicated that acoustic velocity declined in linear fashion with increasing temperature; however, moisture had no appreciable influence [39]. However, the temperature effect was not of consequential significance except when temperatures approached their seasonal extremities. Assuming similar inferences apply to jack pine logs, suggest that acoustic log samplers could treat such variation as a source of random error of minimal importance providing xylem temperatures did not approach their seasonal minimums (<5 °C) or maximums (>30 °C). In cases were temperature effects were of concern, the standardization equation for adjusting acoustic velocities to a reference xylem temperature of 20 °C could be deployed [39].

*4.3. Similarities and Differences between Tree and Log Acoustic-Based Attribute Relationships*

The time-of-flight acoustic approach to estimating the internal attributes within standing trees employs the dilatational stress wave velocity ($v_d$) which transverses the breast-height (1.3 m) region of the main stem [40]. This acoustic velocity measurement when weighted by wood density is also related to the modulus of elasticity as described by a functional specification, similar to that given by

Equation (1) (i.e., replacing $v_l$ by $v_d$). This primary relationship for standing trees can be also potentially expanded to include a similarly structured suite of secondary relationships as that presented in this study for logs (e.g., [16]).

Conceptually, the resonance-based acoustic velocity approach is considered to yield a more representative estimate of the fiber attributes within the xylem tissue than those estimates derived from the time-of-flight approach. This is principally because the mean velocity estimate is derived from many acoustic pulses resonating longitudinally throughout the log rather than the velocity of a single wave front moving between two points within the main stem of a standing tree [41]. The empirical validity of this inference can be partially tested by assessing the differences between the tree and log acoustic-based attribute prediction models parameterized using the same sample tree population. Contrasting the results derived from a companion analysis which assessed the time-of-flight acoustic relationships deploying the same suite of attributes and sample tree population which were measured at the same time [42], revealed that the time-of-flight-based relationships were mostly superior in terms of explanatory power. Specifically, the percentage of attribute variation explained by the dilatational (tree) and longitudinal (log) acoustic relationships were respectively: 71% versus 50% for $m_e$; 30% versus 31% for $w_d$; 19% versus 8% for $m_a$; 66% versus 59% for $w_t$; 42% versus 44% for $c_0$; and 61% versus 47% for $s_a$. However, conversely, in terms of predictive precision, the comparison between the log and tree models revealed minimal differences. The error arising from using a newly sampled acoustic velocity measurement used to estimate $m_e$, $w_d$, $m_a$, $w_t$, $c_0$ and $s_a$ for standing trees and derived logs would be expected at the 95% confidence level to fall within the respective relative error limits of $\pm25\%$ versus $\pm21\%$ for $m_e$, $\pm13\%$ versus $\pm14\%$ for $w_d$, $\pm56\%$ versus $\pm47\%$ for $m_a$, $\pm12\%$ versus $\pm11\%$ for $w_t$, $\pm13\%$ versus $\pm12\%$ for $c_0$, and $\pm10\%$ versus $\pm10\%$ for $s_a$. These similarities in terms of the predictive performance between the tree and log-based acoustic approaches for estimating attributes within the same species (jack pine) while attaining contrasting results between species (jack pine versus red pine for a given time-of-flight- or resonance-based relationship), suggest that acoustic-based attribute relationships may be intrinsic to a given species irrespective of either of these wave types.

### 4.4. Advancing Acoustic-Based Attribute Estimation

The resonance-based acoustic approach to estimating internal fiber attributes within logs and associated demonstrations of its utility in segregation, allocation and merchandizing operations have been presented to the forest management community through various case studies (e.g., [41,43,44] and comprehensive literature reviews (e.g., [11,17]). These contributions have mostly focused on the merits of estimating the modulus of elasticity and using it, or its surrogates, to forecast end-product potential. These include (1) treating wood density as a constant and inferring wood stiffness indirectly from acoustic velocity (e.g., implicitly based on the underlying theoretical relationship expressed by Equation (1)), (2) estimating wood stiffness using parameterized regression models in which the dynamic modulus of elasticity is expressed as a function of acoustic velocity and wood density (e.g., explicitly based on the conceptual formulation (Equation (1)), and (3) directly forecasting wood stiffness using relationships that explicitly relate the static modulus of elasticity of a given end-product (e.g., measured using static bending tests of sawn boards) to acoustic velocity. This last approach while the most informative is also the least developed. Consequentially, the full potential of in-forest acoustic-based forecasting of end-product potential has not been fully realized. Determination and quantification of the explicit relationship between log-based acoustic velocity measures and actual performance-based metrics within the derived manufactured end-product are also required. Preliminary results from several studies in which the static modulus of elasticity within dimensional lumber products have been explicitly related to log-based acoustic velocity measures have been positive. These include statistically viable relationships reported for balsam fir (*Abies balsamea* (L.) Mill.) and white spruce (*Picea glauca* (Moench) Voss.) [45], radiata pine [46], Douglas fir ((*Pseudotsuga menziesii* (Mirb.) Franco) [47], and loblolly pine (*Pinus taeda* L.) [48].

Similarly, previous research has shown that several important performance metrics associated with pulp and paper end-products are also empirically related to acoustic velocity measurements taken on logs. Clark et al. [48] found that fiber length, pulp strength and various handsheet properties derived from radiate pine logs varied systematically with acoustic velocity. Bradley et al. [49] demonstrated that the variation in the pulp quality (freeness) for radiata pine logs could be considerably reduced by stratifying the logs according to their acoustic velocity values before processing. Although beyond the scope of this study, establishment of a broader range of more explicit relationships between log-based acoustic velocity measures and attribute-derived performance metrics within actual manufactured end-products, would be an area worthy of further investigation. As these past studies have shown, in-forest acoustic velocity stratification of log populations not only improves segregation efficiency at the time of harvest but also assists in reducing the variation in end-product quality during the manufacturing and merchandizing stages.

In-forest acoustic log grading has been increasingly used to compliment visual-based approaches in the pursuit of deriving economically efficient sorting networks. This remains an iterative and ongoing process in wood allocation decision-making where the objective is to increase operational efficiency within the upstream portion of the forest products supply chain (sensu [50]). Mechanized acoustic sampling has also advanced to the point where onboard resonance tools have been installed directly on harvesting machines where logs are immediately sorted upon bucking according to their end-product potential (e.g., [51]). Although these innovations have advanced the in-forest non-destructive approach to log grading and sorting, the acoustic approach has been largely limited to providing a single measure of internal wood quality, that being wood stiffness. Thus, the confirmatory empirical results presented in this study for jack pine and previous for red pine logs [15] are collectively supportive of an expanded acoustic-based inferential framework in which estimates for a multitude of end-product-based attribute determinates can be attained: e.g., wood density, microfibril angle, tracheid wall thickness, fiber coarseness, and specific surface area, in addition to the dynamic modulus of elasticity (wood stiffness). Although further research is required to fully realize the benefits of this empirical-based framework, the ability to non-destructively forecast a wide array of commercially relevant attributes could have consequential potential utility in advancing the acoustic approach in forest operations.

## 5. Conclusions

Given the wide spread occurrence of jack pine across the boreal landscape combined with the vast array of potential end-products it can produce, inclusive of solid wood products (e.g., dimensional lumber) and associated mill-work derivatives (window frames, doors, shelving, moldings, and paneling, and composite lumber products such as glulam-based beams, headers and heavy trusses and finger-jointed joists and rafters), and pulp-derived products such as paperboards, newsprint, facial tissues and specialized coated papers [52], the species has become the dominant feedstock species for numerous industrial conversion facilities. However, this diversity of end-products complicates in-forest segregation, allocation, and merchandizing decision-making. Hence the provision of enhanced operational intelligence arising from in-forest forecasts of end-product potential of harvested logs through non-destructive acoustic-based methods, may yield increased efficiencies within the upper portion of the jack pine forest products supply chain. Consequentially, the development and evaluation of a suite of acoustic-based models for predicting the principal attributes governing end-product potential for jack pine as presented in this study, represents an incremental contribution towards more informed decision-making. Specifically, deploying a mixed-effects linear modeling approach combined with cross-validation techniques, viable forecasting models for predicting the dynamic modulus of elasticity, wood density, microfibril angle, cell wall thickness, fiber coarseness and specific surface area were developed. Although these positive results confer additional empirical support for the proposed acoustic-based inferential framework, further research in the areas of accounting for environmentally induced wave variation, specifying end-product-based design thresholds, and explicitly establishing linkages between log-based attribute estimates and those within recoverable end-products, would be beneficial. Collectively, the results presented here for jack pine not only provides the prerequisite parameterized relationships for improving in-forest segregation and allocation decision-making but also contributes to solidifying the empirical foundation of the expanded acoustic-based inferential framework.

**Funding:** Canadian Wood Fiber Centre, Canadian Forest Service, Natural Resources Canada.

**Acknowledgments:** The author expresses his appreciation to: (1) Mike Laporte (retired) of the Canadian Wood Fiber Centre, Canadian Forest Service (CFS), Natural Resources Canada (NRCan), and Gordon Brand of the Great Lakes Forestry Centre, CFS, NRCan for field and laboratory data acquisition support; (2) Dr. Tong and Ny Nelson at FPInnovations Inc., Vancouver, British Columbia, Canada, for completing the SilviScan-3 analysis, (3) Canadian Wood Fiber Centre, CFS, NRCan for fiscal support; and (4) anonymous journal reviewers for constructive and insightful comments and suggestions.

**Conflicts of Interest:** The author declares no conflict of interest.

## References

1. Emmett, B. Increasing the value of our forest. *For. Chron.* **2006**, *82*, 3–4. [CrossRef]
2. Emmett, B. Perspectives on sustainable development and sustainability in the Canadian forest sector. *For. Chron.* **2006**, *82*, 40–43. [CrossRef]
3. Zhang, S.Y.; Chauret, G.; Ren, H.Q.; Desjardins, R. Impact of plantation black spruce initial spacing on lumber grade yield, bending properties and MSR yield. *Wood Fibre Sci.* **2002**, *34*, 460–475.
4. Tsehaye, A.; Buchanan, A.H.; Walker, J.C.F. Sorting of logs using acoustics. *Wood Sci. Technol.* **2000**, *34*, 337–344. [CrossRef]
5. Carter, P.; Briggs, D.; Ross, R.J.; Wang, X. Acoustic testing to enhance western forest values and meet customer wood quality needs. In *Productivity of Western Forests: A Forest Products Focus*; Harrington, C.A., Schoenholtz, S.H., Eds.; Gen. Tech. Rep. PNW-GTR-642; USDA, Forest Service, Pacific Northwest Research Station: Portland, OR, USA, 2005; pp. 121–129.
6. Wang, X.; Carter, P.; Ross, R.J.; Brashaw, B.K. Acoustic assessment of wood quality of raw materials: A path to increased profitability. *For. Prod. J.* **2007**, *57*, 6–14.
7. National Lumber Grades Authority (NLGA). *Standard Grading Rules for Canadian Lumber*; NLGA: Surrey, BC, Canada, 2014.
8. Defo, M. SilviScan-3—A Revolutionary Technology for High-Speed Wood Microstructure and Properties Analysis. Midis de al Foresterie. UQAT. Available online: http://chaireafd.uqat.ca/midiForesterie/pdf/20080422PresentationMauriceDefo.pdf (accessed on 1 October 2018).
9. Wang, X.; Ross, R.J.; Mattson, J.A.; Erickson, J.R. Nondestructive evaluation techniques for assessing modulus of elasticity and stiffness of small-diameter logs. *For. Prod. J.* **2002**, *52*, 79–85.
10. Dickson, R.L.; Raymond, C.A.; Joe, B.; Wilkinson, C.A. Segregation of *Eucalyptus dunnii* logs using acoustics. *For. Ecol. Manag.* **2003**, *179*, 243–251. [CrossRef]
11. Ross, R.J. *Nondestructive Evaluation of Wood*, 2nd ed.; General Technical Report FPL-GTR-238; USDA, Forest Service, Forest Products Laboratory: Madison, WI, USA, 2015; p. 169.
12. Legg, M.; Bradley, S. Measurement of stiffness of standing trees and felled logs using acoustics: A review. *J. Acoust. Soc. Am.* **2016**, *139*, 588–604. [CrossRef] [PubMed]
13. Raymond, C.A.; Joe, B.; Evans, R.; Dickson, R.L. Relationship between timber grade, static and dynamic modulus of elasticity, and Silviscan properties for *Pinus radiata* in New SouthWales. *N. Z. J. For. Sci.* **2007**, *37*, 186–196.
14. Newton, P.F. Development trends of black spruce fibre attributes in maturing plantations. *Int. J. For. Res.* **2016**, 1–12. [CrossRef]
15. Newton, P.F. Predictive relationships between acoustic velocity and wood quality attributes for red pine logs. *For. Sci.* **2017**, *63*, 504–517. [CrossRef]
16. Newton, P.F. Acoustic-based non-destructive estimation of wood quality attributes within standing red pine trees. *Forests* **2017**, *8*, 380. [CrossRef]
17. Wang, X. Acoustic measurements on trees and logs: A review and analysis. *Wood Sci. Technol.* **2013**, *475*, 965–975. [CrossRef]

18. Brashaw, B.K.; Bucur, V.; Divos, F.; Goncalves, R.; Lu, J.; Meder, R.; Yin, Y. Nondestructive testing and evaluation of wood: A worldwide research update. *For. Prod. J.* **2009**, *59*, 7–14.

19. Wang, X.; Senalik, C.A.; Ross, R.J. (Eds.) *20th International Nondestructive Testing and Evaluation of Wood Symposium*; General Technical Report FPL-GTR-249; USDA, Forest Service, Forest Products Laboratory: Madison, WI, USA, 2017; p. 539.

20. McKinnon, L.M.; Kayahara, G.J.; White, R.G. *Biological Framework for Commercial Thinning Evenaged Single-Species Stands of Jack Pine, White Spruce, and Black Spruce in Ontario*; Report TR-046; Ontario Ministry of Natural Resources, Northeast Science and Information Section: Timmins, ON, Canada, 2006; p. 130.

21. Carmean, W.H.; Niznowski, G.P.; Hazenberg, G. Polymorphic site index curves for jack pine in Northern Ontario. *For. Chron.* **2001**, *77*, 141–150. [CrossRef]

22. Rowe, J.S. *Forest Regions of Canada*; Publication No. 1300; Government of Canada, Department of Environment, Canadian Forestry Service: Ottawa, ON, Canada, 1972.

23. Evans, R. Rapid measurement of the transverse dimensions of tracheids in radial wood sections from *Pinus radiata*. *Holzforschung* **1994**, *48*, 168–172. [CrossRef]

24. Siau, J.F. *Wood: Influence of Moisture on Physical Properties*; Virginia Polytechnic Institute and State University, Department of Wood Science and Forest Products: Blacksburg, VA, USA, 1995.

25. Evans, R.; Hughes, M.; Menz, D. Microfibril angle variation by scanning X-ray diffractometry. *Appita* **1999**, *52*, 363–367.

26. Evans, R. Wood stiffness by X-ray diffractometry. In *Characterization of the Cellulosic Cell Wall*; Stokke, D.D., Groom, L.H., Eds.; Wiley: Hoboken, NJ, USA, 2006; pp. 138–146.

27. Evans, R.; Downes, G.; Menz, D.; Stringer, S. Rapid measurement of variation in tracheid transverse dimensions in a radiata pine tree. *Appita* **1995**, *48*, 134–138.

28. Raudenbush, S.W.; Bryk, A.S. *Hierarchical Linear Models: Applications and Data Analysis Methods*, 2nd ed.; Sage: Newbury Park, CA, USA, 2002; p. 485.

29. Raudenbush, S.W.; Bryk, A.S.; Cheong, Y.F.; Congdon, R.T.; Toit, M., Jr. *HLM 7—Hierarchical Linear and Nonlinear Modeling*; Scientific Software International Inc.: Lincolnwood, IL, USA, 2011; p. 360.

30. Gujarati, D.N. *Essentials of Econometrics*, 3rd ed.; McGraw-Hill/Irwin Inc.: New York, NY, USA, 2006; p. 553.

31. Ek, A.R.; Monserud, R.A. Performance and comparison of stand growth models based on individual tree and diameter-class growth. *Can. J. For. Res.* **1979**, *9*, 231–244. [CrossRef]

32. Reynolds, M.R., Jr. Estimating the error in model predictions. *For. Sci.* **1984**, *30*, 454–469.

33. Gribko, L.S.; Wiant, H.V., Jr. A SAS template program for the accuracy test. *Compiler* **1992**, *10*, 48–51.

34. National Lumber Grades Authority (NLGA). *Special Products Standard for Machine Graded Lumber*; NLGA: Surrey, BC, Canada, 2013.

35. Bérubé-Deschênes, A.; Franceschini, T.; Schneider, R. Factors affecting plantation grown white spruce (*Picea glauca*) acoustic velocity. *J. For.* **2016**, *114*, 629–637. [CrossRef]

36. Butler, M.A.; Dahlen, J.; Eberhardt, T.L.; Montes, C.; Antony, F.; Daniels, R.F. Acoustic evaluation of loblolly pine tree-and lumber-length logs allows for segregation of lumber modulus of elasticity, not for modulus of rupture. *Ann. For. Sci.* **2017**, *74*, 1–15. [CrossRef]

37. Kang, H.; Booker, R.E. Variation of stress wave velocity with MC and temperature. *Wood Sci. Technol.* **2002**, *36*, 41–54. [CrossRef]

38. Chauhan, S.S.; Walker, J.C.F. Variations in acoustic velocity and density with age, and their interrelationships in radiata pine. *For. Ecol. Manag.* **2006**, *229*, 388–394. [CrossRef]

39. Newton, P.F. Quantifying the effects of wood moisture and temperature variation on time-of-flight acoustic velocity measures within standing red pine and jack pine trees. *Forests* **2018**, *9*, 527. [CrossRef]

40. Wessels, C.B.; Malan, F.S.; Rypstra, T. A review of measurement methods used on standing trees for the prediction of some mechanical properties of timber. *Eur. J. For. Res.* **2011**, *130*, 881–893. [CrossRef]

41. Wang, X.; Carter, P. Acoustic assessment of wood quality in trees and logs. In *Nondestructive Evaluation of Wood*; Ross, R.J., Ed.; General Technical Report FPL-GTR-238; USDA, Forest Service, Forest Products Laboratory: Madison, WI, USA, 2015; pp. 87–101.

42. Newton, P.F. In-forest acoustic-based prediction of commercially-relevant wood quality attributes within standing jack pine trees. *Forests* **2018**, in preparation.

43. Walker, J.C.F.; Nakada, R. Understanding corewood in some softwoods: A selective review on stiffness and acoustics. *Int. For. Rev.* **1999**, *1*, 251–259.

44.  Harris, P.; Petherick, R.; Andrews, M. Acoustic resonance tools. In *Proceedings of the 13th International Symposium on Nondestructive Testing of Wood*; Forest Products Society: Berkeley, CA, USA, 2003; pp. 195–201.
45.  Ross, R.J.; McDonald, K.A.; Green, D.W.; Schad, K.C. Relationship between log and lumber modulus of elasticity. *For. Prod. J.* **1997**, *47*, 89–92.
46.  Dickson, R.L.; Matheson, A.C.; Joe, B.; Ilic, J.; Owen, J.V. Acoustic segregation of *Pinus radiata* logs for sawmilling. *N. Z. J. For. Sci.* **2004**, *34*, 175–189.
47.  Vikram, V.; Cherry, M.L.; Briggs, D.; Cress, D.W.; Evans, R.; Howe, G.T. Stiffness of Douglas-fir lumber: Effects of wood properties and genetics. *Can. J. For. Res.* **2011**, *41*, 1160–1173. [CrossRef]
48.  Clark, T.A.; Hartmann, J.; Lausberg, M.; Walker, J.C.F. Fibre characterisation of pulp logs using acoustics. In Proceedings of the 56th Appita Annual Conference, Rotorua, New Zealand, 18–20 March 2002; pp. 17–24.
49.  Bradley, A.; Chauhan, S.S.; Walker, J.C.F.; Banham, P. Using acoustics in log segregation to optimise energy use in thermomechanical pulping. *Appita* **2005**, *58*, 306–311.
50.  Murphy, G.; Cown, D. Stand, stem and log segregation based on wood properties: A review. *Scand. J. For. Res.* **2015**, *30*, 757–770. [CrossRef]
51.  Walsh, D.; Strandgard, M.; Carter, P. Evaluation of the Hitman PH330 acoustic assessment system for harvesters. *Scand. J. For. Res.* **2014**, *29*, 593–602. [CrossRef]
52.  Zhang, S.Y.; Koubaa, A. *Softwoods of Eastern Canada: Their Silvics, Characteristics, Manufacturing and End-Uses*; Special Publication SP-526E; FPInnovations: Quebec City, QC, Canada, 2008.

*Article*

# Variations in Orthotropic Elastic Constants of Green Chinese Larch from Pith to Sapwood

**Fenglu Liu** [1,2]**, Houjiang Zhang** [1,2,*]**, Fang Jiang** [1]**, Xiping Wang** [3] **and Cheng Guan** [1,2]

[1]   School of Technology, Beijing Forestry University, Beijing 10083, China; liufenglu3339@bjfu.edu.cn (F.L.); jf0620@bjfu.edu.cn (F.J.); cguan6@bjfu.edu.cn (C.G.)

[2]   Joint International Research Institute of Wood Nondestructive Testing and Evaluation, Beijing Forestry University, Beijing 100083, China

[3]   USDA Forest Products Laboratory, Madison, WI 53726, USA; xiping.wang@usda.gov

*   Correspondence: hjzhang6@bjfu.edu.cn; Tel.: +86-010-62336925 (ext. 401)

Received: 26 April 2019; Accepted: 23 May 2019; Published: 25 May 2019

**Abstract:** Full sets of elastic constants of green Chinese larch (*Larix principis-rupprechtii* Mayr) with 95% moisture content at four different cross-section sampling positions (from pith to sapwood) were determined in this work using three-point bending and compression tests. Variations in the material constants of green Chinese larch from pith to sapwood were investigated and analyzed. The results showed that the sensitivity of each elastic constant to the sampling position was different, and the coefficient of variation ranged from 4.3% to 48.7%. The Poisson's ratios $v_{RT}$ measured at four different sampling positions were similar and the differences between them were not significant. The coefficient of variation for Poisson's ratio $v_{RT}$ was only 4.3%. The four sampling positions had similar Poisson's ratios $v_{TL}$, though the coefficient of variation was 11.7%. The Poisson's ratio $v_{LT}$ had the greatest variation in all elastic constants with a 48.7% coefficient of variation. A good linear relationship was observed between the longitudinal modulus of elastic $E_L$, shear modulus of elasticity $G_{RT}$, Poisson's ratio $v_{RT}$, and sampling distance. $E_L$, $G_{RT}$, and $v_{RT}$ all increased with sampling distance $R$. However, a quadratic relationship existed with the tangential modulus of elasticity $E_T$, radial modulus of elasticity $E_R$, shear modulus of elasticity $G_{LT}$, and shear modulus of elasticity $G_{LR}$. A discrete relationship was found in the other five Poisson's ratios. The results of this study provide the factual changes in the elastic constants of green wood from pith to sapwood for numerical modelling of stress wave propagation in trees or logs.

**Keywords:** orthotropic; elastic constants; green larch; compression; three-point bending

## 1. Introduction

Elastic constants, especially the modulus of elasticity (MOE), which indicate the elastic behavior of wood, are critical parameters for furniture, musical instruments, or wood products, such as plywood, laminated veneer lumber, and cross laminated timber. Numerical simulation is being increasingly used to investigate the propagation of stress wave in standing trees or logs [1–4]. Material elastic constants are required for numerical simulation, especially when defining material properties. Wood, as a complex and anisotropic material, has considerable variations in its mechanical properties from bottom to top, pith to sapwood within a tree. In many studies, wood has been considered an orthotropic material, given its unique and independent material performance in the three principal or orthotropic directions (radial $R$, tangential $T$, and longitudinal $L$) [5–11]. Nine independent elastic constants (reduced from twelve elastic constants according to the symmetry of the stress and strain sensor in orthotropic materials), including three elastic moduli, three shear moduli, and three Poisson's ratios, are required to characterize the elastic behavior of orthotropic materials for mechanical analysis. Wood is also a hygroscopic material, and its mechanical behavior is therefore impacted by variations in moisture content (MC) [12–16]. Hering et al. determined all the independent elastic properties of European beech wood at different moisture conditions (MC ranged from 8.7% to 18.6%), and results indicated that all elastic parameters, except for Poisson's ratios $v_{TR}$ and $v_{RT}$, show a decrease in stiffness with increasing moisture content [13]. Similarly, Jiang et al. examined the influence of moisture content on the elastic and strength anisotropy of Chinese fir (*Cunninghamia lanceolate* Lamb.) wood and found that, except for Poisson's ratios, all investigated elastic and strength parameters decreased with increasing MC (varied from 10.3% to 16.7%), whereby individual moduli and strength values were affected by the MC to different degrees [14].

Resistance strain gauges and ultrasonic are the two of most commonly used methods determining the elastic constants of wood. As a destructive method, resistance strain gauges are usually used to evaluate the material constants of wood. Sliker first proposed the use of compression and bending tests using strain gauges in wood to determine the elastic constants of materials [6–8]. Many studies had been conducted to verify the feasibility and validity of this method. Li successfully measured the full set of elastic constants for *Fraxinus mandshurica* with 13.4% moisture content via the compression test using strain gauges [9]. Gong estimated the elastic moduli parallel to the grain of *Pinus Massoniana* with 15% MC through strain gauges [10]. Wang et al. determined the full set of material constants for White Birch (*Betula platyphylla* Suk.) with 12% MC using the compression and three-point bending tests [17]. Shao et al. measured the seven elastic constants of *Cunninghamia lanceolata* with 12% MC using electric resistance strain gauges [18]. Aira et al. performed compression tests on dry specimens (around 12% MC) to determine the elastic constants of Scots pine (*Pinus sylvestris* L.) and found the MOE values obtained were greater than the average values for softwoods, and Poisson's ratios obtained parallel to the grain were similar to the values in the literature [19].

Ultrasonic, a rapid and efficient non-destructive method for determining material properties, has drawn increasing attention in wood characteristic measurement. Preziosa et al. first determined the stiffness matrix of wood using the ultrasonic technique [20,21]. Then, Bucur measured the elastic constants of six species (pine, spruce, Douglas-fir, oak, beech, and tulip-tree) in the dry condition, applying ultrasonic to different cubic specimens [22]. Francois proposed the use of a polyhedral specimen with 26 faces for the determination of all the elements in the stiffness matrix from a single specimen to measure the elastic constants of dry wood [23]. Many studies have demonstrated the feasibility of using ultrasonic to measure all the elastic constants from a single specimen of dry wood (MC ranged from 7.5% to 12.3%) from different species, such as *Castanea sativa* Mill., ash (*Fraxinus excelsior* L), beech, *Eucalyptus saligna*, *Apuleia leiocarpa*, and *Goupia glabra* [24–27].

Although the full material constants of wood can be measured by both resistance strain gauges and ultrasonic, the full sets of elastic constants for green wood have rarely been reported in the literature. As wood is often used in a dry state, the elastic constants reported in most published papers for wood are in these conditions [28–31]. This would lead to a poor representation when the elastic constants of dry wood are used for modelling standing trees or green logs. Only Davies et al. investigated and obtained the elastic constants of green *Pinus radiata* wood using compression and tension tests [32]. Davies et al. stated that the mathematical modelling would be more realistic with the material constants of green wood rather than those of dry wood. The variation in elastic constants in cross-sections of green wood from pith to sapwood have not been reported in the literature. Davies et al. only measured the material constants of the corewood and outerwood of *Pinus radiata* instead of elastic constants across whole transverse sections of wood. Although Xavier et al. reported that the two stiffness values ($Q_{22}$ and $Q_{66}$) of *P. pinaster* varied across three or four different radial positions using the unnotched Iosipescu test, the moisture content of the specimens was 10.4% and only two stiffness values were investigated [33]. Therefore, to the best of our knowledge, no full sets of elastic constants varying from pith to sapwood have been published for green wood. The variations in the mechanical properties of green wood from pith to sapwood have not yet been studied.

The main purpose of this research was to determine the elastic constants of green Chinese larch from pith to sapwood using compression and three-point bending tests, as well as to investigate and analyze the variations in the mechanical characteristics of green Chinese larch from pith to sapwood. We aimed to obtain basic knowledge about the mechanical properties of wood from pith to sapwood, and to describe standing trees or logs as an orthotropic material in numerical modelling. Use of wood could be optimized in various applications, such as papermaking, furniture, musical instruments, or wood products, according to the measured material performance in different parts of the wood. Standing trees or logs could be modelled more realistically in numerical simulation with the acquired data. The results of this study provide factual elastic constants of green wood from pith to sapwood for numerical modelling of stress wave propagation in trees or logs.

## 2. Materials and Methods

### 2.1. Materials

Two Chinese larches (*Larix principis-rupprechtii* Mayr), a common plantation species in Northern China, were harvested from Maojingba National Plantation Farm, located in Longhua County, Chengde City, Hebei Province, China (118°06′05″ E, 41°28′46″ N at approximately 750 m elevation). The trees (coded A and B) aged 40 years were felled and branches were subsequently removed. The diameter at breast height (DBH) values of tree A and tree B were 32 cm and 36 cm, respectively. Then, two 60-cm-long logs were cut from each selected tree at a height of 0.5 m and 1.25 m above the ground. A 15-cm-thick disc for density and moisture content measurements was cut from each tree at a height of 1.1 m above the ground. A total of four 60-cm-long logs and two 15-cm-thick discs were obtained and immediately sealed in plastic wrap. After, these logs and discs were directly transported to the mechanics laboratory in Beijing Forestry University and kept in a condition room at 15 °C and 95% relative humidity.

### 2.2. Static Testing Method

#### 2.2.1. Specimen Sampling

A schematic of the sawing pattern used to obtain green larch specimens for static testing is presented in Figure 1. Four different sampling positions from pith to sapwood (numbered 1, 2, 3, and 4 in Figure 1 referred to sampling position, defined as P1, P2, P3, and P4 hereinafter, respectively) were chosen to determine the elastic constants at different positions from pith to sapwood in the cross-sections of standing trees and investigate the distribution of elastic constants in the cross-sections of standing trees. As shown in Figure 1, sampling positions P1, P2, and P4 were located at the pith, heartwood, and sapwood, respectively, whereas P3 was located between the heartwood and sapwood. The initial transverse dimension of specimens for static testing in each sampling location was 53 × 53 mm.

Parameter $R$ was used to define the distance between the center of pith and the center of sampling position. Thus, the designed distance $R$ for sampling position P1, P2, P3 and P4 was 13 mm, 56 mm, 76 mm and 132 mm, respectively. It should be noticed that the sampling distance $R$ for P1, P2, P3 and P4 shown in Figure 1 was designed for a log with a DBH ranging from 300 mm to 360 mm. The sampling distance would be different as the DBH of log over 360 mm and need to be changed.

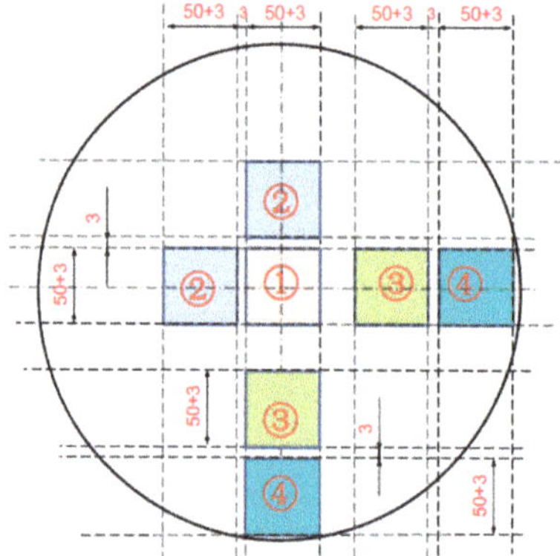

**Figure 1.** Schematic of sawing pattern for testing specimens. Dashed lines represent the sawing line and the units are millimeters. The extra 3 mm width was the machining allowance. The number 1, 2, 3, and 4 in the yellow, gray, green, and blue area, respectively represent four different sampling positions.

One P1 lumber (53 × 53 × 600 mm), two P2 lumbers (53 × 53 × 600 mm), two P3 lumbers (53 × 53 × 600 mm), and two P4 lumbers (53 × 53 × 600 mm) were obtained from each log according to sampling pattern shown in Figure 1. A total of 4 P1- lumbers, 8 P2- lumbers, 8 P3- lumbers, and 8 P4-lumbers were acquired from four 600-mm-long logs. Then, these were used to prepare the specimens for static testing, including the compression test and three-point bending test, performed according to American Society for Testing Materials (ASTM) D5536-94 [34]. Specimens for static testing were instantly sealed with plastic wrap and stored in the condition room (15 °C and 95% relative humidity) before mechanical testing. The dimensions of the specimens used for static testing are provided in Table 1. Each sampling location (referred to as P1, P2, P3, and P4) all used the same size specimens for static testing. Therefore, the elastic constants of these four different positions (from pith to sapwood) in the cross-section of standing trees could be determined.

**Table 1.** Dimensions of specimens for static testing.

| Static Test | Orientation | | Size (mm) (Length × Width × Thickness) | Number of Specimens for One Sampling Position |
|---|---|---|---|---|
| Compression test | Parallel to grain | Longitudinal (L) | 25 × 25 × 100 | 8 |
| | Perpendicular to grain | Radial (R) Tangential (T) With a 45° to tangential | 50 × 50 × 150 | 8 × 3 |
| Three-point bending test | Parallel to grain | Longitudinal (L) | 25 × 25 × 150 | 8 |
| | | | 25 × 25 × 200 | 8 |
| | | | 25 × 25 × 250 | 8 |
| | | | 25 × 25 × 300 | 8 |
| | | | 25 × 25 × 350 | 8 |

### 2.2.2. Compression Test

The four types of test specimens used for the compression test, with wood grain oriented relative to the orthotropic directions and the distribution of strain gauges in specimen, are displayed in Figure 2. Eight clear test specimens of the required shape and orientation were machined from the cut lumber. A total of 128 test specimens (8 replicates × 4 orientations × 4 sampling positions) were used for compression testing. Eight 25 × 25 × 100 mm specimens parallel to the grain (Figure 2a) were used to measure the elastic constants of $E_L$, $\nu_{LR}$, and $\nu_{LT}$. Eight 50 × 50 × 150 mm specimens perpendicular to the grain radially (Figure 2b) were used to test $E_R$, $\nu_{RL}$, and $\nu_{RT}$. Eight 50 × 50 × 150 mm specimens perpendicular to the grain tangentially (Figure 2c) were used to evaluate $E_T$, $\nu_{TR}$, and $\nu_{TL}$.

Eight $50 \times 50 \times 150$ mm specimens inclined at a 45° to the grain (Figure 2d) were used to obtain the shear modulus of elasticity $G_{RT}$.

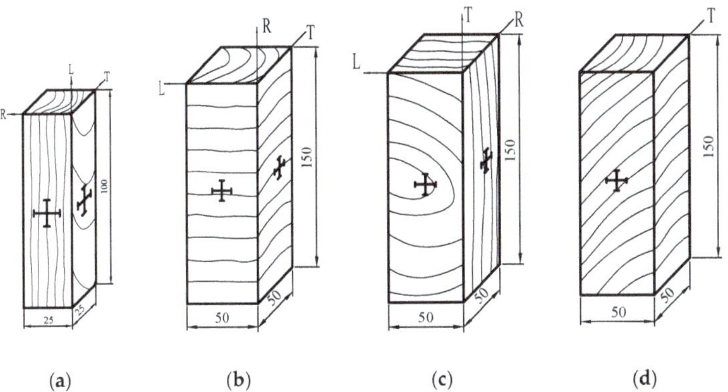

**Figure 2.** Types of specimens and strain gauges placement for compression test: (**a**) parallel to the grain, (**b**) perpendicular to the grain radially, (**c**) perpendicular to the grain tangentially, and (**d**) inclined at 45°.

Resistance strain gauges were directly bonded to the surfaces of the specimen prior to compression testing. Figure 2a–c show that four strain gauges were placed on each specimen. Two resistance strain gauges were glued perpendicularly to a surface of testing sample, and the other two resistance strain gauges were perpendicularly bonded to the adjacent surface. As shown in Figure 2d, only two strain gauges perpendicularly glued to a surface of specimen were used for compression testing. Strain gauges should be placed close to each other so that measurements record the strain state of the same point with no variability in their elastic properties due to the heterogeneity of wood. The adhesive used has a lower longitudinal stiffness than wood to avoid restricting its free deformation when receiving the external load. The adhesive must simultaneously have a high shear stiffness so that the deformation of strain gauge is not damped by the thickness of the adhesive [19]. RGM-4050-100 (made by Reger Instrument Corporation Limited, Shenzhen City, China), a microcomputer-controlled electronic universal testing machine, was used to conduct the compression test according to ASTM D143-09 [35]. Specimens were loaded at a rate of 0.2 mm/min for compression testing. Modulus of elasticity ($E_L$, $E_R$, and $E_T$), Poisson's ratios ($v_{LR}$, $v_{LT}$, $v_{RL}$, $v_{RT}$, $v_{TR}$, and $v_{TL}$) and shear modulus of elasticity ($G_{RT}$) were calculated from Equations (1)–(3), respectively.

$$E = \frac{P_n - P_0}{A_0(\varepsilon_n - \varepsilon_0)} \tag{1}$$

where $E$ is the modulus of elasticity (MPa), $P_n$ is the final load (N), $P_0$ is the initial load (N), $A_0$ is the cross-section area of specimen (mm$^2$), $\varepsilon_n$ is the final strain, and $\varepsilon_0$ is the initial strain.

$$v = -\frac{\Delta\varepsilon'}{\Delta\varepsilon} \tag{2}$$

where $v$ is the Poisson's ratio of specimen, $\Delta\varepsilon'$ is the lateral strain increase, and $\Delta\varepsilon$ is the axial strain increase.

$$G_{RT} = \frac{\Delta P_{45°}}{2A_0\left(\Delta\varepsilon_x - \Delta\varepsilon_y\right)} \tag{3}$$

where $G_{RT}$ is the shear modulus of elasticity in the RT plane (MPa), $A_0$ is the cross-section area of the specimen (mm$^2$), $\Delta P_{45°}$ is the load increase of the elastic deformation phase on the load-strain curve (N), $\Delta\varepsilon_x$ is the strain increase along the axis of the specimen, and $\Delta\varepsilon_y$ is the strain increase perpendicular to the axis of specimen.

Since the moisture content has a significant effect on the mechanical properties of wood, the specimens were taken from the condition room and tested in sequence to ensure less variation in the moisture content of specimens. Measurements for each specimen were performed as soon as possible to reduce the impact of the moisture content on the testing results.

### 2.2.3. Three-Point Bending Test

In the light of ASTM D5536-94, specimens with five different ratios of span to depth were prepared to conduct three-point bending test. We obtained a total of 160 test specimens in total (8 replicates × 5 spans × 4 sampling positions) for the bending tests. The specific size of the specimens, especially the length, are shown in Table 1. The span for each length of specimen was 132 mm, 176 mm, 220 mm, 264 mm, and 308 mm, respectively. Three-point bending tests were conducted using an RGM-4050-100 universal testing machine on the basis of ASTM D143-09. The loading speed of specimens for three-point bending test was 5 mm/min. As above, the specimens for the bending test were successively measured and tested as quickly as possible to reduce the variation in the moisture content of the measured samples. Bending moduli of elasticity (*MOE*) were calculated using Equation (4), and then shear modulus of elasticity (*G*) can be obtained using Equation (5).

$$MOE = \frac{\Delta P l^3}{4\Delta f b h^3} \tag{4}$$

where *MOE* is the bending modulus of elasticity (MPa), $\Delta P$ is loading increase (N), $\Delta f$ is the deflection increase (mm), $b$ is the width of the specimen (mm), $h$ is the thickness of the specimen (mm), and $l$ is the span of the specimen (mm).

$$G = \frac{1.2\Delta(h/l)^2}{\Delta(1/MOE)} \tag{5}$$

where $\Delta(h/l)^2$ is the increase in the square of ratio between the thickness and span and $\Delta(1/MOE)$ is the increase in the reciprocal of the bending modulus of elasticity (mm²/N). The shear modulus of elasticity $G_{LR}$ was obtained by measuring specimen loaded from radial direction, and the shear modulus of elasticity $G_{LT}$ was obtained by measuring the specimen loaded from tangential direction. Both of them can be calculated using Equations (4) and (5).

## 3. Results and Discussion

### 3.1. Twelve Elastic Constants of Green Chinese Larch at Different Sampling Positions

The moisture content (MC) of green Chinese larch was measured in the laboratory using the kiln-dry method, and the average MC of the sample trees was 95%. The average green density of Chinese larch was 625 kg/m³. Twelve elastic constants of green Chinese larch were calculated using Equations (1)–(4) using the experimental data obtained from three-point bending and compression tests. The elastic constants of the four different sampling positions (P1, P2, P3, and P4) are shown in Table 2. The average values of the elastic moduli in the longitudinal, radial, and tangential directions were 7,629 MPa, 773 MPa, and 362 MPa, respectively. Davies et al. reported that the longitudinal, radial, and tangential elastic moduli of the outerwood in green *Pinus radiata* were 4,360 MPa, 490 MPa, and 250 MPa, respectively. The values for the corewood of green *Pinus radiata* were 3,500 MPa, 260 MPa, and 240 MPa, respectively. The average values of the three elastic moduli obtained from this work were higher than those derived from Davies's research due to the differences in the tested species and tree ages. The longitudinal modulus of elasticity ($E_L$) increased from 5016 MPa to 10,137 MPa as the sampling distance ($R$) varied from 13 mm (P1) to 132 mm (P4), respectively. This means that the longitudinal mechanical properties of green Chinese larch increased from pith to sapwood. However, the radial modulus of elasticity ($E_R$) initially increased from 628 MPa to 1,154 MPa as the sampling distance changed from 13 mm (P1) to 76 mm (P3) and then decreased to 342 MPa as sampling distance increased to 132 mm (P4). This may indicate a lower radial modulus of elasticity in sapwood. No significant relationship was found between tangential modulus of elasticity ($E_T$) and sampling distance ($R$).

The average values of shear moduli in the LR, LT, and RT plane were 428 MPa, 393 MPa, and 450 MPa, respectively. Davies et al. presented the lower values for these three shear moduli, 60 MPa, 110 MPa, 50 MPa in LR, LT, and RT plane for outerwood and 40 MPa, 110 MPa, 20 MPa in LR, LT, and RT plane for corewood, respectively. The average values of $v_{LT}$, $v_{LR}$, $v_{TL}$, $v_{TR}$, $v_{RL}$, and $v_{RT}$ Poisson's ratios were 0.30, 0.22, 0.04, 0.60, 0.05, and 0.77, respectively. The corresponding values of these six Poisson's ratios in Davies's paper were 0.60, 0.29, 0.03, 0.33, 0.03, and 0.64 for outerwood and 0.16, 0.46, 0.01, 0.33, 0.05, and 0.54 for corewood, respectively. However, no significant relationship with sampling distance was found for either shear modulus of elasticity or Poisson's ratio.

**Table 2.** Twelve elastic constants at different sampling locations in cross-sections of green Chinese larch.

| Sampling Position | $E_L$ (MPa) | $v_{LT}$ | $v_{LR}$ | $E_T$ (MPa) | $v_{TL}$ | $v_{TR}$ | $E_R$ (MPa) | $v_{RL}$ | $v_{RT}$ | $G_{RT}$ (MPa) | $G_{LR}$ (MPa) | $G_{LT}$ (MPa) |
|---|---|---|---|---|---|---|---|---|---|---|---|---|
| 1 | 5016 | 0.15 | 0.13 | 339 | 0.04 | 0.55 | 628 | 0.04 | 0.74 | 326 | 538 | 403 |
| 2 | 5996 | 0.47 | 0.34 | 423 | 0.05 | 0.66 | 967 | 0.03 | 0.75 | 535 | 339 | 352 |
| 3 | 9365 | 0.21 | 0.15 | 402 | 0.04 | 0.53 | 1154 | 0.06 | 0.79 | 383 | 404 | 371 |
| 4 | 10137 | 0.36 | 0.26 | 282 | 0.04 | 0.66 | 342 | 0.05 | 0.81 | 556 | 430 | 446 |
| Mean | 7629 | 0.30 | 0.22 | 362 | 0.04 | 0.60 | 773 | 0.05 | 0.77 | 450 | 428 | 393 |
| SD [1] | 2503 | 0.15 | 0.10 | 64 | 0.01 | 0.07 | 360 | 0.01 | 0.03 | 113 | 83 | 41 |
| SE [2] | 1252 | 0.07 | 0.05 | 32 | 0.01 | 0.04 | 180 | 0.01 | 0.02 | 57 | 41 | 21 |
| COV [3] (%) | 32.8 | 48.7 | 44.7 | 17.7 | 11.7 | 11.6 | 46.6 | 28.7 | 4.3 | 25.1 | 19.4 | 10.5 |

[1] SD, standard deviation; [2] SE, standard error; [3] COV, coefficient of variation.

Table 2 shows that the longitudinal modulus of elasticity ($E_L$) was higher than the radial modulus of elasticity ($E_R$), and the radial modulus of elasticity ($E_R$) was greater than the tangential modulus of elasticity ($E_T$), i.e., $E_L > E_R > E_T$, for all four sampling positions. Similarly, Poisson's ratio $v_{RT}$ was higher than Poisson's ratio $v_{LT}$, and Poisson's ratio $v_{LT}$ was greater than Poisson's ratio $v_{LR}$, i.e., $v_{RT} > v_{LT} > v_{LR}$ at these four sampling locations. These results align with the findings in dry wood [17,18]. Wood is a highly anisotropic material. Thus, different values would be obtained for the same elastic constant due to different sampling positions. Table 2 also shows that the sensitivity of each elastic constant to the sampling position was different, and the corresponding coefficient of variation ranged from 4.3% to 48.7%. Table 3 provides the results of the analysis of variance (ANOVA) for each elastic constant at different sampling positions. Table 3 shows that sampling position had a significant impact both on $E_L$, $G_{RT}$, $v_{LT}$, and $v_{LR}$ ($p < 0.05$), whereas no significant effect was found in the other elastic constants ($p > 0.05$). For three MOE, only $E_L$ showed significant differences ($p < 0.05$) in sampling positions with a 32.8% coefficient of variation. Poisson's ratios $v_{RT}$ measured at four different sampling positions were similar and the coefficient of variation for Poisson's ratio $v_{RT}$ was only 4.3%, which is in agreement with the insignificant differences between them ($p > 0.05$). The four sampling positions had similar Poisson's ratios $v_{TL}$ and showed an insignificant difference ($p > 0.05$), though the coefficient of variation was 11.7%. Poisson's ratio $v_{LT}$ had the greatest variation in all elastic constants with a 48.7% coefficient of variation and showed a significant difference in sampling position ($p < 0.05$). For shear moduli, only $G_{RT}$ showed significant differences ($p < 0.05$) in sampling positions with a 25.1% coefficient of variation.

**Table 3.** Analysis of variance of elastic constants at different sampling positions.

| | Source | Sum of Squares | Degrees of Freedom | Mean Square | *F*-Value | *p*-Value |
|---|---|---|---|---|---|---|
| $E_{\mathrm{L}}$ | Between groups [1] | $9.92 \times 10^7$ | 3 | $3.31 \times 10^7$ | 8.31 | 0.001 |
| | Within groups | $9.55 \times 10^7$ | 24 | $3.98 \times 10^6$ | | |
| | Total | $1.95 \times 10^8$ | 27 | | | |
| $E_{\mathrm{R}}$ | Between groups | $1.99 \times 10^6$ | 3 | $6.65 \times 10^5$ | 1.62 | 0.21 |
| | Within groups | $9.84 \times 10^6$ | 24 | $4.09 \times 10^5$ | | |
| | Total | $1.18 \times 10^7$ | 27 | | | |
| $E_{\mathrm{T}}$ | Between groups | $1.55 \times 10^6$ | 3 | $5.17 \times 10^5$ | 1.07 | 0.38 |
| | Within groups | $1.16 \times 10^7$ | 24 | $4.82 \times 10^5$ | | |
| | Total | $1.31 \times 10^7$ | 27 | | | |
| $G_{\mathrm{LR}}$ | Between groups | $8.27 \times 10^4$ | 3 | $2.76 \times 10^4$ | 1.29 | 0.322 |
| | Within groups | $2.56 \times 10^5$ | 12 | $2.13 \times 10^4$ | | |
| | Total | $3.38 \times 10^6$ | 15 | | | |
| $G_{\mathrm{LT}}$ | Between groups | $2.04 \times 10^4$ | 3 | $6.79 \times 10^4$ | 0.49 | 0.694 |
| | Within groups | $1.65 \times 10^5$ | 12 | $1.38 \times 10^4$ | | |
| | Total | $1.86 \times 10^5$ | 15 | | | |
| $G_{\mathrm{RT}}$ | Between groups | $2.23 \times 10^5$ | 3 | $7.43 \times 10^4$ | 25.57 | $3.25 \times 10^{-4}$ |
| | Within groups | $5.52 \times 10^4$ | 19 | $2.91 \times 10^3$ | | |
| | Total | $2.78 \times 10^5$ | 22 | | | |
| $v_{\mathrm{LT}}$ | Between groups | 0.439 | 3 | 0.146 | 12.14 | $1.54 \times 10^{-3}$ |
| | Within groups | 0.290 | 24 | 0.012 | | |
| | Total | 0.729 | 27 | | | |
| $v_{\mathrm{LR}}$ | Between groups | 0.221 | 3 | 0.074 | 12.82 | $1.12 \times 10^{-3}$ |
| | Within groups | 0.138 | 24 | 0.006 | | |
| | Total | 0.358 | 27 | | | |
| $v_{\mathrm{TL}}$ | Between groups | 0.006 | 3 | 0.002 | 0.644 | 0.594 |
| | Within groups | 0.072 | 24 | 0.003 | | |
| | Total | 0.078 | 27 | | | |
| $v_{\mathrm{TR}}$ | Between groups | 0.105 | 3 | 0.035 | 1.505 | 0.239 |
| | Within groups | 0.558 | 24 | 0.023 | | |
| | Total | 0.663 | 27 | | | |
| $v_{\mathrm{RL}}$ | Between groups | 0.005 | 3 | 0.002 | 0.974 | 0.421 |
| | Within groups | 0.045 | 24 | 0.002 | | |
| | Total | 0.050 | 27 | | | |
| $v_{\mathrm{RT}}$ | Between groups | 0.024 | 3 | 0.008 | 0.187 | 0.904 |
| | Within groups | 1.040 | 24 | 0.043 | | |
| | Total | 1.065 | 27 | | | |

[1] Groups: there are four groups, i.e., four different sampling positions P1, P2, P3, and P4.

## 3.2. Vadility of Measured Data

Although the elastic constants of green Chinese larch at the four different sampling positions were obtained through experiments and data processing, the validity of testing data needed further verification. According to the mechanics of composite materials, the elastic constants of orthotropic materials should satisfy the limitations of Maxwell's theorem, as shown in Equations (6) and (7). As mentioned, in many studies, wood is considered an orthotropic material in the three main orthotropic directions. Therefore, the modulus of elasticity and the Poisson's ratio of green larch measured in this research should be satisfied the limitations of Maxwell's theorem.

$$\frac{v_{ij}}{E_i} = \frac{v_{ji}}{E_j} \, (i, j = \mathrm{L}, \mathrm{R}, \mathrm{T}) \tag{6}$$

$$\left| v_{ij} \right| < \left| \frac{E_i}{E_j} \right|^{\frac{1}{2}} \tag{7}$$

The modulus of elasticity and the Poisson's ratio obtained from P1 sampling position were taken into Equations (6) and (7), and the results showed that these elastic constants from P1 satisfy the limitation of Maxwell's theorem. Similar results were found in the P2, P3, and P4 sampling locations. Table 4 provides the results of Equation (7) at the P1, P2, P3, and P4 sampling locations. These results indicate that acquired data and calculated elastic constants from compression test were both valid and accurate.

**Table 4.** Results of the limitations of Maxwell's theorem at the four different sampling locations.

| Sampling Position | $\lvert v_{LT}\rvert$ | $\lvert\frac{E_L}{E_T}\rvert^{\frac{1}{2}}$ | $\lvert v_{LR}\rvert$ | $\lvert\frac{E_L}{E_R}\rvert^{\frac{1}{2}}$ | $\lvert v_{TL}\rvert$ | $\lvert\frac{E_T}{E_L}\rvert^{\frac{1}{2}}$ | $\lvert v_{TR}\rvert$ | $\lvert\frac{E_T}{E_R}\rvert^{\frac{1}{2}}$ | $\lvert v_{RL}\rvert$ | $\lvert\frac{E_R}{E_L}\rvert^{\frac{1}{2}}$ | $\lvert v_{RT}\rvert$ | $\lvert\frac{E_R}{E_T}\rvert^{\frac{1}{2}}$ |
|---|---|---|---|---|---|---|---|---|---|---|---|---|
| 1 | 0.15 | 3.85 | 0.13 | 2.83 | 0.04 | 0.26 | 0.55 | 0.73 | 0.04 | 0.35 | 0.74 | 1.36 |
| 2 | 0.47 | 3.76 | 0.34 | 2.49 | 0.05 | 0.27 | 0.66 | 0.66 | 0.03 | 0.40 | 0.75 | 1.51 |
| 3 | 0.21 | 4.83 | 0.15 | 2.85 | 0.04 | 0.21 | 0.53 | 0.59 | 0.06 | 0.35 | 0.79 | 1.69 |
| 4 | 0.36 | 6.00 | 0.26 | 5.44 | 0.04 | 0.17 | 0.66 | 0.91 | 0.05 | 0.18 | 0.81 | 1.10 |

However, Maxwell's theorem was only used to verify the validity of the modulus of elasticity and Poisson's ratio and could not be used to confirm the validity of the shear modulus of elasticity. Correlation analysis, consequently, was conducted on experimental data from the three-point bending tests to verify the validity of calculated shear modulus of elasticity. The relationship between the square of ratio between thickness and span $(h/l)^2$ and the reciprocal of bending modulus of elasticity ($1/MOE$) was analyzed both for radial-loaded and tangential-loaded bending tests. The results of correlation analysis for the P1, P2, P3, and P4 sampling positions are presented in Figures 3–6, respectively.

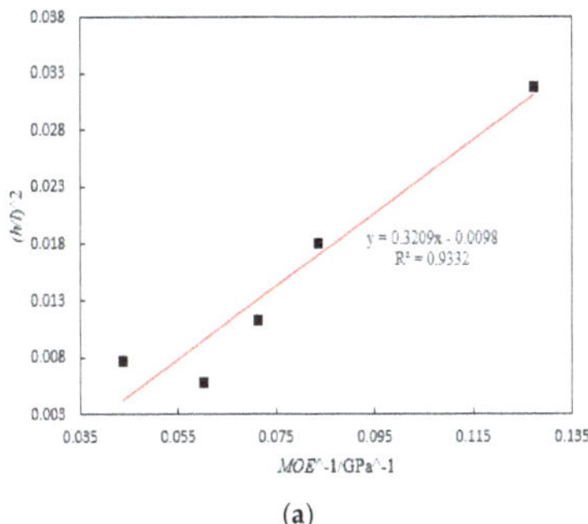


(a)

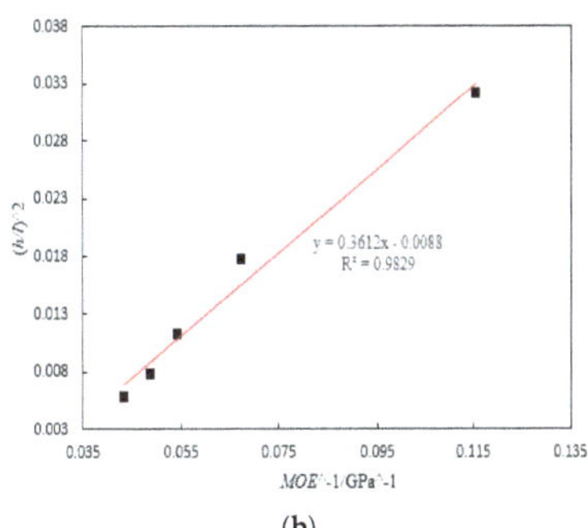


(b)

**Figure 3.** Relationship between the reciprocal of bending elastic modulus and the square of the ratio between depth and length (Sampling position 1) for: (**a**) tangential loading and (**b**) radial loading.

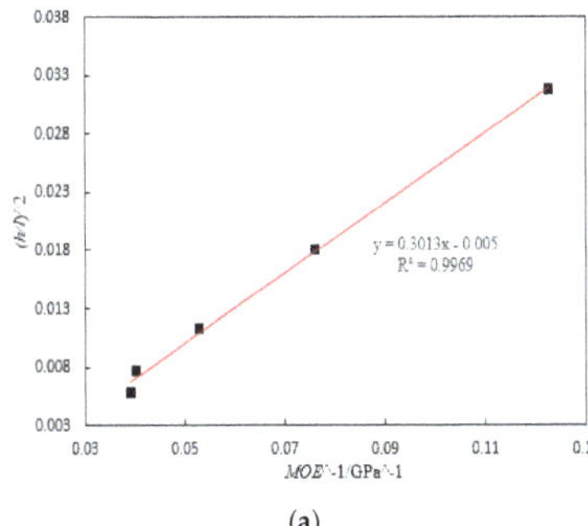


(a)

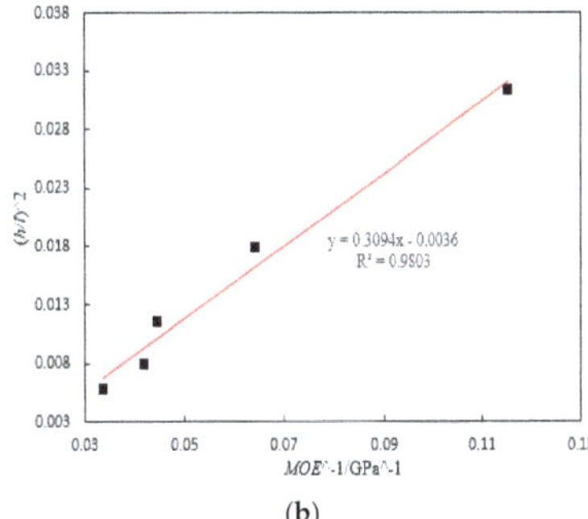


(b)

**Figure 4.** Relationship between the reciprocal of bending elastic modulus and the square of the ratio between depth and length (Sampling position 2) for: (**a**) tangential loading and (**b**) radial loading.

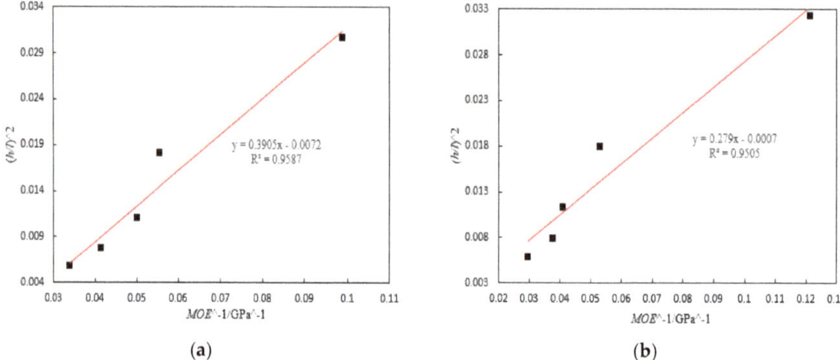

**Figure 5.** Relationship between the reciprocal of bending elastic modulus and the square of the ratio between depth and length (Sampling position 3) for: (**a**) tangential loading and (**b**) radial loading.

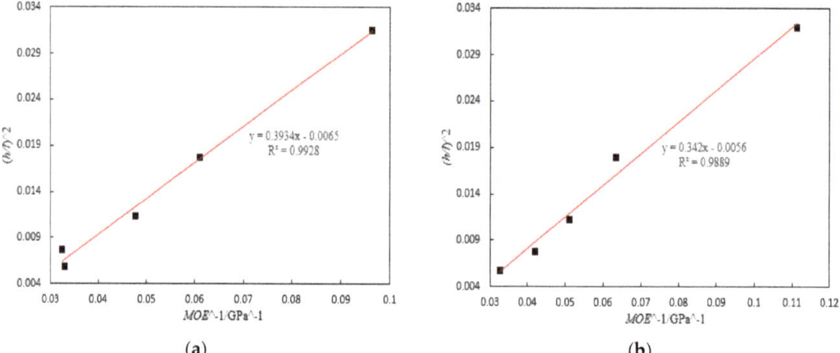

**Figure 6.** Relationship between the reciprocal of bending elastic modulus and the square of the ratio between depth and length (Sampling position 4) for: (**a**) tangential loading and (**b**) radial loading.

Figures 3–6 show that a linear relationship between the square of ratio in span and depth and the reciprocal of bending modulus of elasticity was found in the four sampling positions both for tangential-loaded and radial-loaded bending tests. The correlation coefficients between the square of the ratio in span and depth and the reciprocal of bending modulus of elasticity for the P1, P2, P3, and P4 sampling positions were all over 0.9 in the tangential-loaded and radial-loaded bending tests. These results indicate that the three-point bending test data and calculated shear modulus of elasticity were both effective and reasonable.

### 3.3. Variation in Elastic Constants of Wood Cross-Sections

### 3.3.1. Modulus of Elasticity

Wood is a highly anisotropic material with different mechanical properties throughout its interior. Wood properties change from pith to bark within a tree and differ between trees. Therefore, the mechanical properties, especially the elasticity constants, of wood vary along the cross-section. To investigate the difference and variation in the elastic constants along the cross-section of wood, the relationships between the modulus of elasticity, shear modulus of elasticity, Poisson's ratios, and sampling distance $R$ were analyzed by regression analysis to obtain the variation patterns of the elastic constants along the cross-section of green Chinese larch using the experimental data derived from compression and three-point bending tests.

The results of regression analysis for modulus of elasticity and sampling position are provided in Figure 7 and Table 5. Relationships between sampling distance and the modulus of elasticity for the three principal axes of wood are illustrated in Figure 7. Table 5 displays the corresponding fitting equations and correlation coefficients came from regression analysis.

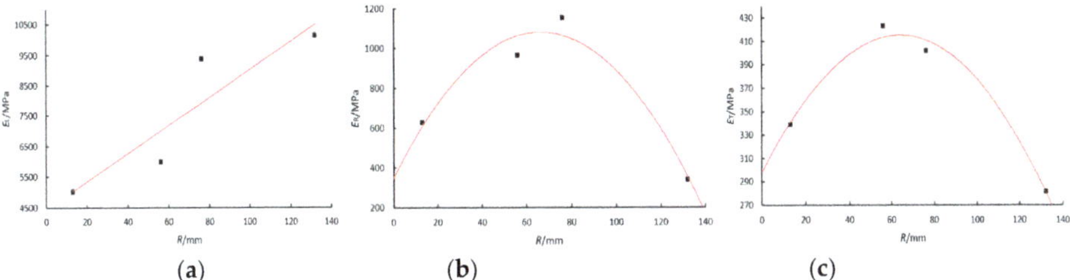

(a)  (b)  (c)

**Figure 7.** Relationships between the distance from pith and elastic moduli in the (**a**) longitudinal, (**b**) radial, and (**c**) tangential directions.

**Table 5.** Mathematical model of elastic moduli in three principle directions and distance from pith.

| Modulus of Elasticity | Fitted Equation | Coefficient of Determination ($R^2$) |
| --- | --- | --- |
| $E_L$ | $E_L = 4436.36 + 46.12R$ | 0.91 |
| $E_R$ | $E_R = 343.25 + 22.27R - 0.17R^2$ | 0.95 |
| $E_T$ | $E_T = 297.55 + 3.68R - 0.03R^2$ | 0.98 |

Figure 7 shows the variation patterns of the three principal moduli of elasticity along the cross-section of the wood. Figure 7a shows that the longitudinal modulus of elasticity ($E_L$) of green Chinese larch linearly increased with sampling distance. However, a quadratic relationship was observed between the radial modulus of elasticity ($E_R$) and the sampling distance, as well as for the tangential modulus of elasticity ($E_T$) and the sampling distance. $E_R$ and $E_T$ both first increased with sampling distance, and then decreased with sampling distances over 70 mm, as shown in Figure 7b,c. $E_R$ and $E_T$ near the bark were significantly lower than in other sampling positions, and even lower than the measured values near the pith. Table 5 shows the linear relationship between the longitudinal modulus of elasticity and sampling distance ($R^2 = 0.91$). Even though a quadratic relationship was found in the tangential and radial moduli of elasticity, both coefficients of determination were higher than 0.95. Little research has been conducted to investigate the variation in elastic constants of wood from pith to sapwood. Only Xavier et al. studied the variation in two stiffness values ($Q_{22}$ and $Q_{66}$) of dry Maritime pine across the radial position using the unnotched Iosipescu test. They found the transverse stiffness ($Q_{22} = E_R/(1 - v_{LR}.v_{RL})$) of dry Maritime pine decreased between the radial position $r_1$ (thirteenth ring, 29% of the radius) and $r_2$ (nineteenth ring, 46% of the radius), and a progressive increase was observed up to $r_4$ (forty-third ring, 81% of the radius) [33]. This means that the transverse stiffness decreased from the center to about the middle radius of stem and increased afterward to the outermost positions. The variation pattern of elastic moduli $E_R$ of green Chinese larch measured in this work was different from that of the transverse stiffness $Q_{22}$. This may be because the transverse stiffness was not only affected by the radial elastic moduli but also by Poisson's ratios $v_{LR}$ and $v_{RL}$. Different moisture contents and tree species may also produce these differences. More data from the same or different species should be acquired to investigate the variation in these three elastic constants in dry or green wood.

### 3.3.2. Shear Modulus of Elasticity

The results of regression analysis for the shear modulus of elasticity and sampling position are provided in Figure 8 and Table 6. The relationships between sampling distance and the shear modulus of elasticity are illustrated in Figure 8. Table 6 provides the corresponding fitting equations and correlation coefficients.

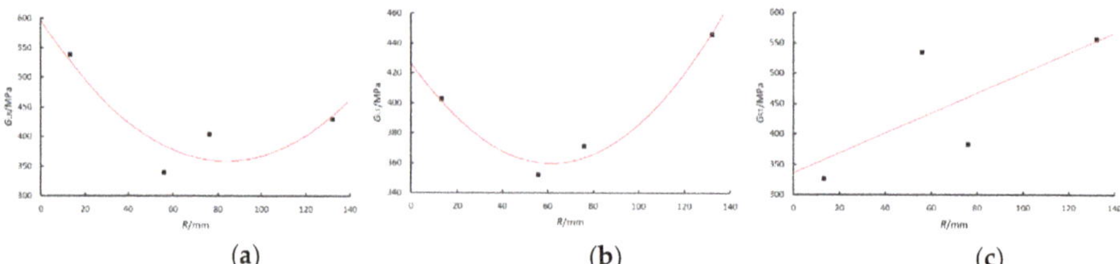

(a) (b) (c)

**Figure 8.** Relationships between distance from pith and shear moduli in: (**a**) Longitudinal-radial (LR) plane, (**b**) Longitudinal-tangential (LT) plane, and (**c**) Radial-tangential (RT) plane.

**Table 6.** Mathematical model of shear moduli and distance from pith.

| Shear Modulus of Elasticity | Fitted Equation | Coefficient of Determination ($R^2$) |
| --- | --- | --- |
| $G_{LR}$ | $G_{LR} = 597.134 + 64.293R + 0.034R^2$ | 0.80 |
| $G_{LT}$ | $G_{LT} = 426.009 - 2.155R + 0.018R^2$ | 0.98 |
| $G_{RT}$ | $G_{RT} = 335.912 + 1.645R$ | 0.72 |

Figure 8 shows the variation patterns of the three shear moduli of elasticity ($G_{LR}$, $G_{LT}$, and $G_{RT}$) along the cross-section of wood. Figure 8a,b demonstrate a quadratic relationship between shear modulus of elasticity $G_{LR}$ and sampling distance, as well as for shear modulus of elasticity $G_{LT}$ and sampling distance. Both $G_{LR}$ and $G_{LT}$ first decreased with increasing sampling distance, and then increased at a sampling distance over 70 mm. $G_{LR}$ and $G_{LT}$ near the bark increased compared to the minimum values. However, the shear modulus of elasticity $G_{RT}$ of green Chinese larch linearly increased with sampling distance. Table 6 shows the coefficient of determination for the shear modulus of elasticity $G_{RT}$ and the sampling distance was 0.72, indicating a robust linear relationship between them. Even though a quadratic relationship was found in the shear modulus of elasticity $G_{LR}$ and $G_{LT}$, their coefficients of determination were 0.80 and 0.98, respectively. The possible interpretation for the relatively lower coefficient of determination ($R^2$) for $G_{RT}$ could be attributed to the large variability in wood performance especially in the cross-sections. Xavier et al. reported the shear stiffness ($Q_{66} = G_{LR}$) of dry Maritime pine decreased from the center (radial position $r_1$, 29% of the radius) to around the middle radius of stem (radial position $r_2$, 46% of the radius), and progressively increased afterward to the outermost positions ($r_4$, 81% of the radius) [33]. Despite the different moisture content and tree species, the variation in shear moduli $G_{LR}$ along the whole cross-section derived in this research was basically in compliance with the results reported in Xavier's study. No study has reported the variation in the shear moduli $G_{LT}$ and $G_{RT}$ along the whole cross-section of wood, whether dry or green. Therefore, the variation patterns of shear moduli $G_{LT}$ and $G_{RT}$ presented in this paper could be used to describe the shear properties of green wood in LT and RT plane. More data from identical or different species need to be obtained to determine the variation in shear properties of dry or green wood, especially for shear moduli $G_{LT}$ and $G_{RT}$.

### 3.3.3. Poisson's Ratio

The results from experimental data (Table 2) showed that the relationship between the values of the Poisson's ratios at the four sampling positions and sampling distance $R$ was relatively discrete, except for Poisson's ratio $\nu_{RT}$. To determine the quantitative relationship between Poisson's ratios and sampling distance $R$, three extra data points were inserted using interpolation for each Poisson's ratio, apart from $\nu_{RT}$. Thus, the relationships between the Poisson's ratios and sampling distance $R$ were obtained as shown in Figure 9a–f. The corresponding fitting equations are provided in Table 7.

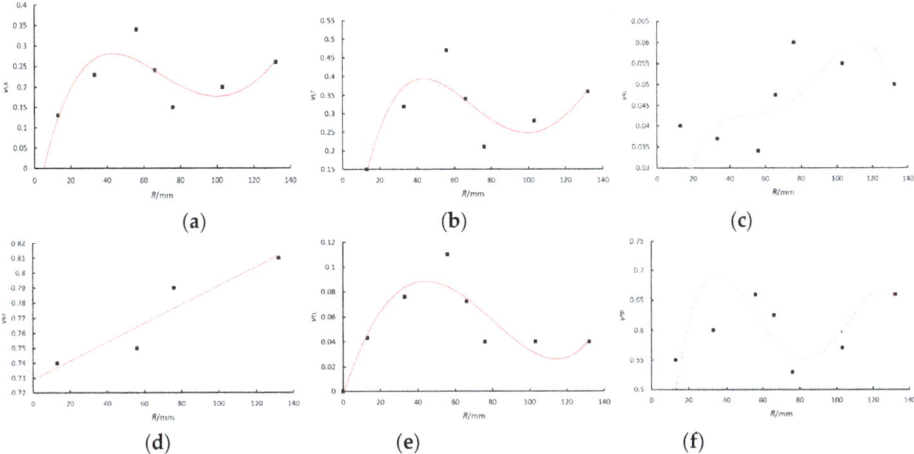

**Figure 9.** Relationships between distance from pith and Poisson's ratio of: (**a**) $v_{LR}$, (**b**) $v_{LT}$, (**c**) $v_{RL}$, (**d**) $v_{RT}$, (**e**) $v_{TL}$, and (**f**) $v_{TR}$.

Figure 9 shows the variation patterns of the six Poisson's ratios along the cross-section of the green wood. Figure 9d depicts the linear relationship between the Poisson's ratio $v_{RT}$ and sampling distance. Poisson's ratio $v_{RT}$ gradually increased with sampling distance. However, for the other five Poisson's ratios, there was a discrete relationship between the Poisson's ratio and sampling distance $R$. For Poisson's ratios $v_{LR}$, $v_{LT}$, and $v_{TR}$, the values first increased at sampling distances lower than 50 mm and then significantly decreased as sampling distance varied from 50 mm to about 80 mm. When the sampling distance was over 80 mm, the values of Poisson's ratio increased with sampling distance again. However, similar results were not found for Poisson's ratios $v_{RL}$ and $v_{TL}$. The values of the Poisson's ratios determined in this study irregularly changed with sampling distance probably due to the variation in moisture content, density, or microfibril angle in different parts of the wood. In general, no significant variation patterns were found in these five Poisson's ratios. Davies et al. estimated six Poisson's ratios both in outerwood and corewood from green *Pinus radiata* and no significant difference between outerwood and corewood was found [32]. Therefore, we still do not understand the variation patterns of the Poisson's ratios along wood cross-sections, and few researchers have evaluated the variation in the Poisson's ratios along the whole cross-section of dry or green wood. Therefore, more efforts are required to investigate the variation patterns of Poisson's ratio in the entire cross-section of dry or green wood.

**Table 7.** Mathematical model of Poisson's ratios and distance from pith.

| Poisson's Ratio | Fitted Equation |
| --- | --- |
| $v_{LR}$ | $v_{LR} = -0.0078 + 0.0138R - 2.112 \times 10^{-4}R^2 + 8.8643 \times 10^{-7}R^3 - 3.155 \times 10^{-8}R^4$ |
| $v_{LT}$ | $v_{LT} = -0.0124 + 0.0172R - 2.2671 \times 10^{-4}R^2 + 5.2757 \times 10^{-7}R^3 + 2.8007 \times 10^{-9}R^4$ |
| $v_{RL}$ | $v_{RL} = 0.0044 + 0.00276R - 7.1388 \times 10^{-5}R^2 + 7.7972 \times 10^{-7}R^3 + 2.8887 \times 10^{-9}R^4$ |
| $v_{RT}$ | $v_{RT} = 0.7312 + 6.2191R$ |
| $v_{TL}$ | $v_{TL} = -0.0023 + 0.0046R - 6.7872 \times 10^{-5}R^2 + 1.9557 \times 10^{-3}R^3 + 5.4162 \times 10^{-10}R^4$ |
| $v_{TR}$ | $v_{TR} = 0.028 + 0.0471R - 0.0011R^2 + 1.0274 \times 10^{-5}R^3 - 3.155 \times 10^{-8}R^4$ |

## 4. Conclusions

The objective of this study was to investigate the variation in the mechanical properties, especially elastic constants, of green Chinese larch from pith to sapwood. The conclusions are as follows:

(1) The relationships between longitudinal modulus of elasticity ($E_L$), radial modulus of elasticity ($E_R$), and tangential modulus of elasticity ($E_T$) were $E_L > E_R > E_T$ for all four sampling positions. Similarly, $v_{RT} > v_{LT} > v_{LR}$ were found for Poisson's ratios $v_{RT}$, $v_{LT}$, and $v_{LR}$ at the four sampling locations. These results align with the reported findings in dry wood.

(2) The sensitivity of each elastic constant to the sampling position was different, and the coefficient of variation ranged from 4.3% to 48.7%. The Poisson's ratios $v_{RT}$ measured at the four different sampling positions were similar and the differences between them were not significant. The coefficient of variation for Poisson's ratio $v_{RT}$ was only 4.3%. The four sampling positions had similar Poisson's ratios $v_{TL}$, though the coefficient of variation was 11.7%. The Poisson's ratio $v_{LT}$ had the greatest variation in all elastic constants with a 48.7% in coefficient of variation.

(3) We found a good linear relationship between the longitudinal modulus of elastic $E_L$, shear modulus of elasticity $G_{RT}$, Poisson's ratio $v_{RT}$ and the distance from pith. $E_L$, $G_{RT}$, and $v_{RT}$ all increased with sampling distance $R$. However, a quadratic relationship existed in the tangential modulus of elasticity $E_T$, radial modulus of elasticity $E_R$, shear modulus of elasticity $G_{LT}$, shear modulus of elasticity $G_{LR}$, and the distance from pith. A discrete relationship was found in the other five Poisson's ratios.

**Author Contributions:** F.L. wrote the original draft of the manuscript and performed most of test and analysis work. H.Z. and C.G. supervised the research team and provided some ideas to research as well as revised the manuscript. F.J. provided the main idea of experiments and assisted with the experiments. X.W. conceived the project and provided technical guidance to the research.

**Funding:** This project was supported by "the Fundamental Research Funds for the Central Universities (NO. BLX201817)", China Postdoctoral Science Foundation (NO. 2018M641225), the Special Research Funds for Public Welfare (NO. 201304512) and the National Natural Science Foundation of China (No. 31328005).

**Acknowledgments:** The authors wish to thank Jianzhong Zhang for his grateful assistance of conducting the experiments.

**Conflicts of Interest:** The authors declare no conflict of interest.

# References


1. Feng, H.L.; Li, G.H. Stress wave propagation model and simulation in non-destructive testing of wood. *J. Syst. Simul.* **2009**, *21*, 2373–2376.
2. Feng, H.L.; Li, G.H.; Fang, Y.M.; Li, J. Stress wave propagation modeling and application in wood testing. *J. Syst. Simul.* **2010**, *22*, 1490–1493.
3. Yang, X.C.; Wang, L.H. Study on the propagation theories of stress wave in log. *Sci. Silvae Sin.* **2005**, *41*, 132–138.
4. Liu, F.L.; Jiang, F.; Wang, X.P.; Zhang, H.J.; Yang, Z.H. *Stress Wave Propagation in Larch Plantation Trees—Numerical Simulation*; USDA Forest Service, Forest Products Laboratory, General Technical Report, FPL-GTR-239; USDA, Forest Products Laboratory: Madison, WI, USA, 22–25 September 2015; pp. 418–427.
5. Bodig, J.; Jayne, B.A. *Mechanics of Wood and Wood Composites*; Van Nostrand Reinhold Company, Inc.: New York, NY, USA, 1982; p. 712.
6. Sliker, A. A method for predicting non-shear compliance in the RT plane of wood. *Wood Fiber Sci.* **1988**, *20*, 44–45.
7. Sliker, A. Orthotropic strains in compression parallel to grain tests. *For. Prod. J.* **1985**, *35*, 19–26.
8. Sliker, A. A measurement of the smaller Poisson's ratios and related compliance for wood. *Wood Fiber Sci.* **1989**, *21*, 252–262.
9. Li, W.J. Elasticity and drying stresses of wood. *J. Nanjing Technol. Coll. For. Prod.* **1983**, *6*, 115–122.
10. Gong, M. A study of wood elastic parameters parallel to grain in compression test by using resistance strain gages. *Sci. Silvae Sin.* **1995**, *31*, 190–191.
11. Zhang, F.; Li, L.; Zhang, L. Study of the determination of the elastic constants and mechanical property parameters of five kinds of wood commonly used in furniture. *For. Mach. Woodwork. Equip.* **2012**, *40*, 16–19.
12. Skaar, C. *Wood-Water Relations*; Springer: Berlin Heidelberg, German, 1988.

13. Hering, S.; Keunecke, D.; Niemz, P. Moisture-dependent orthotropic elasticity of beech wood. *Wood Sci. Technol.* **2012**, *46*, 927–938. [CrossRef]
14. Jiang, J.; Bachtiar, E.V.; Lu, J.; Niemz, P. Moisture-dependent orthotropic elasticity and strength properties of Chinese fir wood. *Eur. J. Wood Wood Prod.* **2017**, *68*, 927–938. [CrossRef]
15. Ozyhar, T.; Hering, S.; Sanabria, S.; Niemz, P. Determining moisture-dependent elastic characteristics of beech wood by means of ultrasonic waves. *Wood Sci. Technol.* **2013**, *47*, 329–341. [CrossRef]
16. Bachtiar, E.V.; Sanabria, S.J.; Mittig, J.P.; Niemz, P. Moisture-dependent elastic characteristics of walnut and cherry wood by means of mechanical and ultrasonic test incorporating three different ultrasound data evaluation techniques. *Wood Sci. Technol.* **2017**, *51*, 47–67. [CrossRef]
17. Wang, L.Y.; Lu, Z.Y.; Shen, S.J. Study on twelve elastic constants values of Betula platyphylla Suk. wood. *J. Beijing For. Univ.* **2003**, *25*, 64–67.
18. Shao, Z.P.; Zhu, S. Study on determining the elastic constants of wood of Cunninghamia lanceolata by means of electric resistance-strain gauges. *J. Anhui Agric. Univ.* **2001**, *28*, 32–35.
19. Aira, J.R.; Arriaga, F.; Íñiguez-González, G. Determination of the elastic constants of Scots pine (*Pinus sylvestris* L.) wood by means of compression tests. *Biosyst. Eng.* **2014**, *126*, 12–22. [CrossRef]
20. Preziosa, C. Méthode de déTermination des Constantes éLastiques du Mátériau Bois Par Utilisation des Ultrasons (Method for Determining Elastic Constants of Wood Material by Use of Ultrasound). Ph.D. Thesis, Université d'Orléans, Orléans, France, 1982.
21. Preziosa, C.; Mudry, M.; Launay, J.; Gilletta, F. Détermination des constantes élastiques du bois par une méthode acoustique goniométrique. (Determination of the elastic constants of wood by an acoustic method using immersion). *CR Acad. Sci. Paris Série II* **1981**, *293*, 91–94.
22. Bucur, V.; Archer, R.R. Elastic constants for wood by an ultrasonic method. *Wood Sci. Technol.* **1984**, *18*, 255–265. [CrossRef]
23. Francois, M.L.M. Identification des symétries matérielles de matériaux anisotropes (Identification of the Material Symmetries of Anisotropic Materials). Ph.D. Thesis, Université Pierre et Marie Curie-Paris VI, Paris, France, 1995.
24. Goncalves, R.; Trinca, A.J.; Pellis, B.P. Elastic constants of wood determined by ultrasound using three geometries of specimens. *Wood Sci. Technol.* **2014**, *48*, 269–287. [CrossRef]
25. Longo, R.; Delaunay, T.; Laux, D.; El Mouridi, M.; Arnould, O.; Le Clézio, E. Wood elastic characterization from a single sample by resonant ultrasound spectroscopy. *Ultrason* **2012**, *52*, 971–974. [CrossRef]
26. Majano-Majano, A.; Fernandez-Cabo, J.L.; Hoheisel, S.; Klein, M. A test method for characterizing clear wood using a single specimen. *Exp. Mech.* **2012**, *52*, 1079–1096. [CrossRef]
27. Vazquez, C.; Gonçalves, R.; Bertoldo, C.; Bano, V.; Vega, A.; Crespo, J.; Guaita, M. Determination of the mechanical properties of Castanea sativa Mill. using ultrasonic wave propagation and comparison with static compression and bending methods. *Wood Sci. Technol.* **2015**, *49*, 607–622. [CrossRef]
28. Yang, N.; Zhang, L. Investigation of elastic constants and ultimate strengths of Korean pine from compression and tension tests. *J. Wood Sci.* **2018**, *64*, 85–96. [CrossRef]
29. Dahmen, S.; Ketata, H.; Ben Ghozlen, M.H.; Hosten, B. Elastic constants measurement of anisotropic Olivier wood plates using air-coupled transducers generated Lamb wave and ultrasonic bulk wave. *Ultrasonics* **2010**, *50*, 502–507. [CrossRef]
30. Nadir, Y.; Nagarajan, P.; Midhun, A.J. Measuring elastic constants of Hevea brasiliensis using compression and Iosipescu shear test. *Eur. J. Wood Wood Prod.* **2014**, *72*, 749–758. [CrossRef]
31. Yoshihara, H.; Ohta, M. Estimation of the shear strength of wood by uniaxial-tension tests of off-axis specimens. *J. Wood Sci.* **2000**, *46*, 159–163. [CrossRef]
32. Davies, N.T.; Altaner, C.M.; Apiolaza, L.A. Elastic constants of green Pinus radiata wood. *N. Z. J. For. Sci.* **2016**, *46*, 399. [CrossRef]
33. Xavier, J.; Avril, S.; Pierron, F.; Morais, J. Variation of transverse and shear stiffness properties of wood in a tree. *Compos. Part A Appl. Sci. Manuf.* **2009**, *40*, 1953–1960. [CrossRef]

34. ASTM D5536-94. *Standard Practice for Sampling Forest Trees for Determination of Clear Wood Properties*; ASTM International: West Conshohocken, PA, USA, 1994.
35. ASTM D143-09. *Standard Test Method for Small Clear Specimens of Timber*; ASTM International: West Conshohocken, PA, USA, 2009.

*Article*

# Intra-Ring Wood Density and Dynamic Modulus of Elasticity Profiles for Black Spruce and Jack Pine from X-Ray Densitometry and Ultrasonic Wave Velocity Measurement [†]

Wassim Kharrat [1], Ahmed Koubaa [2,*], Mohamed Khlif [3] and Chedly Bradai [3]

[1] Centre de Recherche sur les Matériaux Renouvelables, Université Laval,
Ville de Québec, QC G1V 0A6, Canada
[2] Institut de Recherche sur les Forêts, Université du Québec en Abitibi-Témiscamingue,
Rouyn-Noranda, QC J9X 5E4, Canada
[3] Laboratoire des Systèmes électromécaniques (LASEM), École Nationale d'ingénieurs de Sfax,
Sfax 3038, Tunisia
* Correspondence: ahmed.koubaa@uqat.ca; Tel.: +01-819-761-0971 (ext. 2579)
† This manuscript is part of an M.S. thesis by the first author, available online at depositum.uqat.ca.

Received: 30 May 2019; Accepted: 4 July 2019; Published: 9 July 2019

**Abstract:** Currently, ultrasonic measurement is a widely used nondestructive approach to determine wood elastic properties, including the dynamic modulus of elasticity (DMOE). DMOE is determined based on wood density and ultrasonic wave velocity measurement. The use of wood average density to estimate DMOE introduces significant imprecision: Density varies due to intra-tree and intra-ring differences and differing silvicultural treatments. To ensure accurate DMOE assessment, we developed a prototype device to measure ultrasonic wave velocity with the same resolution as that provided by the X-ray densitometer for measuring wood density. A nondestructive method based on X-ray densitometry and the developed prototype was applied to determine radial and intra-ring wood DMOE profiles. This method provides accurate information on wood mechanical properties and their sources of variation. High-order polynomials were used to model intra-ring wood density and DMOE profiles in black spruce and jack pine wood. The transition from earlywood to latewood was defined as the inflection point. High and highly significant correlations were obtained between predicted and measured wood density and DMOE. An examination of the correlations between wood radial growth, density, and DMOE revealed close correlations between density and DMOE in rings, earlywood, and latewood

**Keywords:** ultrasonic wave velocity measurement; nondestructive assessment; wood mechanical properties; intra-ring variation; dynamic modulus of elasticity

## 1. Introduction

"Wood quality is the resultant of physical and chemical characteristics possessed by a tree or a part of a tree that enable it to meet the property requirements for different end products" [1]. Wood density is considered to be the most important wood quality attribute. It is one of the most widely used parameters to predict the mechanical and other physical properties of wood, such as dimensional stability [2]. However, wood density and all its related wood quality attributes are highly variable, with multiple sources of variation, including differences within and between trees, between sites, and between genetic origins. This high variability is due to genetic, environmental, and physiological factors [3,4]. In a same species, variations in wood density also result from variations in anatomical characteristics such as earlywood and latewood width. Wood density is generally defined as the ratio of the wood mass to volume, and is expressed in kilograms per cubic meter ($kg/m^3$). However, this definition does not consider variations in wood density due to biological processes such as earlywood and latewood formation, juvenile wood formation, or environmental conditions. Modern nondestructive measurement methods such as X-ray densitometry are widely used to assess wood quality variations due to biological processes (intra-ring and inter-ring variation), genetic sources, and environmental conditions (e.g., tree-to-tree, site-to-site, and silvicultural treatments).

Intra-ring wood density profiles obtained with X-ray densitometry are generally used to calculate ring density (RD), earlywood density (EWD), latewood density (LWD), ring width (RW), earlywood width (EWW), and latewood width (LWW). These parameters have been determined for many wood species, such as European oak [5], black spruce [2], and *Thuja occidentalis* [6]. Intra-ring wood density profiles are used to determine the use-specific suitability of wood, especially for high value-added applications [5,7]. Intra-ring wood density variation can also indicate wood uniformity and provide information about the wood growth process and the wood fiber yield [7,8].

Earlywood and latewood properties depend on the earlywood–latewood transition point (E/L). Several methods have been reported in the literature to determine the E/L, notably Mork's index [9]. There are at least two interpretations of Mork's index [10]. According to the first, the E/L is obtained when the double wall thickness becomes greater than or equal to the width of the cell lumen. Under the second interpretation, the E/L is obtained when the double cell wall thickness multiplied by two becomes greater than or equal to the lumen width. While this index, using either interpretation, is arbitrary and time-consuming to measure, it allows consistent determination of earlywood and latewood features.

Because Mork's index is based on the double wall thickness and the lumen diameter, these anatomical wood features must be measured in individual growth rings on microscopic slides or using indirect microscopic procedures [11]. In addition, this method is difficult to integrate into X-ray computations. Good agreement was found between earlywood and latewood features determined by three methods: Mork's index, threshold density, and the maximum derivative [12]. However, use of Mork's index and the maximum derivative produced better estimates of physiological variations compared to threshold density. Intra-ring wood density profiles are generally modelled to define the earlywood–latewood transition. Pernestål et al. [8] and Ivkovic et al. [13] used modified spline functions to model intra-ring wood density profiles. The E/L transition was defined using a numerical derivative method. Koubaa et al. [2] demonstrated that high-order polynomial functions consider both profile and intra-ring density variation for E/L estimation. These functions gave consistent estimates of the E/L transition point, with correlation coefficients between measured and predicted density well above 0.90 for the six order polynomial. These results are significant, because modelled intra-ring wood density profiles can simplify the modelling of final wood product properties [13].

While wood density is considered to be the most important wood quality attribute, elastic properties are also important, especially for engineering design purposes [14]. The wood dynamic modulus of elasticity (DMOE), being an elastic constant that describes wood mechanical behavior, is computed from the wood density and the ultrasonic wave velocity [15]. Ultrasonic wave velocity measurement is one of the most widely used nondestructive methods to assess the strength properties of living trees, logs, sawn timbers, and wood-based materials, due to its rapidity, flexibility, portability, cost-effectiveness, and ease of use [16–19]. Wood DMOE has been determined using ultrasonic wave velocity parallel to the grain direction and wood density based on the mass-to-volume ratio of specimens [15,18–21]. However, no study to date has investigated intra-ring wood DMOE profiles to determine variations between earlywood and latewood DMOE. A nondestructive method based on X-ray densitometry and ultrasonic wave velocity measurement was proposed to determine intra-ring wood DMOE profiles.

The objectives of this study were therefore to (1) develop a nondestructive method to determine intra-ring wood DMOE profiles; (2) model intra-ring wood density and DMOE profiles in black spruce and jack pine wood using high-order polynomial functions [2]; (3) determine radial variations in ring wood density and ring DMOE; and (4) analyze correlations between wood radial growth, density, and DMOE.

## 2. Materials and Methods

X-ray densitometry provides intra-ring wood density profiles and determines both annual ring width and wood density components. Intra-ring wood ultrasonic velocity profiles were determined using the developed prototype and a Sonatest Masterscan ultrasonic flaw detector. The superposition of these two profiles is a nondestructive method to obtain the intra-ring wood DMOE profile.

### 2.1. Study Materials

The experimental material used in this study consisted of subsamples from previous studies on the wood quality of jack pine [22] and black spruce [23] sampled from even-aged stands in the Abitibi region of Québec, Canada. Eight black spruce and eight jack pine trees were used. Discs taken at breast height were used in this study. Bark-to-bark samples passing through the pith were extracted from each disc. Thin strips (15 to 20 mm wide and 1.57 to 1.9 mm thick) were sawn from each sample. The sawn strips were extracted with cyclohexane/ethanol (2:1) solution for 24 h and then with distilled water for another 24 h to remove extraneous compounds [24]. After extraction, the strips were air-dried under restraint to prevent warping. Samples were then conditioned to 8% equilibrium moisture content before measurement. The same samples were used to determine wood density and ultrasonic wave propagation time. In this study, a nondestructive method based on X-ray densitometry and ultrasonic wave velocity measurement was used to determine intra-ring wood density and DMOE variation.

### 2.2. Wood Ring Density and Width Measurement

Ring density (RD) and ring width (RW) were measured for each ring using a QTRS-01X Tree-Ring Scanner (Quintek Measurement Systems, Knoxville, TN, USA). The QTRS passes thin strips from increment cores through an accurately collimated soft X-ray beam using a precisely controlled stepping system and linear bearing carriage. Video images of both the wood sample surface and the X-ray density graph are displayed at the same scale on the screen. A linear resolution step size of 40 μm was used for the X-ray densitometry. Rings from pith to bark were scanned in air-dry condition to estimate the basic wood density (ovendry weight/green volume) for each ring. Ring density (RD) and ring width (RW) for each ring were determined based on intra-ring microdensitometer profiles. Incomplete or false rings and rings with compression wood or branch tracers were eliminated from the analysis. Matlab software (R2016a, the MathWorks, Inc., Natick, MA, USA) was used to determine the intra-ring wood density profiles at 40 μm resolution.

### 2.3. Wood Ultrasonic Wave Velocity Measurement

An in-house prototype device was developed for measuring the ultrasonic wave propagation time with the same resolution as that used for the wood density measurement by X-ray densitometry (40 μm). The prototype (Figure 1) consists of a motorized linear translation stage that holds the sample and is controlled by a microcontroller. The ultrasonic wave propagation time in the wood sample is measured between the ultrasonic transmitter (Spot Weld Transducer) and the receiver transducers (Fingertip Contact Transducer CF) at 40 μm resolution. The ultrasonic transducers are mounted in parallel to two mini motorized actuators to ensure constant pressure during measurement.

Following the X-ray densitometry density measurement, nondestructive ultrasonic wave propagation measurement was applied to the same samples using a Masterscan 380 (Sonatest Inc., San Antonio, TX, USA) equipped with 10 MHz frequency transducers. Ultrasonic waves were applied to the samples through two transducers (transmitting and receiving). A coupling agent (Vaseline original petroleum jelly) was used to aid the transmission of the transducer pulses into the test specimens.

A correction factor (C$_f$; s) was applied to calculate the ultrasonic wave velocity in the wood samples to take into consideration the transport time of the electric waves within the measuring circuit.

A Plexiglas sample having the same thickness as the wood samples (2 mm) was used as a reference to determine the correction factor (Equation (1)) [20,25]. The ultrasonic velocity (V; m/s) was then calculated using Equation (2):

$$C_f = t_r - (d_r/v_r) \tag{1}$$

$$V = d/(T - C_f) \tag{2}$$

where d is the thickness of the wood sample (m), T is the ultrasonic wave propagation time (s), $t_r$ is the wave propagation time through the reference Plexiglas core (s), $d_r$ is the thickness of the reference Plexiglas core (m), and $v_r$ is the wave velocity in the reference Plexiglas core (2670 m/s).

The dynamic modulus of elasticity (DMOE; MPa) based on the ultrasonic method was determined using the following one-dimensional wave equation:

$$DMOE = \rho \times V^2 \times 10^{-6}. \tag{3}$$

where $\rho$ is the wood density measured by X-ray densitometry (kg/m$^3$) and V is the ultrasonic wave velocity calculated using Equation (1).

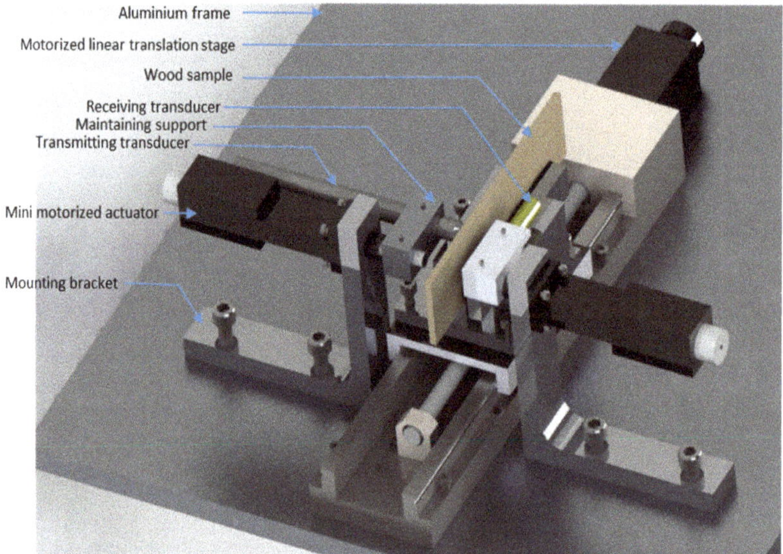


**Figure 1.** Prototype device for measuring the ultrasonic wave propagation time.

*2.4. Modelling Intra-Ring Wood Density and Dynamic Modulus of Elasticity Profiles*

In this study, we used 6th order polynomial functions to model intra-ring wood density and DMOE profiles for black spruce and jack pine wood (Equation (4)).

$$R = a_0 + a_1 RW + a_2 RW^2 + a_3 RW^3 + a_4 RW^4 + a_5 RW^5 + a_6 RW^6 \tag{4}$$

where *R* is the ring density or ring DMOE, *RW* is the ring width in proportion, and $a_i$ are the parameters to be estimated.

The E/L transition was defined as the inflection point obtained from the within-ring density and DMOE profiles. The E/L transition is obtained by equalling the second derivative of the polynomial function to zero (Equation (5)). For a 6th order polynomial function, the second derivative gives 4 solutions, but only one solution is of interest. Certain restrictions were specified in the Matlab program to obtain this unique solution. These restrictions specify that the solution should be included in a positive slope and in the range of 40 to 90% of the ring width proportion. If more than one solution is obtained, the highest value among the solutions is chosen.

$$d^2R/dRW^2 = 2a_2 + 6a_3RW + 12a_4RW^2 + 20a_5RW^3 + 30a_6RW^4 \tag{5}$$

*2.5. Statistical Analysis*

For both softwood species (black spruce and jack pine), the correlations between the wood radial growth, density, and DMOE components were determined using R software (Version 2.15.0 R, R Development Core Team, 2012, Vienna, Austria).

## 3. Results and Discussion

Typical X-ray density and DMOE profiles for jack pine wood are shown in Figure 2. Both the within-ring and radial pattern variation in these properties are shown.

*3.1. Intra-Ring Wood Density and Dynamic Modulus of Elasticity Profiles*

Figure 3 shows the within-ring variation in wood density and DMOE for black spruce and jack pine, revealing similar within-ring density patterns between the two species. Wood density and DMOE increase slowly in earlywood to reach a maximum in latewood. Both properties decrease thereafter at about mid-latewood width to reach a minimum at the boundary between two growth rings. The similarity between intra-ring density and DMOE profiles confirms the close relationship between wood density and wood stiffness, even at the intra-ring level. Some slight differences between the intra-ring density and DMOE profiles appear in earlywood. Thus, the intra-ring wood density profiles increase more slowly in earlywood compared to the DMOE profiles, which show a relatively sharp increase. Similar patterns of within-ring density variation were obtained by Koubaa et al. [2] for black spruce and by Ivkovic et al. [13] for Norway spruce and Douglas fir.

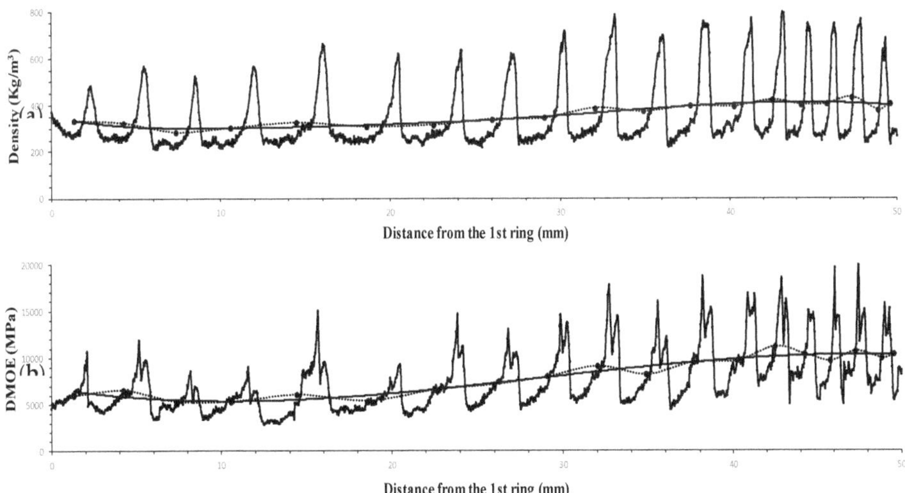


**Figure 2.** Examples of jack pine profiles showing radial variation in: (**a**) Wood density and (**b**) DMOE in (from ring 2 to 19).

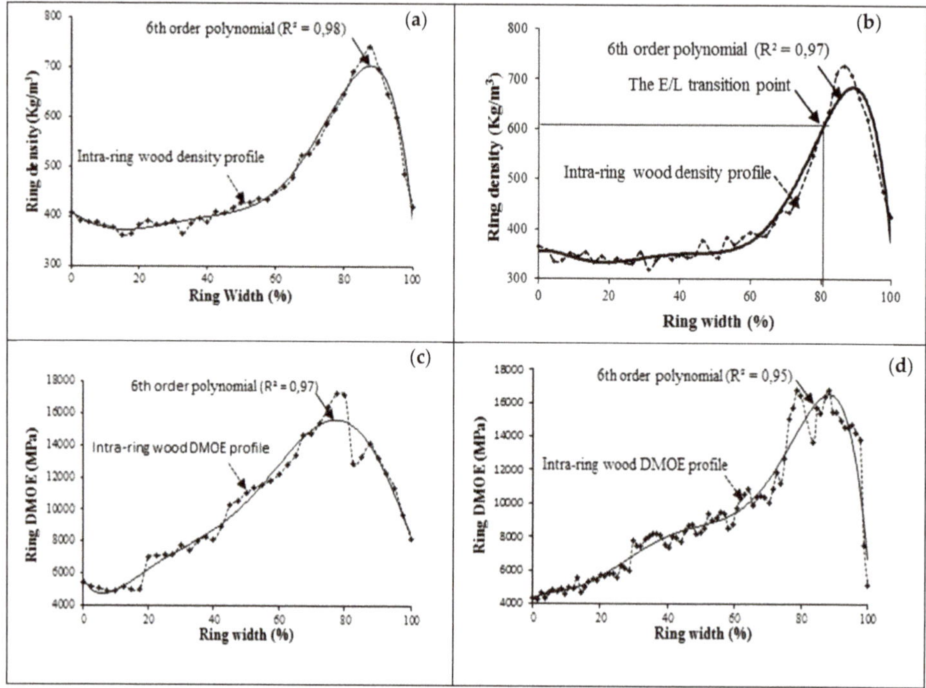

**Figure 3.** Examples of within-ring profiles and the fits obtained with the 6th order polynomials for: (**a**) Ring density in black spruce; (**b**) ring density in jack pine; (**c**) DMOE in Black spruce; and (**d**) DMOE in Jack pine.

The same 6th polynomial function modeling approach suggested by Koubaa et al. [2] was used to model within-ring density and DMOE profiles for the black spruce and jack pine samples in this study. Figure 3 illustrates the fitness of the 6th order polynomials for the intra-ring density and DMOE profiles for both softwoods species: (a) Black spruce and (b) jack pine. Table 1 confirms the fitness. The correlation coefficients obtained between the measured and predicted ring density data range from 0.88 to 1.00, with an average well above 0.95. These results indicate that these models can well describe intra-ring wood density profiles obtained from black spruce and jack pine, and probably other softwood species, as the coefficients are in good agreement with those obtained by Koubaa et al. [2] for black spruce.

Table 1 also indicates that high-order polynomials fit well the intra-ring DMOE profiles for black spruce and jack pine. The correlation coefficients obtained between measured and predicted DMOE data using the 6th order polynomial models range from 0.80 to 0.99, with an average well above 0.90 (Table 1). These results indicate that high-order polynomials can describe DMOE profiles well for these two softwoods.

The measured elastic properties of wood material yield essential information for the understanding of bonding at a very fine structural level [14]. As the elastic properties describe the mechanical behavior of wood, it is mandatory to determine intra-ring wood DMOE profiles. Moreover, modelled intra-ring wood DMOE profiles can serve as effective prediction tools for wood mechanical behavior.

*3.2. The Earlywood–Latewood Transition*

Table 2 shows that wood density at the E/L transition point (Figure 3b) (E/L transition density) as defined by the inflexion point method presents large variation for black spruce (Figure 4a) and jack pine (Figure 4b). The radial variation pattern for the E/L transition density in black spruce (Figure 4a) is similar to that reported by Koubaa et al. [2], and is characterized by large variation with no specific trend. In contrast, the radial variation for the E/L transition in jack pine (Figure 4b) is characterized by a steady increase in juvenile wood and a tendency to level off in mature wood. Similar radial variation patterns for the E/L wood transition were observed by Park et al. [22].

**Table 1.** Average, standard variation (between parenthesis) and range of Pearson's coefficient of determination between measured and predicted within-ring density and DMOE values from the 6th order polynomial models for different rings for black spruce and jack pine.

| | Ring from Pith | | | |
| --- | --- | --- | --- | --- |
| | 5 | 10 | 15 | 20 |
| Wood density profiles | | | | |
| Black spruce | | | | |
| Average profiles | 0.96 (0.02) | 0.97 (0.02) | 0.96 (0.02) | 0.97(0.02) |
| Range | 0.88–0.99 | 0.91–0.99 | 0.91–0.99 | 0.92–1.00 |
| Jack pine | | | | |
| Average profiles | 0.96 (0.02) | 0.95 (0.02) | 0.97 (0.02) | 0.98 (0.01) |
| Range | 0.92–0.98 | 0.90–1.00 | 0.89–0.99 | 0.96–0.99 |
| Dynamic modulus of elasticity profiles | | | | |
| Black spruce | | | | |
| Average profiles | 0.92 (0.03) | 0.94 (0.04) | 0.95 (0.03) | 0.95 (0.02) |
| Range | 0.82–0.99 | 0.88–0.99 | 0.86–0.99 | 0.91–0.99 |
| Jack pine | | | | |
| Average profiles | 0.89 (0.04) | 0.93 (0.02) | 0.93 (0.03) | 0.94 (0.02) |
| Range | 0.80–0.97 | 0.88–0.98 | 0.82–0.99 | 0.91–0.99 |

Within a same ring, the E/L transition also shows substantial variation in the true measures, as indicated by the relatively large standard errors (Figure 4a,b, Table 2). For example, the E/L transition density for the 10th annual ring from the pith varies from 541 to 655 kg/m$^3$ and from 548 to 672 kg/m$^3$ in black spruce and jack pine, respectively (Table 2). These results concur with the findings by Koubaa et al. [26] for black spruce and by Park et al. [22] for jack pine. Earlywood and latewood density defined by this method also show large variation. Thus, for a same annual ring, earlywood density ranges from 383 to 432 kg/m$^3$ for black spruce and from 318 to 367 kg/m$^3$ for jack pine.

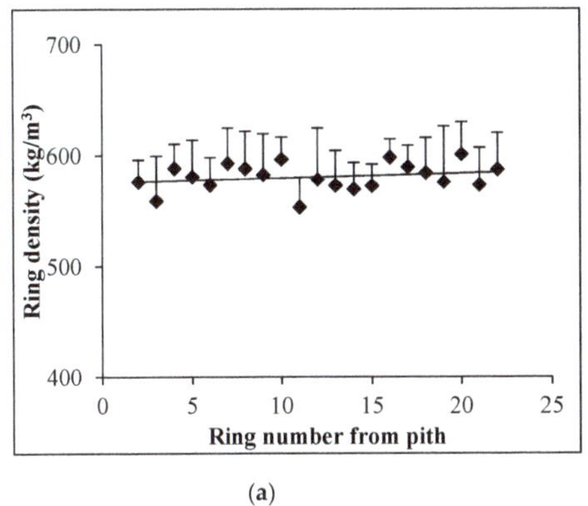
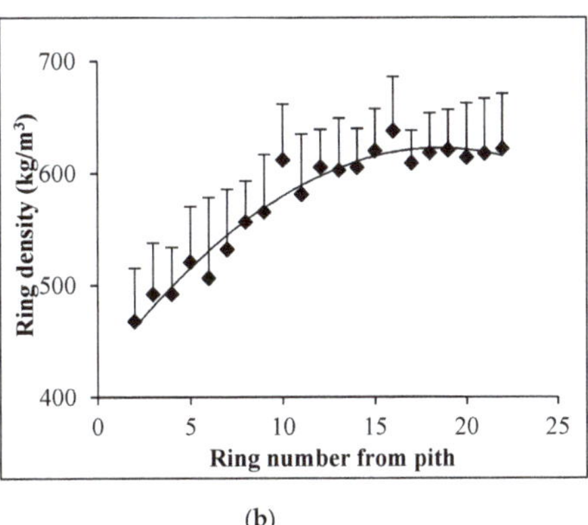

(a)       (b)

**Figure 4.** Radial variation in E/L transition density (bars indicate the standard error) for (**a**) black spruce and (**b**) jack pine.

As shown in Table 2, wood density is variable at the E/L transition point for black spruce and jack pine. For black spruce, the average wood density at the E/L transition point is variable and higher than the 590 kg/m$^3$ reported by Koubaa et al. [2], as well as the threshold wood density of 540 kg/m$^3$ used for black spruce in X-ray densitometry programs. The results for jack pine are similar: The E/L transition density is variable and higher than the threshold density typically used for jack pine in X-ray densitometry programs. As the wood density at the E/L transition point, as defined by the inflexion point method, is variable and higher than the threshold wood density, the average earlywood and latewood width and density as defined by the inflexion point method will differ from those defined by the threshold method, for both black spruce and jack pine. Earlywood width defined by the inflexion point method will be greater, whereas latewood width will be smaller. Consequently, the latewood proportion defined by the inflexion point method will be lower. These results confirm the findings by Koubaa et al. [2] that the E/L transition point varied greatly among individual growth rings and that the use of a predetermined fixed threshold wood density does not reflect the variation in intra-ring wood density profiles across growth rings in a species.

**Table 2.** Average (Av), range (Ra), and standard variation for ring width, wood density, and wood DMOE at the earlywood–latewood transition and in earlywood and latewood, as defined by the inflexion method for different rings.

| | Black Spruce | | | Jack Pine | | |
| --- | --- | --- | --- | --- | --- | --- |
| | **Ring Number from the Pith** | | | | | |
| | **5** | **10** | **20** | **5** | **10** | **20** |
| | *Earlywood width* | | | | | |
| Av (mm) | 1.32 (0.37) | 1.26 (0.35) | 1.02 (0.20) | 2.65 (0.27) | 2.05 (0.31) | 1.23 (0.32) |
| Ra (mm) | 0.69–1.95 | 0.81–1.88 | 0.72–1.37 | 2.05–3.17 | 1.40–2.74 | 0.87–2.09 |
| | *Latewood width* | | | | | |
| Av (mm) | 0.39 (0.12) | 0.30 (0.08) | 0.23 (0.06) | 0.58 (0.06) | 0.50 (0.06) | 0.38 (0.10) |
| Ra (mm) | 0.22–0.61 | 0.15–0.48 | 0.16–0.31 | 0.47–0.67 | 0.44–0.58 | 0.25–0.63 |
| | *Earlywood density* | | | | | |
| Av (kg/m$^3$) | 415 (27) | 403 (14) | 385 (17) | 316 (16) | 340 (13) | 331 (22) |
| Ra (kg/m$^3$) | 376–491 | 383–432 | 344–418 | 296–348 | 318–367 | 295–359 |
| | *Latewood density* | | | | | |
| Av (kg/m$^3$) | 637 (47) | 673 (20) | 692 (36) | 612 (58) | 734 (89) | 726 (62) |
| Ra (kg/m$^3$) | 578–729 | 568–796 | 591–746 | 502–769 | 605–856 | 581–798 |
| | *Density at the earlywood latewood transition* | | | | | |
| Av (kg/m$^3$) | 580 (33) | 596 (20) | 600 (29) | 520 (49) | 612 (49) | 614 (48) |
| Ra (kg/m$^3$) | 536–653 | 541–655 | 547–649 | 433–623 | 548–672 | 482–674 |
| | *Earlywood dynamic modulus of elasticity* | | | | | |
| Av (GPa) | 11.2 (2.5) | 11.1 (1.6) | 11.8 (2.6) | 6.4 (0.8) | 9.9 (1.2) | 10.7 (1.9) |
| Ra (GPa) | 8.3–17.6 | 6.5–13.7 | 8.2–15.3 | 5.2–8.1 | 7.9–1.2 | 8.0–1.4 |
| | *Latewood dynamic modulus of elasticity* | | | | | |
| Av (GPa) | 14.8 (2.4) | 15.2 (2.5) | 17.9 (2.4) | 10.6 (1.6) | 14.6 (1.0) | 16.3 (3.6) |
| Ra (GPa) | 11.9–21.7 | 10.3–18.8 | 14.0–22.1 | 8.1–12.1 | 12.9–16.9 | 11.9–21.0 |
| | *Dynamic modulus of elasticity at the earlywood latewood transition* | | | | | |
| Av (GPa) | 13.4 (2.1) | 13.6 (1.9) | 15.4 (1.9) | 9.8 (1.4) | 12.9 (1.2) | 14.9 (1.9) |
| Ra (GPa) | 10.4–17.7 | 9.3–15.8 | 13.1–18.4 | 6.4–12.2 | 11.2–15.9 | 11.2–17.1 |

The same method was used to determine earlywood DMOE (EWDMOE), latewood DMOE (LWDMOE), and DMOE at the earlywood–latewood transition (Figure 5a,b). Thus, the wood DMOE at the E/L transition for the 10th annual ring varied from 9261 to 15,798 MPa for black spruce and from 11,162 to 15,950 MPa for jack pine. For the same annual ring, EWDMOE ranged from 6500 to 13,652 MPa for black spruce and from 7946 to 11,613 MPa for jack pine. LWDMOE also showed large variation: From 10,327 to 18,792 MPa for black spruce and from 12,873 to 16,916 MPa for jack pine (Table 2). The radial DMOE patterns at the E/L transition for black spruce (Figure 5a) and jack pine (Figure 5b) are shown. As shown in Figure 4a, the radial variation pattern for the E/L transition density in black spruce is characterized by large variation with no specific trend. Similar radial variation patterns for the E/L transition density were observed by Koubaa et al. [2]. In contrast, for jack pine, the radial variation pattern for the E/L transition density is characterized by a steady increase in juvenile wood and a tendency to level off in mature wood (Figure 4b). The radial variation pattern for the E/L transition DMOE in black spruce (Figure 5a) is similar to jack pine (Figure 5b) characterized by a linear increase. These results confirm the importance of measuring ring density and RDMOE separately in order to obtain a more detailed characterization of wood mechanical behavior.

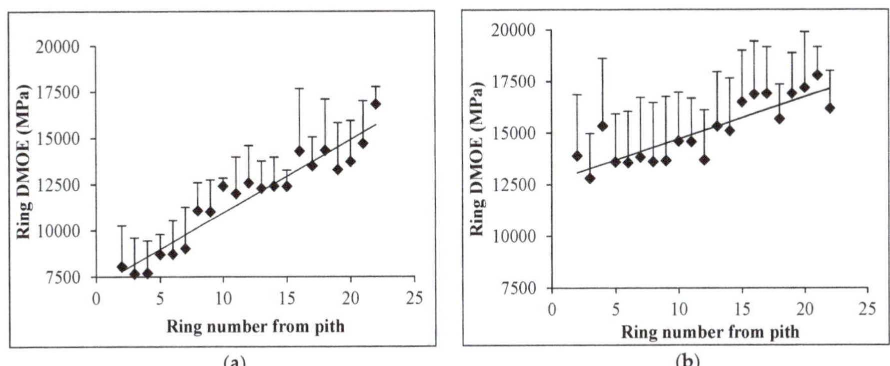


**Figure 5.** Radial variation in E/L transition DMOE (bars indicate the standard error) for (**a**) black spruce ad (**b**) jack pine.

Figure 6 illustrates the close relationship between the transition DMOE measured for the earlywood–latewood transition density and the transition DMOE, as defined by the inflexion point method for all tested samples (black spruce and jack pine). A linear regression curve ($y = 1.06x$) was obtained. Student's *t* test was applied and results indicated no significant differences between the transition DMOE calculated for the earlywood–latewood transition density and the transition DMOE as defined by the inflexion point method calculated from DMOE data. These results reaffirm the close relationship between wood density and wood mechanical properties, and particularly wood DMOE.

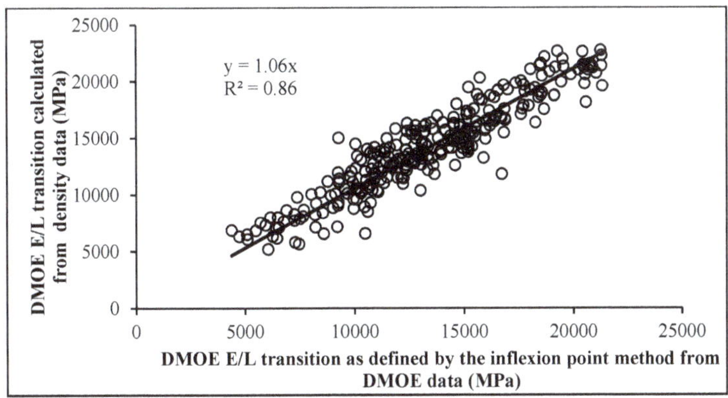


**Figure 6.** Relationship between transition DMOE calculated at the earlywood–latewood (E/L) transition density and transition DMOE, as defined by the inflexion point method for black spruce and jack pine.

### 3.3. Radial Variation in Ring Wood Density and Ring Dynamic Modulus of Elasticity

The mean values for intra-ring density over all samples for both black spruce (a) and jack pine (b) are shown in Figure 7a,b, respectively. The radial variation pattern for wood density is similar to that reported by Park et al. [22] for jack pine, Koubaa et al. [26] for black spruce, and Grabner et al. [24]

for European larch. The ring density is relatively high near the pith and decreases thereafter to reach a minimum in the transition zone leading into the mature wood, where a slow and steady increase is observed. Earlywood density (Figure 7a) decreases rapidly from a maximum near the pith to a low value in the transition zone. The density decreases slowly thereafter with age [26]. Latewood density (Figure 7a) increases almost linearly to a maximum at about ring 13, then levels off in the transition zone and the mature wood [22]. Similar typical variation patterns are seen for DMOE (Figure 8a,b). Ring DMOE increases with tree age, then levels off beyond the 13th ring. A similar radial variation pattern for DMOE was previously reported for hybrid poplar [25]. However, no study to date has investigated radial variation in earlywood and latewood DMOE (EWDMOE and LWDMOE). RDMOE and LWDMOE increase almost linearly in juvenile wood to a maximum at about ring 15, then decrease slowly thereafter through the outer rings in mature wood (Figure 8a,b).

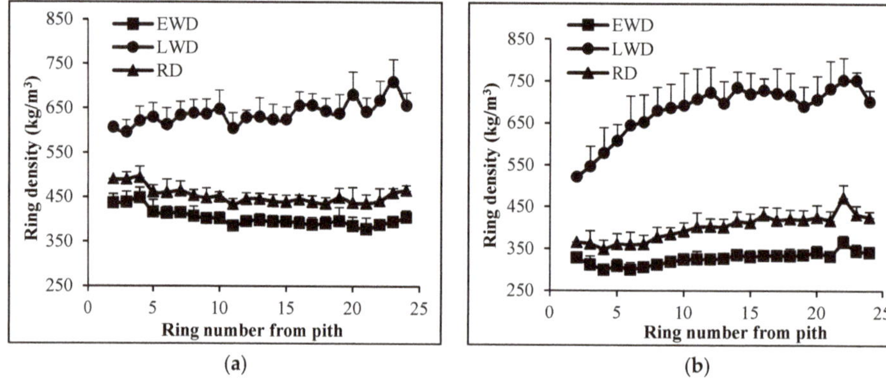

**Figure 7.** Radial variation in ring density (RD), earlywood density (EWD), and latewood density (LWD) for (**a**) black spruce and (**b**) jack pine.

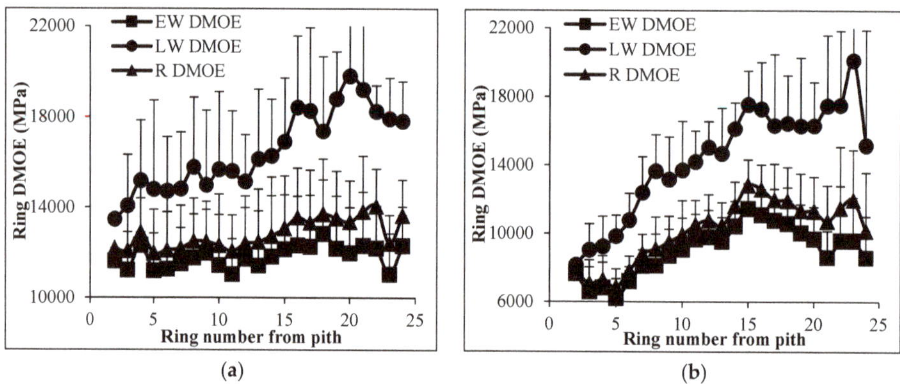

**Figure 8.** Radial variation in ring dynamic modulus of elasticity (RDMOE), earlywood dynamic modulus of elasticity (EWDMOE), and latewood dynamic modulus of elasticity (LWDMOE) for (**a**) black spruce and (**b**) jack pine.

### 3.4. Relationships between Growth, Density, and Elastic Properties

The developed prototype enabled determining relationships between ultrasonic velocity and wood density in rings, earlywood, and latewood (Figures 9 and 10). The coefficient of determination for the linear correlation between RD measured with X-ray densitometry and ring ultrasonic velocity obtained from the developed prototype was $R^2 = 0.66$ using black spruce and jack pine data (Figure 9).

Moderate linear correlations were also obtained between ring, earlywood, and latewood density and ultrasound speed of propagation for Jack pine (Figure 10) and black spruce (not shown).

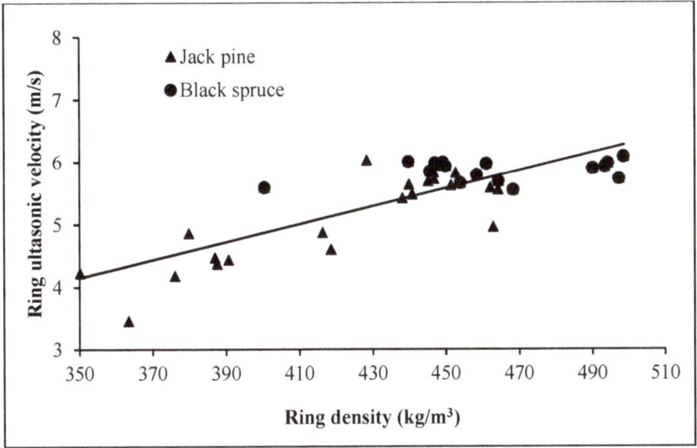

**Figure 9.** Relationship between ring density and ring ultrasonic velocity for black spruce and jack pine.

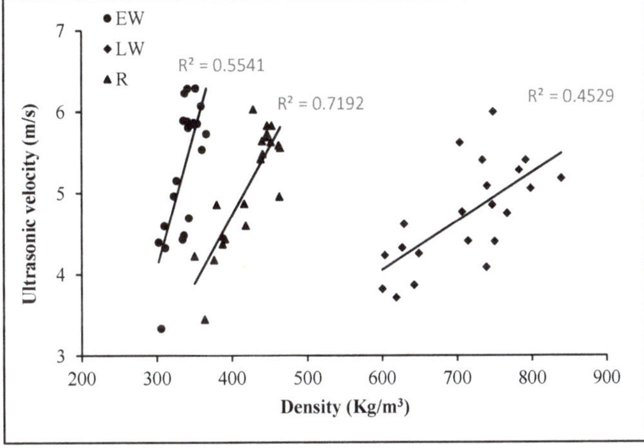

**Figure 10.** Relationships between wood density and ultrasonic velocity in rings (R), earlywood (EW), and latewood (LW) for jack pine.

Table 3 indicates that ring density is positively correlated with earlywood and latewood density. However, for both softwood species, the correlations between ring density and earlywood density are higher than between ring density and latewood density. These results concur with previous studies of black spruce [27]. Ring DMOE shows similar results. Thus, for both black spruce and jack pine, the correlations between ring DMOE and earlywood DMOE are higher than those between ring DMOE and latewood DMOE. The close correlations between DMOE and density shown in Table 3 are due to the fact that the DMOE is obtained from density (Equation (3)). Table 3 also shows a negative and statistically significant correlation between RDMOE and RW. Similar results were found for both earlywood and latewood. In contrast, high positive relationships were found between RDMOE and both earlywood and latewood DMOE, for both softwood species (Table 3). These results have practical implications for a considerably accurate, nondestructive determination of wood density, growth, and stiffness from small samples.

**Table 3.** Pearson's coefficient of correlations between the different traits for black spruce (upper row) and jack pine (lower row).

|  | RW | EWW | LWW | RD | EWD | LWD | RDMOE | EWDMOE | LWDMOE |
|---|---|---|---|---|---|---|---|---|---|
| RW |  | 0.98 *** | 0.81 *** | −0.21 * | 0.00 ns | −0.36 * | −0.37 * | −0.29 * | −0.49 ** |
| EWW | 0.99 *** |  | 0.73 *** | −0.25 * | −0.03 ns | −0.32 * | −0.38 * | −0.30 * | −0.49 ** |
| LWW | 0.82 *** | 0.77 *** |  | −0.02 ns | 0.08 ns | −0.49 ** | −0.31 * | −0.25 * | −0.50 ** |
| RD | −0.69 ** | −0.72 *** | −0.42 ** |  | 0.91 *** | 0.37 * | 0.45 ** | 0.44 ** | 0.33 * |
| EWD | −0.58 ** | −0.58 ** | −0.39 * | 0.85 *** |  | 0.11 ns | 0.40 ** | 0.46 ** | 0.18 ns |
| LWD | −0.50 ** | −0.49 ** | −0.52 ** | 0.75 *** | 0.48 ** |  | 0.18 ns | 0.09 ns | 0.49 ** |
| RDMOE | −0.65 ** | −0.66 *** | −0.48 ** | 0.79 *** | 0.65 ** | 0.66 *** |  | 0.98 *** | 0.85 *** |
| EWDMOE | −0.60 ** | −0.61 ** | −0.44 ** | 0.73 *** | 0.65 ** | 0.57 ** | 0.98 *** |  | 0.75 *** |
| LWDMOE | −0.61 ** | −0.60 ** | −0.55 ** | 0.74 ** | 0.54 ** | 0.80 *** | 0.89 *** | 0.78 *** |  |

* Significant at $\alpha = 0.05$; ** Significant at $\alpha = 0.01$; *** Significant at $\alpha = 0.001$; ns not significant. RW: ring width, EWW: earlywood width, LWW: latewood width, RD: ring density, EWD: earlywood density, LWD: latewood density, RDMOE: ring dynamic modulus of elasticity, EWDMOE: earlywood dynamic modulus of elasticity, LWDMOE: latewood dynamic modulus of elasticity.

## 4. Practical Implications

For this study, we developed a rapid nondestructive method to determine wood density and the dynamic modulus of elasticity (DMOE) based on X-ray densitometry and ultrasonic wave velocity measurement. This method was used to determine earlywood and latewood properties in order to obtain a more detailed characterization of wood mechanical behavior. Only a few studies have investigated earlywood and latewood elastic properties [28–32]. Roszyk et al. [29] reported that latewood modulus of elasticity (MOE) is higher than earlywood MOE for scots pine at low moisture content (8%). Similar results were obtained for Spruce wood [*Picea abies (L.) Karst*] [31] and loblolly pine [28–32] with an important increase in MOE values with the growth of annual rings. However, the preparation of initial and final wood samples was quite complicated and required perfectly parallel annual rings. Different methods have been used to retrieve two adjacent earlywood and latewood bands of 1 mm thick for loblolly pine [28–32]. Moliński et al. [31] reported that two adjacent wood samples were cut out from the region in which the borders of annual rings were straight lines parallel to the longer axis of the plank to obtain two earlywood and latewood samples of 200 μm in thickness for Spruce wood. Thus, it is important to use a rapid nondestructive method with easily prepared samples to determine wood intra-ring mechanical properties, which have direct impacts on wood processing performance. In fact, understanding mechanical properties variations at the earlywood-latewood scale will eventually allow a better knowledge of wood's areas of weakness in order to optimize the performance of wood products. At the wood processing industry scale, this information would be important for mechanical pulping processes where the pulping and refining energies and wood fractioning are closely related to the fiber characteristics including earlywood and latewood mechanical properties [33]. Similarly, oriented strand board (OSB) manufacturing and properties are directly related to earlywood and latewood mechanical properties [28,34]. The results of this study further confirm that ultrasonic measurement can be used to determine the elastic constants of wood with considerable accuracy (0.04 mm). Indeed, the relationships found between ultrasonic wave velocity, density, and wood stiffness demonstrate the experimental efficiency of ultrasonic measurement [35]. Whereas several studies have investigated relationships between static and dynamic MOE in wood [15,19,20,36] and have found significant linear correlations between them, only a few studies have investigated relationships between wood DMOE and wood density [37,38]. For example, linear correlations ($r = 0.70$) were found between DMOE and wood density for *Eucalyptus delegatensis* [39].

The relationships between tree growth and wood properties, especially in terms of mechanical properties, are critical for effective forest management strategies [19,39]. Russo et al. [39] reported that the effects of silvicultural practices (different intensities of thinning) on wood quality can be identified using acoustic measurement to assess the MOED of standing trees with non-destructive method in Calabrian pine. The authors demonstrated that using a low intensity of thinning induced better tree wood quality. In boreal species, enhanced growth through tree improvement programs or intensive forest management strategies can significantly diminish the wood mechanical properties due to several biological factors, including increased earlywood proportion and the production of larger cells with thinner walls. Several anatomical and physical characterization studies have clearly demonstrated the impact of intensive forest management strategies on earlywood, latewood, and overall wood properties [40]. Nevertheless, the impact of intensive forest management on wood mechanical properties has received relatively little attention due to sample size constraints, the destructive nature of characterization tests, and the lack of effective tools for rapid, nondestructive characterization of these properties at the ring level. Nondestructive assessment is essential for understanding the impact of intensive forest management practices on wood mechanical properties as well as the physiological and biological processes involved in wood strength development [39]. The method developed here allows nondestructive measurement of intra-ring wood DMOE and provides deeper insights into wood strength development and its relationships to growth and wood density. As shown in Table 3, radial profiles enable investigating relationships between wood radial growth, density, and elastic properties in rings, earlywood, and latewood.

## 5. Conclusions

Based on the results of this study, the following conclusions can be drawn:

(1) Intra-ring wood dynamic modulus of elasticity (DMOE) profiles can be determined using a nondestructive method based on X ray densitometry and ultrasonic wave velocity measurement.

(2) Sixth order polynomials can well describe intra-ring wood density and dynamic modulus of elasticity profiles in black spruce and jack pine.

(3) The inflexion point method can be used to determine with considerable accuracy the earlywood–latewood transition density and DMOE in black spruce and jack pine.

(4) For black spruce and jack pine, the correlation coefficients between wood density and wood DMOE were positive and statistically significant in rings, earlywood, and latewood. Furthermore, high positive correlations were obtained between ring DMOE and both earlywood and latewood DMOE.

**Author Contributions:** Conceptualization, A.K., W.K., C.B. and M.K.; methodology, W.K. and A.K.; formal analysis, W.K. and A.K.; investigation, W.K. and A.K.; resources, A.K.; data curation, W.K. and A.K.; writing—original draft preparation, W.K.; writing—review and editing, A.K., M.K., C.B.; supervision, A.K., C.B. and M.K.; project administration, A.K.; funding acquisition, A.K.

**Funding:** This research was funded by the Canada Research Chair Program, Grant number: 557752; MITACS Grant number: IT04397 and Natural Resources Canada, CWFC 2017-2018.

**Acknowledgments:** The authors are grateful to Gilles Villeneuve and Williams Belhadef for their invaluable technical assistance. The authors also thank Besma Bouslimi and Zahia Ait-Si-Said for assistance with X-ray density measurements. Thanks also to Margaret McKyes for linguistic editing.



## References

1. Mitchell, H. *A Concept of Intrinsic Wood Quality and Nondestructive Methods for Determining Quality in Standing Timber*; Forest Service, U.S. Report No 2233; Forest Products Laboratory: Madison, WI, USA, 1961.

2. Koubaa, A.; Tony Zhang, S.Y.; Makni, S. Defining the transition from earlywood to latewood in black spruce based on intra-ring wood density profiles from X-ray densitometry. *Ann. For. Sci.* **2002**, *59*, 511–518. [CrossRef]

3. Panshin, A.J.; De Zeeuw, C. *Textbook of Wood Technology*; McGraw-Hill Book Co: New York, NY, USA, 1980; p. 772.

4. Rozenberg, P.; Franc, A.; Bastien, C.; Cahalan, C. Improving models of wood density by including genetic effects: A case study in Douglas-Fir. *Ann. For. Sci.* **2001**, *58*, 385–394. [CrossRef]

5. Zhang, S.Y.; Nepveu, G.; Owoundi, R.E. Intratree and intertree variation in selected wood quality characteristics of European oak (Quercus petraea and Quercus robur). *Can. J. For. Res.* **1994**, *24*, 1818–1823. [CrossRef]

6. Bouslimi, B.; Koubaa, A.; Bergeron, Y. Anatomical properties in *Thuja occidentalis*: Variation and relationship to biological processes. *IAWA J.* **2014**, *35*, 363–384. [CrossRef]

7. Zobel, B.J.; Van Buijtenen, J.P. *Wood Variation: Its Causes and Control*; Springer: Berlin, Germany, 1989; p. 363.

8. Pernestal, K.; Jonsson, B.; Larsson, B. A simple model for density of annual rings. *Wood Sci. Technol.* **1995**, *29*, 441–449. [CrossRef]

9. Mork, E. Die qualität des fichtenholzes unterbesonderer rücksichtnahme auf schleif-und papierholz. *Papier-Fabrikant* **1928**, *26*, 741–747.

10. Denne, M.P. Definition of latewood according to Mork (1928). *IAWA Bull.* **1988**, *10*, 59–62. [CrossRef]

11. Evans, R.; Gartside, G.; Downes, G. Present and prospective use of Silviscan for wood microstructure analysis. In Proceedings of the 49th Appita Annual General Conference proceedings 1995, Hobart, Australia, 2–7 April 1995; pp. 91–96.

12. Barbour, R.J.; Bergqvist, G.; Amundson, C.; Larsson, B.; Johnson, J.A. New methods for evaluating intra-ring X-ray densitometry data: Maximum derivative methods as compared to Mork's index. In Proceedings of the CTIA/IUFRO International Wood Quality Workshop, Quebec, QC, Canada, 18–22 August 1997; p. 61.

13. Ivkovic, M.; Rosenberg, P. A method for describing and modelling of within-ring wood density distribution in clones of three coniferous species. *Ann. For. Sci.* **2004**, *61*, 759–769. [CrossRef]

14. Bucur, V.; Archer, R.R. Elastic constants for wood by an ultrasonic method. *Wood Sci. Technol.* **1984**, *18*, 255–265. [CrossRef]

15. Hassan, K.T.; Horacek, P.; Tippner, J. Evaluation of stiffness and strength of Scots Pine wood using resonance frequency and ultrasonic techniques. Dynamic test of wood. *BioResources* **2013**, *8*, 1634–1645. [CrossRef]

16. Najafi, S.K.; Bucur, V.; Ebrahimi, G. Elastic constants of particleboard with ultrasonic technique. *Mater. Lett.* **2005**, *59*, 2039–2042. [CrossRef]

17. Brashaw, B.K.; Bucur, V.; Divos, F.; Gonçalves, R.; Lu, J.; Meder, R.; Pellerin, R.F.; Potter, S.; Ross, R.J.; Wang, X.; et al. Nondestructive testing and evaluation of wood: A Worldwide Research Update. *For. Prod. J.* **2009**, *59*, 7–13.

18. Chiu, C.M.; Lin, C.H.; Yang, T.H. Application of nondestructive methods to evaluate mechanical Properties of 32-Year Old Taiwan Incense Cedar (*Calocedrus formosana*) wood. *BioResources* **2013**, *8*, 688–700. [CrossRef]

19. Proto, A.R.; Macri, G.; Bernardini, V.; Russo, D.; Zimbalatti, G. Acoustic evaluation of wood quality with a non destructive method in standing trees: A first survey in Italy. *IForest* **2017**, *10*, 700–706. [CrossRef]

20. Yang, J.L.; Fortin, Y. Evaluating strength properties of *Pinus radiata* from ultrasonic measurements on increment cores. *Holzforschung* **2001**, *55*, 606–610. [CrossRef]

21. Horáček, P.; Tippner, J. Nondestructive evaluation of static bending properties of Scots Pine wood using stress wave technique. *Wood Res.* **2012**, *57*, 359–366.

22. Park, Y.I.D.; Koubaa, A.; Brais, S.; Mazerolle, M.J. Effects of cambial age and stem height on wood density and growth of jack pine grown in boreal stands. *Wood Fib. Sci.* **2009**, *41*, 346–358.

23. Ourais, M. Variations intra-arbres de la largeur du cerne, de la masse volumique du bois et des propriétés morphologiques des trachéides de l'épinette noire (*Picea Mariana* (MILL.) B.S.P) avant et après traitements sylvicoles. Master's Thesis, Université du Québec en Abitibi-Témiscamingue, Rouyn-Noranda, QC, Canada, 2012.

24. Grabner, M.; Wimmer, R.; Gierlinger, N.; Evans, R.; Downes, G. Heartwood extractives in larch and effects on X-ray densitometry. *Can. J. Res.* **2005**, *35*, 2781–2786. [CrossRef]

25. Hernández, R.; Koubaa, A.; Beaudoin, M.; Fortin, Y. Selected mechanical properties of fast-growing poplar hybrid clones. *Wood Fib. Sci.* **1998**, *30*, 138–147.

26. Koubaa, A.; Isabel, N.; Zhang, S.Y.; Beaulieu, J.; Bousquet, J. Transition from juvenile to mature wood in black spruce (*Picea Mariana* (Mill.) B.S.P.). *Wood Fib. Sci.* **2005**, *37*, 445–455.

27. Koubaa, A.; Zhang, S.Y.; Isabel, N.; Beaulieu, J.; Bousquet, J. Phenotypic correlations between juvenile-mature wood density and growth in black spruce. *Wood Fib. Sci.* **2000**, *32*, 61–71.

28. Jeong, G.Y.; Zink-Sharp, A.; Hindman, D.P. Tensile properties of earlywood and latewood from lobolly pine (*Pinus taeda*) using digital image correlation. *Wood Fib. Sci.* **2009**, *41*, 51–63.

29. Roszyk, E.; Molinski, W.; Kaminski, M. Tensile properties along the grains of earlywood and latewood of scots pine (*Pinus Sylvestris* L.) in dry and wet state. *BioResources* **2016**, *11*, 3027–3037. [CrossRef]

30. Cramer, S.; Kretschmann, D.; Lakes, R.; Schmidt, T. Earlywood and latewood elastic properties in loblolly pine. *Holzforschung* **2005**, *59*, 531–538. [CrossRef]

31. Moliński, W.; Roszyk, E.; Puszyński, J. Variation in mechanical properties within individual annual ring of the resonance spruce wood [*Picea abies (L.) Karst*]. *Dev. Ind.* **2014**, *65*, 215–223.

32. Mott, L.; Groom, L.; Shaler, S. Mechanical properties of individual southern pine fibers. Part II. Comparison of earlywood and latewood fibers with respect to tree height and juvenility. *Wood Fib. Sci.* **2002**, *34*, 221–237.

33. Vehniäinen, A. Single Fiber Properties—A Key to the Characteristic Defibration Patterns from Wood to Paper Fibers. Ph.D. Thesis, Helsinki University of Technology, Helsinki, Finland, 2008.

34. Jeong, G.Y.; Hindman, D.P. Modeling differently oriented loblly pine strands incorporating variating intraring properties using a stochastic finite element method. *Wood Fib. Sci.* **2010**, *42*, 51–61.

35. Ross, R.J. *Nondestructive Evaluation of Wood*, 2nd ed.; Forest Products Laboratory, Forest Service, WI: U.S, General Technical Report FPL-GTR-238; Department of Agriculture: Madison, WI, USA, 2015.

36. Sales, A.; Candian, M.; De Salles Cardin, V. Evaluation of the mechanical properties of Brazilian lumber (*Goupia glabra*) by nondestructive techniques. *Constr. Build. Mater.* **2011**, *25*, 1450–1454. [CrossRef]

37. Evans, R.; Ilic, J. Rapid prediction of wood stiffness from microfibril angle and density. *For. Prod. J.* **2001**, *51*, 53–57.

38. Lasserre, J.; Mason, E.G.; Watt, M.S.; Moore, J.R. Influence of initial planting spacing and genotype on microfibril angle, wood density, fiber properties and modulus of elasticity in *Pinus radiata D.* Don corewood. *For. Ecol. Manag.* **2009**, *258*, 1924–1931. [CrossRef]

39. Russo, D.; Marziliano, P.A.; Macri, G.; Proto, A.R.; Zimbalatti, G.; Lambardi, F. Does Thinning Intensity Affect Wood Quality? An Analysis of Calabrian Pine in Southern Italy Using a Non-Destructive Acoustic Method. *Forests* **2019**, *10*, 303. [CrossRef]

40. Wang, X.; Ross, R.J.; McClellan, M.; Barbour, R.J.; Erickson, J.R.; Forsman, J.W.; McGinnis, G.D. Nondestructive evaluation of standing trees with a stress wave method. *Wood Fib. Sci.* **2001**, *33*, 522–533.

 *forests*

*Article*

# Use of Time-of-Flight Ultrasound to Measure Wave Speed in Poplar Seedlings

Fenglu Liu [1,2], Pengfei Xu [1,2], Houjiang Zhang [1,2,*], Cheng Guan [1,2], Dan Feng [1,2] and Xiping Wang [3]

[1]  School of Technology, Beijing Forestry University, Beijing 100083, China
[2]  Joint International Research Institute of Wood Nondestructive Testing and Evaluation, Beijing Forestry University, Beijing 100083, China
[3]  USDA Forest Service, Forest Products Laboratory, Madison, WI 53726-2398, USA
*   Correspondence: hjzhang6@bjfu.edu.cn; Tel.: +86-10-6233-6925

Received: 28 June 2019; Accepted: 9 August 2019; Published: 13 August 2019

**Abstract:** In this study, 145 poplar (*Populus × euramericana cv.'74/76'*) seedlings, a common plantation tree species in China, were selected and their ultrasonic velocities were measured at four timepoints during the first growth year. After that, 60 poplar seedlings were randomly selected and cut down to determine their acoustic velocity, using the acoustic resonance method. The effects of influencing factors such as wood green density, microfibril angle, growth days, and root-collar diameter on acoustic speed in seedlings and the relationship between ultrasonic speed and acoustic resonance speed were investigated and analyzed in this work. The number of specimens used for investigating growth days and root-collar diameter was 145 in both cases, while 60 and two specimens were used for investigating wood density and the microfibril angle, respectively. The results of this study showed that the ultrasonic speed of poplar seedlings significantly and linearly increased with growth days, within 209 growing days. The ultrasonic velocity of poplar seedlings has a high and positive correlation with growth days, and the correlation was 0.99. However, no significant relationship was found between the ultrasonic velocity and root-collar diameter of poplar seedlings. Furthermore, a low and negative relationship was found between wood density and ultrasonic speed ($R^2 = 0.26$). However, ultrasonic velocity significantly decreased with increasing microfibril angle (MFA) in two seedlings, and thus MFA may have an impact on ultrasonic speed in poplar seedlings. In addition, ultrasonic velocity was found to have a strong correlation with acoustic resonance velocity ($R^2 = 0.81$) and a good correlation, $R^2 = 0.75$, was also found between the dynamic moduli of elasticity from ultrasonic and acoustic resonance tests. The results of this study indicate that the ultrasonic technique can possibly be used to measure the ultrasound speed of young seedlings, and thus early screen seedlings for their stiffness properties in the future.

**Keywords:** ultrasonic speed; poplar seedlings; acoustic resonance; density; microfibril angle; root-collar diameter

## 1. Introduction

Trees grown in plantation forests are expected to have a high average value of stiffness (i.e., a high modulus of elasticity (MOE)), a low microfibril angle (MFA), and a low shrinkage propensity to distort, thus yielding lumber or other wood products with a high quality, as well as resulting in a high average grade out-turn [1]. The genetics have a significant impact on whether the tree can produce an acceptable yield of wood products, such as lumbers. Therefore, it is extremely desirable to be able to select young seedlings that are more likely to produce better wood products and, hence, can be grown to maturity in the knowledge that they will be more valuable [1]. Many studies show that for juvenile wood, MFA decreases with age [2,3]. There are many studies showing that it is possible to enable an early selection of trees that yield better wood quality. For example, Donaldson and Burdon found that it may be possible to effectively select for desirable MFA, starting from ring 1 [4]. Similarly, Dungey et al. found that it was possible to early select for high stiffness (MOE) in *Pinus radiata* D. Don, around rings 4 to 8 at breast height [5]. In addition, Nakada studied *Cryptomeria japonica* D. Don and found an effective selection method and improvement in average log stiffness at an early age (5 years old). He also suggested that it was worth developing an early selection for tree quality [6]. Moreover, Watt et al. (2010) suggested that *P. radiata* clones with high average stiffness can be selected from trees aged 5 years [7]. Emms et al. stated that the early screening of tree quality has benefits to the forestry industry, potentially yielding better wood quality [1,8].

Acoustic techniques such as longitudinal stress waves, acoustic resonance, and ultrasound have been used to determine the mechanical properties of wood materials for many years [9,10]. Since the acoustic speed of wood materials is related to their stiffness, the measurement of acoustic speed in wood still receives much attention from researchers [11–13]. For instance, the acoustic speeds measured by longitudinal stress waves, acoustic resonance or ultrasound are widely used to evaluate the quality of standing trees and to assess the mechanical performance of logs and lumbers and further sort their grades in terms of measured MOE [14–19]. Therefore, acoustic technologies including longitudinal stress waves, acoustic resonance, and ultrasound have been well developed for standing trees, logs, lumbers, and other wood products. For seedlings, however, few acoustic techniques are applied to evaluate their wood quality. There is a lack of interest in the mechanical properties of seedlings and methods for quality evaluation; hence, there are few studies on acoustic speed measurement and wood quality assessment for seedlings. Only Emms et al. measured the acoustic speeds in 2-year-old *Pinus radiata* seedlings, using a longitudinal-wave time-of-flight prototype that they built and an acoustic resonance technique [1,8]. They found that this longitudinal-wave time-of-flight acoustic technique may be able to become a novel technique for non-damaging measurement of acoustic speed in seedlings, and that the technique shows good promise as a rapid and cost-effective tool for early screening of wood quality. Furthermore, they suggest that the measurement of acoustic speed in seedlings has benefits to the forestry industry, potentially enabling the early selection of trees that yield better quality wood. Huang et al. used the stress-wave technique to measure the stiffness properties of seedlings, and Divos et al. conducted seedling segregation by acoustic velocity using stress-wave devices [20,21]. Therefore, more efforts are still needed to find a proper acoustic technique for non-destructive measurement of sound speed, and to establish a comprehensive evaluation method for the quality of seedlings.

It is well-known that acoustic speed is related to the important quality properties of wood, such as stiffness, referred to as MOE, grain angle and the MFA of the S2 layer in the cell wall. Therefore, it is necessary to accurately measure acoustic speed in seedlings and to investigate the influencing factors on acoustic speed, since this will help to improve the reliability and accuracy of assessment for the quality of seedlings via acoustic techniques. However, there are few reports about acoustic speed measurement in seedlings using acoustic technology, especially for ultrasound, and no reports were found for studying the influencing factors on acoustic speed in seedlings. As mentioned previously, only Emms et al. conducted acoustic speed measurement on *Pinus radiata* seedlings, using longitudinal waves and an acoustic resonance technique [1,8]. Huang et al. used the stress-wave method to test the mechanical properties of seedlings, and Divos et al. performed seedling segregation according to stress-wave acoustic velocity [20,21]. Therefore, to the best of our knowledge, no research on acoustic speed measurement for seedlings using ultrasound was found, and no paper investigating influencing factors, such as wood density, MFA, growth days, and root-collar diameter (commonly used to visually evaluate the physical and mechanical properties of seedlings, and subsequently to grade the quality of seedlings), on acoustic speed has been published for seedlings. It is, hence, highly desirable to measure sound speed in seedlings with the application of an ultrasonic technique and to investigate the influencing factors on ultrasonic propagation speed in seedlings.

The poplar is a common plantation tree species in China, and poplar plantations provide an important raw material for papermaking, plywood, fiberboard, paper matches, sanitary chopsticks, and the packaging industry. The research work presented in this paper aimed to measure the ultrasonic propagation velocity in poplar (*Populus* × *euramericana cv.'74/76'*) seedlings, and to analyze the effect of various influencing factors such as wood density, MFA, growth days, and root-collar diameter on acoustic speed in seedlings. The results of this paper will provide some basic insights for the early screening of seedlings for their stiffness properties, using ultrasonic technology, and for developing a rapid evaluation method of poplar seedling wood quality in the future.

## 2. Materials and Methods

### 2.1. Materials

A total of 22 rows of poplar seedlings (*Populus* × *euramericana cv.'74/76'*) growing in the nursery base of Beijing Forestry University were planted at an initial spacing of 80 cm (row interval) × 30 cm (column interval), on April 10th. Then, 105 days later, 145 poplar seedlings numbered in sequence from P-001 to P-145 were randomly selected from the nursery base to conduct ultrasonic speed testing for the first time. After that, poplar seedlings were cut down, and a $l$ (mm)-long specimen was cut from each green seedling using a portable electric saw. The extracted stem length, $l$ is given by the equation

$$L = 15 \times d + 100 \tag{1}$$

where $d$ (mm) is the root-collar diameter of the poplar seedling, and 100 mm is a reserved length of the specimen for density measurement. A total of 60 $l$ (mm)-long specimens were obtained to perform acoustic resonance tests and density determination. These 60 specimens were immediately sealed by plastic wraps and directly transported to the wood nondestructive evaluation and testing laboratory in Beijing Forestry University, where they were kept in a condition room to maintain the green condition for poplar seedlings prior to acoustic resonance testing. A $15 \times d$ (mm)-long specimen and two 50-mm-long specimens (one from the top and the other one from the bottom) were cut from one $l$ (mm)-long poplar seedling specimen. A total of 60 $15 \times d$ (mm)-long specimen were obtained to conduct acoustic resonance tests. It was found that $15 \times d$ (mm) is the optimal length for specimen to perform acoustic resonance tests, therefore, the ratio of length to diameter of specimen for acoustic resonance tests was 15 in this paper. Moreover, 60 50-mm-long specimens from the top of $l$ (mm)-long poplar seedling specimen and 60 50-mm-long specimens from bottom, in total, were acquired and used to determine the green density of poplar seedlings by the water immersion method [22,23].

Additionally, two poplar seedlings, numbered P-146 and P-147, in the nursery were randomly chosen to determine the microfibril angle of seedlings. Five 30-mm-thick lines, named as A, B, C, D, and E, were marked on the poplar seedlings at the heights of 10, 50, 90, 130, and 170 cm above the base of seedling, respectively, as shown in Figure 1a. The ultrasonic speed between two discs, such as AB, BC, CD, or DE, was tested and recorded prior to being felled using a Fakopp Ultrasonic Timer (Sopron, Hungary) with a frequency of 90 kHz. After that, these two poplar seedlings were felled down and five discs, i.e., A, B, C, D, and E, were cut from each seedling. In order to obtain the specimens used for MFA measurement, at first, a 2-cm-wide strip was symmetrically taken from one disc along the pith from north to south direction. Then, a 2-cm-wide strip was symmetrically cut along the pith from east to west direction to obtain a specimen with a thickness of 0.2 cm (see Figure 1b). Finally, a total of 42 specimens measuring 2 (longitudinal) × 1.5 (tangential) × 0.15 (radial) cm obtained from the P-146 poplar seedling, and 28 specimens from the P-147 poplar seedling, were used to conduct the determination of the MFA of seedlings. These specimens were numbered as, e.g., P-146-A-1, where P-146 is the number of the poplar seedling, the letter A represents the number of the disc, and 1 means the number of the sampling position, as shown in Figure 1b. Specimens obtained from discs A, B, C, and D of P-146 poplar seedlings are shown in Figure 1c. Bruker D8 ADVANCE (Stuttgart, Germany), an X-ray diffractometer, was used in this paper to measure the MFA of specimens.

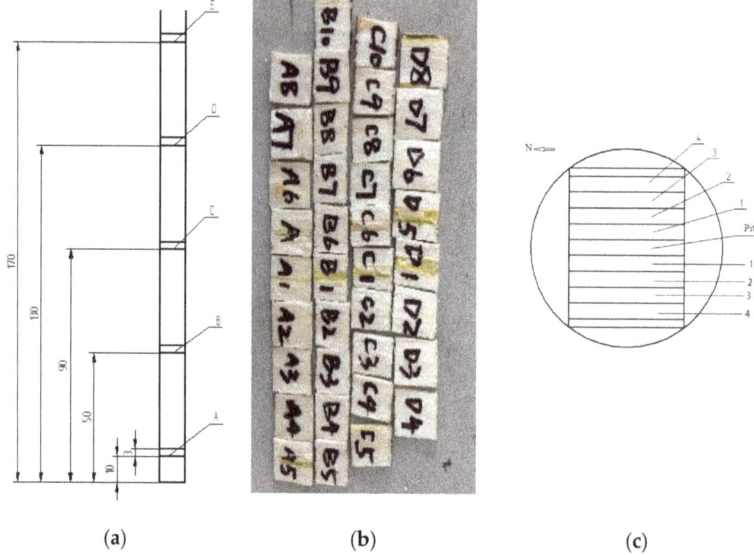

**Figure 1.** Schematic diagram of specimen preparation for MFA measurement: (**a**) disc sample position; (**b**) specimen sampling from cross-section; (**c**) specimens from discs A, B, C, and D of P-146 poplar seedlings.

### 2.2. Methods

#### 2.2.1. Ultrasonic Method

A Fakopp ultrasonic timer, an ultrasonic instrument made in Hungary, was utilized to implement ultrasonic testing on poplar seedlings in situ. This ultrasonic tool is composed of two piezoelectric sensors with cables and one electronic box, as shown in Figure 2. These two sensors, a starter sensor and a receiver sensor, are placed on the same side of the seedling stem. The starter sensor, in field ultrasonic testing, was placed at a height of 200 mm above the ground. The receiver sensor was aligned with the starter sensor, and the distance between them was $L$ (mm), namely measuring distance, seen in Figure 3a. Then, the ultrasonic propagation time for detection distance ($L$) from the starter sensor to the receiver sensor was displayed on the hand-held box. Three readings were recorded for each seedling, and then the average propagation time was used to calculate the corresponding ultrasonic velocity. Figure 3b shows the factual tests in situ using the Fakopp ultrasonic timer.

It should be noted that testing distance, $L$, was still uncertain due to the attenuation of ultrasound propagating in the seedling stems. Generally, a long distance between the starter and receiver sensors was necessary to fully reflect the wood properties of seedlings. However, if the test distance was too large, the ultrasonic pulse signal became weakened as the propagation distance increased. As a result, the receiver sensor did not trigger the timer, and thus no reading was recorded. Therefore, before ultrasonic testing, a pre-experiment was conducted to determine the proper detection distance between the two sensors when measuring ultrasonic velocity in poplar seedling stems. Six poplar seedlings with root-collar diameters ranging from 5 mm to 9 mm were selected to perform the pre-experiment using ultrasonic tools. In the pre-experiment, the detection distance, $L$, was set to a series of values, i.e., 65, 165, 265, 365, 465, 565, 665, 765, 865, and 965 mm. Then, the ultrasonic propagation time was measured for these 10 various testing distances. Moreover, the ultrasonic velocity in the seedling stems was calculated according to Equation (2).

$$v_{\mathrm{u}} = \frac{L}{T - t_0} \times 1000 \tag{2}$$

where $v_{\mathrm{u}}$ is the ultrasonic velocity (m/s), $L$ is the detection distance (mm), $T$ is the transit time appearing on the ultrasonic instrument (s), and $t_0 = 6.1$ s is the time correction, which is the transit time inside the two sensors [24]. $T - t_0$ represents the true ultrasonic wave traveling time at the detection distance $L$ in the seedling stems. Ultimately, the proper detection distance for ultrasonic testing was determined based on wave velocities at different measuring distances.

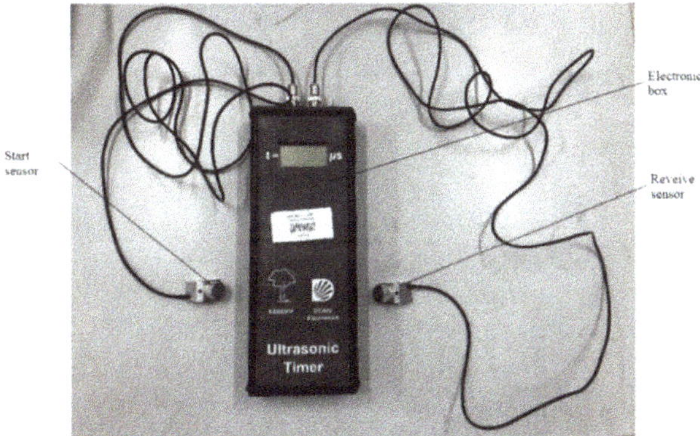


**Figure 2.** Fakopp ultrasonic timer.

**Figure 3.** Ultrasonic propagation time measuring method: (**a**) Schematic diagram, (**b**) Field tests.

Once the proper detection distance ($L$) was identified, the ultrasonic propagation times of 145 seedling samples were measured, and then the ultrasonic propagation velocity of each seedling was calculated using Equation (2). To reduce the effect of environmental factors, the ultrasonic tests for the measurement of 145 seedlings were all completed on the same day. Simultaneously, the root-collar diameters of the seedlings were measured and recorded. A Vernier caliper was used in this step. Two diameters, perpendicular to each other such as east-west and north-south, at the starting point were measured and averaged. Then, the average was denoted as the final root-collar diameter. A total of four timepoints of ultrasonic testing were performed on July 24, September 2, September 20, and November 5 in 2018, respectively—i.e., 105, 145, 165, and 209 days after the seedlings were planted.

### 2.2.2. Acoustic Resonance Method

Acoustic resonance techniques have been used to determine the stiffness of construction materials for a long time. In many respects, acoustic resonance techniques may obtain more useful results, as is evident by the correlations between time-of-flight measurement results on stems and resonance results on logs [25]. Acoustic resonance techniques are more accurate and repeatable than time-of-flight techniques [26], and generally do not require calibration and do not greatly depend on how the operator performs the measurements. The acoustic resonance test, a kind of stress-wave method, was performed in this part to be compared with the ultrasonic experiment described previously.

An acoustic resonance testing system (as shown in Figure 4), comprised of a hammer, two strings, a microphone (Type 2671, Brüel & Kjær, Copenhagen, Denmark), a signal amplifier (Type 1704, Brüel & Kjær, Copenhagen, Denmark) and a data acquisition card (Type USB-6218, National Instruments Corporation, Austin, TN, USA), was used to carry out acoustic resonance tests. As can be seen in Figure 3, the seedling specimens were hung with two strings, and the resonance signal generated by the hammer acted on one end of specimens, with a parallel direction. Then, the acoustic resonance signal was collected by the microphone, and afterwards transmitted to the amplifier and the data acquisition (DAQ) card. Finally, the signal was input to a computer for signal processing and analyzing. 60 15 × $d$ (mm)-long specimens cut from the seedling samples used for the ultrasonic tests were applied for the acoustic resonance tests to measure the periodic frequency of acoustic propagation in seedling specimens. Consequently, the acoustic velocity was determined through Equation (3).

$$V_a = 2l_a f = 30df \tag{3}$$

where $v_a$ is acoustic velocity, $l_a$ is the length of specimen, $d$ is the root-collar diameter of poplar seedlings and $f$ is the frequency of resonance signal traveling between two ends of specimen. Furthermore, to compare with the modulus of elasticity (MOE) from the ultrasonic and acoustic resonance method, the dynamic MOE of these 60 seedling specimens derived from ultrasonic tests and acoustic resonance tests were calculated according to the determined density and measured transmitting velocity using the following equation.

$$E_d = \rho v^2 \tag{4}$$

where $E_d$ is the dynamic MOE of the seedling sample (Pa), $v$ is the wave velocity measured by either the ultrasonic test or the acoustic resonance test (m/s), and $\rho$ is the green density of the sample (kg/m$^3$). It should be noticed that the acoustic test was conducted on only one occasion, on 24 July in 2018, i.e., 105 days after the seedlings were planted.

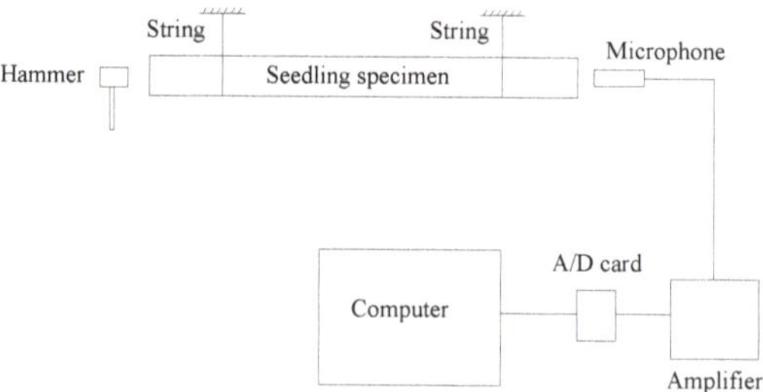


**Figure 4.** Test setup of the acoustic resonance method.

### 2.2.3. Density Measurement

A water immersion method was used to determine the green density of poplar seedlings. A total of 60 100-mm-long specimens were applied for the determination of their density. Firstly, two 50-mm-long specimens used for density measurement were cut from the top and bottom of each l-mm-long specimen, and then these two specimens were marked as top and bottom, respectively. After that, the mass of the top 50-mm-long specimen, marked as $m$ (g), was scaled and recorded using an electronic balance (see Figure 5a). A beaker with water inside was placed on the scale, and then a slim pin was vertically immersed in the water until the marked position, as shown in Figure 5b. Thus, the total mass of the beaker with the water and the immersed slim pin, marked as $m_0$ (g), was scaled and recorded by the electronic balance. Finally, the slim pin was inserted into the top 50-mm-long specimen at a depth of about 1 cm, and then this top specimen was immersed in the beaker until the same position marked on the slim pin, as seen in Figure 5c. Afterwards, the total mass of the beaker with the water, top specimen and immersed slim pin was scaled and recorded, and marked as $m_1$ (g).

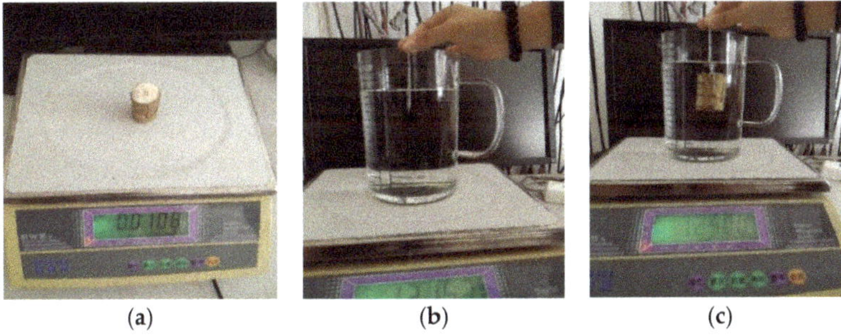

(a)        (b)        (c)

**Figure 5.** Density measurement for a 50-mm-long specimen: (**a**) the mass of $m$, (**b**) the mass of $m_0$, (**c**) the mass of $m_1$.

Since the density of water is 1 g/cm$^3$, the density of the top specimen can be calculated from the Equation (5). Similarly, the density of the bottom specimen can be determined by the above steps and calculated using Equation (5). Therefore, the average density value of the top and bottom specimens was the green density of a poplar seedling specimen. It should be noted that the density measurement method described in this paper only works for wood which has a lower density than water. Another denser liquid, such as mercury, can be used for certain sapwood specimens that have greater densities than water.

$$\rho = \frac{m}{(m_1 - m_0)\cdot 1 \text{ g/cm}^3} \tag{5}$$

## 3. Results and Discussion

### 3.1. Proper Detection Distance for Ultrasonic Tests

Figure 6 shows the relationship between detection distance and ultrasonic velocity for six different seedling samples, i.e., P002, P004, P005, P006, P007, and P008. The root-collar diameters of these six seedling samples, in sequence, were 8.1, 5.9, 6.8, 9.0, 7.1, and 7.2 mm. It can be obviously observed from Figure 6 that no ultrasonic velocity values were obtained when the detection distance was over 765 mm for the P004 seedling. A similar result was also found in the P005 seedling when the detection distance was greater than 865 mm. This may be due to smaller root-collar diameters in the P004 and P005 seedlings (5.9 mm and 6.8 mm, respectively) compared with the other four seedling samples. Ultrasonic propagation time could not be measured as the root-collar diameter was too small.

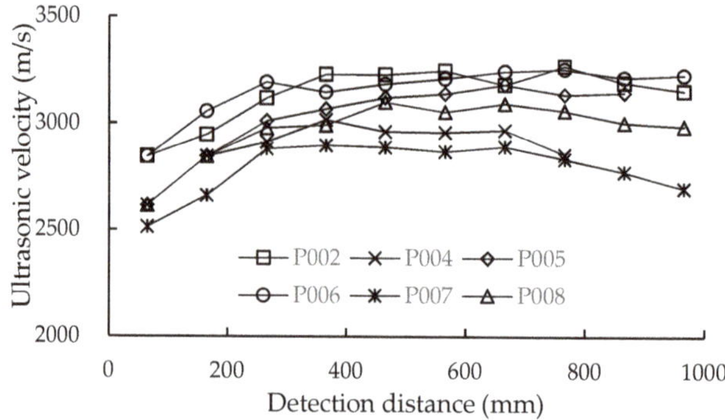

**Figure 6.** Relationship between detection distance and ultrasonic velocity.

It can be seen from Figure 6 that the ultrasonic velocities of the six seedlings all significantly increased when the detection distance changed from 65 mm to 265 mm. For the P004 and P007 seedlings, the ultrasonic velocities both remained basically stable when the detection distance increased from 265 mm to 665 mm, and then linearly decreased once the test distance went beyond 665 mm. For the P002 and P008 seedlings, the ultrasonic velocity generally remained steady when the detection distance was larger than 265 mm. However, for P005 and P006, the ultrasonic velocity slightly increased when the detection distance was larger than 265 mm. Therefore, the proper detection distance for the ultrasonic test should be chosen from 265 mm to 665 mm. Considering the convenience and feasibility of experimental tests, the detection distance used in this paper was ultimately set to 365 mm. The ultrasonic wave speeds shown in the following were all measured at a test distance of 365 mm.

*3.2. Ultrasonic Velocity and Root-Collar Diameters of Poplar Seedlings*

The measured ultrasonic velocities and root-collar diameters of poplar seedlings from tests conducted on July 24, September 2, September 20, and November 5 in the same year, 2018, respectively, are given in Table 1. The values in brackets besides average root-collar diameter and velocity are standard deviations. The average root-collar diameter of 145 seedling samples obtained from four different test dates was 15.98 mm, 20.88 mm, 22.06 mm, and 22.60 mm, respectively. A fast growth rate was observed in average root-collar diameter from July 24 to September 2. In contrast, a lower growth rate was found from September 20 to November 5. This means that poplar seedlings may have a fast growth rate from July to September, and a low growth rate from September to November. Moreover, a low growth rate was also found from September 2 to September 20 due to the shorter growth days.

Ultrasonic velocity, apparently, was increased with growth days. Moreover, in the first (July 24), second (September 2), third (September 20), and fourth (November 5) ultrasonic tests, the maximum values of average ultrasonic velocity—i.e., 1995, 2518, 2703, and 2953 m/s, respectively—were all found in poplar seedlings with a root-collar diameter between 10 mm and 20 mm. The minimum values of average velocity from the first ultrasonic test were found in poplar seedlings with a root-collar diameter between 20 mm and 30 mm. However, for the second and third tests, the minimum values of average velocity were presented in poplar seedlings with a root-collar diameter over 30 mm. Meanwhile, the poplar seedlings with a root-collar diameter of less than 10 mm had the minimum values of average ultrasonic velocity.

Table 1. Results of ultrasonic velocity and root-collar diameter for 145 poplar seedlings.

| Date | Root-Collar Diameter | Number | Average Root-Collar Diameter | Maximum Root-Collar Diameter | Minimum Root-Collar Diameter | Average Velocity | Maximum Velocity | Minimum Velocity |
|---|---|---|---|---|---|---|---|---|
| | (mm) | | (mm) | (mm) | (mm) | (m/s) | (m/s) | (m/s) |
| 24 July | d ≤ 10 | 15 | 8.4(1.5) | 10.0 | 4.9 | 1921(219) | 2241 | 1675 |
| | 10 < d ≤ 20 | 101 | 15.4(2.9) | 20.0 | 10.3 | 1995(123) | 2269 | 1680 |
| | 20 < d ≤ 30 | 29 | 21.9(1.4) | 25.0 | 20.1 | 1919(122) | 2087 | 1748 |
| 2 September | d ≤ 10 | 5 | 8.0(1.5) | 9.4 | 5.7 | 2435(116) | 2520 | 2283 |
| | 10 < d ≤ 20 | 56 | 15.4(2.6) | 20.0 | 10.2 | 2518(153) | 2747 | 2136 |
| | 20 < d ≤ 30 | 79 | 24.9(2.7) | 29.9 | 20.2 | 2315(122) | 2707 | 1996 |
| | d > 30 | 5 | 31.6(1.1) | 33.3 | 30.5 | 2145(88) | 2283 | 2041 |
| 20 September | d ≤ 10 | 5 | 8.0(1.6) | 9.7 | 5.7 | 2509(137) | 2667 | 2404 |
| | 10 < d ≤ 20 | 49 | 15.4(2.6) | 19.5 | 10.2 | 2703(125) | 2923 | 2327 |
| | 20 < d ≤ 30 | 73 | 25.1(2.7) | 29.6 | 20.1 | 2484(134) | 2833 | 2228 |
| | d > 30 | 18 | 31.8(1.6) | 35.0 | 30.1 | 2328(57) | 2404 | 2188 |
| 5 November | d ≤ 0 | 5 | 8.4(1.4) | 9.9 | 6.8 | 2669(281) | 2855 | 2555 |
| | 10 < d ≤ 20 | 48 | 15.6(2.6) | 19.9 | 10.2 | 2953(107) | 3234 | 2707 |
| | 20 < d ≤ 30 | 71 | 25.4(2.8) | 29.8 | 20.5 | 2884(133) | 3353 | 2629 |
| | d > 30 | 21 | 32.5(1.7) | 36.3 | 30.1 | 2741(92) | 2971 | 2573 |

Figure 7 shows the results of the statistical analysis for ultrasonic velocities measured on July 24 and November 5. It can be clearly found from Figure 7 that the ultrasonic velocities of poplar seedlings measured on November 5 were significantly greater than those tested on July 24, maybe due to the longer growth days or the changes in wood density throughout the growing season. In addition, as shown in Figure 7a, the majority of ultrasonic velocities in the first test (July 24) were concentrated near the average velocity (1974 m/s) and ranged from 1825 m/s to 2025 m/s. Moreover, for the fourth test (November 5), the majority of ultrasonic velocities were likewise concentrated near the average ultrasonic velocity (2881 m/s), but ranged from 2680 m/s to 3160 m/s.

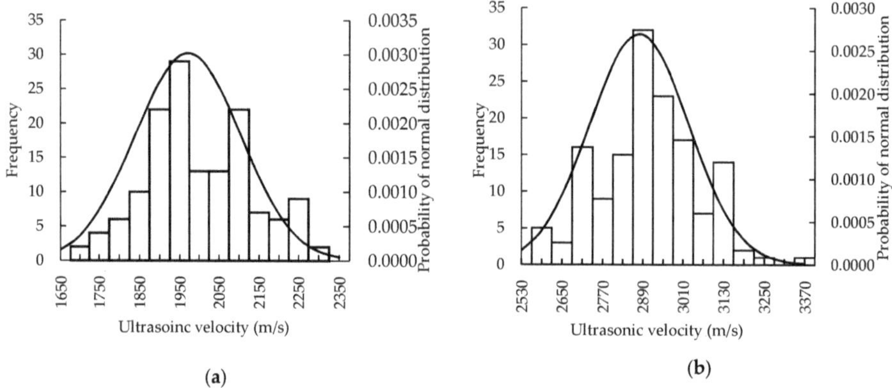

**Figure 7.** Results of statistical analysis for ultrasonic velocities: (**a**) test on July 24; (**b**) test on November 5.

### 3.3. Relationships Between Ultrasonic Velocity and Influencing Factors

### 3.3.1. Growth Days

The effect of growth days on ultrasonic speed in poplar seedlings was analyzed, and the relationships between ultrasonic velocity and growth days are illustrated in Table 2 and Figure 8. The results of the statistical analysis for ultrasonic velocity and growth days are summarized in Table 2. The upper limit of velocity is the wave speed value at the position of positive $3\sigma$ in the ultrasonic velocity probability distribution histogram. Conversely, the lower limit of velocity is the value at the position of negative $3\sigma$ in the ultrasonic velocity probability distribution histogram. As can be seen in Table 2, the average ultrasonic velocities of poplar seedlings were 1972, 2365, 2540, and 2879 m/s, corresponding to 105, 145, 165, and 209 growth days. The ultrasonic speed of poplar seedlings increased with growth days, within 209 growing days. This result means that the growth days may play a positive role in the ultrasonic velocity of poplar seedlings. Similarly, the growth days have a positive influence on the upper limit of velocity and the lower limit of velocity. The upper limit wave speed increased from 2368 to 3323 m/s, and the lower limit wave speed increased from 1576 to 2434 m/s, when growth days increased from 105 to 209 days.

**Table 2.** Results of statistical analysis for ultrasonic velocity and growth days.

| Date | Number of Trees | Growth Days | Average Velocity | Standard Deviation | Upper Limit of Velocity | Lower Limit of Velocity |
|---|---|---|---|---|---|---|
|  |  |  | (m/s) |  | (m/s) | (m/s) |
| 24 July | 145 | 105 | 1972 | 132 | 2368 | 1576 |
| 2 September | 145 | 145 | 2365 | 166 | 2865 | 1866 |
| 20 September | 145 | 165 | 2540 | 183 | 3087 | 1992 |
| 5 November | 145 | 209 | 2879 | 148 | 3323 | 2434 |

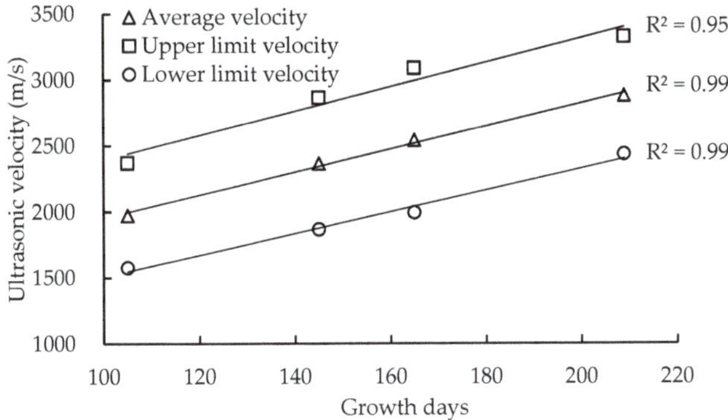


**Figure 8.** Relationship between ultrasonic velocity and growth days.

Figure 8 shows the results of the correlation analysis between the ultrasonic velocities and seedling growth days. It can be seen from Figure 8 that the average velocity, upper limit of velocity and lower limit of velocity were all linearly increased with growth days, within 209 growing days. There are good correlations between these three kinds of ultrasonic velocities and growth days, and the correlations ($R^2$) between the average velocity, upper limit of velocity, lower limit of velocity, and the growth days were 0.99, 0.95, and 0.99, respectively. There was a dramatic difference in the ultrasonic velocities of seedlings at different growth days. This is may be due to the fact that the mechanical properties of poplar seedlings gradually become better as growth days increase, resulting in an increase in ultrasonic wave velocity. Therefore, it could be predicted that the ultrasonic velocity of seedlings would continually increase with growth days due to the underlying changes in density or other wood properties. Growth days, thus, may play an important role in the ultrasonic speed of early stage poplar seedlings, especially within 209 growth days. However, it should be noted that the ultrasonic velocity is likely changing due to the underlying changes in density or other wood properties, and not directly due to more growing days. Underlying factors such as moisture content, MFA, wood density, and other wood properties are also likely to have a role in the development of mechanical properties that affect the ultrasonic velocity.

### 3.3.2. Root-Collar Diameter

It is necessary to investigate the effect of root-collar diameter on ultrasonic velocity and then to determine the relationship between root-collar diameter and ultrasonic speed. In general, the root-collar diameter of seedlings increased with growth days (i.e., the age of seedlings). As showed in Figure 8, the ultrasonic velocity in seedlings linearly increased with increasing growth days. Thereby, it could be speculated that ultrasonic speed may increase with increasing root-collar dimeter of seedlings. However, from the relationships between the ultrasonic velocity and root-collar diameter of poplar seedlings provided in Figure 9, it can be observed that for the four different test dates, the ultrasonic velocity kept relatively stable as the root-collar diameter of poplar seedlings increased. There was no significant correlation between ultrasonic velocity and root-collar diameter in this paper. In other words, it seems that ultrasonic speed does not increase with root-collar diameter.

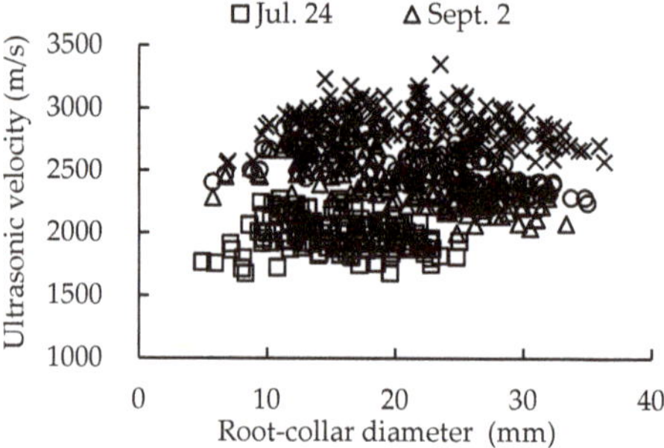


**Figure 9.** Relationship between the ultrasonic velocity and root-collar diameter of seedlings.

In addition, it was found that the overall ultrasonic velocities measured on November 5 were the highest among the four timepoints of ultrasonic tests, while the test on July 24 obtained the lowest velocity. Ultrasonic velocity was overall increased with growth days, which is consistent with the results shown in Figure 8. Therefore, the root-collar diameter of seedlings did not show a significant effect on ultrasonic velocity in this work. More data from the same or diverse seedlings species, however, still need to be acquired to further verify the effect of root-collar diameter on ultrasonic speed and the relationship between them.

### 3.3.3. Density

Wood density is also often utilized to assess the mechanical and physical performance of seedlings, and then to classify the quality of seedlings. Accordingly, it is essential to learn the impact of density on ultrasonic velocity in poplar seedlings and the correlation between wood density and ultrasonic speed.

Figure 10 shows the relationship between the density of seedlings and ultrasonic velocity. It can be seen from Figure 10 that ultrasonic velocity overall tends to decrease with increasing density of seedlings. However, a low value of correlation ($R^2 = 0.26$) was found between ultrasonic velocity and density. Therefore, the relation between ultrasonic speed and the density of seedlings was not significant. This poor relation may be contributed to by the fact that ultrasound speed not only depends on density, but also on other factors such as MFA, grain angle, and the age of seedlings.

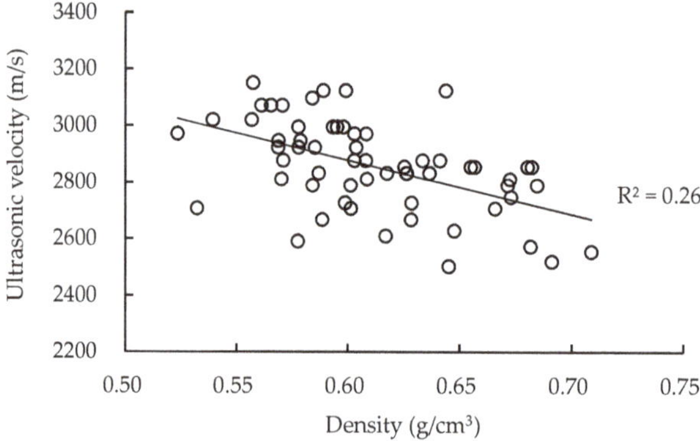


**Figure 10.** Relationship between the ultrasonic velocity and density of poplar seedlings.

The effect of wood density on acoustic speed has been investigated and reported in many studies. However, two contrasting results were found in the relationship between density and acoustic speed. Some studies reported a negative relationship between wood density and acoustic velocity. Isik et al., for instance, found a low but negative correlation for air-dry density and acoustic velocity in *Pinus taeda*, and the corresponding correlation coefficient was −0.2 [27]. Hasegawa et al. showed that the ultrasonic wave velocities of Japanese cedar and Japanese cypress both linearly decreased with air-dry density, and the correlation coefficients were −0.83 and −0.74, respectively [28]. However, other studies indicated a positive relation between wood density and acoustic speed. Krauss et al., for example, reported a low but positive relation for density and ultrasonic velocity in Scots pine, and the correlation was 0.15 [29]. Chen et al. and Lachenbruch et al. both found a moderate and positive relationship for green density and ultrasonic speed, and the correlations were 0.46 in Norway spruce and 0.33 in Douglas fir, respectively [12,30]. Moreover, Blackburn et al. and Ribeiro et al. found that acoustic speed greatly increased with wood basic density, and the correlations were 0.75 for *Eucalyptus nitens* and 0.84 for *Pinus taeda* [31,32]. The results of the present paper were basically in accordance with those reported in Isik et al.'s work. Therefore, density may have an influence on ultrasonic speed. More efforts definitely need to be put into the investigation of the impact of density on ultrasonic speed and the correlations between ultrasonic velocity and density in identical or different seedling species. It may help to early select the better-quality seedlings with high average values of stiffness, if the effect of wood density on ultrasonic velocity were comprehensively understood.

### 3.3.4. Microfibril Angle

The microfibril angle of the S2 cell wall layer is an important parameter of wood. Many studies have showed that the microfibril angle was highly related to the mechanical properties of wood and acoustic speed, for logs and lumbers [33]. The stiffness—i.e., modulus of elasticity—and the acoustic velocity of wood decreased as the microfibril angle increased. Therefore, it is necessary to figure out the effect of microfibril angle on ultrasonic velocity in poplar seedlings and the relationship between MFA and ultrasonic speed.

Microfibril angles measured at different positions (1, 2, 3, 4, ... as shown in Figure 1b) of one disc were averaged and used as the microfibril angle of this disc. Then, the average of the microfibril angle of discs A and B is taken as the microfibril angle between disc A and B. Similarly, the microfibril angles between B and C, C and D, and D and E can be obtained. Thus, the ultrasonic speeds for sections AB, BC, CD, and DE and their corresponding microfibril angles were used to analyze the correlations between them. Figure 11 shows the relationship between the microfibril angle of seedlings and ultrasonic velocity. It can be seen from Figure 11 that ultrasonic velocity significantly decreased with increasing microfibril angle of seedlings. The correlation ($R^2$) between ultrasonic velocity and MFA for poplar seedling P-146 was 0.69, which is lower than that for poplar seedling P-147 ($R^2 = 0.89$). Therefore, the relation between ultrasonic speed and MFA is of great interest for these two seedlings, and MFA may have an impact on ultrasonic speed in poplar seedlings. More seedling specimens are definitely needed to verify this relationship and confirm the effect of MFA on ultrasonic velocity.

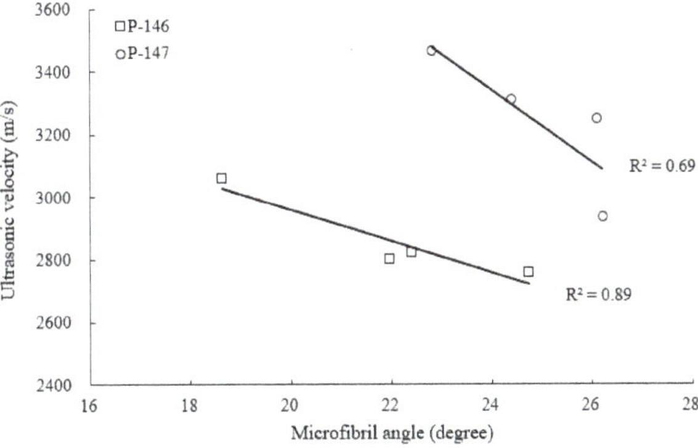


**Figure 11.** Relationship between the ultrasonic velocity and microfibril angle of poplar seedlings.

MFA has been reported to have a significant and negative relation with acoustic velocity for wood materials such as logs and lumbers in many studies. Krauss et al., for example, found a negative relationship ($R^2 = 0.71$) between the ultrasonic wave velocity and MFA of Scots pine [29]. Chen et al. reported a very high but negative relationship ($R^2 = 0.98$) between acoustic velocity and MFA in Norway spruce [30]. Isik et al. and Lachenbruch et al. both observed that acoustic velocity greatly decreased with increasing MFA, and the correlations were 0.70 and 0.69 for *Pinus taeda* and Douglas fir, respectively [12,27]. Moreover, Hasegawa et al. showed that for Japanese cedar, the correlation between ultrasonic wave velocity and MFA was 0.90, and 0.82 for Japanese cypress. They suggested that MFA greatly affects the ultrasonic wave velocity in softwood [28]. The results of the present paper are consistent with their reported results.

### 3.4. Comparison with the Results of Acoustic and Ultrasonic Tests

Figure 12 shows the relationship between ultrasonic velocity and acoustic resonance velocity in the 60 poplar seedlings. The average ultrasonic velocity and acoustic velocity for these 60 poplar seedlings were 3114.8 m/s and 2856.7 m/s, respectively. The standard deviations of ultrasonic velocity and acoustic velocity were 159.53 and 164.68, respectively. The average ultrasonic velocity was approximately 9.1% (i.e., 260 m/s) higher than the average acoustic velocity. In addition, it can be seen from Figure 12 that there was a significant relationship between ultrasonic velocity and acoustic velocity, and the correlation ($R^2$) was 0.81. Similar results have been reported by other researchers as well [1,23]. The acoustic resonance method is generally recognized as a reliable and accurate method for measuring the sound speed of wood material, such as logs and lumbers. Therefore, the prominent relationship between ultrasonic and acoustic velocity may indicate that the ultrasonic method can be used to measure the ultrasonic sound speed of poplar seedlings.

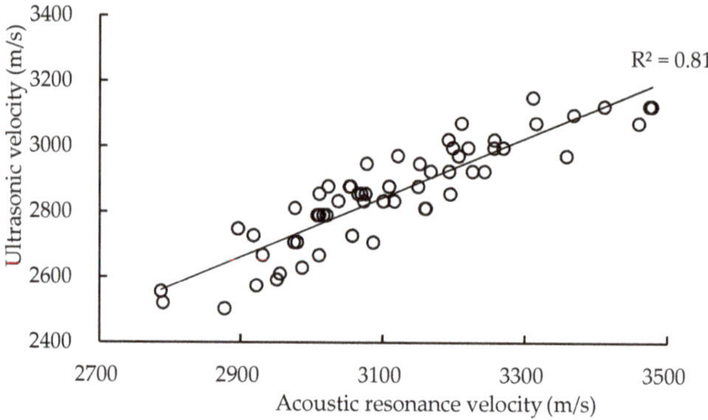


**Figure 12.** Relationship between ultrasonic velocity and acoustic resonance velocity.

Figure 13 presents the relationship between the dynamic MOE results obtained from ultrasonic and acoustic tests carried out in 60 seedlings. The average dynamic MOE values derived from the ultrasonic and acoustic tests for these 60 poplar seedlings were 5.92 and 4.98 GPa, respectively. The standard deviations of dynamic MOE derived from the ultrasonic and acoustic methods were 0.48 and 0.36, respectively. The average dynamic MOE from the ultrasonic test was approximately 18.87% (i.e., 0.94 GPa) higher than that from the acoustic test. Additionally, it can be observed from Figure 13 that there was a good relationship between dynamic MOE from the ultrasonic test and the acoustic test, and the correlation ($R^2$) was 0.75. It is well known that the dynamic elastic modulus measured by the acoustic method can be well used to predict the static elastic modulus of wood. Therefore, this means that the dynamic elastic modulus measured by the ultrasonic method may be able to be used for predicting the static elastic modulus of wood, especially for young seedlings, due to the noticeable relationship between dynamic MOE results from ultrasonic and acoustic tests. However, there is still a lot of work that needs to be done to investigate whether the ultrasonic method could be potentially utilized to evaluate the quality of young seedlings. Moreover, if ultrasound could be applied to the early selection of seedlings, the quality of standing trees and wood-based products may be improved because of the good quality of the young seedlings. Although good quality young seedlings do not guarantee good quality standing trees, they are a good start to cultivate the high properties of plantation trees.

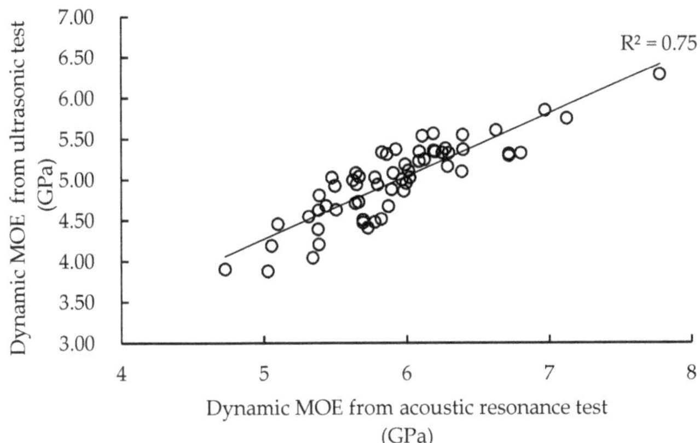


**Figure 13.** Relationship between dynamic MOE values from ultrasonic and acoustic resonance tests.

## 4. Conclusions

The aim of this study was to investigate the application of ultrasonic speed measurement to poplar (*Populus* × *euramericana cv.'74/76'*) seedlings, a common plantation species in China, and to gain some basic insights for the future early selection of poplar seedlings with high wood quality. The conclusions are as follows:

- The proper detection distance for the ultrasonic method to be applied to poplar seedlings is ranged from 265 mm to 665 mm, and 365 mm was used in this study.
- There were good correlations ($R^2 = 0.99$) between the average ultrasonic velocities and growth days. Ultrasonic speed increased with growth days, within 209 growing days. However, almost no relationship was found between the ultrasonic velocities and the root-collar diameters of seedlings, i.e., ultrasonic speed does not seem to increase with increasing root-collar diameter.
- Even though ultrasonic velocity, in general, decreases with increasing density, the density of seedlings showed a weak influence on ultrasonic speed due to the low correlation between them ($R^2 = 0.26$). However, ultrasonic velocity significantly decreased with the increasing microfibril angle of seedlings. The relations between ultrasonic speed and MFA are of great interest for the two sample seedlings, and MFA may have an impact on ultrasonic speed in poplar seedlings. More seedling specimens are definitely needed to verify this relationship and confirm the effect of MFA on ultrasonic velocity.
- There was a significant relationship between ultrasonic velocity and acoustic velocity, and a similar result was also found in the dynamic MOE values derived from the acoustic resonance test and the ultrasonic test, respectively.
- Other influencing factors that were excluded in this paper—such as MFA, temperature, and moisture content—need to be studied in future research to investigate their effect on ultrasonic velocity in seedlings.

**Author Contributions:** F.L. rewrote and revised the manuscript as well as conducted part of data analysis work. P.X. conducted part of analysis work and wrote the original draft of the manuscript. H.Z. and C.G. supervised the research team and provided some ideas to research as well as revised the manuscript. D.F. performed most of test and analysis work. X.W. revised the manuscript.

**Funding:** This project was supported by "the Fundamental Research Funds for the Central Universities (no. BLX201817)", China Postdoctoral Science Foundation (no. 2018M641225) and the National Natural Science Foundation of China (no. 31328005).

**Acknowledgments:** The authors wish to thank Jianhua Hao for his grateful assistance of planting the seedlings.

**Conflicts of Interest:** The authors declare no conflict of interest. The funders had no role in the design of the study; in the collection, analyses, or interpretation of data; in the writing of the manuscript, or in the decision to publish the results.

## References


1.  Emms, G.W.; Nanayakkara, B.; Harrington, J.J. Application of longitudinal-wave time-of-flight sound speed measurement to Pinus radiata seedlings. *Can. J. For. Res.* **2013**, *43*, 750–756.
2.  Huang, Y.H.; Zhao, R.J.; Fei, B.H. Variation patterns of microfibril angle for Chinese fir wood. *J. Northwest For. Univ.* **2007**, *22*, 119–122.
3.  Sun, D.Y.; Yang, W.X.; Liu, Q.L.; Fang, S.Z. A study on geographic variation in wood microfibril angle of *Cyclocarya paliurus*. *J. Nanjing For. Univ. Nat. Sci. Ed.* **2018**, *42*, 81–85.
4.  Donaldson, L.A.; Burdon, R.D. Clonal variation and repeatability of microfibril angle in Pinus radiata. *N. Z. J. For. Sci.* **1995**, *25*, 164–174.
5.  Dungey, H.S.; Matheson, A.C.; Kain, D.; Evans, R. Genetics of wood stiffness and its component traits in Pinus radiata. *Can. J. For. Res.* **2006**, *36*, 1165–1178.
6.  Nakada, R. Within-tree variation of wood characteristics in conifers and the anatomical characteristics specific to very young trees. In *The Compromised Wood Workshop 2007*; University of Canterbury: Christchurch, New Zealand, 2007; pp. 51–67.
7.  Watt, M.S.; Sorensson, C.; Cown, D.J.; Dungey, H.S.; Evans, R. Determining the main and interactive effect of age and clone on wood density, microfibril angle, and modulus of elasticity for Pinus radiata. *Can. J. For. Res.* **2010**, *40*, 1550–1557.
8.  Emms, G.; Harrington, J.J. A novel technique for non-damaging measurement of sound speed in seedlings. *Eur. J. For. Res.* **2012**, *131*, 1449–1459.
9.  Hearmon, R.F.S. The influence of shear and rotatory inertia on the free flexural vibration of wooden beams. *Br. J. Appl. Phys.* **1958**, *9*, 381–388. [CrossRef]
10. Bucur, V. *Acoustics of Wood*; CRC Press: Boca Raton, FL, USA, 1995.
11. Yin, Y.F.; Nagao, H.; Liu, X.L. Mechanical properties assessment of *Cunninghamia lanceolata* plantation wood with three acoustic-based nondestructive methods. *J. Wood Sci.* **2010**, *56*, 33–40.
12. Lachenbruch, B.; Johnson, G.R.; Downes, G.M.; Evans, R. Relationships of density, microfibril angle, and sound velocity with stiffness and strength in mature wood of Douglas-fir. *Can. J. For Res.* **2010**, *40*, 55–64.
13. Todoroki, C.L.; Lowell, E.C.; Dykstra, D. Automated knot detection with visual post-processing of Douglas-fir veneer images. *Comput. Electron. Agric.* **2010**, *70*, 163–171.
14. Chauhan, S.S.; Walker, J.C.F. Variations in acoustic velocity and density with age, and their interrelationships in radiata pine. *For. Ecol. Manag.* **2006**, *229*, 388–394.
15. Toulmin, M.J.; Raymond, C.A. Developing a sampling strategy for measuring acoustic velocity in standing Pinus radiata using the treetap time of flight tool. *N. Z. J. For Sci.* **2007**, *37*, 96–111.
16. Lindstrom, H.; Reale, M.; Grekin, M. Using non-destructive testing to assess modulus of elasticity of Pinus sylvestris trees. *Scand. J. For. Res.* **2009**, *24*, 247–257.
17. Gerhards, C. Longitudinal stress-waves for lumber stress grading: Factors affecting applications: State of the art. *For. Prod. J.* **1982**, *32*, 20–25.
18. Tsehaye, A.; Buchanan, A.H.; Walker, J.C.F. Sorting of logs using acoustics. *Wood Sci. Technol.* **2000**, *34*, 337–344.
19. Carter, P.; Chauhan, S.S.; Walker, J.C.F. Sorting logs and lumber for stiffness using director HM200. *Wood Fibre Sci.* **2006**, *38*, 49–54.
20. Huang, C.L.; Lambeth, C.C. Methods for Determining Potential Characteristics of a Specimen Based on Stress Wave Velocity Measurements. U.S. Patent Application No. 7340958, 11 March 2008.

21. Divos, F. Acoustic tools for seedling, tree and log selection. Presented at the The Future of Quality Control for Wood & Wood Products—COST Action E53, Edinburgh, UK, 4–7 May 2010.

22. ASTM G2395-17. *Standard Test Methods for Density and Specific Gravity (Relative Density) of Wood and Wood-Based Materials*; ASTM International: West Conshohocken, PA, USA, 2017.

23. GB/T1933-2009. *Method for Determination of the Density of Wood*; Standardization Administration of China: Beijing, China, 2009.

24. *Ultrasonic Timer User's Guide*; FAKOPP: Budapest, Hungary, 2008.

25. Wang, X.P.; Ross, R.J.; Carter, P. Acoustic evaluation of wood quality in standing trees. Part I. Acoustic wave behavior. *Wood Fiber Sci.* **2007**, *39*, 28–38.

26. Chauhan, S.S.; Entwistle, K.M.; Walker, J.C.F. Differences in acoustic velocity by resonance and transit-time methods in an anisotropic laminated wood medium. *Holzforschung* **2005**, *59*, 428–434.

27. Isik, F.; Mora, C.R.; Schimleck, L.R. Genetic variation in *Pinus taeda* wood properties predicted using non-destructive techniques. *Ann. For. Sci.* **2011**, *68*, 283–293.

28. Hasegawa, M.; Takata, M.; Matsumura, J.; Oda, K. Effect of wood properties on within-tree variation in ultrasonic wave velocity in softwood. *Ultrasonics* **2011**, *51*, 296–302. [CrossRef] [PubMed]

29. Krauss, A.; Kudela, J. Ultrasonic wave propagation and Young's modulus of elasticity along the grain of Scots pine wood (Pinus Sylvestris L.) varying with distance from the pith. *Wood Res.* **2011**, *56*, 479–488.

30. Chen, Z.Q.; Karlsson, B.; Lundqvist, S.O.; Maria, R.G.G.; Olsson, L.; Wu, H.X. Estimating solid wood properties using Pilodyn and acoustic velocity on standing trees of Norway spruce. *Ann. For. Sci.* **2015**, *72*, 499–508.

31. Blackburn, D.; Farrell, R.; Hamilton, M.; Volker, P.; Harwood, C. Genetic improvement for pulpwood and peeled veneer in *Eucalyptus nitens*. *Can. J. For. Res.* **2012**, *42*, 1724–1732.

32. Ribeiro, P.G.; Goncalez, J.C.; Goncalves, R.; Teles, R.F.; Souza, F.D. Ultrasound waves for assessing the technological properties of *Pinus caribaea* var *hondurensis* and *Eucalyptus grandis* wood. *Maderas Cienc. Tecnol.* **2013**, *15*, 195–204.

33. Cown, D.J.; Hebert, J.; Ball, R. Modelling Pinus radiata lumber characteristics. Part 1: Mechanical properties of small clears. *N. Z. J. For. Sci.* **1999**, *29*, 203–213.

*Article*

# Machinability Study of Australia's Dominate Plantation Timber Resources

**Nathan J. Kotlarewski [1,*], Mohammad Derikvand [1,2], Michael Lee [3] and Ian Whiteroad [4]**

[1]  Australian Research Council Centre for Forest Value, University of Tasmania, Launceston TAS 7250, Australia; Mohammad.Derikvand@utas.edu.au

[2]  Department of Civil Engineering, Aalto University, 02150 Espoo, Finland

[3]  Centre for Sustainable Architecture with Wood, University of Tasmania, Launceston 7250, Australia; M.W.Lee@utas.edu.au

[4]  ARTEC Australia Pty Ltd., Launceston 7250, Australia; Ian.Whiteroad@artec.net.au

*  Correspondence: Nathan.Kotlarewski@utas.edu.au; Tel.: +61-3-6324-4473

Received: 9 August 2019; Accepted: 14 September 2019; Published: 16 September 2019

**Abstract:** This study tested the machinability of three major timber species grown in Tasmania, Australia, under different resource management schemes: plantation fiber-managed hardwood (*Eucalyptus globulus* Labill. and *Eucalyptus nitens* Maiden) and plantation sawlog-managed softwood (*Pinus radiata* D. Don). *P. radiata* was used as a control to identify significant differences in machining fibre-managed plantation timber against sawlog-managed plantation timber with numerically controlled computer technology and manually fed timber production techniques. The potential to fabricate architectural interior products such as moldings with plantation fiber-managed hardwood timber that is high in natural features was the focus of this study. Correlations between wood species, variation in moisture content, and density of individual machinability characteristics were analyzed to determine factors impacting the overall quality of plantation wood machinability. Correlations between species and within species groups from the resulting machinability tests are highlighted and discussed. The results indicate that the machinability of sawlog-managed softwood *P. radiata* is superior in some circumstances to fiber-managed hardwood *E. globulus* and *E. nitens* specimens, according to the American Society for Testing and Materials D1666-11.

**Keywords:** machinability; Eucalyptus; plantation timber; fiber-managed hardwoods

---

## 1. Introduction

Australia has close to one million hectares of plantation hardwood eucalypt species managed for pulplog production. The two major hardwood species grown under this management scheme are *Eucalyptus globulus* Labill., of which 52.7% is predominately grown in Western Australia and the Green Triangle region, followed by *Eucalyptus nitens* Maiden, of which 25.2% is predominately grown in Tasmania (a smaller proportion of the Tasmanian plantation estate for both species is also managed for sawlogs). In addition, there are over one million hectares of planation softwood species managed for sawlog production throughout Australia. *Pinus radiata* D. Don accounts for 74.5% of this estate, which is grown predominately in the Green Triangle region and the Murray Valley (Tasmania also has an established estate of this resource [1]).

In this study, different machinability characteristics of the three major plantation timber species in Australia (*E. globulus*, *E. nitens*, and *P. radiata*) have been evaluated and statistically compared to determine new applications for hardwood plantation resources in machine-manufactured products. With the current supply of planation hardwoods in Australia en masse, and a rise in demand for timber products in the built environment driven by state wood encouragement policies, there is an opportunity to utilize hardwood pulplogs to produce value-added architectural products with advanced manufacturing technologies such as computer numerically controlled (CNC) machinery. The key driver for this research is refocusing hardwood plantation resources into higher-value sawn board applications for furniture and architectural products. To design and manufacture such products, the suitability of processing pulplog with CNC or manually operator-controlled technologies is needed to determine: (i) the machinability of timber derived from pulplogs according to the American Society for Testing and Materials (ASTM) D1666-11 (Standard Test Methods for Conducting Machining Tests on Wood and Wood-Based Materials, 2011) [2] and (ii) the timber properties that most affect the quality of finish for each species.

Utilizing low-quality and low-value plantation logs has been a global topic for a long time [3]. In recent times, *Eucalypts* have attracted much attention for improving the genetics for solidwood production [4] and utilization in value-added materials and product research [5], particularly mass-timber product development such as nail-laminated beams [6] and cross-laminated timber paneling [7]. Traditional wood products (board and veneer), engineered wood products (glulam) and wood-based panels (particleboard and medium-density fiberboard) have revolutionized the way wood is used in the built environment. Wood used in an appearance application relies on a high-quality surface finish to accommodate its final use [8], as well as the application of paints or lamella overlays. The literature consists of various machinability studies that investigate the quality of wood product surface finishes. Not surprisingly, a vast majority of the literature focus on homogeneous wood products such as medium density fiberboard and chipboards due to controllable conditions and less-variable moisture content (MC) and densities [9–11]. Considering these factors, it is of interest to determine the machinability properties of highly variable processed plantation solidwood. How wood specimens are assessed is also a widely presented topic in the literature [12]. Visual assessment has been the standardized procedure for some time now, and new technologies are increasingly being employed to validate quality and compare results [13]. Some researchers go beyond the parameters of ASTM D1666-11, adapting and capturing more data than specified such as the temperature of test specimen after sanding [14] to validate or conclude their findings. Other researchers focus entirely on specific machinability tests such as drilling [15] to advance knowledge. There has also been research conducted in the literature to determine the effects of wood modifications such as thermal treatments on wood machinability [16].

Key variables with any wood machining are cutting speed, feed direction, depth of cut, cutting tool (type and its sharpness) and quality of treatments applied to wood specimens (such as heat or chemical treatment). In addition, the literature states that anatomical characteristics such as species, MC, grain direction, sapwood/hardwood, and density affect the quality of surface machinability [17–19]. New manufacturing knowledge is needed to determine appropriate techniques and commercial processors for the incorporation and potential use of pulplog resources in high-value architectural products, as well as applications to encourage their use in current markets to fulfil demand.

## 2. Materials and Methods

### 2.1. Plantation Timber

The studied timber species included *E. nitens* and *E. globulus* obtained from unthinned and unpruned fiber-managed hardwood plantation resources (from Nook and Trowutta, Tasmania, respectively) for pulplog production in northern Tasmania. These two hardwood pulplog timber species were compared to softwood sawn-board timber obtained from a plantation *P. radiata* resource in Tasmania, Australia. A summary of the three species management schemes, ages and densities is given in Table 1.

**Table 1.** Species sample data.

| Specie | Management Scheme | Age (years) | Average Small End Diameter (mm) | Sample Density Range (kg/m$^3$) | Number of Specimens |
|---|---|---|---|---|---|
| *E. nitens* Maiden | Pulplog | 16 | 345 | 395–741 (523 *) | 54 |
| *E. globulus* Labill. | Pulplog | 26 | 403 | 409–763 (544 *) | 54 |
| *P. radiata* D. Don | Sawlog | 30 | N/A | 444–604 (521 *) | 53 |

* Average specimen density.

The variation in species heterogeneity such as density, presented in Table 1, highlights key characteristics of hardwood species managed under pulplog management schemes. Both hardwood species' density ranges were much wider than *P. radiata*. Specimens prepared for *E. nitens* and *E. globulus* were plainsawn for best recovery. A total of 54 specimens were prepared randomly from ungraded boards for both eucalypt species and varied in origin from each log, deriving from 140-, 120- and 90-mm dressed boards. A maximum of six specimens were machined from individual boards for all species (Figure 1).

**Figure 1.** Six samples machined from individual boards.

For the varying board widths in the hardwood specimens, three sets of six samples were machined from 140-, 120- and 90-mm dressed boards. A total of 53 specimens were prepared for *P. radiata*, all of which derived from utility grade 90-mm dressed boards. All boards designated for sample machining were randomly selected during final processing as run of the mill production to reflect market supply. Prior to testing, boards were stored in a joinery workshop environment at 10 °C (± 4 °C) and 40% (± 5%) relative humidity. This range of environmental conditions (in Tasmania, Australia) was set to test typical joinery workshop environments in which secondary manufacturing commonly takes place, thus allowing a true representation of timber MC in local manufacturing and in service.

## 2.2. Machinability Tests

The machining tests conducted in this study complied with ASTM D1666-11 (2011), with the exception of choice in tooling for CNC operations. All CNC tests were conducted at the Discipline of Architecture and Design, University of Tasmania. The tests conducted included boring, routing, shaping, mortising, and an additional test to determine biscuit boring (doweling) for a more contemporary reference in timber joinery. Each specimen was subjected to boring, routing, and shaping tests. Only a selected number of specimens for each species were subjected to mortising and biscuit boring, as initial results were consistent in machinability quality.

### 2.3. Tooling and Speed/Feed Rates

All boring and routing tests were conducted with a solid carbide, 9.5-mm, 3-flute roughing spiral cutter and finished with a solid carbide, 8-mm, 2-flute compression cutter. A spindle speed of 15,000 revolutions per minute (RPM) and feed rate of 6350 mm/min were used (the standard spindle speed for boring tests is 3600 RPM). The choice in tooling and spindle speeds represented a typical entry-level combination of available tooling in timber joinery workshops for CNC machining. Boring and routing profiles were cut in two passes: first, full depth conventional milling was performed with the roughing tool, leaving 1.6-mm clearance from the finish surface before climb milling with the compression cutter to remove the 1.6 mm overcut. New tools were used for each species. The intention of this change in the standard method was to determine the resilience in timber machinability quality in contemporary industry practice. This was further substantiated and compared with the use of sawlog-managed softwood as a control to determine significant discrepancies between fibre-managed hardwood test results. Mortising tooling complied with the standard (13-mm hollow chisel drill), as did the spindle speed (3600 RPM), which was hand fed by peck drilling on a pedestal drill. Specimen-shaping was conducted on a table router with a no-load spindle speed of 27,000 RPM. The shaping with the table router was done with a 2-flute face-molding carbide tip with a ball bearing guide for profiling. The specimens on the side grain were shaped via hand feeding in two passes due to the depth of the profile. The equipment used to machine biscuit dowels was a Festool DF 500 DOMINO, which cuts the timber stock in a pendulum motion and therefore no spindle speed was recorded. Chip thickness was not measured in any tests. The mortising, shaping, and biscuit-boring introduced a variable of unreliable human-controlled feed rates in comparison with CNC machining. To minimize this variable, one operator conducted each human feed test to maintain consistency in test conditions. This research acknowledges the differences between numerically controlled feed rates and human feed rates, although the intention of this research is still an investigation of the machinability of hardwood plantation resources in commercially mass-produced repetitive machining versus niche one-off productions.

### 2.4. Scoring and Results Analysis

All specimens were graded according to the visual examination classification in ASTM D1666-11 (2011) on the bases of six grades, namely, G0 (defect-free), G1 (excellent), G2 (good), G3 (fair), G4 (poor), and G5 (very poor). While this method of machinability grading is qualitative and context-specific, depending on the product's intended use, two industry-experienced wood machinists with years of sawing, machining, and grading Tasmanian hardwoods were used to visually grade the specimens for each species according to ASTM D1666-11 (2011) and Australian Standard (AS) 2796.3. Grading was conducted using a photo studio lighting kit (Figure 2).

**Figure 2.** Photo studio lighting kit used to grade specimens.

No mechanical or scanning techniques were employed to measure the precision of the visual grading evaluation. A combination of visual and tactile evaluations was employed to determine specimen surface quality as indicated by the existing literature [12,20]. Most of the machining quality was notable by eye and touch, as visual grading provides a rapid and complete analysis of surface quality [21]. The micro-level assessment of surface quality was deemed irrelevant, as products' appearances would normally be graded for Australian markets according to AS 2796. The grade given to each specimen was based on commercially acceptable appearance parameters for high-value architectural products. Commercially acceptable parameters were determined by referencing AS 2796.3 Appendix D, Table D1: "Limits of the machining imperfections and surface finish imperfection on exposed surfaces of hardwood timber for furniture components" [21]. Where specimen tests resulted in surface imperfections, a grade of G2 (good) to G5 (very poor) was given. Specimens with no imperfections were graded as G0 (defect-free) to G1 (excellent). These parameters were also determined by the consistency of surface finish and visually graded using the examples in ASTM D166 and the literature [22]. This process was used to justify allocated grades and to identify significant discrepancies between the resulting specimens within each species and against each species.

*2.5. Statistical Analyses*

The significance of differences between the machinability characteristics of the three wood species in this study were statistically analyzed via Chi-Square testing using IBM SPSS Statistics software (version 23, IBM Corporation, New York, USA). The Analysis of Variance (ANOVA) and Duncan's multiple range test were used for determining the differences between the three species with respect to density and MC. The correlations between machinability characteristics with the variations in density and MC were determined using Pearson's correlation test (interval by interval). All the statistical analyses were conducted at 0.05 significance level.

## 3. Results

*3.1. Statistical Analyses of Density and MC between Species*

The ANOVA results indicated no significant difference between the densities of the three species in this study ($p > 0.05$), which enabled a statistical comparison between the machinability characteristics. The difference between MC values in the three species, however, was significant ($p < 0.05$). The difference between the density and MC of *E. nitens* and *P. radiata* was less than 0.4% and 2.3%, respectively. The average density and MC of *E. globulus* samples were respectively 4.2% and 3.9% higher than that of *E. nitens*, and 4.6% and 1.5% higher than *P. radiata*. The variations in the density and MC values of the test samples within each species can be seen in Figures 3–5.

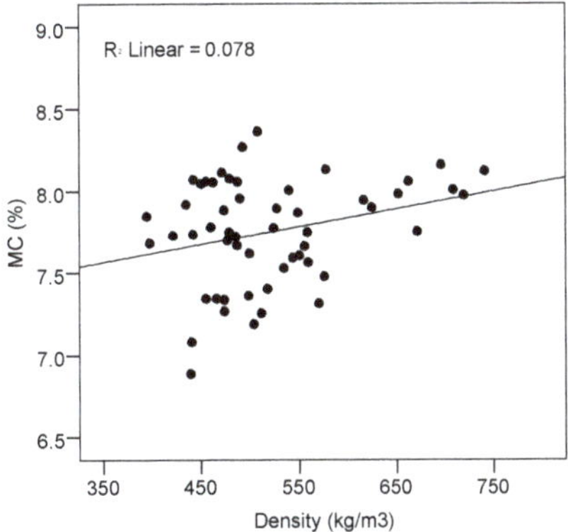


**Figure 3.** Variation in moisture content and density of *E. nitens* samples.

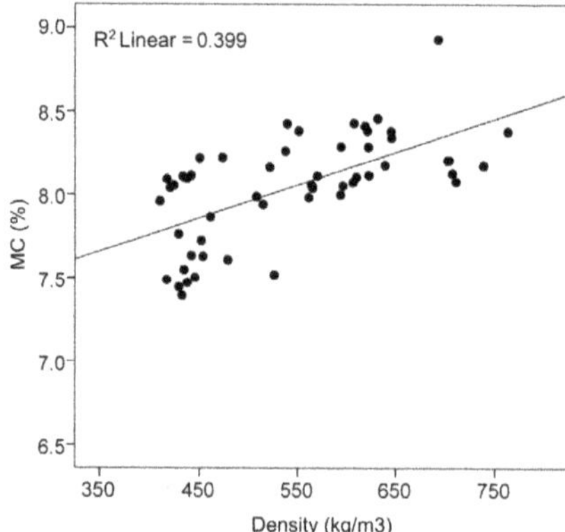

**Figure 4.** Variation in moisture content and density of *E. globulus* samples.

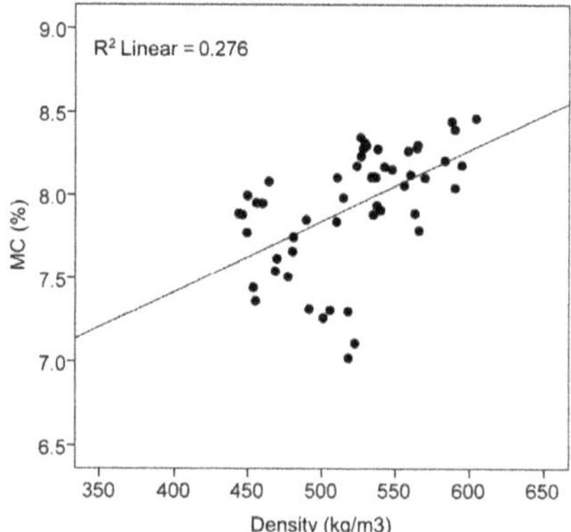

**Figure 5.** Variation in moisture content and density of *P. radiata* samples.

*3.2. Statistical Analyses of Machinability Results between Species*

The test results obtained with respect to routing end grain (fuzzy and raised) and boring (crushing, fuzzy, and smoothness) indicated that the test samples from the three species were all defect-free with consistent quality (having a grade of G0). These test results are therefore not presented in this study. The results of the statistical analyses for the remaining machinability characteristics are presented. No statistically significant difference was found between the three species with respect to raised routing side grain, chipped routing end grain, chipped shaping side grain, boring tear-out and biscuit bore (crushed and chipped) ($p > 0.05$). Statistically significant differences were found between the three species for routing side grain (fuzzy and chipped), shaping side grain (raised and fuzzy), mortising (crushing, tearing and smoothness) and fuzzy biscuit-bore grain ($p < 0.05$).

The grading results of the machinability characteristics for each species sample are shown in Tables 2–7. The values with the most important contributions to the statistical significance for each characteristic are highlighted in grey where applicable (indicating the differences within species and between species). All the *E. nitens* and *P. radiata* samples received a G0 grade (defect-free) for routing side grain (Table 2). The fuzzy routing side grain (Figure 6), was more variable within the *E. globulus* samples, with more than 37% of the samples having a grade between G1 to G4. There was no significant difference between the *E. nitens* and *P. radiata* samples with respect to the chipped routing side grain. The number of samples with a grade worse than G0 were significantly higher in *E. globulus* compared to *E. nitens* and *P. radiata*.

**Table 2.** Grading results according to routing side grain.

| Routing Side Grain Results | | Routing Side Grain (Raised) | | | Routing Side Grain (Fuzzy) | | | | | Routing Side Grain (Chipped) | | | | Total |
|---|---|---|---|---|---|---|---|---|---|---|---|---|---|---|
| | | Grade 0 | Grade 1 | Grade 2 | Grade 0 | Grade 1 | Grade 2 | Grade 3 | Grade 4 | Grade 0 | Grade 1 | Grade 2 | Grade 3 | |
| Species | *E. nitens* | Count | | | | | | | | | | | | | 54 |
| | | 53 | 1 | 0 | 54 | 0 | 0 | 0 | 0 | 49 | 1 | 1 | 3 | 54 |
| | | Standardised residual | | | | | | | | | | | | |
| | | 0.2 | −0.3 | −1 | 1 | −1.3 | −1.9 | −1 | −0.6 | 0 | 1.1 | −1.2 | 1 | |
| | *E. globulus* | Count | | | | | | | | | | | | | 54 |
| | | 49 | 2 | 3 | 34 | 5 | 11 | 3 | 1 | 44 | 0 | 8 | 2 | 54 |
| | | Standardised residual | | | | | | | | | | | | |
| | | −0.4 | 0.6 | 2 | −1.9 | 2.6 | 3.8 | 2 | 1.1 | −0.7 | −0.6 | 2.9 | 0.2 | |
| | *P. radiata* | Count | | | | | | | | | | | | | 53 |
| | | 52 | 1 | 0 | 53 | 0 | 0 | 0 | 0 | 53 | 0 | 0 | 0 | 53 |
| | | Standardised residual | | | | | | | | | | | | |
| | | 0.2 | −0.3 | −1 | 1 | −1.3 | −1.9 | −1 | −0.6 | 0.7 | −0.6 | −1.7 | −1.3 | |
| Total | | Count | | | | | | | | | | | | | |
| | | 154 | 4 | 3 | 141 | 5 | 11 | 3 | 1 | 146 | 1 | 9 | 5 | 161 |

**Table 3.** Grading results according to routing end grain.

| Routing End Grain Results | | | Routing End Grain Chipped | | Total |
|---|---|---|---|---|---|
| | | | Grade 0 | Grade 2 | |
| Species | *E. nitens* | Count | 53 | 1 | 54 |
| | | Standardised residual | −0.1 | 1.1 | |
| | *E. globulus* | Count | 54 | 0 | 54 |
| | | Standardised residual | 0 | −0.6 | |
| | *P. radiata* | Count | 53 | 0 | 53 |
| | | Standardised residual | 0 | −0.6 | |
| Total | | Count | 160 | 1 | 161 |

**Table 4.** Grading results according to shaping side grain.

| Shaping Side Grain Results | | | Shaping Side Grain (Raised) | | | | Shaping Side Grain (Fuzzy) | | | | Shaping Side Grain (Chipped) | | | | | | Total |
|---|---|---|---|---|---|---|---|---|---|---|---|---|---|---|---|---|---|---|
| | | | Grade 0 | Grade 1 | Grade 2 | Grade 3 | Grade 0 | Grade 1 | Grade 2 | Grade 3 | Grade 0 | Grade 1 | Grade 2 | Grade 3 | Grade 4 | Grade 5 | |
| Species | *E. nitens* | Count | 42 | 6 | 6 | 0 | 54 | 0 | 0 | 0 | 47 | 1 | 1 | 2 | 1 | 2 | 54 |
| | | Standardised residual | −0.9 | 1.4 | 2.4 | −0.6 | 1.2 | −2.2 | −1.7 | −0.6 | −0.2 | 0.4 | 0 | 0.6 | −0.5 | 1 | |
| | *E. globulus* | Count | 48 | 4 | 1 | 1 | 47 | 0 | 6 | 1 | 44 | 1 | 2 | 2 | 4 | 1 | 54 |
| | | Standardised residual | 0 | 0.4 | −0.9 | 1.1 | 0.2 | −2.2 | 1.7 | 1.1 | −0.6 | 0.4 | 1 | 0.6 | 1.8 | 0 | |
| | *P. radiata* | Count | 53 | 0 | 0 | 0 | 36 | 14 | 3 | 0 | 53 | 0 | 0 | 0 | 0 | 0 | 53 |
| | | Standardised residual | 0.9 | −1.8 | −1.5 | −0.6 | −1.4 | 4.4 | 0 | −0.6 | 0.8 | −0.8 | −1 | −1.1 | −1.3 | −1 | |
| Total | | Count | 143 | 10 | 7 | 1 | 137 | 14 | 9 | 1 | 144 | 2 | 3 | 4 | 5 | 3 | 161 |

**Table 5.** Grading results according to boring.

| Boring Results | | | Boring Tear-Out | | | Total |
|---|---|---|---|---|---|---|
| | | | Grade 0 | Grade 2 | Grade 3 | |
| Species | E. nitens | Count | 54 | 0 | 0 | 54 |
| | | Standardised residual | 0.1 | −0.6 | −0.6 | |
| | E. globulus | Count | 52 | 1 | 1 | 54 |
| | | Standardised residual | −0.2 | 1.1 | 1.1 | |
| | P. radiata | Count | 53 | 0 | 0 | 53 |
| | | Standardised residual | 0.1 | −0.6 | −0.6 | |
| Total | | Count | 159 | 1 | 1 | 161 |

**Table 6.** Grading results according to mortising.

| Mortising Results | | | Mortising (Crushing) | | | | Mortising (Tearing) | | | | Mortising (Smoothness) | | Total |
|---|---|---|---|---|---|---|---|---|---|---|---|---|---|
| | | | Grade 2 | Grade 3 | Grade 4 | Grade 5 | Grade 2 | Grade 3 | Grade 4 | Grade 5 | Grade 4 | Grade 5 | |
| Species | E. nitens | Count | 0 | 20 | 10 | 0 | 0 | 20 | 9 | 1 | 0 | 30 | 30 |
| | | Standardised residual | −1.4 | 1.2 | 1.5 | −2.5 | −1.4 | 0.8 | 0.2 | −1.2 | −1.4 | 0.4 | |
| | E. globulus | Count | 0 | 2 | 9 | 19 | 0 | 6 | 16 | 8 | 0 | 30 | 30 |
| | | Standardised residual | −1.4 | −3.4 | 1.1 | 5 | −1.4 | −2.6 | 2.7 | 2.9 | −1.4 | 0.4 | |
| | P. radiata | Count | 6 | 24 | 0 | 0 | 6 | 24 | 0 | 0 | 6 | 24 | 30 |
| | | Standardised residual | 2.8 | 2.2 | −2.5 | −2.5 | 2.8 | 1.8 | −2.9 | −1.7 | 2.8 | −0.8 | |
| Total | | Count | 6 | 46 | 19 | 19 | 6 | 50 | 25 | 9 | 6 | 84 | 90 |

**Table 7.** Grading results according to biscuit boring.

| Biscuit Bore Results | | | Biscuit Bore (Crushed) | | Biscuit Bore (Fuzzy) | | | | Biscuit Bore (Chipped) | | Total |
|---|---|---|---|---|---|---|---|---|---|---|---|
| | | | Grade 0 | Grade 3 | Grade 0 | Grade 1 | Grade 2 | Grade 3 | Grade 0 | Grade 4 | |
| Species | E. nitens | Count | 30 | 0 | 0 | 0 | 0 | 30 | 29 | 1 | 30 |
| | | Standardised residual | 0.1 | −0.6 | −1.5 | −0.8 | −2.5 | 2.1 | −0.1 | 0.4 | |
| | E. globulus | Count | 29 | 1 | 7 | 0 | 4 | 19 | 29 | 1 | 30 |
| | | Standardised residual | −0.1 | 1.2 | 3.1 | −0.8 | −0.9 | −0.4 | −0.1 | 0.4 | |
| | P. radiata | Count | 30 | 0 | 0 | 2 | 15 | 13 | 30 | 0 | 30 |
| | | Standardised residual | 0.1 | −0.6 | −0.5 | 1.6 | 3.4 | −1.7 | 0.1 | −0.8 | |
| Total | | Count | 89 | 1 | 7 | 2 | 19 | 62 | 88 | 2 | 90 |

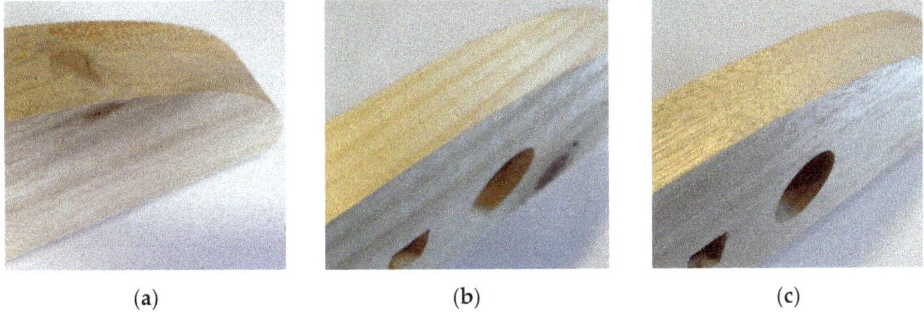

(a)          (b)          (c)

**Figure 6.** Examples of *E. globulus* (**a**) with a grade of G4 (poor) and *P. radiata* (**b**) and *E. nitens* (**c**) with a grade of G0 (excellent) for fuzzy routing side grain.

Routing end grain (Figure 7) divided the *E. nitens* samples into grades of G0 and G2; however, no statistically significant difference was found between the three species with respect to the chipped routing end grain (Table 3). This machinability characteristic was less sensitive among the species.

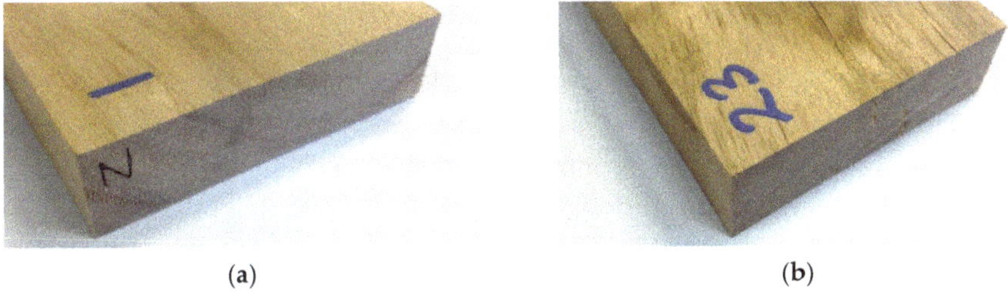

(a)          (b)

**Figure 7.** Examples of *E. nitens* with grades of G0 (excellent) (**a**) and G2 (good) (**b**) for routing end grain.

The grading results of the samples with respect to shaping side grain are shown in Table 4. For both shaping side grains (raised or chipped), *P. radiata* displayed a better finish quality than *E. nitens* and *E. globulus*, with 100% defect-free results. The *E. nitens* samples, however, had the highest fuzzy shaping side grain quality compared to *P. radiata* and *E. globulus*, with 100% of the samples being graded as G0. None of the test samples from the three species had any grade worse than G3 with respect to shaping side grain (raised or fuzzy) (Figure 8).

(a)          (b)

**Figure 8.** Examples of *E. globulus* with a grade of G2 (good) for fuzzy shaping side grain (**a**) and *E. nitens* with a grade of G5 (very poor) for chipped shaping side grain (**b**).

All the *E. nitens* and *P. radiata* samples were graded as G0 with respect to boring tear-out (Table 5). Only two samples from *E. globulus* had a grade worse than G0, a difference that was not statistically significant (Figure 9).

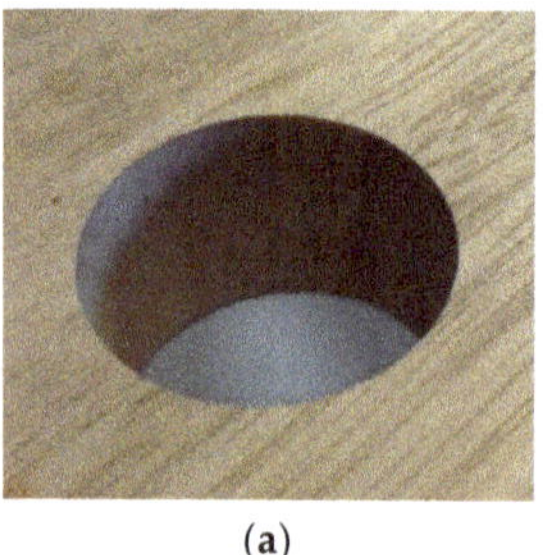 

(**a**)  (**b**)

**Figure 9.** Examples of *E. nitens* with a grade of G0 (excellent) (**a**) and *E. globulus* with a grade of G2 (good) (**b**) for boring tear-out.

The results indicated that mortising quality is significantly correlated to the wood species (Table 6), with *P. radiata*, *E. nitens*, and *E. globulus* having the best to the worst overall mortising qualities, respectively (Figure 10). There was no statistically significant difference between *E. nitens* and *E. globulus* in respect to mortising smoothness, whereas *P. radiata* had significantly better mortising smoothness than both eucalypt species. All *E. nitens* and *E. globulus* samples had a G5 grade for mortising smoothness.

(**a**)  (**b**)  (**c**)

**Figure 10.** Examples of *E. nitens* (**a**) and *E. globulus* (**b**) with grades of G5 (very poor), and *P. radiata* (**c**) with a grade of G2 (good) for mortising (tearing).

The *E. nitens* samples had the worst quality when graded based on biscuit bore (Table 7), with 100% of the samples being graded as G3 (Figure 11). More than 30% of the *E. globulus* samples were defect-free (G0), which made a high contribution to the statistical significance between the three species. The three species had almost the same quality when the biscuit bore (crushed and chipped) was used as the grade-determining parameter. The *E. globulus* samples showed more variations in the fuzzy biscuit-bore grain compared to the other two species.

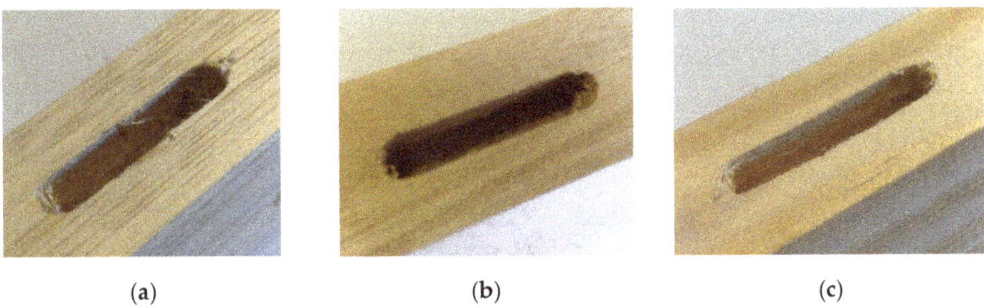

(a)                 (b)                 (c)

**Figure 11.** Examples of *E. nitens* (**a**) with a grade of G3 (fair) for fuzzy biscuit-bore grain and *E. globulus* (**b**) and *P. radiata* (**c**) with a grade of G2 (good).

## 4. Discussion

The findings in this research intend to demonstrate appropriate new applications for hardwood plantation resources in machine-manufactured products. The following sections validate opportunities where hardwood plantation resources could serve as appropriate materials of choice.

### 4.1. Statistical Analyses of Density and MC Within Species

The results indicate a higher variability in the machinability of the *E. nitens* and *E. globulus* specimens compared to that of *P. radiata*. Part of this is because of the variation in the density of the samples and its influence on the results obtained. A possible physical phenomenon that explains the variation in machinability results—due to the degree of changes in densities—could relate to the management of the resource that was initially intended for pulplog production. The variations in MC for *E. nitens* specimens showed no significant correlation with any of the studied machinability characteristics ($p > 0.05$). The variations in specimen densities, however, had significant correlations with chipped routing end grain, raised shaping side grain and mortising (crushing and tearing) ($p < 0.05$). For *E. globulus*, the variation in MC values had significant correlations with fuzzy routing side grain and fuzzy shaping side grain ($p < 0.05$). In addition, the variations in specimen densities also had significant correlations with fuzzy routing side grain, fuzzy shaping side grain and mortising (tearing) ($p < 0.05$). For *P. radiata*, the variations in MC values had significant correlations with mortising (crushing, tearing, and smoothness) ($p < 0.05$) and the variations in specimen densities showed no statistically significant correlation with the studied machinability characteristics.

### 4.2. Routing End Grain and Boring

Despite being managed for pulplog production, machinability tests of *E. nitens* and *E. globulus*—as well as sawlog *P. radiata*—for routing end grain (fuzzy and raised), boring (crushing, fuzzy, and smoothness), resulted in defect-free specimens. Both pulplog resources were out-graded in quality of finish by *P. radiata* due to chipping in the end grain (*E. nitens*) and tear-out from boring (*E. globulus*). Expectedly, chipping and tear-out were present in the fiber-managed plantation species given that the nature of the resource to break apart is a direct reason for its use in pulp production. The chipping observation could be due to the long fiber lengths when machined perpendicular to the grain, and the tear-out evident in boring (also perpendicular to the grain) may be caused by pulling fibers. Regardless, the results suggest that either species would be an acceptable choice for products such as cabinetry or acoustic panels that require end-grain routing or a degree of good and better surface boring for fixtures or perforations, particularly where high-quality surface finishes are essential. Extra care in machining could mitigate chipping or tear-out from the pulplog resources. As suggested in ASTM D1666-11 (2011), a roughing cut offset by 1.6 mm then finished in a final pass can ensure that any edge damaged in roughing is removed for a better surface finish. As previously highlighted in this study, two solid carbide tools—one for roughing and the other for finishing—were used to ensure that the best surface quality could be achieved. In addition, the roughing cut was conventional milling and the finishing cut was climb milling. Generally, optimal surface qualities were achieved directly parallel and perpendicular to the wood grain, and most raised grain, fuzzing, and chipping were at tangent angles following a parabola specimen shape as set out in ASTM D-1666.

*4.3. Routing Side Grain and Shaping*

*E. nitens* and *P. radiata* samples showed better routing qualities on the side grain than *E. globulus*. Once again, *P. radiata* out-graded both pulplog resources, and *E. nitens* out-graded *E. globulus*. In line with routing end grain, chipping appears to have been dominant for both hardwood species, particularly towards the edge of a specimen. This could have been caused by the length of wood fiber in plantation *Eucalyptus*, which typically results in pulling out or tearing more stock material than intended by machining. Similarly, fuzzy routing side grain for *E. globulus* was the greatest reason for the downgrading of these specimens. The results of shaping on the side grain also suggest that *E. globulus* and *P. radiata* are less desirable for architectural applications such as moldings where high-quality surface finishes are necessary. In comparison, shaping on the side grain of *E. nitens* resulted in raised grain.

*4.4. Mortising and Biscuit Boring*

*P. radiata* out performed *E. nitens* and *E. globulus*, however the results for all species were far from perfect, with no tests resulting in a defect-free grade (G0). This may suggest that following the defined test method set out in ASTM D1666-11 (2011) for mortising is not an ideal form of joinery. The results also suggest that the surface hardness of the tested species could have been low, and therefore the observed crushing by compression and tearing could have been mitigated by a change in choice of tooling. Consideration should be made, however, of the fact that grading of the mortise refers to an internal surface that is not seen in final products such as assembled furniture. As an alternative to mortise and tenon joinery, boring via CNC fabrication would be an acceptable alternative, as substantiated in the boring tests. In keeping with the grading standards set out in ASTM D1666-11 (2011), this study considered the quality in surface finish from biscuit boring, a more contemporary approach to joinery with dowels. All species performed exceptionally well against crushing and chipping. In this test, it was fuzzy grain that was the dominate grade-reducing feature. This may have been due to the pendulum motion of the biscuit dowel cutter. Regardless, the extrusion generated in this test—like a mortise—is internal, and not seen in a product's finally assembly. Moreover, the fuzzy grain caused by the biscuit borer could advantageously improve the retention of the biscuit dowel and glue for furniture or table tops.

*4.5. Other Considerations*

Another reason for the higher variability in the machinability of the *E. nitens* and *E. globulus* specimens compared to that of *P. radiata* could be that the pulplog specimens were selected randomly from ungraded timber boards high in natural features. Considering this, there could be a high potential to improve the machinability of these plantation species by making use of an appropriate timber grading system that would allow proper resources to be selected for appropriate higher-value products. Although *E. nitens* and *E. globulus* specimens were derived from pulplog resources, the results suggest that in some applications these species are appropriate alternatives for products where hardwood species are desirable or in demand.

The physical phenomena observed in *E. globulus* and *E. nitens*, such as the tearing, fuzzing, and chipping throughout the test specimens, could be related to fiber length, elasticity, hardness, and ductility of the pulp resources [23]. *E. globulus* is known for its high density, high coarseness, and high fiber-length [23], which contribute to its use in pulp production; however, these properties also render the resource useful for raw forest products as well as other composite and engineering forest products [23]. Carefully considered machining processes to mitigate chipping and tearing parallel and perpendicular to wood grains may improve and reduce the quantity of machining defects in these hardwood species. This could be as simple as including lead-ins and -outs, making helical cuts, and programming multiple steps to mitigate visible fibre damage to value-added timber products. Future research could investigate these multiple variables to identify processes or techniques to avoid when machining plantation *Eucalyptus* species.

All species demonstrated both a high-quality and less-than-favorable surface finished with both CNC and manual machining techniques. Where possible, any automated system that is replicable and controllable is ideal for consistency in quality. This research suggests that both forms of machining can produce acceptable finishes for architectural value-added products.

Further research on the machinability of *Eucalyptus* pulplogs could consider the origin of specimens from log and tree positions, as well as the orientations of cuts. The impact of surface and internal checking could also be investigated to determine if these characteristics impact chipping and tear-out from various machinability tests conducted on fiber-managed hardwood resources. Furthermore, the impact of live and dead knots on machinability properties could be investigated to determine the acceptable presence of different types of knots on the appearance of products that are routed, shaped, or bored.

## 5. Conclusions

The aim of this study was to determine the machinability of Tasmanian plantation fibre-managed hardwood *Eucalyptus globulus* and *Eucalyptus nitens* and to evaluate their potential use in architectural interior products such as moldings, as well as other timber products such as furniture. Plantation sawlog-managed softwood *Pinus radiata* was used as a control reference given its acceptable quality and use in a wide range of architectural and product applications.

- The results in this study suggest that fibre-managed plantation hardwood *E. nitens* and *E. globulus* have the same machinability qualities (with no statistically significant difference) as sawlog-managed *P. radiata* for routing end grain (fuzzy, raised, and chipped), boring (crushing, fuzzy, smoothness, and tear-out), raised routing side grain, chipped shaping side grain, and biscuit bore (crushed and chipped). In products and applications where secondary manufacturing involves routing (end grain and side grain), boring, shaping, and biscuit boring, producers can expect acceptable machinability qualities that would allow the use of fibre-managed plantation hardwoods as an alternative to sawlog-managed plantations softwoods.
- No statistically significant difference was observed between *E. nitens* and *E. globulus* except when routing side grain (fuzzy and chipped), shaping side grain (raised and fuzzy), mortising (crushing and tearing), and fuzzy biscuit-bore grain were used as the grade-determining parameters.
- The correlations between the variations in density and machinability characteristics of *E. nitens* and *E. globulus* were statistically significant, whereas no significant correlations existed between the variations in density and machinability of *P. radiata*.

### 5.1. E. nitens

- The machinability characteristics of *E. nitens* were statistically comparable to *P. radiata* in all cases except for shaping side grain (raised and fuzzy), mortising (crushing, tearing, and smoothness) and fuzzy biscuit-bore grain.
- The studied *E. nitens* samples received the worst grades among the studied timber species when graded against raised shaping side grain and fuzzy biscuit-bore grain as the grade-determining parameters.
- The machinability of the *E. nitens* samples was significantly better than both *E. globulus* and *P. radiata* in respect to fuzzy shaping side grain.
- The *E. nitens* had better quality than *E. globulus* in respect to routing side grain (fuzzy and chipped), fuzzy shaping side grain, mortising (crushing and tearing), and fuzzy biscuit-bore grain.

- Unlike *P. radiata* and *E. globulus*, the variations in the MC of the samples had no important correlation with the machinability characteristics of *E. nitens*.

### 5.2. E. globulus

- The machinability characteristics of *E. globulus* were statistically comparable to *P. radiata* in all cases except for routing side grain (fuzzy and chipped), fuzzy shaping side grain, mortising (crushing, tearing, and smoothness) and fuzzy biscuit-bore grain.
- The only cases in which the *E. globulus* samples showed significantly better qualities than *P. radiata* were fuzzy shaping side grain and fuzzy biscuit-bore grain.

**Author Contributions:** Conceptualization, N.K.; formal analysis, N.K. and M.D.; investigation, N.K. and M.D.; methodology, N.K., M.D., M.L., and I.W.; resources, N.K. and M.L.; validation, N.K., M.D., M.L. and I.W.; writing—original draft preparation, N.K.; writing—review and editing, N.K, M.D. and M.L.

**Funding:** This research was funded by the Australian Research Council, Centre for Forest Value, University of Tasmania, TAS, Australia, grant number IC150100004. The APC was funded by the Australian Research Council, Centre for Forest Value, University of Tasmania, TAS, Australia.

**Acknowledgments:** The authors would like to thank Trevor Innes (TimberLink Australia) for supplying the sawn-board *P. radiata* for this study; Forico Pty Ltd. for providing the fibre-managed hardwood logs; and Britton Timbers for milling, drying, and dressing the fiber-managed hardwood resources.

**Conflicts of Interest:** The authors declare no conflict of interest. The funders had no role in the design of the study; in the collection, analyses, or interpretation of data; in the writing of the manuscript, or in the decision to publish the results.

## References

1. Australian Bureau of Agricultural and Resource Economics and Sciences (ABARES). Australian Plantation Statistics 2017 Update, Australian Bureau of Agricultural and Resource Economics and Sciences, Canberra, August 2017. CC BY 3.0. Available online: https://data.gov.au/dataset/ds-dga-a1fcbbec-807c-438e-b8fd-db1a9a93f887 (accessed on 4 October 2018).
2. American Society for Testing and Materials (ASTM). *Standard Test Methods for Conducting Machining Tests of Wood and Wood-Base Materials*; ASTM D1666-11; ASTM International: West Conshohocken, PA, USA, 2011; 23p.
3. Zobel, B. The changing quality of the world wood supply. *Wood Sci. Technol.* **1984**, *181*, 1–17. [CrossRef]
4. Hamilton, M.G. The Genetic Improvement of Eucalyptus Globulus and Eucalyptus Nitens for Solidwood Production. Ph.D. Dissertation, University of Tasmania, Newnham, Australia, 2007.
5. Dugmore, M.; Nocetti, M.; Brunetti, M.; Naghizadeh, Z.; Wessels, C.B. Bonding quality of cross-laminated timber: Evaluation of test methods on Eucalyptus grandis panels. *Constr. Build. Mater.* **2019**, *211*, 217–227. [CrossRef]
6. Derikvand, M.; Kotlarewski, N.; Lee, M.; Jiao, H.; Chan, A.; Nolan, G. Short-term and long-term bending properties of nail-laminated timber constructed of fast-grown plantation eucalypt. *Constr. Build. Mater.* **2019**, *211*, 952–964. [CrossRef]
7. Pangh, H.; Hosseinabadi, H.Z.; Kotlarewski, N.; Moradpour, P.; Lee, M.; Nolan, G. Flexural performance of cross-laminated timber constructed from fibre-managed plantation eucalyptus. *Constr. Build. Mater.* **2019**, *208*, 535–542. [CrossRef]
8. Aguilera, A.; Martin, P. Machining qualification of solid wood of *Fagus silvatica* L. and *Piceaexcelsa* L.: Cutting forces, power requirements and surface roughness. *Holz Als Roh-Und Werkst.* **2001**, *59*, 483–488. [CrossRef]
9. Aguilera, A.; Meausoone, P.J.; Martin, P. Wood material influence in routing operations: The MDF case. *Eur. J. Wood Wood Prod.* **2000**, *58*, 278–283. [CrossRef]
10. Lin, R.J.; van Houts, J.; Bhattacharyya, D. Machinability investigation of medium-density fibreboard. *Holzforschung* **2006**, *60*, 71–77. [CrossRef]
11. Szymanowski, K.; Gorski, J.; Czarniak, P.; Wilkowski, J.; Podziewski, P.; Cyrankowski, M.; Morek, R. Machinability index of certain types of chipboards. In *Forestry and Wood Technology; Annals of Warsaw*; Warsaw University of Life Sciences Press: Warsaw, Poland, 2015; p. 90.

12. Goli, G.; Sandak, J. Proposal of a new method for the rapid assessment of wood machinability and cutting tool performance in peripheral milling. *Eur. J. Wood Wood Prod.* **2016**, *74*, 867–874. [CrossRef]

13. Sütçü, A. Investigation of parameters affecting surface roughness in CNC routing operation on wooden EGP. *BioResources* **2012**, *8*, 795–805. [CrossRef]

14. Moya-Roque, R.; Tenorio-Monge, C.; Salas-Garita, C.; Berrocal-Jiménez, A. Evaluation of wood properties from six native species of forest plantations in Costa Rica. *Bosque* **2016**, *37*, 71–84. [CrossRef]

15. Podziewski, P.; Szymanowski, K.; Górski, J.; Czarniak, P. Relative Machinability of Wood-Based Boards in the Case of Drilling–Experimental Study. *BioResources* **2018**, *13*, 1761–1772. [CrossRef]

16. Ratnasingam, J.; Ioras, F. Effect of heat treatment on the machining and other properties of rubberwood. *Eur. J. Wood Wood Prod.* **2012**, *70*, 759–761. [CrossRef]

17. Laina, R.; Sanz-Lobera, A.; Villasante, A.; López-Espí, P.; Martínez-Rojas, J.A.; Alpuente, J.; Sánchez-Montero, R.; Vignote, S. Effect of the anatomical structure, wood properties and machining conditions on surface roughness of wood. Maderas. *Cienc. Y Tecnol.* **2017**, *19*, 203–212. [CrossRef]

18. Thoma, H.; Peri, L.; Lato, E. Evaluation of wood surface roughness depending on species characteristics. Maderas. *Cienc. Y Tecnol.* **2015**, *17*, 285–292. [CrossRef]

19. Aguilera, A.; Zamora, R. Surface roughness in sapwood and heartwood of Blackwood (*Acacia melanoxylon* R. Br.) machined in 90-0 direction. *Eur. J. Wood Wood Prod.* **2009**, *67*, 297–301. [CrossRef]

20. Ramanakoto, M.F.; Andrianantenaina, A.N.; Ramananantoandro, T.; Eyma, F. Visual and visuo-tactile preferences of Malagasy consumers for machined wood surfaces for furniture: Acceptability thresholds for surface parameters. *Eur. J. Wood Wood Prod.* **2017**, *75*, 825–837. [CrossRef]

21. Standards Australia. *Australian Standard 2796 Timber—Hardwood—Sawn and Milled Products Part 3: Timber for Furniture Components*; Standards Associations of Australia: Sydney, NSW, Australia, 1999; reconfirmed in 2016.

22. Goli, G.; Marchal, R.; Negri, M.; Costes, J.P. Surface quality: Comparison among visual grading and 3D roughness measurements. In Proceedings of the 15th International Wood Machining Seminar, Los Angeles, CA, USA, 30 July–1 August 2001.

23. Carrillo-Varela, I.; Valenzuela, P.; Gacitúa, W.; Mendonca, R.T. An Evaluation of Fiber Biometry and Nanomechanical Properties of Different Eucalyptus Species. *BioResources* **2019**, *14*, 6433–6446. [CrossRef]

*Article*

# Artificial Neural Network Modeling for Predicting Wood Moisture Content in High Frequency Vacuum Drying Process

**Haojie Chai [1], Xianming Chen [2], Yingchun Cai [1] and Jingyao Zhao [1,***

[1] Ministry of Education, Key Laboratory of Bio-Based Material Science and Technology, College of Material Science and Engineering, Northeast Forestry University, Harbin 150040, China; nefuchj@163.com (H.C.); caiyingchunnefu@163.com (Y.C.)

[2] College of Information and Computer Engineering, Northeast Forestry University, Harbin 150040, China; chenxianming@nefu.edu.cn

* Correspondence: zjy_20180328@nefu.edu.cn; Tel./Fax: +86-451-8219-1002

Received: 5 December 2018; Accepted: 26 December 2018; Published: 29 December 2018

**Abstract:** The moisture content (MC) control is vital in the wood drying process. The study was based on BP (Back Propagation) neural network algorithm to predict the change of wood MC during the drying process of a high frequency vacuum. The data of real-time online measurement were used to construct the model, the drying time, position of measuring point, and internal temperature and pressure of wood as inputs of BP neural network model. The model structure was 4-6-1 and the decision coefficient $R^2$ and Mean squared error (Mse) of the training sample were 0.974 and 0.07355, respectively, indicating that the neural network model had superb generalization ability. Compared with the experimental measurements, the predicted values conformed to the variation law and size of experimental values, and the error was about 2% and the MC prediction error of measurement points along thickness direction was within 2%. Hence, the BP neural network model could successfully simulate and predict the change of wood MC during the high frequency drying process.

**Keywords:** neural network; high frequency drying; moisture content; wood

## 1. Introduction

Wood MC (moisture content) is one of the crucial indicators in the drying process as it has a direct impact on the stability of wood drying quality, and a reasonable control of MC can help in meeting the various quality requirements of actual wood products [1]. High frequency vacuum drying is a joint drying technology with a fast drying rate, low energy consumption, and low environmental pollution [2], and is in widespread use throughout the wood drying industry [3]. However, due to the interference of high frequency electromagnetic fields, the traditional MC online monitoring device cannot be used normally, which makes the online prediction and effective detection of wood MC problematic [4]. Therefore, the research on the prediction model of wood MC is of great significance in the high frequency drying process.

The wood structure is complex and it is difficult to establish a precise mathematical model through mathematical mechanism. An accurate control of MC requires precise mathematical models. The high frequency vacuum drying of wood is a non-linear, complex drying process, which is difficult to accurately express, control, or implement by using general mathematical methods [5]. The concept of BP (Back Propagation) neural network comes from the biological system of brain, which is composed of numerous neurons that are connected to each other through synapses that process information. The neural network has decent characteristics for predicting nonlinear complex systems [6,7], and the model reflects the intrinsic connection of experimental data after a finite number of iterative calculations. It is not only strong at processing nonlinearity, self-organizing adjustment, adaptive learning, and fault-tolerant anti-noise [8–10] but also can effectively deal with nonlinear and complex fuzzy processes. An effective network prediction model can be established without any assumption or theoretical relationship analysis, based on the historical data and powerful self-organization integration capabilities [11,12].

Artificial neural networks are increasingly being used for modeling in the field of wood science. For instance, in the field of wood drying, Avramidis (2006) [13] predicted the drying rate of wood based on neural network construction model; Zhang Dongyan (2008) [14] constructed a neural network model for predicting wood MC during conventional drying; İlhan Ceylan (2008) [15] used neural network models to study wood drying characteristics; Watanabe (2013, 2014) [16,17] employed artificial neural network model to predict the final moisture content of Sugi (*Cryptomeria japonica*) during drying and evaluate the drying stress on the wood surface. Ozsahin (2017) [18] utilized artificial neural networks to successfully predict the equilibrium moisture content and specific gravity of heat-treated wood. The artificial neural networks are widely used in the study of conventional drying characteristics, stress monitoring, and MC prediction of wood [19]; however, the use of neural networks to predict changes in the wood MC during high frequency drying has been rarely studied.

Hence, in order to provide a predictive model for the control of wood MC during high frequency drying, based on the BP neural network algorithm and using the real-time online measurement data, drying time, location of measuring point, and internal temperature and pressure of wood as the input to neural network model, the changes in the wood MC can be predicted. Also, the feasibility and prediction accuracy of the model was analyzed.

## 2. Materials and Methods

### 2.1. On-Line Monitoring of Wood Internal Temperature and Pressure

Some uniform and defect-free Mongolian pine (*Pinus sylvestris* var. *mongholica* Litv.) were selected. The 200 mm ends were removed at both ends of the test piece, and the specifications were 120 × 120 × 500 mm specimens after sawing and planning, and the initial moisture content was 50%. As shown in Figure 1, five temperature pressure measuring points were uniformly preset at the center of the sample in the thickness direction. Drilling holes on the side of specimen with a 4 mm drill bit to depth of 60 mm (seeing Figure 1 for specific locations). Each measuring point was embedded with one of the pressure and temperature fiber sensors, and the locations where the sensors were in contact with the surface of wood were coated with silica gel to ensure good sealing. The data was recorded online through the optical fiber sensors.

As shown in Figure 2, 1 is drying tank of high frequency vacuum with the diameter of 650 mm and length of 1350 mm; 2 and 4 are upper and lower plates respectively; and 3 is test material. The high frequency generator oscillates at the frequency of 27.12 MHz and outputs the effective power of 1 kW, which is powered by the center of electrode plate length.

During the drying process, the wood control temperature was set to 55 °C, the ambient pressure was set to 8 kPa, and the control of high frequency output time was set to stop for 2 min after a continuous oscillation for 7 min. In the early stage of drying, wood was quickly taken out and weighed after every 4 h, the real-time MC of wood was calculated, and the pressure and temperature values of five measuring points before the sample was taken out were recorded. In the middle stage of drying, the data were recorded once every 8 h; while, in the later stage of drying, the data were recorded once every 12 h.

The drying was carried out for 204 h until wood MC was dried to 11.56%. The experiment was stopped and a total of 135 data were recorded.

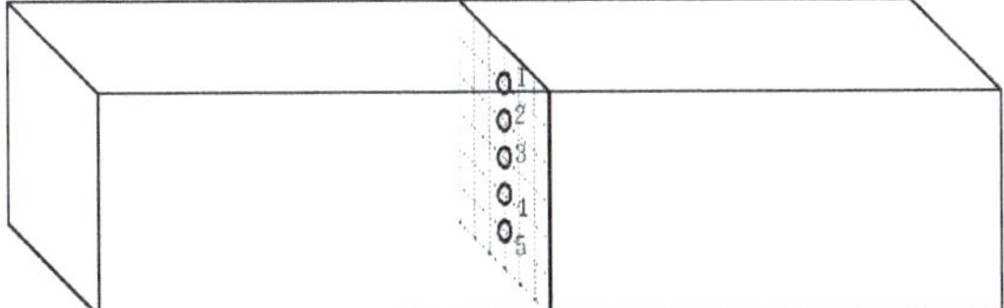

**Figure 1.** Diagram of the wood tested sample and location of the sensors. (1 is the upper layer measuring point; 2 is the upper middle layer measuring point; 3 is the core layer measuring point; 4 is the lower middle layer measuring point; and 5 is the lower layer measuring point).

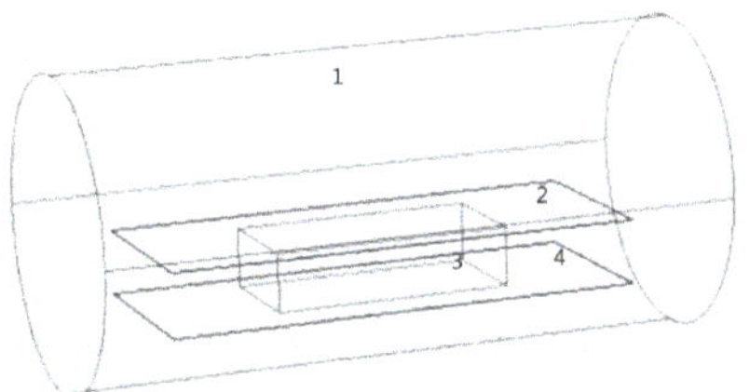

**Figure 2.** Drying tank of high frequency vacuum.

## 2.2. BP Neural Network Model

The BP (Back Propagation) model is currently the most studied and widely used artificial neural network model [20]. It has a powerful nonlinear mapping ability and the qualities of human intelligence such as self-learning, adaptive, associative memory, and parallel information processing. It can imitate the human brain nervous system to store, retrieve, and process the information with an excellent fault tolerance, and is extremely suitable for modeling and control of complex systems [21]. The Python language has a rich and powerful class library. It is an interpreted, interactive, and pure object-oriented scripting programming language that combines the best design principles and ideas of several different languages and is widely used in various fields of software development and application programming. Therefore, this paper built the BP neural network model using Python language programming.

### 2.2.1. Determination of Neuron Number

The neural network prediction model in this paper uses a three-layer feedforward network structure, which includes an input layer, a hidden layer, and an output layer [22]. The hidden layer can be further divided into a single hidden layer and multiple hidden layer according to the layer number. The multiple hidden layer is composed of multiple single hidden layers. Compared with a single hidden layer, a multiple hidden layer has a stronger generalization ability and higher prediction accuracy, but the training time is longer. The selection of hidden layers should be considered comprehensively based on the network accuracy and training time. For a simple mapping relationship, in case the network accuracy meets the requirements, the single hidden layer can be selected to speed up the process. For a complex mapping relationship, the multiple hidden layer can be selected to enhance the network prediction accuracy. Therefore, according to the research requirements, the study chose a single hidden layer.

The number of hidden neurons also has a certain impact on the network [23]. The neuron number in the hidden layer is directly related to the predictive power of network model. If the number is too high, it will not only increase the network training time but also the network will not converge to the target error, resulting in an over-fitting. If the number is too small, the model training will be insufficient and would not be able to completely express the relationship between the input variables and output parameters, thus affecting the predictive ability of the model. Therefore, the determination of neuron number in hidden layer is particularly critical [24].

The optimal neurons number was determined via trial and error method [5]. The neuron number in hidden layer was set to 4~10, and the learning error and epoch of different nodes were tested by network training. The optimal node was obtained by comparison analysis.

### 2.2.2. Data Normalization

The data obtained during the experiment were randomly divided into two data sets: a training group and a test group. The 101 test data of the training group accounted for 75% of the total data, while 34 data of the test group accounted for 25% of the total data.

Each input sample usually has different physical meanings and dimensions; hence, in order to make each input sample have an equally important position and also to prevent the adjustment of the weight into the flat area of error, the input sample needs to be normalized [5]. In addition, as the neurons of the BP neural network adopt the Sigmoid transfer function and the output is between [0, 1], it is also necessary to normalize the output samples (Equation (1)).

$$X' = \frac{X - X_{min}}{X_{max} - X_{min}} \tag{1}$$

where $X'$ is the $X$ normalization value; $X_{max}$ and $X_{min}$ are the maximum and minimum values of $X$, respectively.

The neurons in each layer are only connected to the neurons in the adjacent layer and there is no connection between the neurons in each layer. Also, there is no feedback connection between the neurons in each layer. The input signal first propagates forward to the hidden node and then through the transformation function. The output information of the hidden node is propagated to the output node and the output result is given after processing. In general, the Sigmoid transfer function (Equation (2)) is used on all nodes of hidden layer. In the output layer, all nodes use the linear transfer function Pureline.

$$f = \frac{1}{1 + e^{-x}} \tag{2}$$

where $f$ represents the neuron output value and x represents the neuron input value.

### 2.2.3. Model Performance Analysis

In the model correlation test, the model was evaluated by using the determination coefficient $R^2$ and Mse (Mean squared error) of the training sample [25].

The determination coefficient $R^2$ is defined as:

$$R^2 = \frac{\sum_{i=1}^{n} (t_i - p_i)^2}{\sum_{i=1}^{n} t_i^2 \sum_{i=1}^{n} p_i^2}. \tag{3}$$

The Mean square error (Mse) is calculated as [12]:

$$Mse = \frac{1}{n} \sum_{n=1}^{n} (t_i - p_i)^2 \tag{4}$$

where $t_i$ ($i$ = 1, 2, . . . , $n$) is the predicted value of the ith sample, $p_i$ ($i$ = 1, 2, . . . , $n$) is the true value of the ith sample, and n is the total number of all samples. The decision coefficient is in [0, 1], and the closer the value to 1, the better the model performance, and the closer to 0, the worse the model performance. The smaller the sample Mean square error, the better the prediction performance and the better the model performance. The learning efficiency is set to 0.01.

## 3. Results and Discussion

### 3.1. Determination of Neuron Number

The corresponding relationship between the neuron number of hidden layer and the training error and epoch of neural network is shown in Figure 3. When the node number of hidden layer is 6, the training error is the smallest at 0.07355, and the epoch is 17, the network training is faster. These results show that the neural network model has superb generalization ability at this time [26]; hence, the node number of hidden layer is determined to be 6. According to the node number of hidden layer, the structure of neural network is shown in Figure 4.

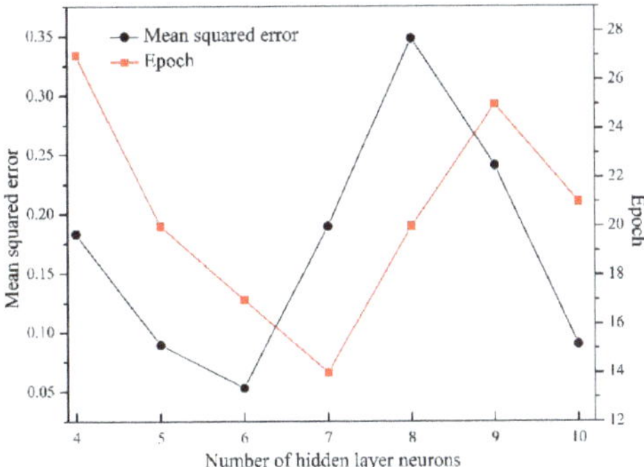


**Figure 3.** Correspondence between the network error and the number of hidden layer neurons.

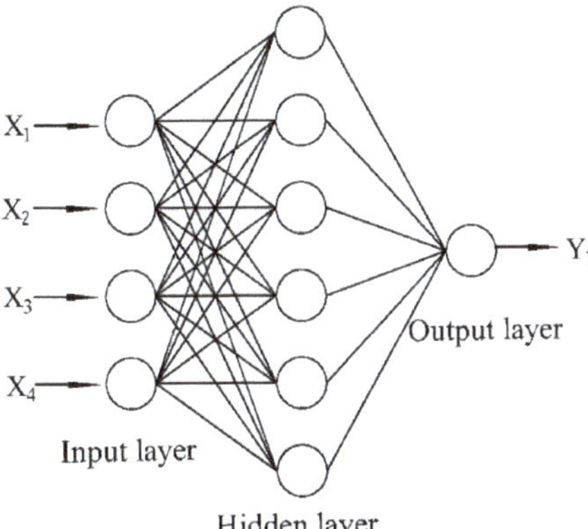


**Figure 4.** BP (Back Propagation) neural network structure diagram ($X_1$: drying time; $X_2$: measuring point position; $X_3$: temperature; $X_4$: pressure; $Y_1$: MC (moisture content)).

### 3.2. Model Performance Analysis

The training regression map for the BP neural network is shown in Figure 5. The linear regression equation between experimental and the predicted value is y = 0.948x + 1.24 while the determination coefficient $R^2$ is 0.974. These results indicate that the experimental and predicted values fit well. The BP neural network model has a good performance and can explain 97% of the above experimental values [24].

The predicted fitted curve for the neural network is shown in Figure 6. The remaining 25% of the samples are predicted and compared with the experimental values. The predicted values are consistent with the variation and size of the experimental values. Initially, the BP neural network model can simulate and predict the change of wood MC during high frequency drying.

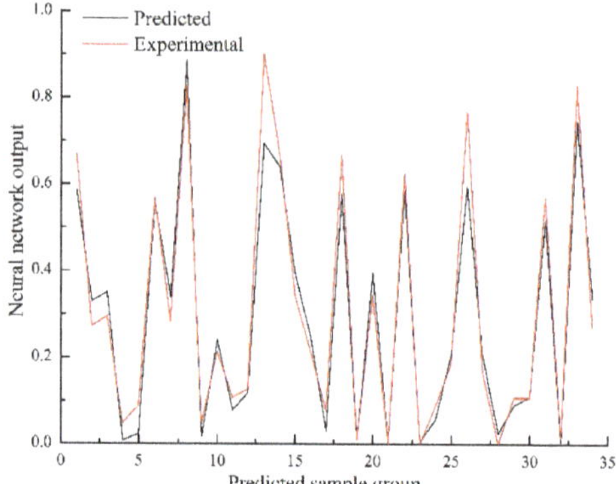


**Figure 5.** Training regression graph of BP neural network.

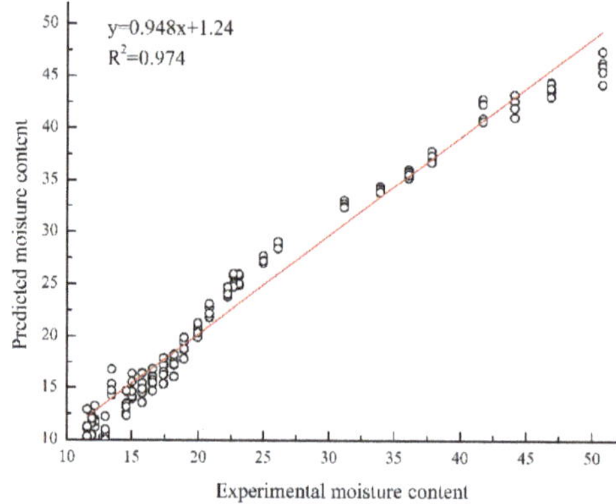


**Figure 6.** Prediction fitting curve of BP neural network.

*3.3. Prediction of Moisture Content Change*

During the drying process of wood, the free water is primarily discharged along the large capillary system above the fiber saturation point. The bound water in cell wall is mainly discharged along the microcapillary system below the fiber saturation point. The bound water is affected by the hydroxyl interaction force in the amorphous region of cell wall [27]. In the early stage of drying, there is a short accelerated drying section. The energy of high frequency radiation is basically used to raise the temperature of wood, and the drying rate is gradually increased from zero. The middle stage of drying is constant-speed drying section. The energy of high frequency radiation is basically used to evaporate the moisture in wood. The MC decreases rapidly and exhibits constant-speed drying tendency. This stage basically completes the evaporation process of moisture in wood. In the later stage of drying, there is less water in wood, and the evaporation rate of moisture and the drying rate of wood gradually decrease [28].

The experimental data is input into the trained model for simulation verification. Figure 7 presents the curve of predicted and experimental values with time. In the early stage of drying, the predicted values are slightly lower than the experimental values; in the middle stage of drying, the predicted values are slightly higher than the experimental values; and in the later stage of drying, the predicted values have a slight wave motion, but overall the value is basically the same as the experimental values.

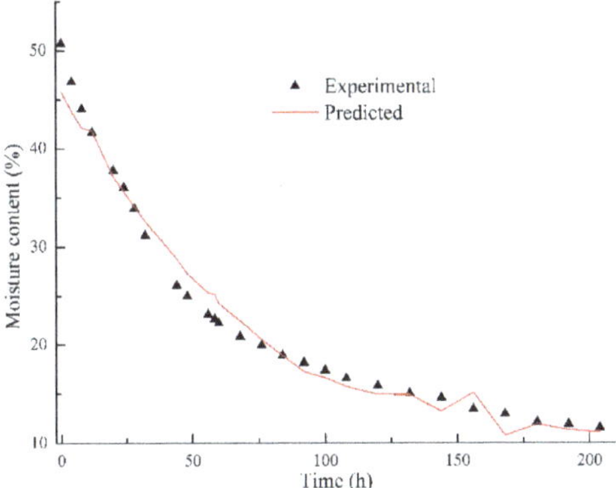



**Figure 7.** Simulation results of BP neural network.

Figure 8 displays the predicted error curve for the neural network (error = experimental value − predicted value [29]). The overall error range is −4%~6% and most of the data is concentrated between −2%~2%, which can basically meet the requirements of prediction accuracy in wood drying.

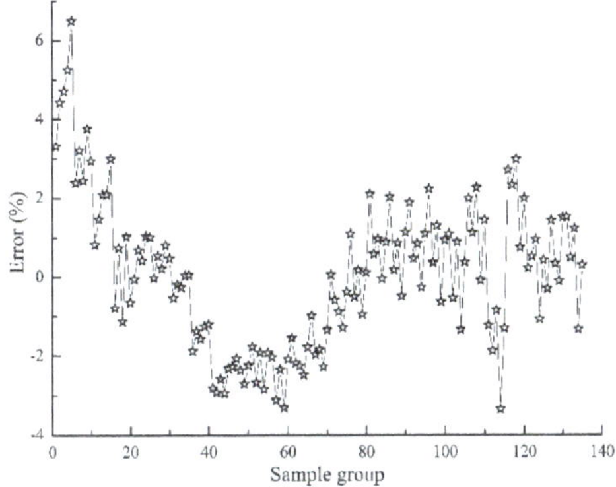



**Figure 8.** Prediction error of BP neural network.

Overall, the predicted data could basically reflect the change trend of MC during the high frequency drying process. The prediction error is about 2%, which proves the feasibility of BP neural network model in MC prediction. Moreover, if the external environmental parameters in the high frequency drying process and the relevant parameters of wood itself are known, the trained neural network model can be used to predict the MC change, thereby eliminating the complicated experimental detection process and saving time and cost [30].

*3.4. Analysis of Stratified Moisture Content Prediction Error*

Figure 9 shows the MC prediction error of measurement points along thickness direction. In the early and later stage of drying, the error is positive and the predicted values are slightly less than the experimental values. In the middle stage of drying, the error is negative and the predicted values are slightly larger than the experimental values. Among these, the error in the middle stage of drying is the largest, followed by the early and later stage of drying. In the early and middle stage of drying, along the thickness direction of test material, from top to bottom, the error increases firstly, then decreases, then increases, and then decreases. The results show M-type trend with no clear law, and the error of the upper surface measurement point is the smallest. In the later stage of drying, along the thickness direction of the test material, from top to bottom, the error is firstly reduced, then increased, and then decreased, while the error at the upper intermediate layer is the smallest.

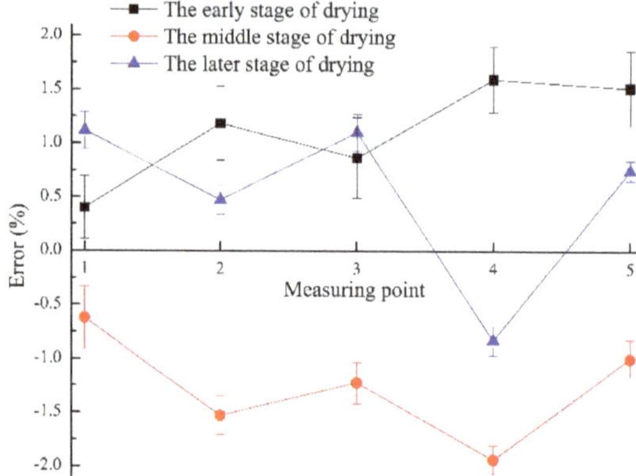


**Figure 9.** Error analysis of stratified moisture content prediction.

Due to the difference in material properties at different locations of the wood and the degree of electromagnetic radiation, the prediction accuracy of each measurement point is different. But overall, the prediction error of MC of each layer is less than 2%, indicating that the prediction accuracy of each measurement point is good and can meet the demand for stratified moisture content prediction.

## 4. Conclusions

The BP neural network was used to simulate the wood MC during the high frequency drying process. The drying time, the location of measuring point, and the internal temperature and pressure of the wood were taken as input variables, while the wood MC was the output variable, 101 test data of the training group accounted for 75% of the total data, while 34 data of the test group accounted for 25% of the total data. The results showed that when the number of hidden layer of neurons was six, the neural network training error was the smallest and the BP neural networks had better stability. The error between the predicted and the experimental values was about 2% and the stratified moisture content prediction error was within 2%, which the model could well simulate the change trend of wood MC during the drying process. In general, although the performance of wood varies greatly and the complex relationship has not been completely elucidated, the proposed neural network model is reliable and has a good predictive power.

**Author Contributions:** H.C. and X.C. conceived and designed the experiments; J.Z. performed the experiments; H.C. and Y.C. analyzed the data and wrote the paper.

**Funding:** The Fundamental Research Funds for the Central Universities (Grant No. 2572018BB08) and the National Natural Science Foundation of China (Grant No. 31670562), financially supported this research.

**Acknowledgments:** The authors thank Zhiqiang Huang for the technical support in the computer field.

**Conflicts of Interest:** The authors declare no conflict of interest.

## References

1. Xie, J. Study on Larch Wood Drying Models by Artificial Neural Network. Master's Thesis, Northeast Forestry University, Harbin, China, 2013.
2. Chai, H.J. Development and Validation of Simulation Model for Temperature Field during High Frequency Heating of Wood. *Forests* **2018**, *9*, 327. [CrossRef]
3. Kong, F.X. Study on Technique of Radio-Frequency Vacuum Drying for Oak Veneer. *J. Northeast. For. Univ.* **2018**, *6*, 46.

4.  Li, X.J. *Characteristics of Microwave Vacuum Drying of Wood and Mechanism of Thermal and Mass Transfer*; China Environmental Science Press: Beijing, China, 2009; p. 65. ISBN 9787802099890.

5.  Yu, H.M. Study on Drying Mathematical Model of Hawthorn Using Microwave Coupled with Hot Air and Drying Machine Design. Ph.D. Thesis, Jilin University, Changchun, China, 2015.

6.  Zhou, P. *MATLAB Neural Network Design and Application*; Tsinghua University Press: Beijing, China, 2013; p. 153. ISBN 9787302313632.

7.  Poonnoy, P.; Tansakul, A. Artificial Neural Network Modeling for Temperature and Moisture Content Prediction in Tomato Slices Undergoing Microwave-Vacuum Drying. *J. Food Sci.* **2007**, *72*, E042–E047. [CrossRef] [PubMed]

8.  Sablani, S.S.; Kacimov, A. Non-iterative estimation of heat transfer coefficients using artificial neural network models. *Int. J. Heat Mass Transf.* **2005**, *48*, 665–679. [CrossRef]

9.  Rai, P.; Majumdar, G.C. Prediction of the viscosity of clarified fruit juice using artificial neural network: A combined effect of concentration and temperature. *J. Food Eng.* **2005**, *68*, 527–533. [CrossRef]

10. Shyam, S.; Sablani, M. Using neural networks to predict thermal conductivity of food as a function of moisture content, temperature and apparent porosity. *Food Res. Int.* **2003**, *36*, 617–623.

11. Hagan, M.T. *Neural Network Design*; PWS Publishing Company: Boston, MA, USA, 1996.

12. Fu, Z. Artificial neural network modeling for predicting elastic strain of white birch disks during drying. *Eur. J. Wood Wood Prod.* **2017**, *75*, 949–955. [CrossRef]

13. Wu, H.W.; Stavros, A. Prediction of Timber Kiln Drying Rates by Neural Networks. *Dry Technol.* **2006**, *24*, 1541–1545. [CrossRef]

14. Zhang, D.Y. Neural network prediction model of wood moisture content for drying process. *Sci. Silva Sin.* **2008**, *44*, 94–98.

15. İlhan, C. Determination of Drying Characteristics of Timber by Using Artificial Neural Networks and Mathematical Models. *Dry Technol.* **2008**, *26*, 1469–1476.

16. Watanabe, K. Artificial neural network modeling for predicting final moisture content of individual Sugi (*Cryptomeria japonica*) samples during air-drying. *J. Wood Sci.* **2013**, *59*, 112–118. [CrossRef]

17. Watanabe, K. Application of Near-Infrared Spectroscopy for Evaluation of Drying Stress on Lumber Surface: A Comparison of Artificial Neural Networks and Partial Least Squares Regression. *Dry Technol.* **2014**, *32*, 590–596. [CrossRef]

18. Ozsahin, S. Prediction of equilibrium moisture content and specific gravity of heat treated wood by artificial neural networks. *Eur. J. Wood Wood Prod.* **2018**, *76*, 563–572. [CrossRef]

19. Aghbashlo, M. Application of Artificial Neural Networks (ANNs) in Drying Technology: A Comprehensive Review. *Dry Technol.* **2015**, *33*, 1397–1462. [CrossRef]

20. Zhang, Z.Y. Research on Load Forecasting Analysis of Wuhai Power Grid Based on Artificial Neural Network BP Algorithm. Master Thesis, University of Electronic Science and Technology of China, Sichuan, China, 2008.

21. Bi, W. Research and Design of a Wood Drying Control System Based on Artificial Neural Network. Master's Thesis, Jilin University, Changchun, China, 2011.

22. MATLAB Chinese Forum. *30 Case Analysis of MATLAB Neural Network*; Beijing University of Aeronautics and Astronautics Press: Beijing, China, 2010.

23. Liu, T.Y. Improvement Research and Application of BP Neural Network. Master's Thesis, Northeast Agricultural University, Harbin, China, 2011.

24. Fu, Z.Y. Study on Dry Stress and Strain of Birch Tree in Conventional Drying Process. Ph.D. Thesis, Northeast Forestry University, Harbin, China, 2017.

25. Zhang, T. Prediction Model of Moisture Content of Ginger Hot Air Drying Based on BP Neural Network and SVM. Master's Thesis, Inner Mongolia Agricultural University, Hohhot, China, 2014.

26. Xu, X. Nonlinear fitting calculation of wood thermal conductivity using neural networks. *Zhejiang Univ. (Eng. Sci.)* **2007**, *41*, 1201–1204.

27. Gao, J.M. *Wood Drying*, 1st ed.; Science Press: Beijing, China, 2008; p. 10. ISBN 9787030205179.

28. Allegretti, O. Nonsymmetrical drying tests—Experimental and numerical results for free and constrained spruce samples. *Dry Technol.* **2018**, *36*, 1554–1562. [CrossRef]

29. Cai, Y.C. *Wood High Frequency Vacuum Drying Mechanism*; Northeast Forestry University Press: Harbin, China, 2007.
30. Wang, Y. Prediction of Wood Moisture Content Based on BP Neural Network Optimized by SAGA. *Tech. Autom. Appl.* **2013**, *32*, 4–6.

*Article*

# Multi-Scale Evaluation of the Effect of Phenol Formaldehyde Resin Impregnation on the Dimensional Stability and Mechanical Properties of *Pinus Massoniana* Lamb.

Xinzhou Wang [1], Xuanzong Chen [1], Xuqin Xie [2], Shaoxiang Cai [1], Zhurun Yuan [1] and Yanjun Li [1,*]

[1]    College of Materials Science and Engineering, Nanjing Forestry University, Nanjing 210037, China
[2]    Dehua Tubao New Decoration Material Co., Ltd., Zhejiang 313200, China
*    Correspondence: lalyj@njfu.edu.cn; Tel.: +86-025-85428507

Received: 1 July 2019; Accepted: 27 July 2019; Published: 31 July 2019

**Abstract:** The local chemistry and mechanics of the control and phenol formaldehyde (PF) resin modified wood cell walls were analyzed to illustrate the modification mechanism of wood. Masson pine (*Pinus massoniana* Lamb.) is most widely distributed in the subtropical regions of China. However, the dimensional instability and low strength of the wood limits its use. Thus, the wood was modified by PF resin at concentrations of 15%, 20%, 25%, and 30%, respectively. The density, surface morphology, chemical structure, cell wall mechanics, shrinking and swelling properties, and macro-mechanical properties of Masson pine wood were analyzed to evaluate the modification effectiveness. The morphology and Raman spectra changes indicated that PF resin not only filled in the cell lumens, but also penetrated into cell walls and interacted with cell wall polymers. The filling and diffusing of resin in wood resulted in improved dimensional stability, such as lower swelling and shrinking coefficients, an increase in the elastic modulus ($E_r$) and hardness ($H$) of wood cell walls, the hardness of the transverse section and compressive strength of the wood. Both the dimensional stability and mechanical properties improved as the PF concentration increased to 20%; that is, a PF concentration of 20% may be preferred to modify Masson pine wood.

**Keywords:** *Pinus massoniana* Lamb.; phenol formaldehyde resin; wood impregnation; wood properties; cell-wall mechanics

---

## 1. Introduction

*Pinus massoniana* Lamb., commonly known as Masson pine, is one of the most widely distributed tree species in the subtropical regions of China [1]. The wood from Masson pine planted forests has become an important industrial raw material for wide commercial use such as wood construction, wood-based panels, and polymer composites due to the beautiful wood grain, good adaptability to the environment, and shorter growth cycle [2,3]. However, as one of the common fast-growing tree species, Masson pine wood also presents drawbacks which limit its practical application, such as low dimensional stability, softness, and low bio-durability [4,5]. In the past few decades, a number of modification methods have been proposed to enhance the quality and high value-added utilization of plantation wood, including thermal or densification treatment, surface coating, and chemical impregnation [6–10]. Above all, chemical impregnation under vacuum or pressure has been proven to be an effective method to improve the properties of the wood. Among the existing chemical modification methods, impregnating and curing the wood with resins appears to especially promote the industrial utilization of wood [11,12].

Phenol formaldehyde (PF) resin is a popular thermosetting agent that forms a three dimensional structure via cross-linking reactions after curing, which are extensively used in exterior grade wood-based panels for its excellent performance, including water resistance and chemical stability [13–15]. It has therefore been widely used to improve dimensional stability and strength and prolong the service life of wood for indoor and outdoor use [16–18]. To date, changes in physical, mechanical, and chemical properties of chemically modified wood have been intensively analyzed at the macro scale [19,20]. The modified wood achieved high dimensional and stiffness stability and biological resistance. Monomers such as methyl methacrylate, styrene-methyl methacrylate and styrene-glycidyl methacrylate have been utilized for wood modification and have proven that the monomers not only filled in the cell lumens, but also penetrated into the cell walls [21,22]. However, for a pre-polymer like PF resin with a higher molecular weight, it is still unclear whether it can penetrate into cell wall or not, or what the accompanied influence on the cell wall would be. In particular, only limited attempts have been made to find out the correlations between chemical, physical, and mechanical performance between the cell wall- and macro-level to indicate the contribution of cell wall modification. Nanoindentation (NI) has been successfully applied for measuring the mechanics of wood cell walls including modulus of elasticity, hardness, etc. [23–25] and the Raman spectra technique can detect the local chemical structure of wood cell walls [26], which facilitates in situ characterization of the effects of PF resin impregnation on wood cell walls.

The purpose of this study was to characterize the properties of PF resin modified Masson pine wood for enriching the fundamental theory of wood chemical modification, which may benefit the modification process. For this purpose, the changes in morphology, local chemical structure and mechanics at the cell wall level and the density, dimensional stability, and mechanical properties at the macro-scale of Masson pine wood after PF resin impregnation were analyzed using scanning electron microscopy (SEM), Raman, NI, and conventional physical and mechanical test instruments, respectively.

## 2. Materials and Methods

### 2.1. Materials

Wood samples were obtained from 40-year-old Masson pine (*Pinus massoniana* Lamb.) wood harvested from plantation forestry located in Fujian Province, China. Wood blocks with the dimensions 20 mm$^3$ × 20 mm$^3$ × 160 mm$^3$ and 50 mm$^3$ × 50 mm$^3$ × 70 mm$^3$ (longitudinal × tangential × radial) were cut from the sapwood around the 21st growth ring. The initial moisture content was about 11%. A commercial phenol formaldehyde (PF) resin (Dynea Co., Ltd., Guangdong, China) with a solid content of 48% and a viscosity of 150 mPa·s at 25 °C was used in this experiment.

### 2.2. Impregnation Treatment

The wood samples were oven-dried at 103 ± 2 °C until a constant weight was achieved, and the weight was determined before impregnation. The PF resin was diluted with distilled water into resin concentrations of 15%, 20%, 25%, and 30% for separate treatments. Twenty replicate samples in each treatment group were conducted in a stainless-steel chamber under a vacuum of 0.09 MPa for 30 min and then at 0.8 MPa for 2 h. After impregnation, the samples were air-dried at room temperature for 48 h after removing the excess resin on the surface, and then were cured at a temperature of 130 °C for 2 h in an oven.

### 2.3. Determination of Weight Percent Gain and Density

The oven-dried weight and volume of the 20 replicate samples with the dimensions 20 mm$^3$ × 20 mm$^3$ × 20 mm$^3$ (L × T × R) cut from the PF resin modified wood were determined to calculate the oven-dried density and weight percent gain (WPG) of the samples. The WPG was calculated according to Equation (1):

$$WPG\ (\%) = \frac{W_1 - W_0}{W_0} \times 100\% \tag{1}$$

where $W_1$ is the oven-dried weight of the modified wood and $W_0$ is the oven-dried weight of the control wood.

*2.4. Morphology Observation*

Both the cross-section and tangential section of the surfaces of the control and modified wood samples were observed using scanning electron microscopy (SEM, JSM-7600F, JEOL Japan Electronics Co., Ltd., Japan) at an accelerating voltage of 20 kV.

*2.5. Raman Measurement*

The local chemical distribution analysis in the transversal section of the control and modified wood samples was analyzed by a laser Raman spectrometer (DXR532, Thermo Fisher Scientific Inc., USA) equipped with a linear-polarized 780 nm laser. The cross-sections of the wood samples were sliced by an ultra-microtome (Leica MZ6, Germany) with a thickness of 20 μm and then placed on glass slides covered with glass coverslips. All spectra were collected in the range of 1800 cm$^{-1}$ to 600 cm$^{-1}$.

*2.6. Dimensional Stability Analysis*

The dimensions (L × T × R) of the control and modified wood samples were measured under three different moisture contents (> fiber satruated point, air-dried, and oven-dried) to analyze the dimensional stability according to the testing procedure of Chinese National Standards (GB/T 1932-2009 and GB/T 1934.2-2009). For determination of shrinkage, samples were soaked in distilled water at 20 °C until the dimension was constant, and then the wet samples were conditioned at 20 °C and relative humidity (RH) of 65%, and finally the air-dried samples were dried in an oven at 103 °C until a constant weight was achieved. In contrast, the determination of swelling was processed in reverse order. The swelling coefficient ($\alpha$), shrinkage coefficient ($\beta$), anti-swelling efficiency (ASE), and anti-shrinking efficiency (ASE′) can be calculated by Equations (2)–(13):

$$\alpha_w \ (\%) = \frac{l_w - l_0}{l_0} \times 100\% \tag{2}$$

$$\alpha_{max} \ (\%) = \frac{l_{max} - l_0}{l_0} \times 100\% \tag{3}$$

$$\alpha_{Vw} \ (\%) = \frac{V_w - V_0}{V_0} \times 100\% \tag{4}$$

$$\alpha_{Vmax} \ (\%) = \frac{V_{max} - V_0}{V_0} \times 100\% \tag{5}$$

$$ASE_w \ (\%) = \frac{\alpha_{Vw(c)} - \alpha_{Vw(m)}}{\alpha_{Vw(c)}} \times 100\% \tag{6}$$

$$ASE_{max} \ (\%) = \frac{\alpha_{Vmax(c)} - \alpha_{Vmax(m)}}{\alpha_{Vmax(c)}} \times 100\% \tag{7}$$

where $\alpha_w$ and $\alpha_{max}$ are the linear swelling coefficient from oven-dried to air-dried and from oven-dried to wet, respectively; $l_{max}$, $l_w$, and $l_0$ are the length in the tangential and radial directions of wet, air-dried, and oven-dried samples, respectively; similarly, $\alpha_{Vw}$ and $\alpha_{Vmax}$ are the volumetric swelling coefficients; $V_{max}$, $V_w$, and $V_0$ are the volume of samples at different moisture conditions; $\alpha_{Vw(c)}$ and $\alpha_{Vmax(m)}$ are the volumetric swelling coefficient of the control and modified wood samples, respectively.

$$\beta_{max} \ (\%) = \frac{l_{max} - l_0}{l_{max}} \times 100\% \tag{8}$$

$$\beta_w \ (\%) = \frac{l_{max} - l_w}{l_{max}} \times 100\% \tag{9}$$

$$\beta_{Vmax}\,(\%) = \frac{V_{\max} - V_0}{V_{\max}} \times 100\%  \tag{10}$$

$$\beta_{Vw}\,(\%) = \frac{V_{\max} - V_{\mathrm{w}}}{V_{\max}} \times 100\%  \tag{11}$$

$$ASE'_{max}\,(\%) = \frac{\beta_{Vmax(c)} - \beta_{Vmax(m)}}{\beta_{Vmax(c)}} \times 100\%  \tag{12}$$

$$ASE'_{w}\,(\%) = \frac{\beta_{Vw(c)} - \beta_{Vw(m)}}{\beta_{Vw(c)}} \times 100\%  \tag{13}$$

where $\beta_{\max}$ and $\beta_{\mathrm{w}}$ are the linear shrinking coefficient from oven-dried to air-dried and from oven-dried to wet, respectively; $l_{\max}$, $l_{\mathrm{w}}$, and $l_0$ are the length in the tangential and radial directions of oven-dried, air-dried, and wet samples, respectively; similarly, $\beta_{Vmax}$ and $\beta_{Vw}$ are the volumetric shrinking coefficients; $V_{\max}$, $V_{\mathrm{w}}$, and $V_0$ are the volume of samples at different moisture conditions; $\beta_{Vmax(m)}$ and $\beta_{Vw(c)}$ are the volumetric shrinking coefficients of the control and modified wood samples, respectively.

### 2.7. Mechanical Property Testing

Wood samples with the dimensions 5 mm$^3$ × 5 mm$^3$ × 10 mm$^3$ (T × R × L) were obtained for nanoindentation (NI) to evaluate the effect of modification on the wood cell wall. The transverse section of the samples was polished by an ultra-microtome with a diamond knife (Micro Star Tech Inc., Huntsville, AL, USA). As shown in Figure 1, the cell wall mechanics of both the control and modified wood, which had been equilibrated at 20 °C and 65% RH for 48 h, were measured by using a Hysitron TriboIndenter system (Hysitron Inc., USA) equipped with scanning probe microscopy (SPM). Testing was operated with the load function: loading, holding at the peak load of 400 µN, and unloading for 5 s, respectively. About 30 valid indents were obtained to calculate the reduced elastic modulus ($E_r$) and hardness ($H$) based on Equations (14) and (15) introduced by Oliver and Pharr [27]:

$$H = \frac{P_{\max}}{A}  \tag{14}$$

where $P_{\max}$ is the peak load, and $A$ is the projected contact area of the tips at peak load.

$$E_r = \frac{\sqrt{\pi}}{2\beta}\frac{S}{\sqrt{A}}  \tag{15}$$

where $E_r$ is the combined elastic modulus; $S$ is initial unloading stiffness; and $\beta$ is a correction factor correlated to indenter geometry ($\beta = 1.034$).

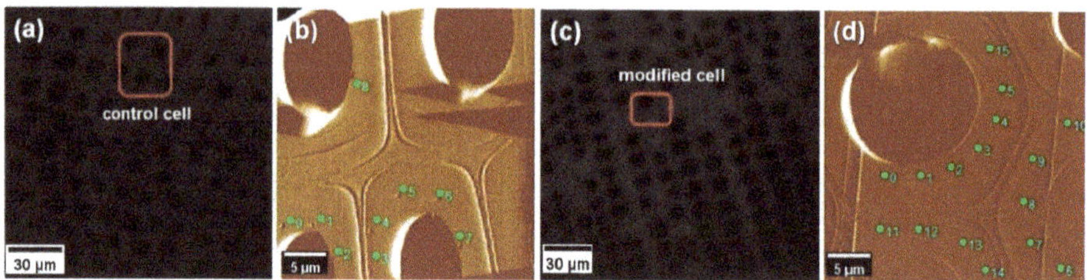


**Figure 1.** Microscope images showing the positioning of indents: (**a**) and (**c**) optical micrograph of the transverse section of wood samples; (**b**) and (**d**) scanning probe microscopy (SPM) images of wood cell walls after nanoindentation (NI).

The Janka hardness of a transverse section and the compressive strength parallel to the grain of the wood was determined in accordance with GB/T 1941-2009 and GB/T 1935-2009 standards, respectively. The dimensions of the samples were 30 mm$^3$ × 20 mm$^3$ × 20 mm$^3$ (L × T × R). A total of 20 replicate samples for each treatment were conditioned until they reached a moisture content of approximately 12% before testing.

## 3. Results and Discussion

### 3.1. Weight Percent Gain and Density

Figure 2 shows the weight percent gain (WPG) and density of the control and modified wood samples. It can be observed that the WPG and density increase with increasing PF concentration. Density was positively correlated with WPG, which gradually increased by 34.7% and 39.6% as compared to the control when the PF concentration was 30%, respectively. The increased density of the samples was mainly attributed to the filling of the cell lumens with PF resin. Meanwhile, the lower increased rate of density than that of WPG may due to the swelling of cell wall filled with PF resin. However, both the WPG and density increased slowly when the PF concentration was above 20%. In a previous study, the viscosity of the solution has been found to affect the penetration in wood [28,29]. That is, higher concentration PF resin could decrease the permeability of the resin in wood.

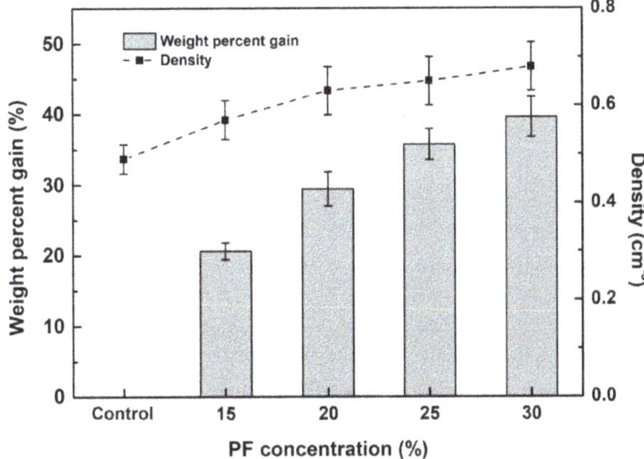


**Figure 2.** Weight percent gain (WPG) and density of the control and modified wood. PF = phenol formaldehyde.

### 3.2. Morphological Analysis

The scanning electron microscopy images of the cross-sectional and tangential-sectional surfaces of the wood samples are shown in Figure 3. The presence of the polymeric structure of PF resin can be easily noted in many cell lumens and pits (Figure 3d,f). During impregnation in Masson pine wood, the chemicals entered the interior of the wood primarily through the wood tracheids and then circulated through the pits in an axial and transverse direction [10,26]. Under the action of exterior pressure, the PF molecules freely diffused into the intercellular spaces of the wood. It also can be observed that there is no obvious boundary between the wood cell wall and the filled PF resin, which may indicate that some PF resin has penetrated into the cell wall and that they interact with each other well. Furthermore, some cracks appear on the cell walls that are not filled with resin; however, the surface of the cell walls filled with resin was smoother and more compact after drying. This finding illustrates that the penetration of resin into the cell wall can enlarge the differences in shrinkage properties of the filled and unfilled cell walls and might be responsible for the cracks between them (Figure 3e).

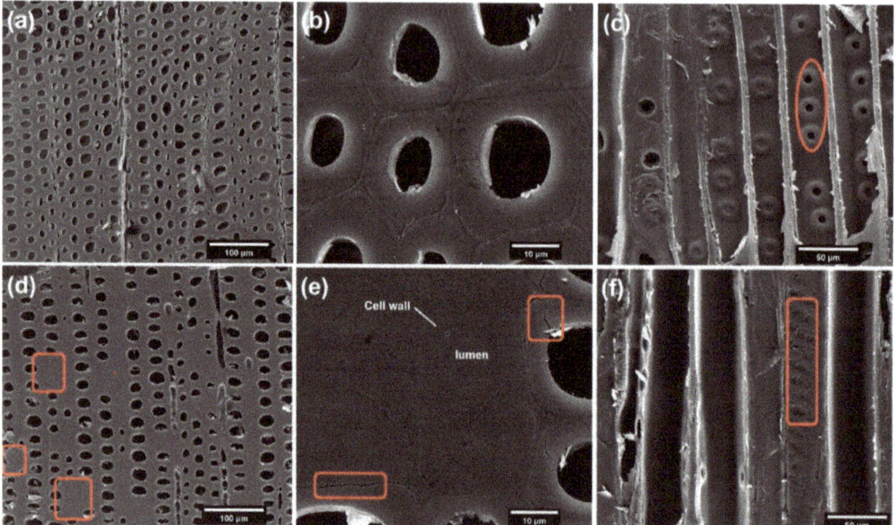

**Figure 3.** Micrographs of the cross-sections of (**a,b**) the control wood, (**d,e**) the wood modified by PF resin, and tangential-section of (**c**) the control wood, (**f**) the modified wood.

### 3.3. Local Chemical Analysis by Raman

Raman was used to detect the local chemical groups to confirm the possible interactions between the wood cell wall materials and the resin. Figure 4 shows the typical spectra for the control wood cell wall, PF resin within the lumen, and cell wall filled with resin. The spectra with black and blue color in Figure 4 indicate that the chemical nature of the native cell wall material is clearly different from the cured PF resin. Wood cell walls are mainly composed of cellulose, hemicelluloses, and lignin. The bands at 1091 cm$^{-1}$ , 1336 cm$^{-1}$ , and 1376 cm$^{-1}$ on the black IR spectra corresponded to the C-O stretching and flexural vibrations in cellulose and hemicellulose, and the absorption at 1595 cm$^{-1}$ and 1656 cm$^{-1}$ arises from the non-conjugated and conjugated C=C stretching vibrations in the aromatic ring of the phenol in lignin [30–32]. For phenol formaldehyde resin, the intensive band at 1607 cm$^{-1}$ originates from the C=C stretching vibration in the aromatic ring of the phenol, while the bands at 1287 cm$^{-1}$ and 778 cm$^{-1}$ are attributable to the biphenyl C-C bridge stretching vibration and C-H flexural vibration in the aromatic ring. It is intriguing that a number of spectral modifications appeared on the spectra with red color for the modified cell wall, although the general aspect of the spectra remained unchanged. The intensity of the C=C stretching and C-O stretching at 1656 cm$^{-1}$ , 1376 cm$^{-1}$, and 1336 cm$^{-1}$ decreased significantly, while some new bands at 1287 cm$^{-1}$ and 778 cm$^{-1}$ appeared as compared to the control cell wall, which can be attributed to the penetration of PF resin into the cell wall [33,34]. Moreover, a relatively broader band between 1150 cm$^{-1}$ and 1100 cm$^{-1}$ appeared in the spectra of modified wood, corresponding to the asymmetric stretching vibration of C-O-C aliphatic ether, which is in agreement with the literature that the chemical reactions of the -OH groups of wood and PF resin occurred at the cell wall level [35,36].

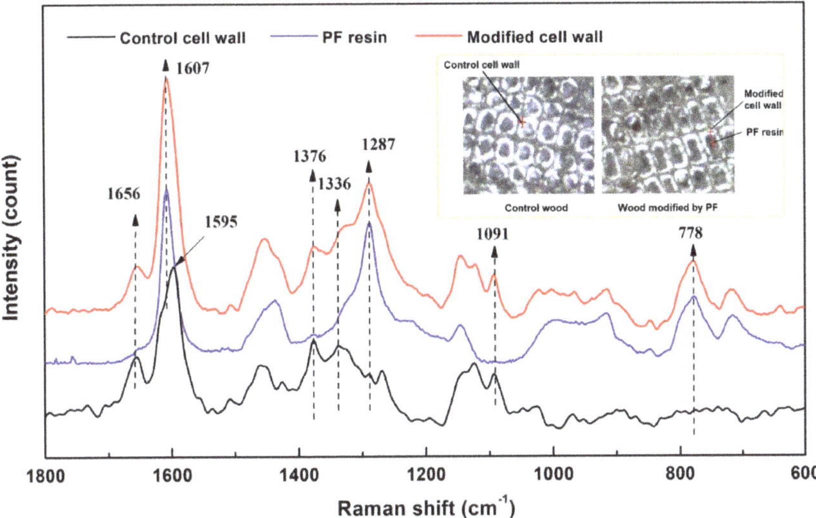

**Figure 4.** Raman spectra of the control and modified wood.

## 3.4. Dimensional Stability

The swelling and shrinking coefficients ($\alpha$ and $\beta$) of the control and chemically modified wood samples under different moisture conditions are presented in Figure 5. The swelling and shrinking coefficients of wood with the changing moisture content (MC) between oven-dried and air-dried conditions ($\alpha_w$ and $\beta_w$) were primarily measured to evaluate the dimensional stability of wood applied on locations with low equilibrium moisture content. Both the $\alpha_w$ and $\beta_w$ in the radial direction were about half of that in the tangential direction, resulting from the anatomical structures such as the limitation of xylem ray, the difference of lignin content in the radial and tangential cell wall, etc. [10]. The $\alpha_w$ and $\beta_w$ decreased significantly at first and then kept stable with an increase in PF resin concentration in comparison with the control, indicating that chemical treatment could effectively improve wood dimensional stability [19]. The radial, tangential, and volumetric $\alpha_w$s of wood modified by PF resin with 20% concentration were about 45%, 58%, and 54% lower than that of the control, respectively. However, the $\beta_w$ became stable when the PF concentration was beyond 15%, indicating that PF resin concentrations ranging from 15% to 20% are better for wood impregnation.

As applied in areas with a larger moisture content range, such as outdoors, the swelling and shrinking coefficients of wood with the changing MC between oven-dried and wet conditions ($\alpha_{max}$ and $\beta_{max}$) need to be analyzed too. It can be observed from Figure 5c and 5d that the effect of PF resin impregnation on the $\alpha_{max}$ and $\beta_{max}$ is similar to that of $\alpha_w$ and $\beta_w$; that is, the dimensional stability of wood can be modified effectively by the PF resin at concentrations below 20%. The radial, tangential, and volumetric $\beta_{max}$s of wood modified by PF resin at 20% concentration decreased by about 49%, 50%, and 51%, respectively. Moreover, both the $\alpha_{max}$ and $\beta_{max}$ were almost twice as high as the $\alpha_w$ and $\beta_w$, respectively, which can be attributed to the different moisture content. The MC of air-dried wood is approximately 12–15%, which is half of the fiber saturation point (FSP). Generally, the hygroscopic water located within the cell wall plays the most important role on the dimension stability rather than the free water occupied in the cell lumen or other macro-voids.

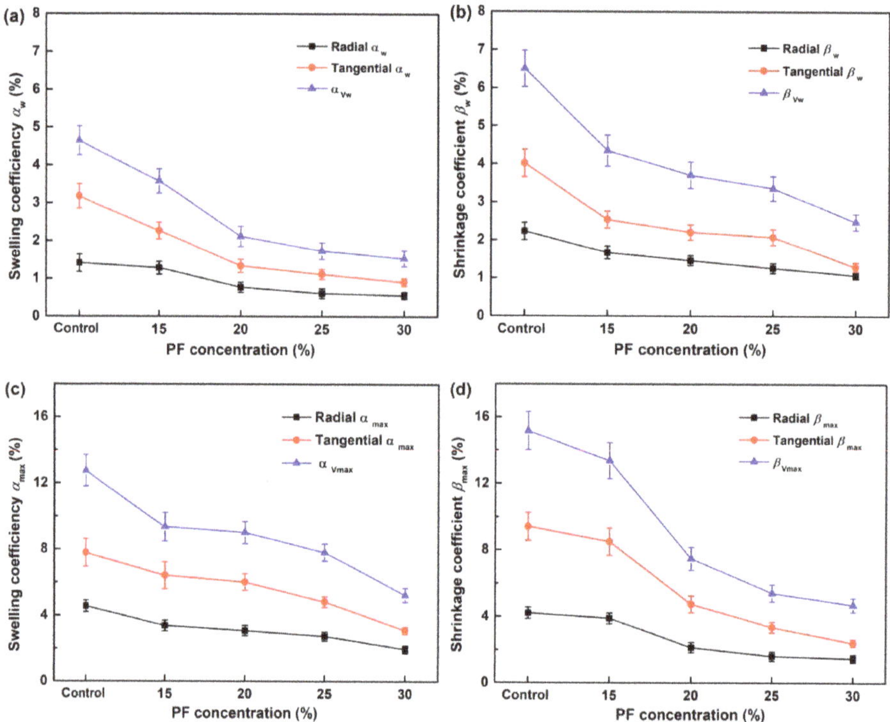

**Figure 5.** The swelling and shrinking coefficients of the control and modified wood: (**a**) and (**b**) swelling and shrinking coefficients from oven-dried to wet; (**c**) and (**d**) swelling and shrinking coefficients from oven-dried to air-dried.

The anti-swelling efficiency (ASE) and anti-shrinking efficiency (ASE′) were positively affected by the PF resin concentration (Figure 6). The $ASE_w$ and $ASE'_w$ initially increased at a concentration of 20% and then kept stable or only increased slightly with a further increase in concentration. The $ASE_w$ and $ASE'_w$ of the wood treated with 20% PF resin reached 54% and 50%, respectively. The deposition of PF in the cell walls reduced the space within the cell walls, which could be occupied by water in untreated wood. In addition, the reduction in swelling and shrinking of the modified wood could be partly attributed to cross-linking of particle cell wall polymers [37,38]. However, the $ASE_{max}$ and $ASE'_{max}$ of the wood kept increasing with the increased concentration and reached the maximum of 59% and 62% as PF resin concentration increased to 30%. As the MC of wood exceeded the FSP, the free voids in the wood provided space for free water, which also affected the swelling and shrinking. Thus, the higher WPG attributed to the higher PF resin concentration could occupy more free space and finally increase the $ASE_{max}$ and $ASE'_{max}$. The $ASE_w$ and $ASE'_w$ of the wood treated with 20% PF resin is close to the maximum $ASE_{max}$ and $ASE'_{max}$, indicating that the lower concentration PF resin is suitable for treating the wood utilized in areas with small MC range.

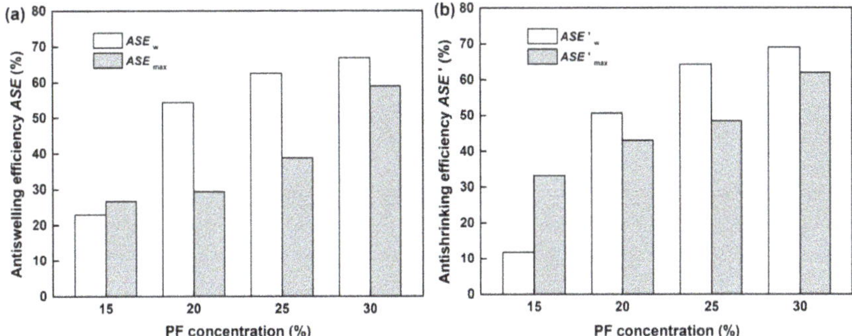

**Figure 6.** The anti-swelling and anti-shrinking efficiency of the modified wood: (**a**) anti-swelling efficiency; (**b**) anti-shrinking efficiency.

### 3.5. Micro-Mechanics of Wood Cell Walls

The longitudinal reduced elastic modulus ($E_r$) and hardness ($H$) of the control and modified wood cell walls are presented in Figure 7, respectively. It is remarkable that the $E_r$ and $H$ values increased after impregnation by PF resin. For instance, the $E_r$ and $H$ values of the wood cell walls modified with PF resin at 15% concentration increased by about 24.9% and 47.3%, which further confirmed the results of the Raman measurements that some PF molecules penetrated into the cell wall successfully. The filling of the voids and the cross-linking of -OH groups of the wood polymer with the resins may reinforce the cell wall [39,40]. However, the cell wall mechanics increased slowly and even decreased accompanying the increase in PF resin concentration. The $E_r$ and $H$ of the wood cell walls modified by 30% PF resin were 8.7% and 11.6% lower than that of the cell wall modified by 20% PF, which can be easily interpreted as a result of the increasing bulking effects attributed to the deposition of resin in the cell walls at higher WPG.

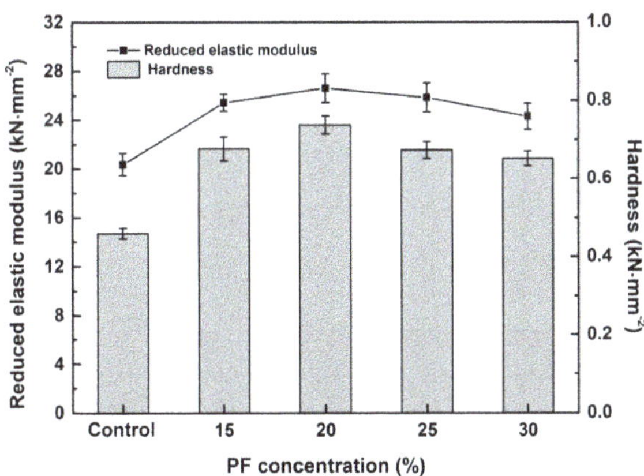

**Figure 7.** The reduced elastic modulus and hardness of the control and modified wood cell walls.

### 3.6. Macro-Mechanics of Wood

Figure 8 shows the hardness of the transverse section and the compressive strength parallel to grain of the control and modified wood at the macro-level. The initial hardness and compressive strengths of the control wood are 39.5 N·mm$^{-2}$ and 49.1 N·mm$^{-2}$, respectively. Both the hardness and compressive strength gradually increased with an increase in PF resin concentration, reaching the maximum of 54.9 N·mm$^{-2}$ and 59.7 N·mm$^{-2}$ with a concentration of 30%. The higher filling degree of the wood voids led to a higher density, which mainly contributed to the improved hardness and compressive strength after curing of the PF resin [41]. In addition, the increased mechanics of wood cell walls induced by the chemical modification also played an important role to achieve the desired modification effect. However, as the PF concentration was above 20%, the hardness and compressive strength kept stable, which was in agreement with the results of dimensional stability and cell wall mechanics. That is, the PF concentration of 20% may be preferred to modify Masson pine wood.

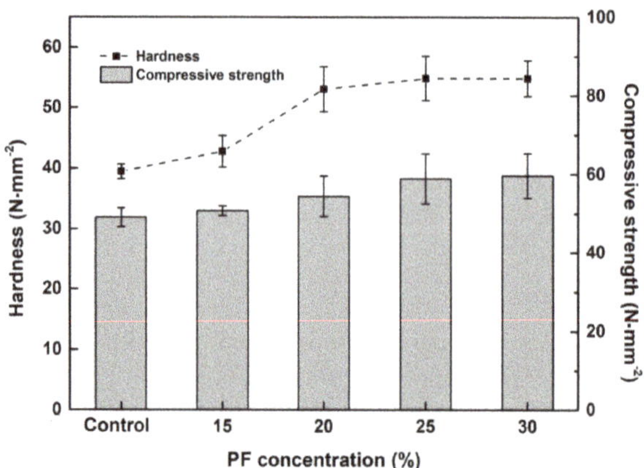


**Figure 8.** The compressive strength and hardness of the control and modified wood.

## 4. Conclusions

The effects of PF resin impregnation on the density, dimensional stability, mechanical strength, and microscopic chemical and mechanical properties of Masson pine wood were determined in this paper. PF resin was impregnated into the wood cell lumen and diffused into the cell walls, as verified by scanning electron microscopy and Raman spectrum. Swelling and shrinking coefficients were significantly reduced while the anti-swelling and anti-shrinking efficiency of wood were improved accompanying the increase in PF resin concentration. The inter-reaction between the resin and cell walls made a positive contribution to the cell wall mechanics of wood cell walls. The elastic modulus ($E_r$) and hardness ($H$) of the wood cell walls modified by 15% PF resin increased by about 24.9% and 47.3% as compared to the control. Both the increased density attributed to the filling of resin in cell lumens and cell walls and the improved cell wall mechanics resulted in the remarkable increase in hardness and the compress strength of wood. However, both the dimensional stability and mechanical properties improved slowly as the PF concentration was above 20%; that is, the PF concentration of 20% may be preferred to modify Masson pine wood.

**Author Contributions:** X.W. conceived and designed the experiments; X.C. and S.C. performed the experiments; X.C., X.X., and Z.Y. analyzed the data; X.W. wrote the paper, with revisions by Y.L.

**Acknowledgments:** The authors would like to gratefully acknowledge the financial support from the Natural Science Foundation of Jiangsu Province (BK20180774), Natural Science Foundation of China (31570552), and Key University Science Research Project of Jiangsu Province (17KJA220004).

**Conflicts of Interest:** The authors declare no conflict of interest.

## References


1.  Gu, H.; Wang, J.; Ma, L.; Shang, Z.; Zhang, Q. Insights into the BRT (Boosted Regression Trees) method in the study of the climate-growth relationship of Masson pine in subtropical China. *Forests* **2019**, *10*, 228. [CrossRef]
2.  He, Q.; Zhan, T.; Ju, Z.; Zhang, H.; Hong, L.; Brosse, N.; Lu, X. Influence of high voltage electrostatic field (HVEF) on bonding characteristics of Masson (*Pinus massoniana* Lamb.) veneer composites. *Eur. J. Wood Wood Prod.* **2019**, *77*, 105–114. [CrossRef]

3.  Ge, S.; Ma, J.; Jiang, S.; Liu, Z.; Peng, W. Potential use of different kinds of carbon in production of decayed wood plastic composite. *Arab. J. Chem.* **2018**, *11*, 838–843. [CrossRef]

4.  Wang, X.; Chen, X.; Xie, X.; Wu, Y.; Zhao, L.; Li, Y.; Wang, S. Effects of thermal modification on the physical, chemical and micromechanical properties of Masson pine wood (*Pinus massoniana* Lamb.). *Holzforschung* **2018**, *72*, 1063–1070. [CrossRef]

5.  Li, W.; Wang, H.; Ren, D.; Yu, Y.; Yu, Y. Wood modification with furfuryl alcohol catalysed by a new composite acidic catalyst. *Wood Sci. Technol.* **2015**, *49*, 845–856. [CrossRef]

6.  Altgen, M.; Hofmann, T.; Militz, H. Wood moisture content during the thermal modification process affects the improvement in hygroscopicity of Scots pine sapwood. *Wood Sci. Technol.* **2016**, *50*, 1–15. [CrossRef]

7.  Gaff, M.; Babiak, M.; Kačík, F.; Sandberg, D.; Turčani, M.; Hanzlík, P.; Vondrová, V. Plasticity properties of thermally modified timber in bending—The effect of chemical changes during modification of European oak and Norway spruce. *Compos. Part B Eng.* **2019**, *165*, 613–625. [CrossRef]

8.  Shi, J.; Lu, Y.; Zhang, Y.; Cai, L.; Shi, S.Q. Effect of thermal treatment with water, $H_2SO_4$ and NaOH aqueous solution on color, cell wall and chemical structure of poplar wood. *Sci. Rep.* **2018**, *8*, 17735. [CrossRef]

9.  Wu, Y.; Wu, J.; Wang, S.; Feng, X.; Chen, H.; Tang, Q.; Zhang, H. Measurement of mechanical properties of multilayer waterborne coatings on wood by nanoindentation. *Holzforschung* **2019**. [CrossRef]

10. Cai, M.; Fu, Z.; Cai, Y.; Li, Z.; Xu, C.; Xu, C.; Li, S. Effect of impregnation with maltodextrin and 1,3-dimethylol-4,5-dihydroxyethyleneurea on Poplar wood. *Forests* **2018**, *9*, 676. [CrossRef]

11. Gérardin, P. New alternatives for wood preservation based on thermal and chemical modification of wood-a review. *Ann. For. Sci.* **2016**, *73*, 559–570. [CrossRef]

12. Sandberg, D.; Kutnar, A.; Mantanis, G. Wood modification technologies—A review. *iForest* **2017**, *10*, 895–908. [CrossRef]

13. Li, W.; Jan, V.D.B.; Dhaene, J.; Zhan, X.; Mei, C.; Van Acker, J. Investigating the interaction between internal structural changes and water sorption of mdf and OSB using X-ray computed tomography. *Wood Sci. Technol.* **2018**, *52*, 701–716. [CrossRef]

14. Yue, K.; Chen, Z.; Lu, W.; Liu, W.; Li, M.; Shao, Y.; Tang, L.; Wan, L. Evaluating the mechanical and fire-resistance properties of modified fast-growing Chinese fir timber with boric-phenol-formaldehyde resin. *Constr. Build. Mater.* **2017**, *154*, 956–962. [CrossRef]

15. Sun, S.; Zhao, Z. Influence of acid on the curing process of tannin-sucrose adhesives. *BioResources* **2018**, *13*, 7683–7697. [CrossRef]

16. Shams, M.I.; Morooka, T.; Yano, H. Compressive deformation of wood impregnated with low molecular weight phenol formaldehyde (PF) resin V: Effects of steam pretreatment. *J. Wood Sci.* **2006**, *52*, 389–394. [CrossRef]

17. Khalil, H.A.; Amouzgar, P.; Jawaid, M.; Abdullah, C.; Issam, A.; Zainudin, E.; Paridah, M.T.; Hassan, A. Physical and thermal properties of microwave-dried wood lumber impregnated with phenol formaldehyde resin. *J. Compos. Mater.* **2013**, *47*, 3565–3571. [CrossRef]

18. Biziks, V.; Bicke, S.; Militz, H. Penetration depth of phenol-formaldehyde (PF) resin into beech wood studied by light microscopy. *Wood Sci. Technol.* **2018**, *53*, 165–176. [CrossRef]

19. Deka, M.; Saikia, C.N. Chemical modification of wood with thermosetting resin: Effect on dimensional stability and strength property. *Bioresour. Technol.* **2000**, *73*, 179–181. [CrossRef]

20. Özçifçi, A. Impacts of impregnation with boron compounds on the bonding strength of wood materials. *Constr. Build. Mater.* **2008**, *22*, 541–545. [CrossRef]

21. Devi, R.R.; Ali, I.; Maji, T. Chemical modification of rubber wood with styrene in combination with a crosslinker: Effect on dimensional stability and strength property. *Bioresour. Technol.* **2003**, *88*, 185–188. [CrossRef]

22. Yildiz, U.C.; Yildiz, S.; Gezer, E.D. Mechanical properties and decay resistance of wood-polymer composites prepared from fast growing species in Turkey. *Bioresour. Technol.* **2005**, *96*, 1003–1011. [CrossRef]

23. Gindl, W.; Schoberl, T. The significance of the elastic modulus of wood cell walls obtained from nanoindentation measurements. *Compos. Part A* **2004**, *35*, 1345–1349. [CrossRef]

24. Wimmer, R.; Lucas, B.N.; Tsui, T.Y.; Oliver, W.C. Longitudinal hardness and young's modulus of spruce tracheid secondary walls using nanoindentation technique. *Wood Sci. Technol.* **1997**, *31*, 131–141. [CrossRef]

25. Yu, Y.; Fei, B.; Wang, H.; Tian, G. Longitudinal mechanical properties of cell wall of *Masson pine* (Pinus massoniana Lamb) as related to moisture content: A nanoindentation study. *Holzforschung* **2011**, *65*, 121–126. [CrossRef]

26. Ali, E.M. Modification of spruce wood by UV-crosslinked PEG hydrogels inside wood cell walls. *React. Funct. Polym.* **2018**, *131*, 100–106.

27. Oliver, W.C.; Pharr, G.M. An improved technique for determining hardness and elastic modulus using load and displacement sensing indentation experiments. *J. Mater. Res.* **1992**, *7*, 1564–1583. [CrossRef]

28. Xie, Y.; Xiao, Z.; Grüneberg, T.; Militz, H.; Hill, C.A.; Steuernagel, L.; Mai, C. Effects of chemical modification of wood particles with glutaraldehyde and 1,3-dimethylol-4,5-dihydroxyethyleneurea on properties of the resulting polypropylene composites. *Compos. Sci. Technol.* **2010**, *70*, 2003–2011. [CrossRef]

29. Xie, Y.; Fu, Q.; Wang, Q.; Xiao, Z.; Militz, H. Effects of chemical modification on the mechanical properties of wood. *Eur. J. Wood Wood Prod.* **2013**, *71*, 401–416. [CrossRef]

30. Tjeerdsma, B.F.; Militz, H. Chemical changes in hydrothermal treated wood: FTIR analysis of combined hydrothermal and dry heat-treated wood. *Holz Roh. Werkst.* **2005**, *63*, 102–111. [CrossRef]

31. González-Peña, M.M.; Curling, S.F.; Hale, M.D. On the effect of heat on the chemical composition and dimensions of thermally-modified wood. *Polym. Degrad. Stab.* **2009**, *94*, 2184–2193. [CrossRef]

32. Kotilainen, R.A.; Toivanen, T.J.; Alén, R.J. FTIR monitoring of chemical changes in softwood during heating. *J. Wood Chem. Technol.* **2000**, *20*, 307–320. [CrossRef]

33. Krajnc, M.; Poljanšek, I. Characterization of phenol-formaldehyde prepolymer resins by in line FTIR spectroscopy. *Acta Chim. Slov.* **2015**, *52*, 238–244.

34. De, D.; Adhikari, B.; De, D. Grass fiber reinforced phenol formaldehyde resin composite: Preparation, characterization and evaluation of properties of composite. *Polym. Adv. Technol.* **2007**, *18*, 72–81. [CrossRef]

35. Wang, X.; Deng, Y.; Li, Y.; Kjoller, K.; Roy, A.; Wang, S. In situ identification of the molecular-scale interactions of phenol-formaldehyde resin and wood cell walls using infrared nanospectroscopy. *RSC Adv.* **2016**, *6*, 76318–76324. [CrossRef]

36. Paris, J.L.; Kamke, F.A.; Xiao, X. X-ray computed tomography of wood-adhesive bondlines: Attenuation and phase-contrast effects. *Wood Sci. Technol.* **2015**, *49*, 1185–1208. [CrossRef]

37. Yelle, D.J.; Ralph, J. Characterizing phenol-formaldehyde adhesive cure chemistry within the wood cell wall. *Int. J. Adhes. Adhes.* **2016**, *70*, 26–36. [CrossRef]

38. Wang, J.; Laborie, M.P.G.; Wolcott, M.P. Correlation of mechanical and chemical cure development for phenol–formaldehyde resin bonded wood joints. *Thermochimica Acta* **2011**, *513*, 20–25. [CrossRef]

39. Jakes, J.E.; Hunt, C.G.; Yelle, D.J.; Lorenz, L.; Hirth, K. Synchrotron-based X-ray fluorescence microscopy in conjunctionwith nanoindentation to study molecular-scale interactions of phenol-formaldehyde in wood cell walls. *ACS Appl. Mater. Interfaces* **2015**, *7*, 6584–6589. [CrossRef]

40. Kuai, B.; Wang, X.; Lv, C.; Xu, K.; Zhang, Y.; Zhan, T. Orthotropic tension behavior of two typical chinese plantation woods at wide relative humidity range. *Forests* **2019**, *10*, 516. [CrossRef]

41. Olaniran, S.O.; Michen, B.; Mora Mendez, D.F.; Wittel, F.K.; Bachtiar, E.V.; Burgert, I.; Markus, R. Mechanical behaviour of chemically modified *Norway spruce* (Picea abies L. karst.): Experimental mechanical studies on spruce wood after methacrylation and in situ polymerization of styrene. *Wood Sci. Technol.* **2019**, *53*, 425–445. [CrossRef]

*Article*

# The Impact of Anatomical Characteristics on the Structural Integrity of Wood

**Lukas Emmerich *, Georg Wülfing and Christian Brischke**

Wood Biology and Wood Products, University of Goettingen, Buesgenweg 4, D-37077 Goettingen, Germany; georg.wuelfing@stud.uni-goettingen.de (G.W.); christian.brischke@uni-goettingen.de (C.B.)
* Correspondence: lukas.emmerich@uni-goettingen.de

Received: 8 February 2019; Accepted: 21 February 2019; Published: 24 February 2019

**Abstract:** The structural integrity of wood is closely related to its brittleness and thus to its suitability for numerous applications where dynamic loads, wear and abrasion occur. The structural integrity of wood is only vaguely correlated with its density, but affected by different chemical, physico-structural and anatomical characteristics, which are difficult to encompass as a whole. This study aimed to analyze the results from High-Energy Multiple Impact (HEMI) tests of a wide range of softwood and hardwood species with an average oven-dry wood density in a range between 0.25 and 0.99 g/cm$^3$ and multifaceted anatomical features. Therefore, small clear specimens from a total of 40 different soft- and hardwood species were crushed in a heavy vibratory ball mill. The obtained particles were fractionated and used to calculate the 'Resistance to Impact Milling (RIM)' as a measure of the wood structural integrity. The differences in structural integrity and thus in brittleness were predominantly affected by anatomical characteristics. The size, density and distribution of vessels as well as the ray density of wood were found to have a significant impact on the structural integrity of hardwoods. The structural integrity of softwood was rather affected by the number of growth ring borders and the occurrence of resin canals. The density affected the Resistance to Impact Milling (RIM) of neither the softwoods nor the hardwoods.

**Keywords:** brittleness; density; dynamic strength; High-Energy Multiple Impact (HEMI)–test; Resistance to Impact Milling (RIM)

---

## 1. Introduction

Most elasto-mechanical and rheological properties of wood are closely related to wood density and are therefore rather easily predictable. However, the anatomical features of wood, which can be wood species-specific, further affect especially dynamic strength properties such as the impact bending strength and shock resistance [1–3]. For instance, the large earlywood pores in ring-porous hardwoods such as English oak (*Quercus robur* L.), Sweet chestnut (*Castanea sativa* L.), Black locust (*Robinia pseudoacacia* L.) or Wych elm (*Ulmus glabra* Huds.) can serve as predetermined breaking points. Further deviations from an ideal homogeneous xylem structure such as large rays in European beech (*Fagus sylvatica* L.) or Alder (here: false rays, *Alnus* spp.), distinct parenchyma bands in Bongossi (*Lophira alata* Banks ex C. F. Gaertn.) or agglomerates of resin canals in Red Meranti (*Shorea* spp.), also have the potential to either strengthen or reduce the structural integrity of wood.

Similarly, wood cell wall modification affects different mechanical properties including the wood hardness and abrasion resistance, but also its brittleness and consequently its structural integrity. This has been shown previously with the help of High-Energy Multiple Impact (HEMI)-tests, where small wood specimens are subjected to thousands of dynamic impacts by steel balls in the bowl of a heavy vibratory mill. The fragments obtained are analyzed afterwards [4]. For instance, the weakening of cell walls by heat during thermal modification processes, especially in the middle lamella region, leads to a steady decrease in the structural integrity of wood with increasing treatment intensity. The HEMI-test has further been used to detect fungal decay by soft rot, brown rot and white rot fungi (even in very early stages), the effect of gamma radiation, wood densification, wood preservative impregnation, wax and oil treatments, and different chemical wood modification processes [5].

It has previously been shown that the Resistance to Impact Milling (RIM), which serves as a measure of wood's structural integrity is very insensitive to varying densities, natural ageing, and the occurrence of larger cracks [5]. Furthermore, the RIM varies only little within one wood species, as shown for Scots pine sapwood (*Pinus sylvestris* L.) samples from trees in six Northern European countries [6]. However, the results from previous studies indicated that the structural integrity determined in HEMI-tests is not well correlated with wood density, since further variables such as wood species-specific anatomical characteristics of the xylem tissue interfere with the effect of density [7].

*Objective*

The aim of this study was to analyze the results from HEMI-tests of a wide range of softwood and hardwood species with an average oven-dry wood density in a range between 0.25 and 0.99 g/cm$^3$ and with multifaceted anatomical features.

## 2. Materials and Methods

One hundred replicate specimens of 10 (ax.) $\times$ 5 $\times$ 20 mm$^3$ were prepared from a total of 40 different wood species and separated between sapwood and heartwood, as listed in Tables 1 and 2.

To determine the oven-dry density (ODD), $n = 10$ replicate specimens of 10 (ax.) $\times$ 5 $\times$ 20 mm$^3$ per wood species were oven dried at 103 °C until a constant mass, weighed to the nearest 0.0001 g; the dimensions were then measured to the nearest 0.001 mm. The oven dry density was calculated according to the following equation:

$$\rho_0 = \frac{m_0}{V_0} \ [\text{g cm}^{-3}] \tag{1}$$

where:

$\rho_0$ is the oven-dry density, in g·cm$^{-3}$;
$m_0$ is the oven-dry mass, in g;
$V_0$ is the oven-dry volume, in cm$^3$.

**Table 1.** The oven-dry density (ODD), Resistance to Impact Milling (RIM), degree of integrity (I), and fine percentage (F) of different softwood species. The standard deviation (SD) is in parentheses.

| Name [1] | Botanical Name | ODD [g cm$^{-3}$] | | RIM [%] | | I [%] | | F [%] | |
|---|---|---|---|---|---|---|---|---|---|
| Scots pine sw | *Pinus sylvestris* | 0.41 | (0.02) | 88.2 | (0.9) | 67.4 | (1.1) | 13.5 | (1.1) |
| Scots pine hw | | 0.58 | (0.04) | 84.5 | (0.8) | 41.9 | (3.1) | 1.3 | (0.4) |
| Radiata pine sw | *Pinus radiata* | 0.43 | (0.02) | 88.8 | (0.5) | 55.4 | (2.1) | 0.0 | (0.0) |
| Carribean pine hw | *Pinus carribaea* | 0.39 | (0.04) | 87.3 | (0.4) | 52.4 | (1.8) | 1.1 | (0.3) |
| European Larch sw | *Larix decidua* | 0.56 | (0.02) | 85.2 | (0.4) | 44.5 | (2.2) | 1.2 | (0.3) |
| European Larch hw | | 0.51 | (0.02) | 80.8 | (1.5) | 35.5 | (4.8) | 4.1 | (0.4) |
| Douglas fir sw | *Pseudotsuga menziesii* | 0.63 | (0.02) | 86.3 | (0.4) | 45.6 | (1.8) | 0.2 | (0.2) |
| Douglas fir hw | | 0.51 | (0.02) | 82.2 | (0.5) | 34.8 | (1.3) | 1.9 | (0.3) |
| Norway spruce | *Picea abies* | 0.43 | (0.03) | 82.9 | (1.7) | 35.9 | (6.1) | 1.5 | (0.4) |
| Coastal fir | *Abies grandis* | 0.40 | (0.06) | 80.6 | (0.5) | 26.5 | (1.3) | 1.4 | (0.4) |
| Western hemlock | *Tsuga heterophylla* | 0.42 | (0.03) | 83.8 | (0.7) | 40.0 | (2.1) | 1.6 | (0.3) |
| Yew | *Taxus baccata* | 0.60 | (0.03) | 84.5 | (0.9) | 43.9 | (3.2) | 1.9 | (0.3) |

[1] sw = sapwood, hw = heartwood; heartwood if not otherwise indicated.

**Table 2.** The oven-dry density (ODD), Resistance to Impact Milling (RIM), degree of integrity (I), and fine percentage (F) of different hardwood species. The standard deviation (SD) is in parentheses.

| Name [1] | Botanical Name | ODD [g cm$^{-3}$] | | RIM [%] | | I [%] | | F [%] | |
|---|---|---|---|---|---|---|---|---|---|
| English oak sw | *Quercus robur* | 0.49 | (0.02) | 83.3 | (0.5) | 44.2 | (1.7) | 3.7 | (0.4) |
| English oak hw | | 0.59 | (0.01) | 87.3 | (1.2) | 59.0 | (4.3) | 3.3 | (0.4) |
| Black locust | *Robinia pseudoacacia* | 0.68 | (0.05) | 83.5 | (1.2) | 41.0 | (3.9) | 2.3 | (0.2) |
| Sweet chestnut | *Castanea sativa* | 0.50 | (0.03) | 78.1 | (2.3) | 36.0 | (4.2) | 7.9 | (1.8) |
| Ash | *Fraxinus excelsior* | 0.62 | (0.02) | 83.1 | (0.8) | 40.4 | (2.6) | 2.7 | (0.3) |
| Locust | *Gleditsia sp.* | 0.66 | (0.02) | 86.7 | (1.1) | 52.6 | (3.4) | 1.9 | (0.4) |
| Common walnut | *Juglans regia* | 0.63 | (0.02) | 85.2 | (0.5) | 49.8 | (2.1) | 2.9 | (0.3) |
| Wild cherry | *Prunus avium* | 0.55 | (0.01) | 86.7 | (0.7) | 53.0 | (2.2) | 2.0 | (0.3) |
| Black cherry | *Prunus serotina* | 0.64 | (0.04) | 87.7 | (0.6) | 54.9 | (2.1) | 1.4 | (0.2) |
| Beech | *Fagus sylvatica* | 0.66 | (0.02) | 88.0 | (0.4) | 55.9 | (2.2) | 1.4 | (0.3) |
| Maple | *Acer sp.* | 0.61 | (0.01) | 89.1 | (0.6) | 58.0 | (2.3) | 0.5 | (0.1) |
| Lime | *Tilia sp.* | 0.44 | (0.01) | 90.1 | (0.8) | 61.1 | (2.6) | 0.2 | (0.3) |
| Birch | *Betula pendula* | 0.57 | (0.02) | 87.9 | (0.4) | 54.2 | (1.6) | 0.8 | (0.1) |
| Hazel | *Corylus avellana* | 0.68 | (0.02) | 86.9 | (1.0) | 52.8 | (3.9) | 1.8 | (0.2) |
| Boxwood | *Buxus sempervirens* | 0.96 | (0.01) | 90.3 | (0.9) | 64.1 | (3.7) | 0.9 | (0.0) |
| Poplar | *Populus nigra* | 0.39 | (0.02) | 86.3 | (0.3) | 50.5 | (0.9) | 1.8 | (0.3) |
| Alder | *Alnus glutinosa* | 0.48 | (0.01) | 86.9 | (0.9) | 54.6 | (3.3) | 2.3 | (0.5) |
| Kiri | *Paulownia tomentosa* | 0.25 | (0.02) | 80.9 | (1.5) | 40.0 | (4.0) | 5.5 | (0.9) |
| Shining gum | *Eucalyptus nitens* | 0.74 | (0.11) | 83.2 | (1.5) | 46.7 | (4.5) | 4.6 | (0.9) |
| Teak | *Tectona grandis* | 0.63 | (0.09) | 84.1 | (0.7) | 48.0 | (2.1) | 3.9 | (0.8) |
| Ipe | *Handroanthus sp.* | 0.93 | (0.02) | 86.0 | (0.5) | 51.8 | (1.2) | 2.6 | (0.7) |
| Merbau | *Intsia spp.* | 0.74 | (0.03) | 68.1 | (2.4) | 27.9 | (1.8) | 18.6 | (2.7) |
| Bangkirai | *Shorea laevis* | 0.79 | (0.05) | 87.7 | (0.7) | 54.9 | (1.9) | 1.4 | (0.4) |
| Balau | *Shorea spp.* | 0.92 | (0.03) | 84.3 | (1.1) | 51.7 | (2.8) | 4.8 | (1.0) |
| Bongossi | *Lophira alata* | 0.97 | (0.03) | 85.9 | (1.0) | 51.9 | (2.7) | 2.8 | (0.7) |
| Amaranth | *Peltogyne sp.* | 0.88 | (0.01) | 88.6 | (0.7) | 57.9 | (2.7) | 1.1 | (0.0) |
| Basralocus | *Dicorynia sp.* | 0.81 | (0.02) | 84.8 | (0.6) | 50.9 | (1.9) | 4.0 | (0.4) |
| Garapa | *Apuleia sp.* | 0.76 | (0.04) | 86.7 | (1.1) | 53.0 | (3.3) | 2.1 | (0.5) |
| Limba | *Terminalia superba* | 0.50 | (0.03) | 83.2 | (1.2) | 45.1 | (2.7) | 4.1 | (0.9) |
| Kambala | *Milicia sp.* | 0.62 | (0.03) | 79.7 | (0.7) | 45.2 | (2.7) | 8.8 | (0.4) |
| Massaranduba | *Manilkara bidentata* | 0.99 | (0.04) | 85.9 | (0.6) | 53.2 | (2.5) | 3.2 | (0.2) |
| Greenheart | *Chlorocardium rodiei* | 0.96 | (0.02) | 85.9 | (1.5) | 49.9 | (5.3) | 2.1 | (0.8) |

[1] sw = sapwood, hw = heartwood.

Afterwards, selected density specimens were cut with a traversing microtome and used for digital reflected-light microscopy with a Keyence Digital microscope VHX 5000 (Keyence Corporation, Osaka, Japan). Cross section photographs were taken at a magnification of 30×, and the diameter of the earlywood vessels, the vessel density, and the wood ray density were determined at a magnification of 200× for both the soft- and hardwoods. For the tropical species, the listed anatomical features were determined at a magnification of 100×. Therefore, $n = 10$ replicate measurements were conducted per wood species to determine the ray density and vessel density. The earlywood vessel diameter was determined on $n = 30$ vessels.

Five times 20 specimens of 10 (ax.) × 5 × 20 mm$^3$ were submitted to High-Energy Multiple Impact (HEMI)–tests. The development and optimization of the HEMI-test have been described by [4] and [8]. In the present study, the following procedure was applied: 20 oven-dried specimens were placed in the bowl (140 mm in diameter) of a heavy-impact ball mill (Herzog HSM 100-H; Herzog Maschinenfabrik, Osnabrück, Germany), together with one steel ball of 35 mm diameter for crushing the specimens. Three balls of 12 mm diameter and three of 6 mm diameter were added to avoid small fragments from hiding in the angles of the bowl, thus ensuring impact with smaller wood fragments. The bowl was shaken for 60 s at a rotary frequency of 23.3 s$^{-1}$ and a stroke of 12 mm. The fragments of the 20 specimens were fractionated on a slit sieve according to [9], with a slit width of 1 mm, using an orbital shaker at an amplitude of 25 mm and a rotary frequency of 200 min$^{-1}$ for 2 min. The following values were calculated:

$$I = \frac{m_{20}}{m_{all}} \times 100 \; [\%] \tag{2}$$

where:

    I is the degree of integrity, in %;

    $m_{20}$ is the oven-dry mass of the 20 biggest fragments, in g;

    $m_{all}$ is the oven-dry mass of all the fragments, in g.

$$F = \frac{m_{fragments<1mm}}{m_{all}} \times 100 \; [\%] \tag{3}$$

where:

    F is the fine percentage, in %;

    $m_{fragments<1mm}$ is the oven-dry mass of fragments smaller than 1 mm, in g;

    $m_{all}$ is the oven-dry mass of all the fragments, in g.

$$RIM = \frac{(I - 3 \times F) + 300}{400} \; [\%] \tag{4}$$

where:

    RIM is the Resistance to Impact Milling, in %;

    I is the degree of integrity, in %;

    F is the fine percentage, in %.

## 3. Results and Discussion

### 3.1. Structural Integrity

The Resistance to Impact Milling (RIM) varied between 68.1% (Merbau) and 90.3% (Boxwood). In contrast, the degree of integrity (I) varied significantly more, i.e., between 26.5% (Coastal fir) and 67.4% (Scots pine sapwood), as did the fine percentage (F): i.e., between 0.0% (Radiata pine) and 18.6% (Merbau). The data for the RIM, I, and F are summarized in Table 1 for the tested softwood species and in Table 2 for the hardwood species. Besides differences between the wood species, the three indicators showed differences in the variation within one species, here expressed as the standard deviation (SD). The highest variation was obtained for F, followed by I and RIM. This supports previous findings pointing out the benefit of using the combined measure RIM, which is of higher sensitivity to differences in the structural integrity paired with less scattering of data compared to I and F [4,7]. In total, the SD of the RIM was between 0.3% (Poplar) and 2.4% (Merbau), corresponding to coefficients of variation (COV) between 0.4% and 3.5%, which is very low compared to mechanical properties such as the bending or impact bending strength (e.g., [7]).

### 3.2. Impact of Oven-Dry Density on Structural Integrity

A clear relationship between the ODD and structural integrity did not become evident, as shown for all the examined wood species and separately for the softwoods, ring- and semi-ring-porous hardwoods and diffuse-porous hardwoods in Figure 1. The RIM seemed to be at least superposed by further parameters such as structural features and anatomical characteristics. This coincides with the data for the Ash, Scots pine and Beech previously reported by [8], who showed that the density and RIM were not even correlated within one wood species. More recently, [7] reported that the density and RIM were also poorly correlated when considering ten different wood species representing a range of ODD between 0.37 and 0.77 g/cm$^3$. However, according to [7] the RIM was fairly well correlated with the impact bending strength (IBS, $R^2 = 0.67$) and modulus of rupture (MOR, $R^2 = 0.56$), as determined on axially matched specimens, which indicates that these strength properties are also at least partly affected by similar anatomical characteristics as the RIM is.

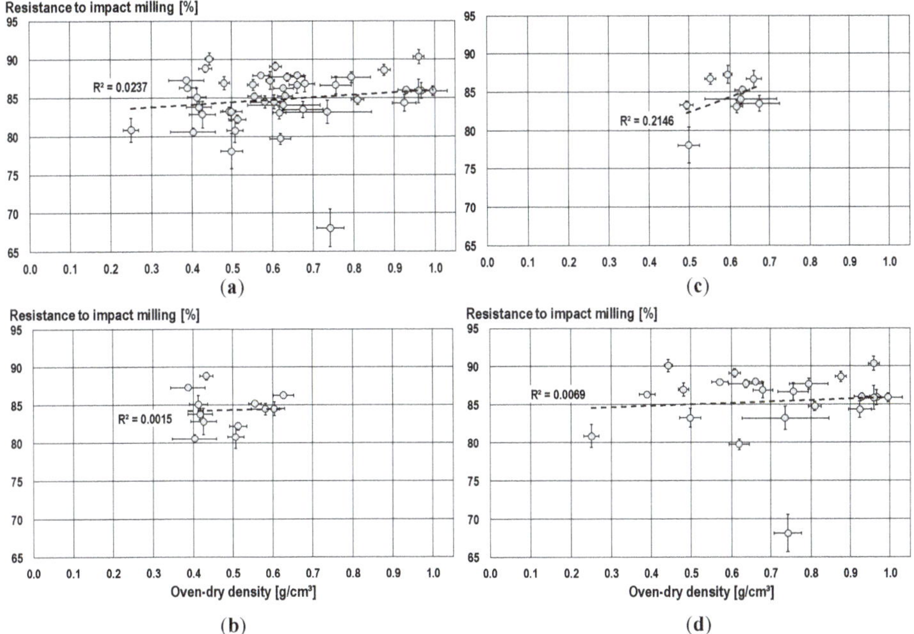

**Figure 1.** The relationship between the average oven-dry density and Resistance to Impact Milling (RIM): (**a**) all wood species (y = 3.1629x + 82.887); (**b**) softwoods (y = 1.1035x + 83.791); (**c**) ring- and semi-ring-porous hardwoods (y = 19.634x + 72.545); and (**d**) diffuse-porous hardwoods (y = 1.8475x + 84.086).

### 3.3. Impact of Anatomical Characteristics on Structural Integrity

The tested softwood species had a rather homogeneous and uniform anatomical appearance compared to the different hardwood species. However, even within this group the RIM varied between 80.6% and 88.8%. As summarized in Table 3, the softwood species differed also in the average tracheid diameter and in wood ray density. Nevertheless, the fracture patterns observed during the HEMI-tests were rather uniform, and fractures occurred predominantly along the growth ring borders in a tangential direction and along the wood rays and resin canals in a radial direction. The wood species showing an abrupt transition between the earlywood and latewood, such as the Larch and Scots pines, did not show a lower structural integrity compared to the species with a more gradual transition, such as the Norway spruce and Douglas fir, as one might expect due to a more sudden change of density within the tracheid tissue of one annual ring. Consequently, no fractures were observed along the transition line between the earlywood and latewood. In contrast to other softwood species, the Carribean and Radiata pines showed fractures in a tangential direction not only at the growth ring borders, but also where the resin canals ran in an axial direction.

As exemplarily shown for the heartwood of the Scots pine and Douglas fir in Figure 2, the major weak points, where fractures predominantly occurred, were the following: (a) the growth ring borders, where the less dense earlywood follows the dense latewood, and (b) the wood rays, which (1) consist of parenchyma cells, and (2) are running orthogonal to the main cell orientation in the tracheid tissue.

**Table 3.** The anatomical characteristics (tracheid diameter, ray density) and description of fractures during the HEMI-tests of different softwood species (standard deviation in parentheses).

| Wood Species | Tracheid Ø [μm] | | Wood Ray Density [mm$^{-1}$] | | Fracture Behaviour tang. | rad. | Remarks |
|---|---|---|---|---|---|---|---|
| Scots pine sw | 29 | (6) | 4.6 | (1.2) | GR | RC | wider rings compared to hw |
| Scots pine hw | 25 | (5) | 3.7 | (1.5) | GR | RC | - |
| Radiata pine sw | 22 | (4) | 4.4 | (1.3) | GR | R, RC | - |
| Carribean pine hw | 28 | (4) | 5.2 | (1.2) | GR | R, RC | - |
| European larch sw | 35 | (7) | 5.9 | (1.7) | GR | R | wider rings compared to hw |
| European larch hw | 35 | (6) | 4.4 | (1.1) | GR | R | - |
| Douglas fir sw | 25 | (6) | 4.3 | (1.2) | GR | R, RC | - |
| Douglas fir hw | 23 | (5) | 3.9 | (1.2) | GR | R | - |
| Norway spruce | 25 | (5) | 4.5 | (1.0) | GR | R | - |
| Coastal fir | 28 | (5) | 5.7 | (1.3) | GR | R | - |
| Western hemlock | 25 | (5) | 5.0 | (1.3) | GR | R | - |
| Yew | 10 | (3) | 7.1 | (1.4) | (GR) | (R) | Irregular fracture pattern |

GR = along growth rings, R = along rays; RC = along resin canals; tang. = tangential growth direction; rad. = radial growth direction.

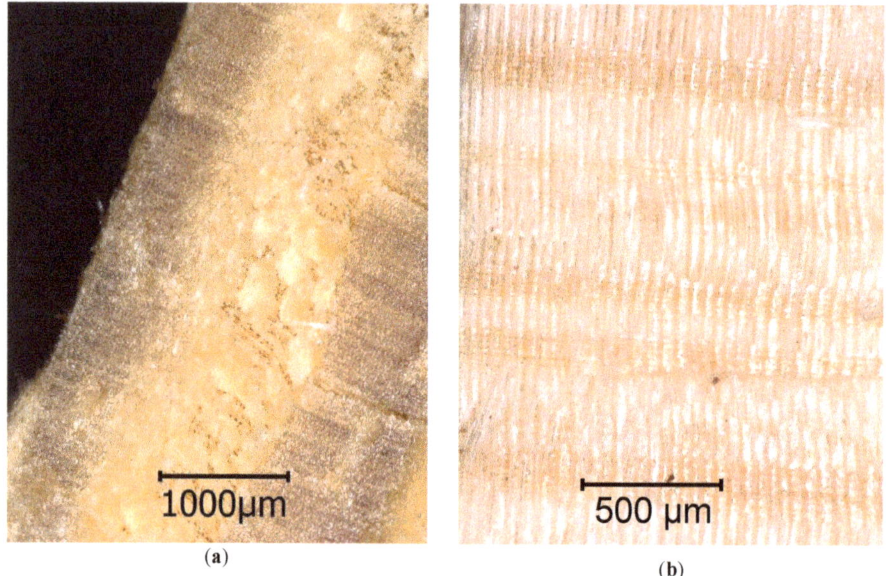

**Figure 2.** The fracture pattern in the softwoods: (**a**) Cross section of the Scots pine heartwood, fracture along a growth ring border; (**b**) The radial fracture section of the Douglas fir heartwood, fracture along the rays.

The fractures in the ring-porous hardwood species often followed the wide-luminous earlywood vessels, such as in the English oak, Sweet chestnut, Ash, Locust, and Black locust (Table 4). The specimens consequently broke apart in a tangential direction. In addition, the fractures occurred along the latewood vessel fields where high portions of paratracheal parenchyma were present (Figure 3). The ring-porous hardwoods with broad wood rays, such as the English oak, also showed fractures running parallel to the latter. Finally, the average diameters of the earlywood vessels were not correlated with the structural integrity, although, in the earlywood of all the ring-porous hardwoods, the fractures occurred preferentially in a tangential direction following the vessel rings.

**Table 4.** The anatomical characteristics (earlywood vessel diameter, vessel density, ray density) and description of fractures during the HEMI-tests of different hardwood species (standard deviation in parentheses).

| Wood Species | Earlywood Vessel Ø [μm] | | Vessel Density [mm$^{-2}$] | | Wood Ray Density [mm$^{-1}$] | | Fracture Behaviour [1] | |
|---|---|---|---|---|---|---|---|---|
| | | | | | | | tang. | rad. |
| English oak sw [2] | 247 | (51) | 7.7 | (1.6) | 8.4 | (2.0) | EW | P |
| English oak hw | 202 | (49) | 10.1 | (2.0) | 10.7 | (1.8) | EW | P |
| Black locust | 190 | (40) | 11.5 | (1.0) | 6.9 | (1.1) | EW | R * |
| Sweet chestnut | 209 | (30) | 7.7 | (1.5) | 11.8 | (1.5) | EW | V-V |
| Ash [3] | 169 | (21) | 13.8 | (1.7) | 6.6 | (0.8) | EW | n.a. |
| Locust | 165 | (25) | 16.1 | (2.0) | 4.4 | (1.2) | EW | R, P * |
| Common walnut | 134 | (32) | 7.7 | (2.0) | 5.7 | (1.3) | V-V | V-V |
| Wild cherry | 33 | (8) | 171.6 | (31.3) | 6.1 | (1.4) | GR | R |
| Black cherry [4] | 33 | (9) | 67.9 | (22.9) | 5.6 | (1.3) | n.a. | R |
| European beech | 40 | (8) | 131.9 | (15.4) | 3.0 | (1.3) | GR * | n.a. |
| Maple [5] | 46 | (7) | 54.5 | (3.4) | 7.9 | (1.7) | GR | R * |
| Lime [6] | 39 | (9) | 104.7 | (14.0) | 4.8 | (1.1) | n.a. | n.a. |
| Birch | 54 | (13) | 45.2 | (8.5) | 8.3 | (2.4) | n.a. | R |
| Hazel | 28 | (6) | 98.9 | (20.2) | 11.6 | (2.5) | GR | n.a. |
| Boxwood [6] | 10 | (4) | 213.9 | (14.0) | 11.0 | (2.5) | n.a. | R * |
| Poplar [7] | 58 | (13) | 33.7 | (6.4) | 11.0 | (1.5) | n.a. | R |
| Alder | 41 | (10) | 108.0 | (16.7) | 11.7 | (2.0) | GR * | R |
| Kiri | 164 | (55) | 5.2 | (2.0) | 2.4 | (0.8) | V-V | V-V, R * |
| Shining gum [2] | 144 | (25) | 7.6 | (3.2) | 11.3 | (1.1) | V-V | V-V |
| Teak [8] | 184 | (57) | 6.3 | (1.7) | 4.1 | (0.7) | V-V | R * |
| Ipé [2] | 103 | (9) | 23.2 | (2.7) | 7.8 | (0.9) | P * | V-V |
| Merbau | 250 | (40) | 4.0 | (1.4) | 4.2 | (0.9) | V-V, P * | V-V, P * |
| Bangkirai | 207 | (32) | 7.3 | (1.7) | 3.7 | (1.3) | P | V-V, R * |
| Balau | 137 | (13) | 11.9 | (2.9) | 9.1 | (1.2) | P | V-V, R * |
| Bongossi | 232 | (41) | 2.9 | (1.1) | 9.9 | (1.2) | P | V-V, P * |
| Amaranth | 109 | (16) | 4.4 | (1.7) | 6.9 | (1.7) | P, V * | R |
| Basralocus [2] | 190 | (33) | 2.8 | (1.0) | 7.9 | (1.0) | P * | V-V |
| Garapa | 121 | (19) | 15.2 | (3.0) | 8.3 | (1.3) | P | V-V |
| Limba | 139 | (28) | 4.4 | (1.7) | 10.2 | (1.0) | n.a. | R |
| Kambala | 193 | (41) | 2.8 | (0.8) | 4.4 | (1.0) | (P) | R |
| Massaranduba | 113 | (18) | 13.1 | (3.3) | 10.5 | (1.5) | (P) | R |
| Greenheart [2] | 90 | (16) | 14.0 | (2.0) | 7.5 | (0.9) | n.a. | V-V |

[1] n.a. = not available (no clear pattern evident), GR = along growth rings, R = along rays, RC = along resin canals, EW = along earlywood vessels, P = in parenchyma tissue, V-V = vessel to vessel, V = at vessels, * = characteristic plays minor role; remarks related to fracture patterns: [2] radial, parallel to rays; [3] no clear radial pattern; [4] very often parallel to rays; [5] parallel to growth rings; [6] irregular fracture pattern; [7] samples often compressed; [8] often at growth ring border.

This stands to some extent in contrast to findings by [2], who studied the perpendicular-to-grain properties of eight North-American hardwood species and found that the earlywood vessel area fraction negatively influenced the radial maximum stress and strain, whereas the ray width and area fraction were positively related to the maximum radial properties. The rays also affected the transverse stiffness significantly.

Studies conducted by [10] showed that wood rays have a positive effect on the tensile strength of English oak and European ash wood. However, as shown for the fragments obtained in the HEMI-tests, the latewood vessel fields turned out to be weak spots when it comes to dynamic loads in different anatomical directions. Therefore, the potentially positive effect of the wood rays on the structural integrity might be superposed by other anatomical features.

Finally, the RIM of the heartwood of the English oak (87.3%) was significantly higher than that of its sapwood (83.3%), which is to some extent surprising since sapwood is often considered to be less brittle than heartwood [11]. While the fine percentage (F) of both English oak materials was almost equal, the degree of integrity (I) of the heartwood was remarkably higher than that of the sapwood, which might be related to the potential 'gluing' effects of the tylosis which were present in the earlywood vessels in the heartwood (Figure 3b), but were absent in the sapwood. Whether and to what extent the formation of tylosis has a positive effect on structural integrity would need to be further investigated using different generally tylosis-forming wood species.

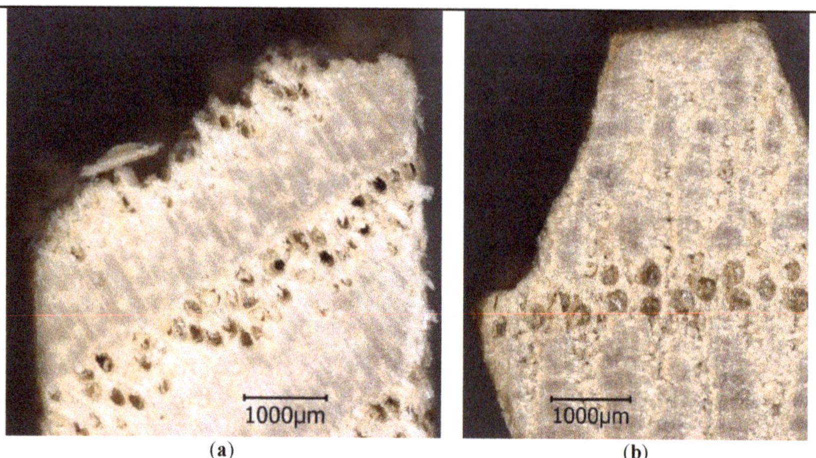

(a)           (b)

**Figure 3.** The fracture pattern in the ring-porous hardwoods: (**a**) Cross section of the Ash, fracture within a ring of the earlywood vessels; (**b**) Cross section of the English oak heartwood, fracture along the field of the latewood pores and the adjacent parenchyma cells.

By far, the Sweet chestnut showed the lowest RIM among the ring-porous hardwoods, which might be related to its high wood ray density (Table 4), but no clear correlation between the ray density and structural integrity became evident (Figure 4). Furthermore, the radial fractures in the Sweet chestnut were also running from one vessel to the next. More likely, the higher percentage of vessels and axial parenchyma leads to a higher number of weak points within the xylem of the Sweet chestnut compared to the other ring-porous species within this study.

The group of semi-ring-porous hardwood species, which was represented by the Teak, Wild cherry and Walnut in this study, takes an intermediate position between the ring- and the diffuse-porous species. This also became evident when analyzing the fracture patterns obtained through the HEMI-test. As shown in Figure 5a for the Wild cherry, the fractures occurred along the growth ring borders but did not run through the earlywood vessel rings.

In the diffuse-porous hardwoods, the RIM varied most, i.e., between 80.9% (Kiri) and 90.3% (Boxwood), respectively. Although these two species also represent the extremes in ODD, the latter was not correlated with the structural integrity, as shown in Figure 1. Nevertheless, in contrast to the ring-porous hardwood species, the average earlywood vessel diameter of the diffuse-porous hardwood species was correlated with the RIM ($R^2 = 0.4704$), as shown in Figure 6. [12] studied angiosperm wood species and concluded that the tissue density outside the vessel lumens must predominantly influence wood density. Furthermore, they suggest that both the density and the vessel lumen fraction affect the mechanical strength properties.

It became also obvious that in different wood species such as the Kiri, Walnut, Shining gum and further tropical species, the fractures occurred between the vessels, both in the radial and tangential directions (Table 4). Consequently, the vessels turned out to be general weak points in the fiber tissue of the hardwoods, where the weakness increases with an increasing vessel diameter. Figure 7a shows, as an example for the Bongossi, that the vessels served as a starting point for the fractures independently from its anatomical orientation. Tropical species with comparatively small vessels such as the Amaranth, Bangkirai, Garapa, and Ipé showed a rather high RIM. On the extreme end of the scale, the Merbau showed the lowest RIM and also the largest vessel diameters of all the species. Furthermore, distinct parenchyma bands and wood rays appeared to be weak (and therefore starting points for fractures) in tropical species as well, as also shown in Figure 7. The fractures cutting the wood rays appeared only where the rays were deflected by the vessels from their straight radial orientation.

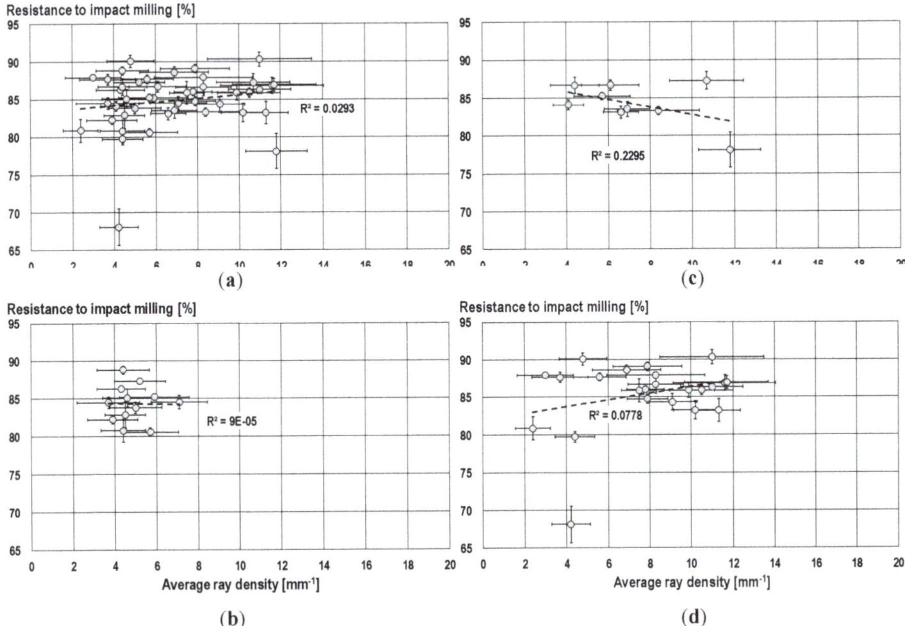

Figure 4. The relationship between the average ray density and the Resistance to Impact Milling (RIM): (a) all wood species (y = 0.2354x + 83.247); (b) softwoods (y = −0.0252x + 84.454); (c) ring- and semi-ring-porous hardwoods (y = −0.5083x + 87.875); (d) and diffuse-porous hardwoods (y = 0.4365x + 81.997).

(a)  (b)

Figure 5. The fracture pattern in semi-ring-porous and diffuse-porous hardwoods: (a) the cross section of the Wild cherry, the fracture along a growth ring border; (b) the radial fracture section of the Alder, the fracture along the rays.

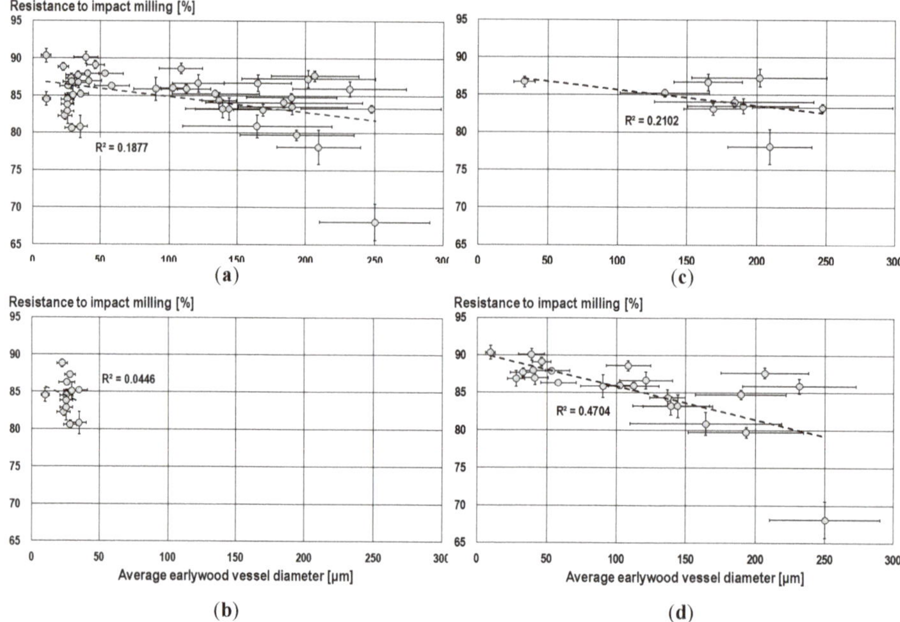

**Figure 6.** The relationship between the average earlywood vessel diameters and the resistance to impact milling (RIM): (**a**) all wood species (y = −0.0213x + 86.982); (**b**) softwoods (y = −0.0814x + 86.425); (**c**) ring- and semi-ring-porous hardwoods (y = −0.0213x + 87.857); and (**d**) diffuse-porous hardwoods (y = −0.0445x + 90.309).

**Figure 7.** The fracture pattern in the diffuse-porous hardwoods: (**a**) the cross section of the Bongossi, the tangential fractures; (**b**) the cross section of the Amaranth—the radial fractures along the rays.

## 4. Conclusions

In this study, we showed that the differences in the structural integrity of wood and thus in the brittleness are predominantly affected by anatomical characteristics. The size, density and distribution of the vessels as well as the ray density of the wood were found to have a significant impact on the structural integrity of the hardwoods. The structural integrity of the softwoods was, on the other hand, affected by the number of growth ring borders and the occurrence of resin canals. The density affected the Resistance to Impact Milling (RIM) of neither the softwoods nor the hardwoods.

Consequently, for applications where the brittleness of wood is more relevant than its elasto-mechanical properties, which are generally strongly correlated with wood density, other anatomical characteristics need to be considered for assessing wood quality. In particular, where dynamic loads impact on wooden components, the brittleness of wood becomes a critical issue. Dynamic loads paired with long-term wear and abrasion can be expected, for instance, on outdoor flooring. Furthermore, during wood processing, machining and handling during industrial processes, numerous dynamic impacts occur and affect the structural integrity of wood.

Wood quality is consequently strongly purpose-specific and cannot be simply derived from wood density data. Anatomical features showed a high potential to serve as better indicators for the structural integrity of wood. Additional influences such as the occurrence of reaction wood, alternating rotational growth and other types of fiber deviations likely affect the structural integrity of wood to a similarly extent. In summary, the findings from this study confirmed the need for test methods other than standard strength tests. As long as the common knowledge about wood anatomy and its effects on mechanical wood properties is incomplete, methods are needed that are sensitive, reliable, and accurate enough to characterize the structures of wood in a comprehensive manner. As shown with the HEMI-method applied in this study, indicators can be delivered for instance of the structural integrity of wood. However, further tests are needed, paired with more detailed analyses of the anatomical and chemical constitution of the wood samples being tested, to achieve a fully satisfactory insight on the relationship between wood anatomy and its structural integrity.

**Author Contributions:** Mainly responsible for the conceptualization, methodology used for these investigations and also the data evaluation, data validation and formal analysis was L.E. together with C.B. Investigations and data curation were conducted by G.W. together with L.E. The original draft of this article was prepared by L.E. together with C.B. who was involved in the review and editing process of this article. L.E. and G.W. did care for the visualization, supervised by C.B. who had the project's administration.

**Funding:** This research received no external funding.

**Acknowledgments:** The authors gratefully acknowledge Florian Zeller (GD Holz, Germany), Uwe Herrmann (Vandecasteele Houtimport, Belgium), Miha Humar (University Ljubljana, Slovenia) and Martin Rosengren (Rönnerum, Sweden) for providing wood samples.

**Conflicts of Interest:** The authors declare no conflict of interest.

## References

1. Ghelmeziu, N. Untersuchungen über die Schlagfestigkeit von Bauhölzern. *Holz Roh. Werkst.* **1938**, *1*, 585–601. [CrossRef]
2. Beery, W.H.; Ifju, G.; McLain, T.E. Quantitative wood anatomy–Relating anatomy to transverse tensile strength. *Wood Fiber Sci.* **1983**, *15*, 395–407.
3. Alteyrac, J.; Cloutier, A.; Ung, C.H.; Zhang, S.Y. Mechanical properties in relation to selected wood characteristics of black spruce. *Wood Fiber Sci.* **2007**, *38*, 229–237.
4. Rapp, A.O.; Brischke, C.; Welzbacher, C.R. Interrelationship between the severity of heat treatments and sieve fractions after impact ball milling: A mechanical test for quality control of thermally modified wood. *Holzforschung* **2006**, *60*, 64–70. [CrossRef]
5. Brischke, C. Agents affecting the structural integrity of wood. In Proceedings of the 10th Meeting of the Northern European Network on Wood Science and Engineering, Edinburgh, UK, 13–14 October 2014; pp. 43–48.

6.  Zimmer, K.; Brischke, C. Within-species variation of structural integrity—A test on 204 Scots pine trees from Northern Europe. In Proceedings of the 12th Meeting of the Northern European Network on Wood Science and Engineering, Riga, Latvia, 12–13 September 2016.

7.  Brischke, C. Interrelationship between static and dynamic strength properties of wood and its structural integrity. *Drvna Ind.* **2017**, *68*, 53–60. [CrossRef]

8.  Brischke, C.; Rapp, A.O.; Welzbacher, C.R. High-energy multiple impact (HEMI)-test—Part1: A new tool for quality control of thermally modified timber. In Proceedings of the The International Research Group on Wood Preservation, Tromsø, Norway, 18–22 June 2006; IRG Document No: IRG/WP/06-20346.

9.  ISO 5223. *Test Sieves for Cereals*; ISO (International Organization for Standardization): Geneva, Switzerland, 1996.

10. Reiterer, A.; Burgert, I.; Sinn, G.; Tschegg, S. The radial reinforcement of the wood structure and its implication on mechanical and fracture mechanical properties-a comparison between two tree species. *J. Mater. Sci.* **2002**, *37*, 935–940. [CrossRef]

11. Bamber, R.K. Sapwood and heartwood. In *Forestry Commission of New South Wales*; Wood Technology and Forest Research Division: Beecroft, Australia, 1987; Volume 2, pp. 1–7.

12. Zanne, A.E.; Westoby, M.; Falster, D.S.; Ackerly, D.D.; Loarie, S.R.; Arnold, S.E.J.; Coomes, D.A. Angiosperm wood structure: Global patterns in vessel anatomy and their relation to wood density and potential conductivity. *Am. J. Bot.* **2010**, *97*, 207–215. [CrossRef] [PubMed]

*Article*

# Orthotropic Tension Behavior of Two Typical Chinese Plantation Woods at Wide Relative Humidity Range

Bingbin Kuai [1], Xuan Wang [1], Chao Lv [1], Kang Xu [2], Yaoli Zhang [1] and Tianyi Zhan [1,*]

[1]   College of Materials Science and Engineering, Nanjing Forestry University, Nanjing 210037, China; bbkuai@njfu.edu.cn (B.K.); wangxuan9210@outlook.com (X.W.); lvchao@njfu.edu.cn (C.L.); zhangyaoli@126.com (Y.Z.)

[2]   Hunan Provincial Collaborative Innovation Center for High-Efficiency Utilization of Wood and Bamboo Resources, Central South University of Forestry and Technology, Changsha 410004, China; xkang86@126.com

*   Correspondence: tyzhan@njfu.edu.cn; Tel.: +86-025-85427058

Received: 2 June 2019; Accepted: 18 June 2019; Published: 19 June 2019

**Abstract:** *Research Highlights:* Orthotropic tension behaviors of poplar and Chinese fir were investigated at a wide relative humidity (RH) range. *Background and Objectives:* Poplar and Chinese fir are typical plantation tree species in China. Mechanical properties of plantation-grown wood varies from naturally-grown one. To utilize poplar and Chinese fir woods efficiently, fully understanding their moisture content (MC) and orthotropic dependency on tension abilities is necessary. *Materials and Methods:* Plantation poplar and Chinese fir wood specimens were prepared and conditioned in series RH levels (0–100%). Tensile modulus ($E$) and strength ($\sigma$) were tested in longitudinal ($L$), radial ($R$), and tangential ($T$) directions. *Results:* The $E$ and $\sigma$ results in transverse directions confirmed the general influence of the MC that decreased with increasing MC. However, both $E$ and $\sigma$ in $L$ direction showed a trend that increased at first, and then decreased when MC increased. The local maximums of stiffness and strength may be associated with straightened non-crystalline cellulose, induced by the penetration of water into the wood cell wall. Using the visualization method for compliance, the tension abilities of poplar and Chinese fir exhibited clear moisture and orthotropic dependency. *Conclusion:* Both poplar and Chinese fir showed a significantly higher degree of anisotropy in the $L$, $R$, and $T$ directions. The results in this study provided first-hand data for wooden construction and wood drying.

**Keywords:** orthotropic; tensile modulus; tensile strength; moisture content; relative humidity

---

## 1. Introduction

Wood supply gradually shifts from natural forest to plantation forest, with the worldwide trend toward intensive forest management to meet the wood demand and to protect the environment [1]. China serves as a typical example of this change, and China has planted more than 4 million hectares of new forest each year since the 1990s [2]. Poplar (*Poplus spp.*) and Chinese fir (*Cunninghamia lanceolate* [Lamb.] Hook.) are common plantation species in China. Considerable differences in physical and mechanically properties can be found between naturally-grown (NG) and plantation-grown (PG) woods [3–5].

The properties of wood from plantation trees are inferior to those from natural forests, primarily because the proportion of juvenile part increases as rotation age is reduced [6]. It is widely accepted that density is the most important parameter in determining the mechanical properties of wood [7–11]. Due to the fast-grown characteristics of plantation wood, especially its juvenile wood, the density of PG wood is lower than NG wood [1]. Besides density, the inferior properties of PG wood are also influenced by its anatomical structures and chemical compositions. PG wood has shorter fibers with considerably thinner cell walls. Compared to NG wood, PG wood has larger micro-fibril angles and the tendency for larger amounts of reaction wood in predominantly juvenile parts [12,13]. Large fibril angles, short fibers with thin walls, and low percentages of latewood in the annual ring all contribute to unsatisfactory strength and stiffness in PG wood [6].

Since wood by nature is anisotropic, it is usually described as a symmetric orthotropic material with three principal axes (longitudinal *L*, radial *R*, and tangential *T*). Mechanical properties in *L*, *R*, and *T* directions vary significantly. The mechanical characterization of wood, based on a three-dimensional approach, is needed as input for ambitious calculations in such fields like civil engineering and material science [14]. When used as truss or grille in a wooden construction, tension behavior in *L* direction is of importance, which has been published for numerous wood species [15–20]. Compared to the *L* direction, the tension behavior in *R* and *T* directions is limited to a few references. In mortise-tenon joint or glue joint area, wood and its composites would be loaded with transverse tensile force. During wood drying, when drying stress exceeds the ultimate tensile strength in transverse direction, crack, or split occur [21]. Hence, tension behavior in transverse direction also deserves to be investigated.

Wood moisture content (MC) influences nearly all its mechanical properties. The effect of MC on the mechanical properties of wood has been an extensively researched topic over the last decades [7,22–28]. In addition, the moisture-dependent orthotropic characteristics of some selected wood species have been published in Keunecke et al. [29], Hering et al. [14], Clauss et al. [30], Bachtiar et al. [31], and Jiang et al. [32]. With the exception of Poisson's ratios, stiffness and strength decrease with increasing MC [14,32], and the influence degree of MC on stiffness/strength in different directions are various. In addition, the decreasing stiffness or strength as a function of MC is observed at given MC ranges in the above mentioned reports.

Wood MC significantly varies at relative humidity (RH) ranges, from 0 to 100%. Fully understanding the MC-dependent mechanical properties at a wide RH range is necessary for utilizing wood and its composites efficiently. To obtain first-hand data of tension behavior of typical Chinese PG woods–poplar and Chinese fir, and to elucidate the orthotropic tensile behavior at wide relative humidity range, the tensile modulus and strength were investigated in RH range from 0 to 100%. The influences of MC on the elastic and strength were studied, and the moisture dependent stiffness was visualized in wide RH range as well.

## 2. Materials and Methods

### 2.1. Materials

Poplar (*Populus tomentosa* Carrière) and Chinese fir (*Cunninghamia lanceolata* [Lamb.] Hook.), two typical Chinese plantation resources, were selected as testing species. Polar and Chinese fir were taken from Henan, and Zhejiang provinces, China, respectively. For each wood species, at least ten logs were cut. Once cut, the logs were air-dried for more than six months to a stable MC around 10 to 12%. The air-dried density of poplar and Chinese fir was $0.50 \pm 0.04$ g/cm$^3$, and $0.37 \pm 0.02$ g/cm$^3$, respectively. After air-dried, specimens without any visual defects were prepared with dimensions of $370 \times 20 \times 15$ ($L \times R \times T$), $30 \times 150 \times 20$ ($L \times R \times T$), and $30 \times 20 \times 150$ ($L \times R \times T$). Tensile tests were conducted along length direction of the specimens, hence, the three dimensions of specimens were named as *L*, *R*, and *T*, respectively. All the specimens were processed as dog-bone shapes (Figure S1) according to Chinese national standards (GB/T 1938-2009 and GB/T 14107-2009).

A total number of 480 specimens were prepared and divided into eight groups with 20 specimens per load axes *L*, *R* and *T*. Each group was conditioned in sealed containers over $P_2O_5$, or saturated solutions (LiCl$_2$, MgCl$_2$, NaBr, NaCl, KCl, and KNO$_3$) or distilled water, which provided RH of 0, 11, 33, 58, 75, 85, 94 or 100% at room temperature (25 °C).

### 2.2. Tensile Test

After the conditioned period, the tensile test was performed using a Zwick Z100 universal testing machine with a 100 kN load cell. A displacement-controlled test was performed with a testing speed of about 1mm/min.

Tensile modulus ($E$, GPa) was obtained from the ratio of the stress ($\sigma$, N) to the corresponding strain ($\varepsilon$, mm$^2$) in the linear elastic range:

$$E = \frac{\Delta\sigma}{\Delta\varepsilon} = \frac{\sigma_2 - \sigma_1}{\varepsilon_1 - \varepsilon_2} \tag{1}$$

where the subscripts "0" and "1" designate the corresponding data at 30 and 70% of the actual force of 50% elastic limit. The orientation and moisture-dependent elastic limits for poplar and Chinese fir were determined in the preliminary tests. To measure the tensile strength (ultimate stress) $\sigma$, the specimens were loaded with a constant loading rate to make sure the specimen failure was reached in $90 \pm 30$s. $\sigma$ was tested using the following relationship

$$\sigma = \frac{P_{\max}}{A} \tag{2}$$

where $P_{\max}$ is the maximum load at the failure point (N), and $A$ is the cross section area at 1/2 length of the unloaded specimen (mm$^2$). For $L$, $R$ and $T$ specimens, the value of $A$ is 60, 36, and 36 mm$^2$, respectively.

### 2.3. Determination of Moisture Content (MC)

MC was determined by the mass of the specimens after conditioned at target RH ($m_i$) and oven-dry ($m_0$) on a dry basis:

$$MC = \frac{m_i - m_0}{m_0} \tag{3}$$

### 2.4. Statistical Analysis

The statistical software, SPSS version 17.0 (SPSS Inc, Chicago, IL, USA) was used for data analysis. Significant effects of MC on orthotropic modulus and strengths were analyzed by Duncan's multiple comparison test ($p = 0.05$).

## 3. Results and Discussion

### 3.1. Moisture Content

MCs of poplar and Chinese fir wood, at different RHs, are shown in Figure 1a,b. At each RH level, the average value of MC was calculated, based on the results of $L$, $R$, and $T$ specimens. Non-significant differences was found among specimens in three directions, regardless of RH levels. At a given RH level, the MC of Chinese fir was higher than poplar. Statistical analyses was conducted, as shown in Table S1. Both poplar and Chinese fir did not reached the fiber saturation point even at the most RH level. Wood MC was influenced by chemical components in cell wall, and associated with bulk density of wood. The lower density of Chinese fir delivered much more specific surface area and void in cell wall. Consequently, Chinese fir absorbed much moisture than poplar, especially at high RH levels.

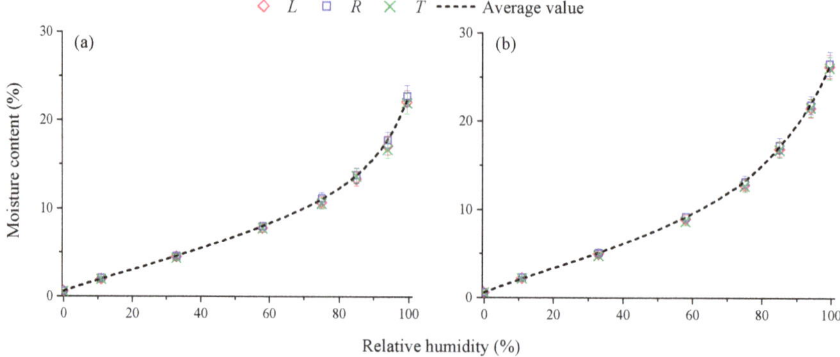



**Figure 1.** Moisture contents of poplar (**a**) and Chinese fir (**b**) at a series of relative humidity.

### 3.2. Moisture-Dependent Orthotropic Elasticity

#### 3.2.1. Tensile Modulus $E$

Changes of $E$, at different RH levels, are shown in Figure 2a,b, for poplar, and Chinese fir, respectively. $E$ or $L$ specimens was significantly greater than $R$ or $T$ specimens for both poplar and Chinese fir. The orientations of tracheids (in Chinese fir) or wood fiber (in poplar) in the L direction provided higher stiffness for $L$ specimens than $R$ or $T$. The stiffness of $R$ specimens was higher than $T$ for both poplar and Chinese fir. Anisotropic behavior, within the transverse plane, was evaluated by the value of $E_R$:$E_T$, while the MC-dependent $E_R$:$E_T$ was still controversial. Decreased trend of $E_R$:$E_T$ with increasing MC was concluded in Bachtiar [31], while Ozyhar et al. [33,34] and Jiang et al. [32] reported that $E_R$:$E_T$ remained unchanged. In this study, $E_R$:$E_T$ of poplar and Chinese fir increased from 1.33 to 1.56, and 1.18 to 1.45, respectively within the whole RH region (0–100%). Regardless of RH level, a higher value of $E_R$:$E_T$ could be found for poplar than Chinese fir. Anatomical structure of poplar is more heterogeneous than that of Chinese fir, which was mainly consisted by tracheids. More cell types and arrangements donated much anisotropic behavior in the transverse plane of poplar.

Based on RH-dependent MC (Figure 1a,b) and $E$ (Figure 2a,b), the relations between MC and $E$ could be established. In Figure 2c,d, changes of $E$ as a function of MC are displayed. Within the whole MC range, $E_L$ increased first and then decreased. The local maximum value of $E_L$ was 10.7, and 8.4 GPa, respectively for poplar at 2.0% MC and Chinese fir at 2.4% MC. The similar phenomena were also reported when wood was subjected to bending force by Takahashi et al. [35], Lu et al. [36], and Jiang et al. [37]. When water molecules penetrated into the wood cell wall, a slight amount of water induced a bridge effect involving hydrogen bonds forming a relative ordered cohesion state between molecular chains [35]. In the previous studies [14,30,32,38], the moisture ranges were limited, in hence, no local maximum value of $E_L$ were reported.

Different from $E_L$, modulus in transverse directions ($E_R$ and $E_T$) decreased monotonically with increasing MC. Water acted as a softener for the wood polymers. Almost all the mechanical properties of wood were affected by moisture [39]. The plasticization of the amorphous polymers enhanced the flexibility of the polymer network when MC increased. A greater decrease in the extent of $E$ of transverse specimens ($R$ and $T$) could be found than $L$ specimens (Table S2). When MC increased from 0 to 26.3%, the decreasing rate of $E_L$, $E_R$, and $E_T$ was 36.5, 50.8, and 60.0%, respectively for Chinese fir. The matrix of lignin and hemicelluloses was much more sensitive to moisture than cellulose fibrils [40,41], and the matrix played a more important role than cellulose fibrils in the interpretation of transverse mechanical properties [42]. Hence, a greater decrease in the stiffness in the transverse direction could be found.

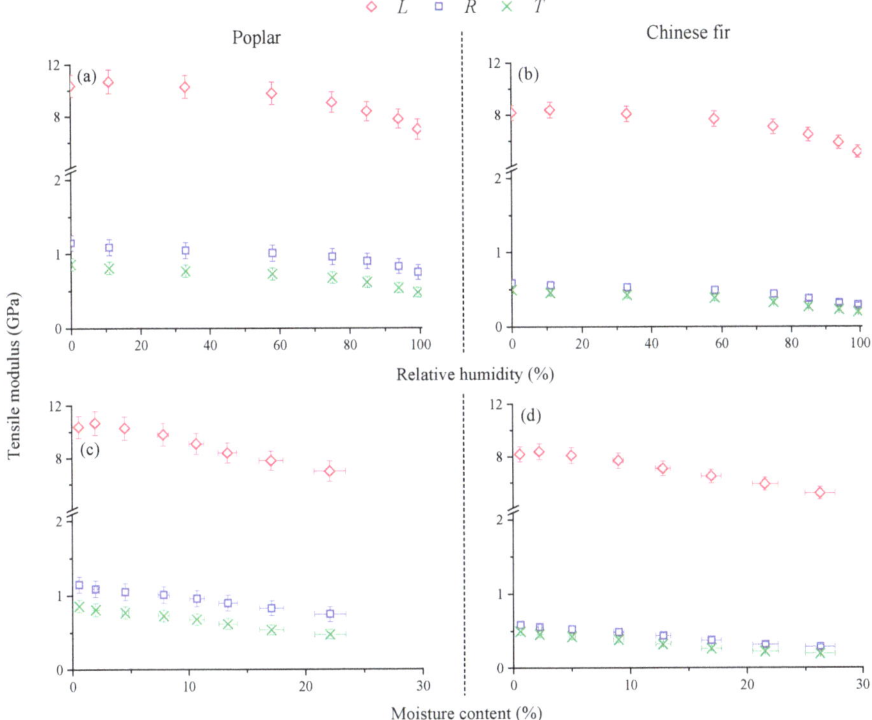

**Figure 2.** Influence of relative humidity on tensile modulus of poplar (**a**) and Chinese fir (**b**) and moisture dependence relations (**c,d**).

### 3.2.2. Visualization of the Orthotropic Elasticity

Elasticity at three orthotropic directions ($L$, $R$ and $T$) could also be used to evaluate the elastic response when stress and strain deviated from these axes. Grimsel [43] proposed a first-time spatial illustration as "deformation body" and described the detailed transformation procedures. To demonstrate the moisture dependency of compliance, two-dimensional illustrations in the principle planes ($L$, $R$ and $T$) of poplar and Chinese fir are presented, based on stiffness values ($E_L$, $E_R$, and $E_T$) (Figure 2a,b), and values of poisson ratio and shear modulus investigated by Bao et al. [1] and Cheng [44]. In Figure 3a,b, the relationships between compliance value and RH level were clearly visible. For both poplar and Chinese fir, increases in compliance as a function of increasing RH level can be observed in axis and off-axis directions. Moreover, the moisture-dependent compliance is spatially illustrated in Figure 4a,b by comparing the hemi deformation bodies at four RH levels. The deformation body can be interpreted as follows [29]: To any arbitrarily chosen axis in the three-dimensional coordinate system, representing the $L$, $R$, and $T$ directions of a wood species, an identical tensile load was applied. The bodies illustrated the degree of deformation depending on the load direction. By the deformation bodies, the influence of the MC on the overall deformation capability of poplar and Chinese fir wood, at a wide humidity range can be summarized effectively. Keunecke et al. [29], Hering et al. [14], Ozyhar et al. [33], Clauss et al. [30], and Jiang et al. [32] presented that the deformation body was suitable to describe the degree of the deformation dependent on the load direction.

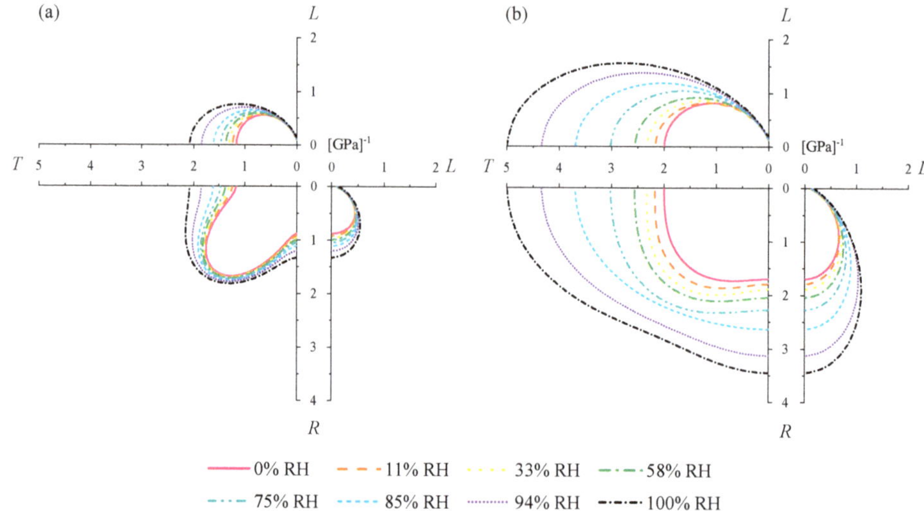

Figure 3. Load-directional dependence of poplar (**a**) and Chinese fir (**b**) wood compliance via polar diagrams for the principal planes of anisotropy.

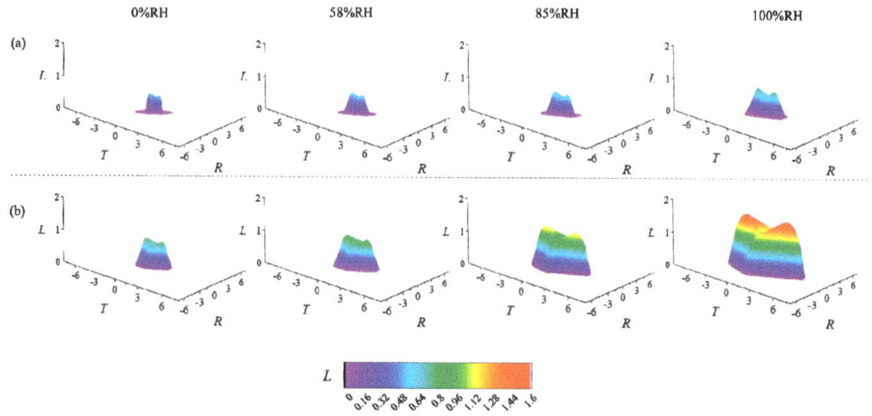

**Figure 4.** Load-directional dependence of poplar (**a**) and Chinese fir (**b**) wood compliance by the three-dimensional representation at four RH levels (0, 58, 85 and 100%).

### 3.3. Moisture-Dependent Orthotropic Strength

Figure 5a,b exhibit the changes of orthotropic $\sigma$ at a series of RHs. Furthermore, the relationships of $\sigma$ and MC are displayed in Figure 5c,d. At a given RH level (or MC), $\sigma$ in the longitudinal direction ($\sigma_L$) was significantly higher than in the transverse directions ($\sigma_R$ and $\sigma_T$) (Table S3). The theoretical tensile strength for cutting the main chains of cellulose molecular was about 8000 MPa, and the typical rupture was attributed to the shearing slip within the crystalline regions in $L$ direction [45]. However, in $R$ and $T$ directions, the middle lamella was the most vulnerable area when resisting transverse tensile force. Within the transverse plane, the capacity of the bearing load in the $R$ direction was superior than in the $T$ direction. Taking 75% RH as an example, the value of $\sigma_R$:$\sigma_T$ was 2.0:1, and 1.5:1, respectively for poplar and Chinese fir. This variation was associated with the orientation of wood ray cells. Wood ray played as reinforcement when tensile load applied in $R$ direction [46]. According to Cheng [44], rays percentage of poplar and Chinese fir ranged 7–13, and 3–6 mm$^{-1}$, respectively.

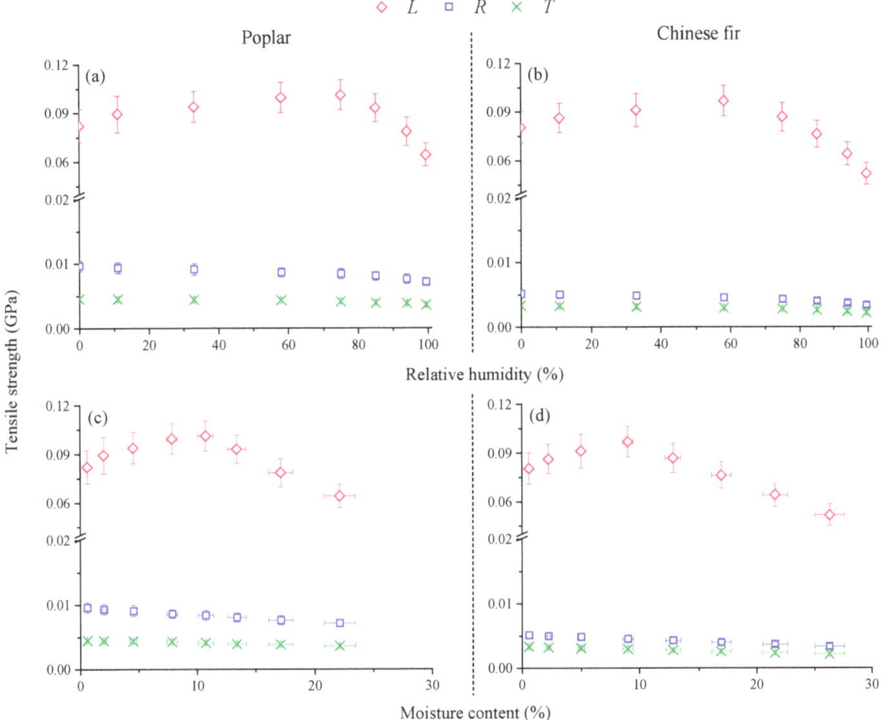

**Figure 5.** Influence of relative humidity on tensile strength of poplar (**a**) and Chinese fir (**b**) and moisture dependence relations (**c,d**).

Within the wide MC ranges, local maximum value of $\sigma_L$ could be found (Figure 5c,d). For poplar and Chinese fir, the local maximum value of $\sigma_L$ was 1.01 and 0.97 GPa at corresponding MC of 10.72 and 8.99%. Commonly, most of the mechanical properties of wood decreased with increasing MC, while the changing trend of $\sigma_L$ within the wide MC ranges was different. In the wood cell wall, both crystalline and non-crystalline cellulose was embedded in a viscoelastic matrix [47]. Typically, slip and rupture in non-crystalline cellulose occurred prior to deformation of crystalline cellulose. When penetrated into the wood cell wall, water molecules induced the non-crystalline cellulose from curl states into straight states. The straightened non-crystalline cellulose may add the crystalline index of wood to some extent [48]. The increased crystalline index allowed more potential energy inside wood cell wall. Hence, local maximum value of $\sigma_L$ can be found at certain MC condition. While, the molecule mechanism of the occurrence of the local maximum of $\sigma_L$ deserves a depth investigation.

## 4. Conclusions

The moisture-dependent orthotropic tension behavior (modulus and strength) of two typical Chinese plantation woods (poplar and Chinese fir) were qualified at wide humidity range (0–100 RH). Both poplar and Chinese fir showed significantly higher degrees of anisotropy in longitudinal (*L*), radial (*R*), and tangential (*T*) directions. The stiffness (*E*) and strength ($\sigma$) results in transverse directions, and confirmed that *E* and $\sigma$ decrease when MC increases. However, both *E* and $\sigma$ in *L* direction showed a trend that increased at first, and then decreased when MC increased. The local maximums of stiffness and strength may be associated with the straightened non-crystalline cellulose induced by the penetration of water into the wood cell wall. Using the visualization method for compliance, tension ability of poplar and Chinese fir exhibited clear moisture and orthotropic dependency. The results in this study provided first-hand data for wooden construction and wood drying.

**Supplementary Materials:** The following are available online at http://www.mdpi.com/1999-4907/10/6/516/s1, Figure S1: Schematic of *L* (a), *R* (b) or *T* (c) specimens, Table S1: Moisture content of poplar and Chinese fir, Table S2: Relative humidity-dependent stiffness of poplar and Chinese fir, Table S3: Relative humidity-dependent strength of poplar and Chinese fir.

**Author Contributions:** T.Z. conceived and designed the experiment. B.K., X.W., C.L., and K.X. performed the experiment and analyzed data. Y.Z. supervised the project. All the authors wrote the manuscript.

**Acknowledgments:** This research was financially supported by the National Key Research and Development Program of China (2017YFD0600202), the National Natural Science Foundation of China (No. 31700487), the Natural Science Foundation of Jiangsu Province (CN) (No. BK20170926).

**Conflicts of Interest:** The authors declare no conflict of interest.

## References

1. Bao, F.; Jiang, Z.; Jiang, X.; Lu, X.; Luo, X.; Zhang, S. Differences in wood properties between juvenile wood and mature wood in 10 species grown in China. *Wood Sci. Technol.* **2001**, *35*, 363–375. [CrossRef]
2. Xu, J. China's new forests aren't as green as they seem. *Nature News* **2011**, *477*, 371. [CrossRef] [PubMed]
3. Phelps, J.; Chen, P. Lumber and wood properties of plantation-grown and naturally grown black walnut. *Forest Prod. J.* **1989**, *39*, 58–60.
4. Bosman, M.T.; de Kort, I.; van Genderen, M.K.; Baas, P. Radial variation in wood properties of naturally and plantation grown light red meranti (Shorea, Dipterocarpaceae). *IAWA J.* **1994**, *15*, 111–120. [CrossRef]
5. Bosman, M.T. Longitudinal variation in selected wood properties of naturally and plantation grown Light Red Meranti (Shorea leprosula and S. parvifolia, Dipterocarpaceae). *IAWA J.* **1996**, *17*, 5–14. [CrossRef]
6. Bendtsen, B.; Senft, J. Mechanical and anatomical properties in individual growth rings of plantation-grown eastern cottonwood and loblolly pine. *Wood Fiber Sci.* **2007**, *18*, 23–38.
7. Miyoshi, Y.; Kojiro, K.; Furuta, Y. Effects of density and anatomical feature on mechanical properties of various wood species in lateral tension. *J. Wood Sci.* **2018**, *64*, 509–514. [CrossRef]
8. Niklas, K.J.; Spatz, H.C. Worldwide correlations of mechanical properties and green wood density. *Am. J. Bot.* **2010**, *97*, 1587–1594. [CrossRef]
9. Kutnar, A.; Kamke, F.A.; Sernek, M. Density profile and morphology of viscoelastic thermal compressed wood. *Wood Sci. Technol.* **2009**, *43*, 57. [CrossRef]
10. Liu, F.; Zhang, H.; Jiang, F.; Wang, X.; Guan, C. Variations in Orthotropic Elastic Constants of Green Chinese Larch from Pith to Sapwood. *Forests* **2019**, *10*, 456. [CrossRef]
11. Emmerich, L.; Wülfing, G.; Brischke, C. The Impact of Anatomical Characteristics on the Structural Integrity of Wood. *Forests* **2019**, *10*, 199. [CrossRef]
12. Cave, I.; Walker, J. Stiffness of wood in fast-grown plantation softwoods: The influence of microfibril angle. *Forest Prod. J.* **1994**, *44*, 43.
13. Barnett, J.R.; Bonham, V.A. Cellulose microfibril angle in the cell wall of wood fibres. *Biol. Rev.* **2004**, *79*, 461–472. [CrossRef] [PubMed]
14. Hering, S.; Keunecke, D.; Niemz, P. Moisture-dependent orthotropic elasticity of beech wood. *Wood Sci. Technol.* **2012**, *46*, 927–938. [CrossRef]
15. Navi, P.; Rastogi, P.K.; Gresse, V.; Tolou, A. Micromechanics of wood subjected to axial tension. *Wood Sci. Technol.* **1995**, *29*, 411–429. [CrossRef]
16. Kojima, Y.; Yamamoto, H. Properties of the cell wall constituents in relation to the longitudinal elasticity of wood. *Wood Sci. Technol.* **2004**, *37*, 427–434. [CrossRef]
17. Hering, S.; Niemz, P. Moisture-dependent, viscoelastic creep of European beech wood in longitudinal direction. *Eur. J. Wood Wood Prod.* **2012**, *70*, 667–670. [CrossRef]
18. Sonderegger, W.; Martienssen, A.; Nitsche, C.; Ozyhar, T.; Kaliske, M.; Niemz, P. Investigations on the physical and mechanical behaviour of sycamore maple (Acer pseudoplatanus L.). *Eur. J. Wood Wood Prod.* **2013**, *71*, 91–99. [CrossRef]
19. Niemz, P.; Clauss, S.; Michel, F.; Hansch, D.; Hansel, A. Physical and mechanical properties of common ash (Fraxinus excelsior L.). *Wood Res.* **2014**, *59*, 671–682.

20. Oscarsson, J.; Olsson, A.; Enquist, B. Strain fields around knots in Norway spruce specimens exposed to tensile forces. *Wood Sci. Technol.* **2012**, *46*, 593–610. [CrossRef]

21. Wang, X.; Song, L.; Cheng, D.; Liang, X.; Xu, B. Effects of saturated steam pretreatment on the drying quality of moso bamboo culms. *Eur. J. Wood Wood Prod.* **2019**. [CrossRef]

22. Gerhards, C.C. Effect of moisture content and temperature on the mechanical properties of wood: An analysis of immediate effects. *Wood Fiber Sci.* **2007**, *14*, 4–36.

23. Rémond, R.; Passard, J.; Perré, P. The effect of temperature and moisture content on the mechanical behaviour of wood: A comprehensive model applied to drying and bending. *Eur. J. Mech A - Solids* **2007**, *26*, 558–572. [CrossRef]

24. Cave, I. Modelling moisture-related mechanical properties of wood Part I: Properties of the wood constituents. *Wood Sci. Technol.* **1978**, *12*, 75–86. [CrossRef]

25. Yu, Y.; Fei, B.; Wang, H.; Tian, G. Longitudinal mechanical properties of cell wall of Masson pine (Pinus massoniana Lamb) as related to moisture content: A nanoindentation study. *Holzforschung* **2011**, *65*, 121–126. [CrossRef]

26. Cave, I. Modelling moisture-related mechanical properties of wood Part II: Computation of properties of a model of wood and comparison with experimental data. *Wood Sci. Technol.* **1978**, *12*, 127–139. [CrossRef]

27. Ishimaru, Y.; Arai, K.; Mizutani, M.; Oshima, K.; Iida, I. Physical and mechanical properties of wood after moisture conditioning. *J. Wood Sci.* **2001**, *47*, 185–191. [CrossRef]

28. Cousins, W. Young's modulus of hemicellulose as related to moisture content. *Wood Sci. Technol.* **1978**, *12*, 161–167. [CrossRef]

29. Keunecke, D.; Hering, S.; Niemz, P. Three-dimensional elastic behaviour of common yew and Norway spruce. *Wood Sci. Technol.* **2008**, *42*, 633–647. [CrossRef]

30. Clauss, S.; Pescatore, C.; Niemz, P. Anisotropic elastic properties of common ash (Fraxinus excelsior L.). *Holzforschung* **2014**, *68*, 941–949. [CrossRef]

31. Bachtiar, E.V.; Sanabria, S.J.; Mittig, J.P.; Niemz, P. Moisture-dependent elastic characteristics of walnut and cherry wood by means of mechanical and ultrasonic test incorporating three different ultrasound data evaluation techniques. *Wood Sci. Technol.* **2017**, *51*, 47–67. [CrossRef]

32. Jiang, J.; Bachtiar, E.V.; Lu, J.; Niemz, P. Moisture-dependent orthotropic elasticity and strength properties of Chinese fir wood. *Eur. J. Wood Wood Prod.* **2017**, *75*, 927–938. [CrossRef]

33. Ozyhar, T.; Hering, S.; Niemz, P. Moisture-dependent elastic and strength anisotropy of European beech wood in tension. *J. Mater. Sci.* **2012**, *47*, 6141–6150. [CrossRef]

34. Ozyhar, T.; Hering, S.; Sanabria, S.J.; Niemz, P. Determining moisture-dependent elastic characteristics of beech wood by means of ultrasonic waves. *Wood Sci. Technol.* **2013**, *47*, 329–341. [CrossRef]

35. Takahashi, C.; Nakazawa, N.; Ishibashi, K.; Iida, I.; Furuta, Y.; Ishimaru, Y. Influence of variation in modulus of elasticity on creep of wood during changing process of moisture. *Holzforschung* **2006**, *60*, 445–449. [CrossRef]

36. Lu, J.; Jiang, J.; Wu, Y.; Li, X.; Cai, Z. Effect of moisture sorption state on vibrational properties of wood. *Forest Prod. J.* **2012**, *62*, 171–176. [CrossRef]

37. Jiang, J.; Lu, J.; Cai, Z. The vibrational properties of Chinese fir wood during moisture sorption process. *BioResources* **2012**, *7*, 3585–3596.

38. Kretschmann, D.E.; Green, D.W. Modeling moisture content-mechanical property relationships for clear southern pine. *Wood Fiber Sci.* **2007**, *28*, 320–337.

39. Salmén, L. Micromechanical understanding of the cell-wall structure. *C R Biol.* **2004**, *327*, 873–880. [CrossRef]

40. Engelund, E.T.; Thygesen, L.G.; Svensson, S.; Hill, C.A. A critical discussion of the physics of wood–water interactions. *Wood Sci. Technol.* **2013**, *47*, 141–161. [CrossRef]

41. Kulasinski, K.; Guyer, R.; Derome, D.; Carmeliet, J. Water adsorption in wood microfibril-hemicellulose system: Role of the crystalline–amorphous interface. *Biomacromolecules* **2015**, *16*, 2972–2978. [CrossRef] [PubMed]

42. Åkerholm, M.; Salmén, L. The oriented structure of lignin and its viscoelastic properties studied by static and dynamic FT-IR spectroscopy. *Holzforschung* **2003**, *57*, 459–465. [CrossRef]

43. Grimsel, M. *Mechanisches Verhalten von Holz: Struktur-und Parameteridentifikation eines anisotropen Werkstoffes*; w.e.b.-Univ.-Verl.: Dresden, Germany, 1999; ISBN 3-933592-66-6.

44. Cheng, J. *Wood Science*; China Forestry Pub.: Beijing, China, 1985.

45. Bodig, J.; Jayne, B.A. *Mechanics of Wood and Wood Composites*; Van Nostrand Reinhold Company Inc.: New York, NY, USA, 1982.
46. Reiterer, A.; Burgert, I.; Sinn, G.; Tschegg, S. The radial reinforcement of the wood structure and its implication on mechanical and fracture mechanical properties—a comparison between two tree species. *J. Mater. Sci* **2002**, *37*, 935–940. [CrossRef]
47. Zhan, T.; Jiang, J.; Peng, H.; Lu, J. Dynamic viscoelastic properties of Chinese fir (Cunninghamia lanceolata) during moisture desorption processes. *Holzforschung* **2016**, *70*, 547–555. [CrossRef]
48. Ketoja, J.; Paavilainen, S.; McWhirter, J.L.; Róg, T.; Järvinen, J.; Vattulainen, I. Mechanical properties of cellulose nanofibrils determined through atomistic molecular dynamics simulations. *Nord. Pulp Pap. Res. J.* **2012**, *27*, 282–286. [CrossRef]

*Article*

# Bending Stiffness, Load-Bearing Capacity and Flexural Rigidity of Slender Hybrid Wood-Based Beams

**Barbara Šubic [1,*], Gorazd Fajdiga [2] and Jože Lopatič [3]**

[1]   M SORA d.d., Trg Svobode 2, 4226 Žiri, Slovenia
[2]   Biotechnical Faculty, University of Ljubljana, Jamnikarjeva 101, 1000 Ljubljana, Slovenia; Gorazd.Fajdiga@bf.uni-lj.si
[3]   Faculty of Civil and Geodetic Engineering, University of Ljubljana, Jamova 2, 1000 Ljubljana, Slovenia; Joze.Lopatic@fgg.uni-lj.si
*   Correspondence: Barbara.Subic@m-sora.si; Tel.: +386-31-541-681

Received: 19 October 2018; Accepted: 9 November 2018; Published: 13 November 2018

**Abstract:** Modern architecture suggests the use of opened spaces with large transparent envelope surfaces. Therefore, windows of long widths and large heights are needed. In order to withstand the wind loads, such wooden windows can be reinforced with stiffer materials, such as aluminium (Al), glass-fibre reinforced polymer (GFRP), and carbon-fibre reinforced polymer (CFRP). The bending stiffness, load-bearing capacity, and flexural rigidity of hybrid beams, reinforced with aluminium, were compared through experimental analysis, using a four-point bending tests method, with those of reference wooden beams. The largest increases in bending stiffness (29%–39%), load-bearing capacity (33%–45%), and flexural rigidity (43%–50%) were observed in the case of the hybrid beams, with the highest percentage of reinforcements (12.9%—six reinforcements in their tensile and six reinforcements in their compressive zone). The results of the experiments confirmed the high potential of using hybrid beams to produce large wooden windows, for different wind zones, worldwide.

**Keywords:** wood based composites; hybrid beams; bending stiffness; flexural rigidity; aluminium reinforcements; wooden windows

---

## 1. Introduction

The share of wood-based windows (i.e., wooden and aluminium-wooden profiled windows) is the largest (66%) among certified windows for low-energy houses, due to their high energy efficiency and their beneficial life cycle assessment characteristics [1]. Polyvinyl chloride (PVC) and aluminium windows have 19% and 13% shares, respectively. Contemporary architectural design of buildings integrate wooden products in the building envelope as well as in the interior. One of the reasons for this is that it was proven that the use of natural wood in the buildings has positive psychological, emotional, and health impacts on the people living in such an environment [2]. Windows have an important role (solar gains through glazing), effecting the energy efficiency of the building. Therefore, there has been a gradual increase in the proportion of the transparent part of the building envelope, over the last decade. Windows are more and more frequently placed from the bottom to the top of individual storeys, with heights exceeding 3.0 m, or even over several storeys, with heights exceeding 5.0 m (an example of such building is presented in Figure 1).

Two most important window elements are window frame and window glazing. From a thermal point of view, the window frame is a weak point of the low energy windows, since its thermal transmittance varies from 0.8–1.6 W/m$^2$K, whereas thermal transmittance of the glazing is almost half this value and varies from 0.5–0.8 W/m$^2$K [1]. To improve thermal transmittance of the window we need to either improve thermal transmittance of the frame or increase the surface of the glazing, compared to the whole window area. The architectural trends prefer the open view through windows, with as little visible window frame as possible. Therefore, for this study small cross sectioned (i.e., slender) profiles were analysed, to explore the limits of improving their bending stiffness and strength, when being reinforced with stiffer materials.

**Figure 1.** An example of window elements with a height of 5.8 m.

In Europe, windows are mostly manufactured from softwood species—90% (e.g., Norway spruce—*Picea abies* (L.) H. Karst., Pine—*Pinus sylvestris* L. and Siberian larch—*Larix sibirica* Ledeb.), whereas hardwoods are seldom used—10% (e.g., oak—*Quercus* spp. L., Red grandis—*Eucalyptus grandis* W. Hill ex Maiden, walnut—*Juglans regia* L., beech—*Fagus sylvatica* L.) [3]. The main reason for this is that softwoods (especially Norway spruce and Pine, which in Europe are a domestic wood species), are easy to process and have up to 38% better thermal properties than hardwoods [4]. This is an important parameter when trying to comply with today's high thermal efficiency demands for modern buildings.

Over the last twenty years, many studies have been performed with regard to the flexural rigidity and load-bearing capacity of hybrid beams; not only experimentally [5–8], but also analytically [9–14] and numerically [15–19]. In most of the mentioned studies much larger cross sections than those used for windows were investigated (e.g., up to 500 mm in height) [12,15]. Only a few researchers have studied the reinforcing effect on specimens with cross sections smaller than 100 mm × 100 mm [14,19,20], where limited space for reinforcements is available. In most cases the reinforcements are visible at least on one side of the tested specimens and mostly oriented horizontally [5,6,12]. Windows are influencing the overall architectural appearance of the building from the outside, as well as the interior design, and therefore no visible reinforcements are preferred on either side of the window.

Different reinforcing materials have been used in many studies to improve the bending stiffness and load-bearing capacity of wooden beams. Steel profiles [5,9,12], glass-fibre reinforced polymer profiles (GFRP (glass-fibre reinforced polymer) profiles) [6,14,18], and carbon-fibre reinforced polymer profiles (CFRP (carbon-fibre reinforced polymer) profiles) [7,11,15,19] are the most common, but the effect of many others has also been investigated; e.g., basalt fibres [13], hemp, flax, basalt, and bamboo fibre reinforcements [21], as well as concrete in combination with wood [22].

The aim of this study is to explore possibilities of using domestic wood species for technologically challenging big-sized window elements, with high added value. For this reason, the Norway spruce, as the basic wood species, was used in this study.

During the first stage of the study, the mechanical characteristics of the used materials were defined. Slender hybrid beams were manufactured using aluminium reinforcements and four-point bending tests were performed in order to compare the bending stiffness and load-bearing capacities of beams having different percentages, orientations, and positions of the reinforcing material.

All decisions about the type of adhesive, the number and position of the reinforcements and the choice of material were made taking into account not only the required mechanical characteristics but also future production feasibility and cost-effectiveness.

## 2. Materials and Methods

Norway spruce was used as the basic material for all the beam specimens which were prepared for this study. Beams were manufactured from lamellas with thicknesses from 8 to 28 mm, with a measured averaged moisture content of 12% and an average density of 460 kg/m$^3$. The mechanical characteristics of the spruce, that correspond to structural grade C30 [23], used in window production, compared to those of the reinforcing materials, are presented in Table 1. The shear strength was determined experimentally, according to standard EN 205 [24].

**Table 1.** Material characteristics.

|  | Norway Spruce | Aluminium | GFRP | CFRP |
|---|---|---|---|---|
| Density $\varrho$ (kg/m$^3$) | 460 | 2700 | 2000 | 1420 |
| Poisson's ratio $\nu$ | 0.37 | 0.30 | 0.27 | 0.3 |
| Modulus of elasticity $E_0$ (MPa) | 12,000 | 70,000 | 38,000 | 120,000 |
| Shear modulus G (MPa) | 750 | 25,000 | 5000 | 4000 |
| Tensile strength $f_t$ (MPa) | 19 | 195 | / | / |
| Compressive strength $f_c$ (MPa) | 24 | / | / | / |
| Shear strength $f_v$ (MPa) | 8.4 * | / | / | / |
| Thermal conductivity $\lambda$ (W/mK) | 0.11 | 210 | 0.3 | 5 |
| Price factors compared to wood (per m$^3$) | 1 | 28 | 71 | 180 |

* Experimentally defined value.

Three different types of reinforcing material were planned to be used later on within the wider scope of this study; aluminium, GFRP, and CFRP profiles. The longitudinal modulus of elasticity of the CFRP profiles, aluminium, and GFRP profiles compared to that of Norway spruce are higher by a factor of 10.0, 5.8, and 3.2, respectively (Table 1). Thus, CFRP profiles should provide the best improvements in bending stiffness and load-bearing capacity [10,14,25]. Despite this, we decided to perform the first phase of the tests with aluminium reinforcements only, in order to define their positioning and orientation effect on the observed parameters and to analyse the possibilities of manufacturing window profiles with non-visible reinforcements. The advantages of aluminium are the uniformity of its mechanical properties, its high tensile capacity, and the easy processing, accessibility, and low costs of the material itself. On the other hand, its high thermal conductivity decreases the low energy efficiency of the windows, so it would be expected that in a further step of the research it would be replaced with GFRP or CFRP profiles.

The aluminium profiles used for the analysis were made of aluminium alloy 6060-T66 and technical data are collected from technical sheets provided by the manufacturer [26]. For GFRP and CFRP profiles the data were also obtained from technical data sheets from the manufacturers [27].

For this study pre-analysis of the adhesive selection for connection between aluminium and wood was performed. Tensile shear strength of three two-component epoxy adhesives (2K-EP) and three two-component polyurethane adhesives (2K-PU) were defined according to EN 1465:2009 [28], with tensile lap-joint tests on aluminium specimens. For each adhesive, 10 specimens were tested and the shear strength was measured. The mean shear strength of the Novasil P-SP adhesive was the lowest and of COSMO PU was the highest (Table 2).

**Table 2.** Adhesives characteristics and experimental results of their shear strength.

| Brand Name | Type | Mean Shear Strength (MPa) | Standard Deviation of Shear Strength (MPa) | Technical Sheet Shear Strength (MPa) |
|---|---|---|---|---|
| Körapox 565/GB | 2K-EP | 9.9 | 0.96 | 24 |
| Permabond ET515 | 2K-EP | 4.4 | 0.53 | 10 |
| COSMO EP-200.110 | 2K-EP | 6.1 | 0.75 | 18 |
| Körapur 790/30 | 2K-PU | 8.9 | 1.13 | 18 |
| Novasil P-SP 6944 | 2K-PU | 4.0 | 0.96 | / |
| COSMO PU-200.280 | 2K-PU | 11.2 | 0.75 | 18 |

Due to its largest shear strength, it was decided to use the two-component polyurethane adhesive—COSMO PU 200.280—for further experimental analysis. All of the mean shear strengths defined through experimental analysis had a value that was lower compared to the declared ones on the manufacturers' technical sheets. Nevertheless, the shear strength of 11.2 MPa (of the selected 2K-PU adhesive), was higher than the shear strength of the wood itself (8.4 MPa), which should present sufficient strength to avoid the cohesive failure of the joint.

In the second step the four different window beam sections (scantlings), designated types A, B, C, and D, were selected for the experimental study (Figure 2). Norway spruce was used as the basic material, and aluminium as the reinforcing material.

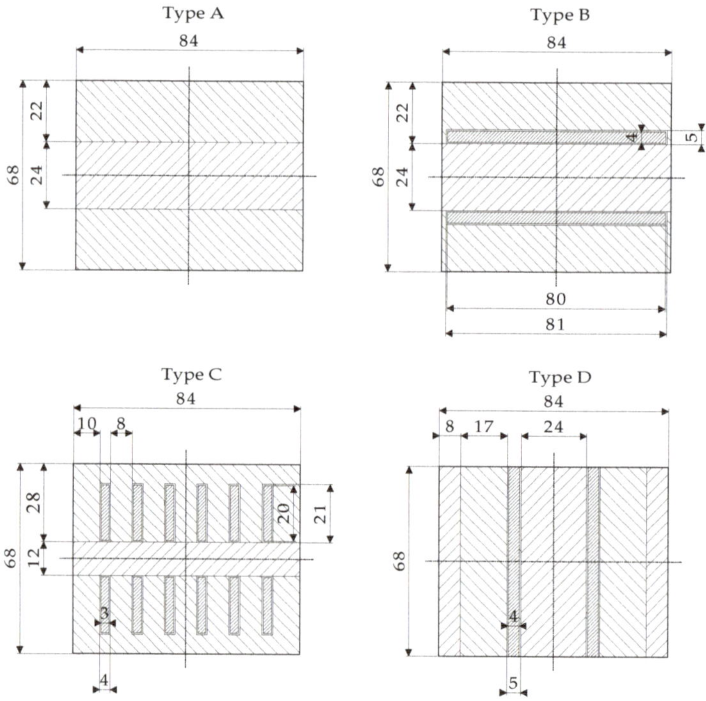


**Figure 2.** Cross sections of the type A, B, C, and D specimens (measurements in mm).

The position and orientation of the reinforcements in the specimens, B, C, and D, were defined, taking into account number of different aspects:

- To improve bending stiffness, load-bearing capacity and flexural-rigidity the reinforcements need to be positioned near tensile and compressive faces, as far away from the neutral axis as possible [9,15,25,29].
- The moment of inertia of the cross section should be maximized by proper orientation of the reinforcements and the provision of a sufficient number of them.
- An aesthetic requirement—no reinforcement should be visible when the window is completed.
- The dimensions of the reinforcements should correspond to those of standard products available on the market, in order to decrease production costs.
- The production feasibility aspect—the basic thickness of wooden lamellas, to produce window scantlings, is 24 mm.
- Cost-effectiveness of the hybrid beams.

The outside measurements of the specimens were prepared with the following dimensions: width (w) × height (h) × length ($L_e$) = 84 mm × 68 mm × 1950 mm. Basic wooden beams for specimens were manufactured from lamellas with thicknesses from 8 to 28 mm. The type A specimens were unreinforced (i.e., reference specimens). The type B specimens had two horizontal reinforcements (with width × height dimensions of 80 mm × 4 mm) placed in the pre-milled grooves of the contact regions between the two neighbouring wooden lamellas. The type C specimens had twelve vertical reinforcements (with width × height dimensions of 3 mm × 20 mm), which were located in each of the two outside wooden lamellas (six in each lamella). Type D specimens were prepared with two vertical reinforcements with dimensions of 68 × 4 mm. All the pre-milled grooves were in width and height 1 mm bigger than reinforcements, leaving 0.5 mm space for the adhesive around the reinforcements. The adhesive was applied to the grooves and the reinforcements were then manually inserted into them. Specimens C had the highest number of reinforcements with its highest area (12.6%), while specimens D had smallest number of reinforcements with the smallest area (9.5%) (Table 3).

**Table 3.** Characteristics of the tested specimens.

| Specimen | Number of Specimens (-) | Series | Dimensions of Reinforcement b × h (mm) h | Number of Reinforcements (-) | $A_{real}$ Percentage of Reinforcement (%) |
|---|---|---|---|---|---|
| A1–A5 | 5 | S1 | - | - | - |
| * B1–B5 | 4 | S1 | 80 × 4 | 2 | 11.2 |
| C1–C5 | 5 | S1 | 3 × 20 | 12 | 12.6 |
| A6–A10 | 5 | S2 | - | - | - |
| B6–B10 | 5 | S2 | 80 × 4 | 2 | 11.2 |
| C6–C10 | 5 | S2 | 3 × 20 | 12 | 12.6 |
| D1–D5 | 5 | S2 | 4 × 68 | 2 | 9.5 |

* B4 was excluded from the analysis due to its different dimension of the cross section (84 mm × 65 mm).

A decision about the preparation of type D specimens was made after the first tests of Series S1 on specimens A, B, and C had been completed, in order to be able to make additional comparisons of the effect of different reinforcement orientations on the bending stiffness, load-bearing capacity, flexural rigidity, and type of failure mechanism. The type D specimens did not fulfil the aesthetic requirement for non-visibility of the reinforcements.

Two test fields, for four-point bending tests, were prepared for two series of hybrid specimens; Series S1 (Figure 3) and Series S2 (Figure 4). The difference between them was in different spans L, lengths of the middle areas l, and the distance of the applied load from the support a. The simply supported beams of Series S1 were 1950 mm long and had a span of L = 1750 mm. Loads of P/2, which were 750 mm apart, were applied at a distance a = 500 mm from the beam supports (Figure 3). The simply supported beams of Series S2 were 1950 mm long and had a span of L = 1850 mm. In this case loads of P/2, which were 700 mm apart, were applied at a distance a = 625 mm from the beam supports (Figure 4). The difference between both series was that the distance between the loads l decreased and the span L increased, in the case of Series S2. With this change, the bending moment with specimens of Series S2 increased, at the same loading force, compared to specimens of Series S1. This step was made in order to decrease the probability that the failure would be caused by shear strength, prior the bending strength would be fully exploited.

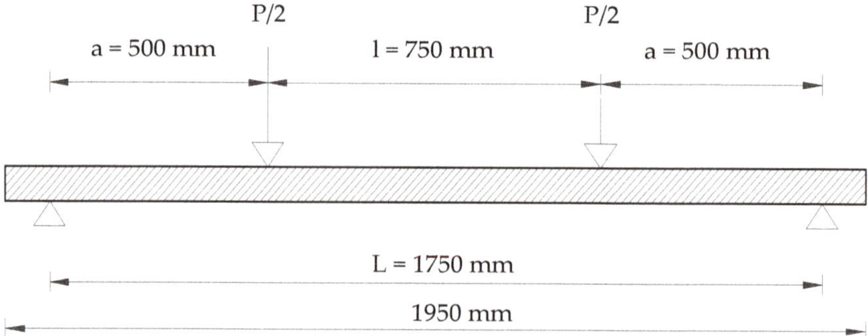


**Figure 3.** Schematic diagram of the four-point bending test field configuration for the specimens of Series S1.

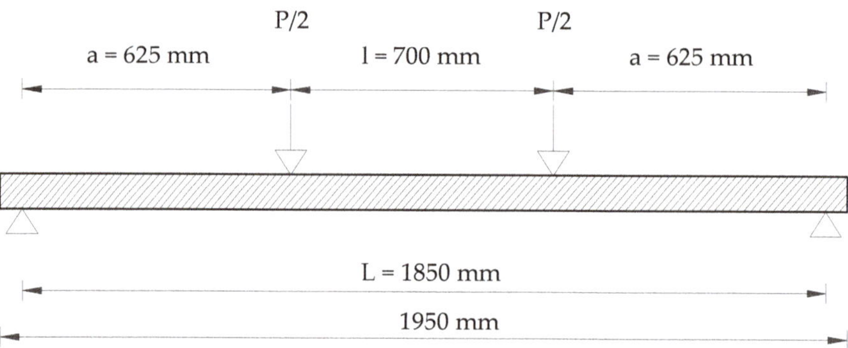


**Figure 4.** Schematic diagram of the four-point bending test field configuration for the specimens of Series S2.

The four-point static bending tests, up to failure, were performed using a Roel Amsler HA 100 universal servo-hydraulic testing machine (Zwick GmbH & Co. KG, Ulm, Germany) (Figure 5).

The flexural loading tests were performed in displacement (i.e., stroke) control mode, with an actuator movement rate of 0.1 mm/s, up to failure. In the region between the two applied loads the bending moment was constant, the shear force was zero, and the maximum relative deflection was caused by the bending moments only. On each specimen the deflections were measured at two points using linear voltage displacement transducers (LVDTs) marked as IND in Figure 6.

The absolute deflection $w_1$ was calculated as the average value of the deflections measured at the midspan by the IND 1 and IND 2. All the physical properties (i.e., the displacements, the strains, and the load P) were measured and recorded by a Dewesoft DEWE 2500 data acquisition system (DEWESoft d.o.o., Trbovlje, Slovenija).

**Figure 5.** The testing equipment for the performance of the four-point bending tests.

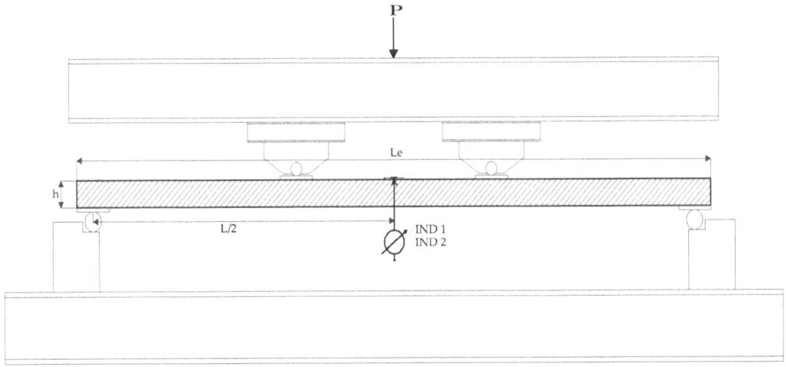

**Figure 6.** Basic configuration of the test instrumentation.

Effective flexural rigidity $(EI)_{eff}$ was calculated on the basis of the average absolute midspan deflection $w_1$, using an expression of technical mechanics (Equation (1)) that corresponds to the four-point bending test load arrangement. The equation neglects a small part of deflections caused by shear stresses.

$$(EI)_{eff} = 0.33\, P_u \cdot a \cdot (3L^2 - 4a^2)/48 \cdot w_1 \tag{1}$$

The ultimate load $(P_u)$ is the load where the failure occurred, a is the distance between the support and the load, L is the span length and $w_1$ is the absolute midspan deflection.

## 3. Results

The bending stiffness was calculated for each specimen type at three different loading stages: at a load corresponding to 33% of the ultimate load $(0.33\, P_u)$; at the limit of proportionality $(P_p)$, which represents the end of the region of linear behaviour; and at the ultimate load $(P_u)$ (Table 4). Beside the corresponding load, the deflection occurring at midspan $(w_{1,i})$ and the bending stiffness $(k_{1,i})$ are given at each loading step (i = 1/3 for the loading step $0.33\, P_u$, i = p for the loading step $P_p$ and i = u for the loading step $P_u$). The bending stiffness was calculated as $k_{1,i} = P_i/w_{1,i}$. The flexural rigidity was calculated from the displacement $w_1$, and was the highest at C specimens and the lowest at A specimens (Table 4).

**Table 4.** Summary of the experimental results of the specimens of both series of tests (i.e., S1 and S2). The loads ($P_i$) and midspan deflections ($w_{1,i}$) are presented as averaged measured values of all specimens, of the same type and series, and the bending stiffness was calculated as $k_i = P_i/w_i$. In the last six columns comparisons of the different loads ($P_i$), the midspan deflections ($w_{1,i}$), the bending stiffnesses, and flexural rigidity are presented.

| | Specimens No. | At 33% of Ultimate Load | | | At the Proportional Limit | | | At Ultimate Load | | | Comparisons | | | | | Flexural Rigidity |
|---|---|---|---|---|---|---|---|---|---|---|---|---|---|---|---|---|
| | | $P_{1/3}$ (kN) | $w_{1,1/3}$ (mm) | $k_{1,1/3}$ (kN/mm) | $P_p$ (kN) | $w_{1,p}$ (mm) | $k_{1,p}$ (kN/mm) | $P_u$ (kN) | $w_{1,u}$ (mm) | $k_{1,u}$ (kN/mm) | $P_{u,(B,C,D)}/P_{u,A}$ (-) | $P_p/P_u$ (-) | $w_{1,p}/w_{1,u}$ (-) | $k_{1,p}/k_{1,u}$ (-) | $k_{1,1/3}/k_{1,u}$ (-) | $(EI)_{eff}$ (kN*m²) |
| Series S1 | A1–A5 | 6.2 | 19.5 | 0.32 | 12.7 | 39.4 | 0.32 | 18.6 | 67.4 | 0.30 | - | 0.68 | 0.58 | 1.09 | 1.13 | 27.5 |
| | B1–B5 * | 5.7 | 12.1 | 0.48 | 14.3 | 30.9 | 0.46 | 17.1 | 37.5 | 0.46 | 0.92 | 0.84 | 0.82 | 1.01 | 1.04 | 38.7 |
| | C1–C5 | 8.2 | 17.0 | 0.47 | 15.9 | 33.2 | 0.48 | 24.7 | 58.2 | 0.44 | 1.33 | 0.64 | 0.57 | 1.09 | 1.10 | 41.2 |
| Series S2 | A6–A10 | 4.8 | 20.0 | 0.24 | 10.3 | 43.3 | 0.24 | 14.5 | 78.6 | 0.19 | - | 0.71 | 0.55 | 1.26 | 1.29 | 27.3 |
| | B6–B10 | 5.7 | 17.3 | 0.33 | 14.0 | 42.5 | 0.33 | 17.2 | 54.6 | 0.32 | 1.18 | 0.82 | 0.78 | 1.04 | 1.04 | 37.5 |
| | C6–C10 | 7.0 | 20.8 | 0.34 | 13.9 | 40.6 | 0.34 | 21.1 | 80.4 | 0.27 | 1.45 | 0.66 | 0.51 | 1.28 | 1.27 | 39 |
| | D1–D5 | 6.5 | 20.0 | 0.32 | 11.8 | 36.2 | 0.33 | 19.5 | 81.7 | 0.24 | 1.35 | 0.60 | 0.44 | 1.35 | 1.34 | 37.6 |

* B4 was excluded from further analysis due to the change in cross section dimension.

The results show that the ultimate load ($P_u$) and bending stiffness ($k_i$) were the highest in the case of the type C beam specimens, which had the highest percentage of reinforcements, whereas it was the lowest, as expected, in the case of the type A beam specimens, which had no reinforcement (Figure 7). The comparisons of the bending stiffnesses $k_{1,p}$ and $k_{1,u}$ show the plasticity capacity of the tested beams. In the case of the type B beam specimens, both parameters have almost equal values, which corresponds to an almost linear load-deflection relationship up to failure (Figure 7). The biggest difference between these two parameters (i.e., 34%) occurred in the case of the beams D1–D5, which indicates their nonlinear behaviour between the limit of the proportionality and the ultimate load.

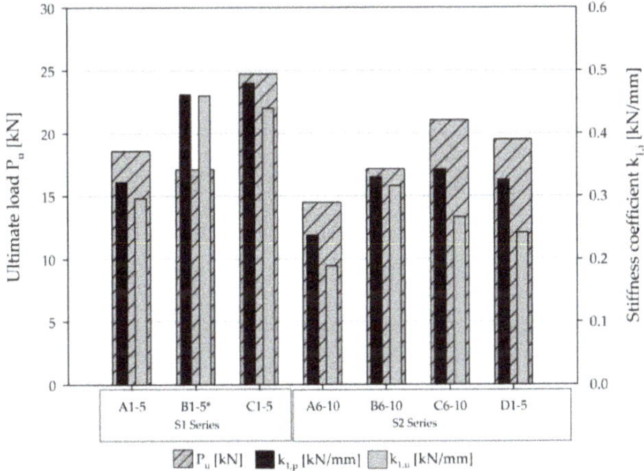



**Figure 7.** Ultimate load-bearing capacity and stiffness coefficient of the tested specimens of types A, B, C, and D.

The average values of the measured midspan deflections were compared at an applied load equal to P = 10 kN, where the load-deflection relationships are still linear (Figure 8).

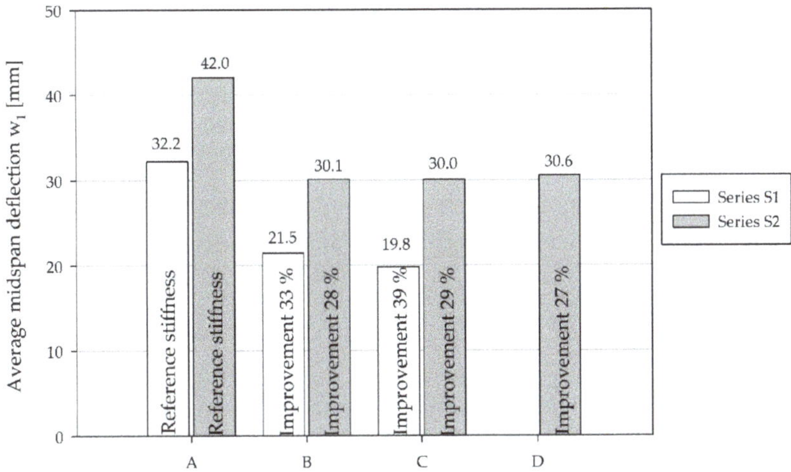



**Figure 8.** Average midspan deflection ($w_1$) of the beam specimens A, B, C, D at the load P = 10 kN and the bending stiffness improvement (%) of the beams of types B, C, D compared to the type A beams.

The average measured midspan deflections ($w_1$—calculated as the mean values of IND 1 and 2) of the beam specimens of types A, B, and C, of Series S1, were 32.2 mm, 21.5 mm, and 19.8 mm, respectively (Figure 8). The bending stiffness improvements of the beam specimens of types B and C when compared to the type A reference specimens, were 33% and 39%, respectively. The average midspan deflections of the Series S2 beams of types A, B, C and D amounted to 42.0 mm, 30.1 mm, 30.0 mm, and 30.6 mm, respectively. The improvement in bending stiffness of the hybrid beams of types B, C, and D, compared to that of type A beams, amounted to 28%, 29%, and 27%, respectively. In both series of tests, the midspan deflection of the reinforced hybrid beams was decreased compared to the reference specimens of type A. The largest standard deviation of the midspan deflection was observed with the type A beams, in the case of both series. This was expected, since these specimens were made solely of wood, which is a non-homogeneous material, with significant variations in density and different grain orientations.

On average, the hybrid beams have improved load-bearing capacity compared to the reference beams (Figure 9). Only the type B hybrid beams, of Series S1, had 8% lower average load-bearing capacity than the reference type A specimens. The reason for this lies in the high shear stresses, which caused a shear failure (Figure 10b), before the flexural load-bearing capacity was reached. The failure happened in the adhesion layer between the wood and the reinforcement. The type C specimens had a load-bearing capacity which was 33% greater than that of the type A reference specimens (Figure 9). The load-bearing capacity of the Series S2 tested specimens of types A, B, C, and D amounted to 14.5 kN, 17.2 kN, 21.1 kN, and 19.5 kN, respectively (Figure 9). The load-bearing capacity was increased in all cases of the reinforced beams by more than 18%. The highest load-bearing capacity was observed in the case of the type C hybrid beams. It was higher than the load-bearing capacity of the type A, B, and D beams by 45%, 23%, and 8%, respectively. Type A, C, and D specimens had a typical tensile failure and type B specimens a typical shear failure (Figure 10).

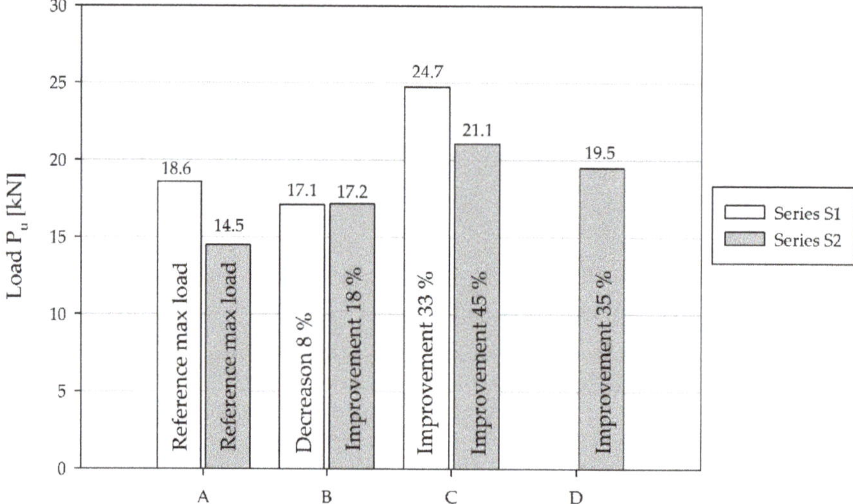


**Figure 9.** Average load-bearing capacity ($P_u$) of the beams of type A, B, C, and D, and the improvement in this capacity of the beams of types B, C, and D compared to the type A beams.

(**a**) tensile failure         (**b**) shear failure

**Figure 10.** Typical failure mechanisms of specimen types A, C, D (**a**) and specimen type B (**b**).

Effective flexural rigidity $(EI)_{eff}$ was calculated on the basis of the average absolute midspan deflection $w_1$, using a well-known expression of technical mechanics (Equation (1)) that corresponds to the four-point bending test load arrangement. The equation neglects a small part of deflections caused by shear stresses.

The hybrid beams of type C had the highest flexural rigidity, followed by the beams of type B, the beams of type D, and lastly the beams of type A (Figure 11). For Series S1 the flexural rigidity of type B and C beams was higher in comparison with beams type A by 41% and 50%, respectively. For Series S2, the flexural rigidity of type B, C, and D beams was higher in comparison with beam type A for 37%, 43%, and 38%, respectively.

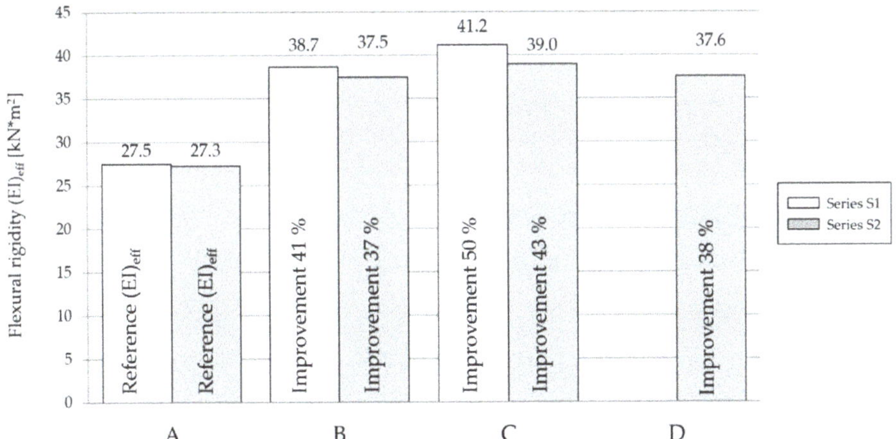

**Figure 11.** Effective flexural rigidity of type A, B, C, and D specimens.

Due to the high ratio between the modulus of elasticity $(E_0)$ and the shear modulus $(G)$ of wood (i.e., $E_0/G = 16$, Table 1) and a relatively high span and specimen height $(L/h)$ ratio (which amounts to $L/h = 25.7$ and $L/h = 27.2$ for Series S1 and Series S2, respectively), the deflection caused by shear stresses should not be neglected. Kretschmann [30] suggested that the flexural rigidity should be increased by 10% if the deflection due to the shear stresses is neglected, as it is in the case of Equation (1). The results of a study reported by Eierle and Bös [31] confirmed, that the shear deflection depends on the length and height ratio (i.e., $L/h$) of the beam, and on the ratio between the shear and elastic moduli (i.e., $G/E_0$). The smaller this ratio is, the larger is the effect of shear stresses on the deflection.

At rectangular shaped simply supported beams with L/h ratio above 25, the shear deflection should not represent more than 2% of the total deflection.

Midspan deflection ($w_1$) was measured in dependence of the load (P), for all beam types of Series S1 and Series S2 (Figures 12 and 13). The angle of the linear part of the curves from the abscissa is defining the bending stiffness; the higher the angle the higher the bending stiffness. Load-bearing capacity is defined by the load at which the curve ends from the exception of specimens marked with * sign, which are specimens, where the full range of at least one LVDT was reached, before the failure occurred.

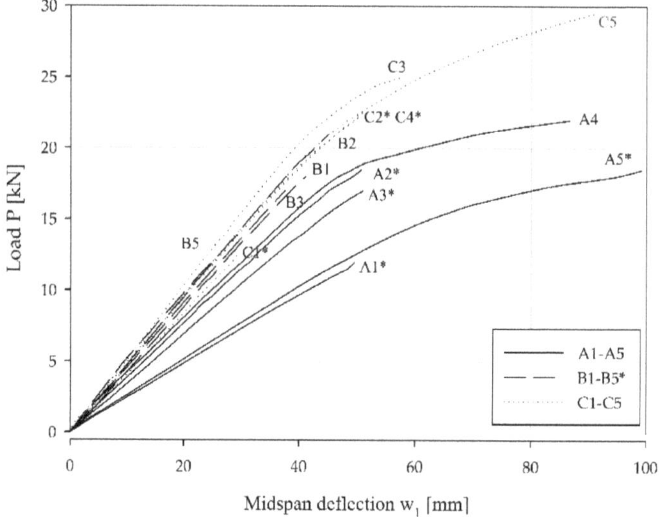

**Figure 12.** Load-midspan deflection relationship of Series S1. * B4 was excluded from further analysis due to the change in cross section dimension.

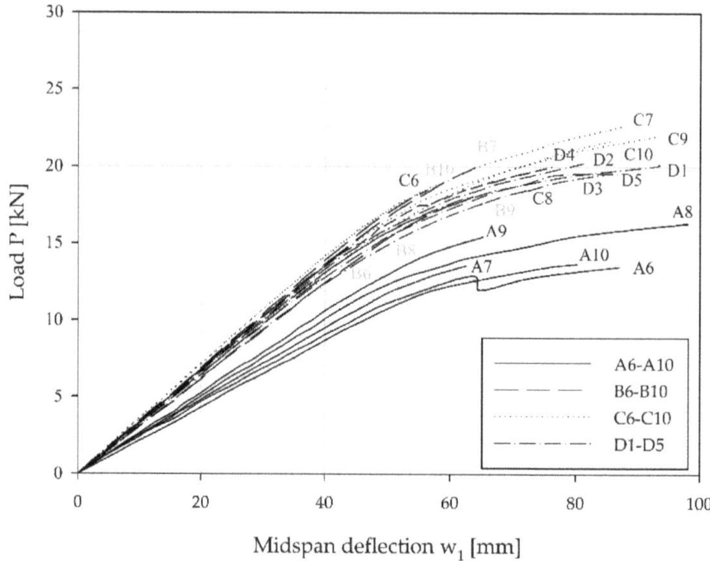

**Figure 13.** Load-midspan deflection relationship of Series S2.

Type A specimens had the smallest bending stiffness and load-bearing capacity but showing the highest dispersion of the results at the same time (Figure 12). The highest values (bending stiffness and ultimate load) were measured at type C specimens. The ultimate load was recorded and taken into consideration in the analysis of the results in Table 4.

In both figures (Figures 12 and 13) the same deflection range and load range had been used in order to be able to directly compare also the behaviour of specimens being tested on two different testing fields. With the increase of the span L and the increase of the distance between the support and the load a, the bending moment has increased (at the same load level) at specimens of Series S2, compared to specimens of Series S1. The increase of the bending moment by 25% in Series S2, has decreased the ultimate load $P_u$ (Series S2) for only 12% compared to ultimate load of specimens of Series S1. The reason lies in a prevailing failure mechanism, which was with Series S2 specimens a typical tensile failure.

## 4. Discussion

Aluminium reinforcements used in wooden beams have improved bending stiffness, load-bearing capacity and flexural rigidity. The experiments showed almost linear dependency between reinforcement-wood ratio and bending stiffness and nonlinear dependency between reinforcement-wood ratio and load-bearing capacity. This was proven also by Kim and Harries [16], who showed that there is a limit of the reinforcement/wood ratio, beyond which the load-bearing capacity can no longer increase. The reason for the smallest increase of load-bearing capacity with specimens B, lies in the high number of shear failure mechanisms (89% of all B specimens tested have failed in shear). Borgin et al. [32] have described the positive effect of the knots in the wood on the shear stress resistance. In the window industry all wood must be free of knots, so the shear stress depends on the wood shear properties only. The shear strength of the specimens could be additionally increased with different wood species in the middle part of the cross section [33], but the most effective it is with the additional reinforcements. There are not many attempts to use aluminium as the reinforcing material. Due to higher moduli of elasticity and better thermal properties, steel is used much more often, [5,9,12]. Nevertheless, in our experimental analysis, it was proven that aluminium is a good material to be used in a first phase of the wood hybrid beam analysis, to define the positioning, orientation of the reinforcements, and the selection of the proper adhesives. One of the biggest advantages of this approach is the ability to process aluminium-wooden hybrid beams with standard machinery for window manufacturing. During the shear strength tests of the adhesives it was observed that surface treatment of the aluminum is essential in order to achieve the adequate adhesion between adhesive and aluminium. The importance of surface treatment of the reinforcement was also declared by Jasienko and Nowak [5], who sanded the steel surface before adhesion. For proper adhesion, aluminium embedded in our specimens was anodized.

The presented study was made as a starting point for further analysis, more focused on the reinforcements in window profiled elements. Mullion profile (middle window profile in the double sashed window, Figure 14) is a combination of one of the most frequently used window profiles and one of the most exposed window elements to wind load.

From the four-point bending tests results and observed failure mechanisms it can be concluded, that the optimal reinforcement arrangement would be a combination of the reinforcement used in the type C and D specimens. With reinforcements positioned in both, the tensile and compressive zones, the tensile and compressive stress strength of the beams (as in the case of the type C specimens) can be increased. A vertically positioned reinforcement over the almost whole cross sectional height (as in the case of the type D specimens) prevents a shear failure before the tensile and compressive stress limits are reached. A preposition of such a window hybrid beam is presented in Figure 15.

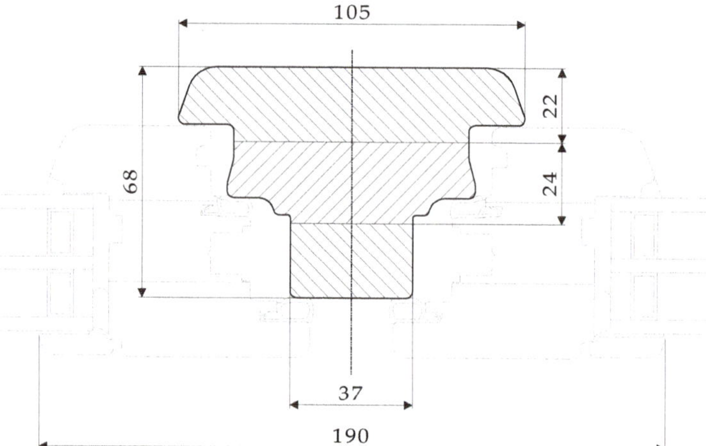

**Figure 14.** Mullion-middle fixed element of the double sashed window (measurements in mm).

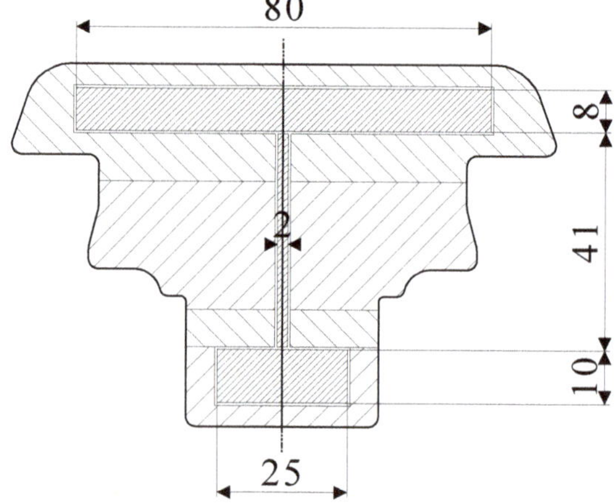

**Figure 15.** Example of window profile with horizontally and vertically positioned reinforcements (measurements in mm).

But on the other hand, a vertical metal reinforcement through the whole specimen height can lead to thermal transmission and therefore decrease thermal efficiency of the window or even worse, can lead to a local decrease of inner temperature that can further on lead to condensation and mold growth [34]. If we could achieve the proper bending stiffness of the window elements with aluminium reinforcements that would not go through the whole height of the profile, this would be the most cost-effective and applicable solution in window production.

In further analysis, the mullion profile will be investigated to define maximal possible bending stiffness and load-bearing improvements, considering positioning, orientation, material, and number of reinforcements.

## 5. Conclusions

An experimental analysis of the effect of reinforcements on the bending stiffness, load-bearing capacity, and flexural rigidity of small cross section hybrid window beams, reinforced with aluminium reinforcements was analysed. The results show that all three analysed parameters were the best in the case of the type C specimens with 12 (3 × 20 mm) vertical reinforcements in two rows. The type B specimens had a very low load-bearing capacity, due to their low resistance to shear stress, whereas the type D beams never failed due to a lack of shear strength. The shape of the window profile and the results of this analysis indicate that the reinforcement should be arranged by combining the reinforcing methods used in the type C and type D beam specimens. Based on the results of this study, further numerical and experimental analyses will need to be performed, using real shaped window profiles (Figure 15).

**Author Contributions:** B.Š., J.L. and G.F. conceived and designed experiments; B.Š. and J.L. performed the experiments; B.Š. analysed the data; J.L. and G.F. conducted the validation and formal analysis; B.Š., writing—original draft preparation; J.L. and G.F., writing—review and editing.

**Funding:** This work was supported by Ministry of Education, Science and Sport of the Republic of Slovenia, and by the European Regional Development Fund, European Commission (project TIGR4smart, grant number 5441-1/2016/116) and by Ministry of Education, Science and Sport of the Republic of Slovenia within the framework of the Program P2-0182.

**Conflicts of Interest:** The authors declare no conflict of interest.

## References


1. PHI Certified Component Database. Available online: https://database.passivehouse.com/en/components/list/window (accessed on 30 September 2018).
2. Jiménez, P.; Dunkl, A.; Eibel, K.; Denk, E.; Grote, V.; Kelz, C.; Moser, M. Wood or Laminate? Psychological Research of Customer Expectations. *Forests* **2016**, *7*, 275. [CrossRef]
3. Verband Fenster + Fassade, G.F. und H.V. *Holzarten für den Fensterbau—Teil 1: Eigenschaften, Holzarttabellen*; Verband Fenster + Fassade, G.F. und H.V: Frankfurt am Main, Germany, 2018.
4. ISO 10077-2: 2017. *Thermal Performance of Windows, Doors and Shutters—Calculation of Thermal Transmittance—Part 2: Numerical Method for Frames*; ISO: Geneva, Switzerland, 2017.
5. Jasieńko, J.; Nowak, T.P. Solid timber beams strengthened with steel plates—Experimental studies. *Constr. Build. Mater.* **2014**, *63*, 81–88. [CrossRef]
6. Alhayek, H.; Svecova, D. Flexural Stiffness and Strength of GFRP-Reinforced Timber Beams. *J. Compos. Constr.* **2012**, *16*, 245–252. [CrossRef]
7. Andor, K.; Lengyel, A.; Polgár, R.; Fodor, T.; Karácsonyi, Z. Experimental and statistical analysis of spruce timber beams reinforced with CFRP fabric. *Constr. Build. Mater.* **2015**, *99*, 200–207. [CrossRef]
8. Raftery, G.M.; Kelly, F. Basalt FRP rods for reinforcement and repair of timber. *Compos. Part B Eng.* **2015**, *70*, 9–19. [CrossRef]
9. Borri, A.; Corradi, M. Strengthening of timber beams with high strength steel cords. *Compos. Part B Eng.* **2011**, *42*, 1480–1491. [CrossRef]
10. Fiorelli, J.; Dias, A.A. Analysis of the strength and stiffness of timber beams reinforced with carbon fiber and glass fiber. *Mater. Res.* **2003**, *6*, 193–202. [CrossRef]
11. Premrov, M.; Dobrila, P. Experimental analysis of timber–concrete composite beam strengthened with carbon fibres. *Constr. Build. Mater.* **2012**, *37*, 499–506. [CrossRef]
12. Winter, W.; Tavoussi, K.; Pixner, T.; Parada, F.R. Timber-Steel-Hybrid Beams for Multi-Storey Buildings. In Proceedings of the World Conference on Timber Engineering, Auckland, New Zealand, 15–19 July 2012; Volume 2012.
13. Thorhallsson, E.R.; Hinriksson, G.I.; Snæbjörnsson, J.T. Strength and stiffness of glulam beams reinforced with glass and basalt fibres. *Compos. Part B Eng.* **2017**, *115*, 300–307. [CrossRef]
14. Nadir, Y.; Nagarajan, P.; Ameen, M.; Arif, M. Flexural stiffness and strength enhancement of horizontally glued laminated wood beams with GFRP and CFRP composite sheets. *Constr. Build. Mater.* **2016**, *112*, 547–555. [CrossRef]
15. Valipour, H.R.; Crews, K. Efficient finite element modelling of timber beams strengthened with bonded fibre reinforced polymers. *Constr. Build. Mater.* **2011**, *25*, 3291–3300. [CrossRef]
16. Kim, Y.J.; Harries, K.A. Modeling of timber beams strengthened with various CFRP composites. *Eng. Struct.* **2010**, *32*, 3225–3234. [CrossRef]

17. Taheri, F.; Nagaraj, M.; Khosravi, P. Buckling response of glue-laminated columns reinforced with fiber-reinforced plastic sheets. *Compos. Struct.* **2009**, *88*, 481–490. [CrossRef]

18. Raftery, G.M.; Harte, A.M. Nonlinear numerical modelling of FRP reinforced glued laminated timber. *Compos. Part B Eng.* **2013**, *52*, 40–50. [CrossRef]

19. de Jesus, A.M.P.; Pinto, J.M.T.; Morais, J.J.L. Analysis of solid wood beams strengthened with CFRP laminates of distinct lengths. *Constr. Build. Mater.* **2012**, *35*, 817–828. [CrossRef]

20. Khelifa, M.; Lahouar, M.A.; Celzard, A. Flexural strengthening of finger-jointed *Spruce* timber beams with CFRP. *J. Adhes. Sci. Technol.* **2015**, *29*, 2104–2116. [CrossRef]

21. Borri, A.; Corradi, M.; Speranzini, E. Reinforcement of wood with natural fibers. *Compos. Part B Eng.* **2013**, *53*, 1–8. [CrossRef]

22. Ferrier, E.; Agbossou, A.; Michel, L. Mechanical behaviour of ultra-high-performance fibrous-concrete wood panels reinforced by FRP bars. *Compos. Part B Eng.* **2014**, *60*, 663–672. [CrossRef]

23. EN 338:2016. *Structural Timber—Strength Classes*; BSI: London, UK, 2016.

24. EN 205:2016. *Adhesives—Wood Adhesives for Non-Structural Applications—Determination of Tensile Shear Strength of Lap Joints*; BSI: London, UK, 2016.

25. Yang, Y.; Liu, J.; Xiong, G. Flexural behavior of wood beams strengthened with HFRP. *Constr. Build. Mater.* **2013**, *43*, 118–124. [CrossRef]

26. Voßedelstahlhandel GmbH & Co. KG. *Technical Datasheet for Aluminium Alloy—6060-T66*; Voßedelstahlhandel GmbH & Co. KG: Neu Wulmstorf, Germany, 2014.

27. Exel Composites Benefites of Composites. Available online: http://www.exelcomposites.com/en-us/english/composites/composites/ (accessed on 9 March 2017).

28. EN 1465:2009. *Adhesives—Determination of Tensile Lap-Shear Strength of Bonded Assemblies*; BSI: London, UK, 2009.

29. De Luca, V.; Marano, C. Prestressed glulam timbers reinforced with steel bars. *Constr. Build. Mater.* **2012**, *30*, 206–217. [CrossRef]

30. Kretschmann, E.D. *Mechanical Properties of Wood*; General Technical Report FPL-GTR-190; Forest Products Laboratory, United States Department of Agriculture: Washington, DC, USA, 2010; p. 46.

31. Eierle, B.; Bös, B. Schubverformungen von Stabtragwerken in der praktischen Anwendung. *Bautechnik* **2013**, *90*, 747–752. [CrossRef]

32. Borgin, K.B.; Loedolff, G.F.; Asce, M.; Saunders, G.R. Laminated wood beams reinforced with steel strips. *Proc. Am. Soc. Civ. Eng.* **1968**, *94*, 1681–1706.

33. Tomasi, R.; Parisi, M.A.; Piazza, M. Ductile Design of Glued-Laminated Timber Beams. *Pract. Period. Struct. Des. Constr.* **2009**, *14*, 113–122. [CrossRef]

34. Ruediger, D.; Jehl, W.; Koos, F.; Seggebruch, F.; Strasser, G.; Detlef, T.; Wiegand, L.; Zimmermann, H.-H. *Wärmetechnische Anforderungen an Baukörperanschlüsse für Fenster*; Verband der Fenster-und Fassadenhersteller e.V. Gutgemeinschaften Fenster und Haustüren: Frankfurt am Main, Germany, 2016; Volume 2001.

 *forests*

*Article*

# Nondestructive Characterization of Dry Heat-Treated Fir (*Abies Alba* Mill.) Timber in View of Possible Structural Use

**Aleš Straže *, Gorazd Fajdiga and Bojan Gospodarič**

Biotechnical Faculty, University of Ljubljana, Jamnikarjeva 101, 1000 Ljubljana, Slovenia;
gorazd.fajdiga@bf.uni-lj.si (G.F.); bojan.gospodaric@bf.uni-lj.si (B.G.)
* Correspondence: ales.straze@bf.uni-lj.si; Tel.: +386-1-320-3635

Received: 14 November 2018; Accepted: 11 December 2018; Published: 15 December 2018

**Abstract:** The use of heat-treated timber for building with wood is of increasing interest. Heat treatment improves the durability and dimensional stability of wood; however, it needs to be optimized to keep wood's mechanical properties in view of the possible structural use of timber. Therefore, dry vacuum heat treatment varying the maximum temperature between 170 °C and 230 °C was used on fir (*Abies alba* Mill.) structural timber, visually top graded according to EN 338, to analyze its final weight loss, hygroscopicity, CIELAB color, and dynamic elastomechanical properties. It turned out that weight loss and total color difference of wood positively correlates with the increasing intensity of the heat treatment. The maximum 40% reduction of the hygroscopicity of wood was already reached at 210 °C treatment temperature. The moduli of elasticity in longitudinal and radial direction of wood, determined by ultrasound velocity, increased initially up to the treatment temperature of 210 °C, and decreased at higher treatment temperature. Equally, the Euler-Bernoulli modulus of elasticity from free-free flexural vibration of boards in all five vibration modes increased with the rising treatment temperature up to 190 °C, and decreased under more intensive treatment conditions. The Euler-Bernoulli model was found to be valid only in the 1st vibration mode of heat-treated structural timber due to the unsteady decrease in the evaluated moduli of elasticity related to the increasing mode number.

**Keywords:** heat treatment; wood; structural changes; nondestructive testing; ultrasound; Euler-Bernoulli; modulus of elasticity

---

## 1. Introduction

Thermal treatment at high temperature, i.e., between 160 °C to 260 °C, is one of the eco-friendly methods for the enhancement of the biological durability of wood and lignocellulosic composites. Heat treatment processes vary in terms of furnace design, type and condition of heating medium, and treatment schedules, and mostly depend on final usage of heat-treated material. The common factor of these processes is a modification of the chemical structure of timber, which has consequences on the physical and mechanical properties of wood [1–7].

With the improved hygroscopicity and dimensional stability of heat-treated wood, there is a desire to use it for structural purposes, especially in more demanding climates. However, the important aspects in a case of thermally treated wood are strength reduction and stiffness alteration, which vary with the anatomical direction of wood, testing method, and wood species. Many studies have shown a reduction in the bending stiffness and strength of heat-treated wood, combined with the reduced wood density [8–14], since the latter is the main influencing factor in the mechanical properties of wood [15]. However, exceptions are found to be related to the significant decrease in modulus of elasticity only when the weight loss of wood exceeds a particular value [16]. The latter is related significantly to the treatment conditions, since material cracking and degradation of the cell structure of heat-treated wood can be induced as well [12]. The important role of material changes during heat treatment mostly concerns the initial structure and density inhomogeneity, which is almost always present in real size solid wood. In the case of the use of such heat-treated solid wood for structural purposes, it is necessary to ensure reliable quality control based on non-invasive techniques [17], widely present in the management of wood quality in the whole forest-wood chain [18,19].

Therefore, the main goal of the study was to use non-destructive mechanical and physical testing methods to investigate possible internal structural changes of fir (*Abies alba* Mill.) real size quarter-sawn timber after vacuum heat treatment, having varying intensity. Additionally, the machine stress grading and dynamic mechanical response of structural timber before and after heat treatment was analyzed and compared with the weight loss of boards and their color changes.

## 2. Materials and Methods

### 2.1. Material

Forty-five radially-oriented fir wood boards (*Abies alba* Mill.) of 45 mm thickness ($L_T$), 120 mm wide ($L_R$) and 4 m long ($L_L$), Figure 1, were selected from the conditioned warehouse (T = 20 °C; RH (Relative humidity) = 65%) of a local construction timber trade company. In the sample population, we included boards without present fissures, deformations, wane, rot, insect damages, or other abnormal defects. We only allowed the presence of single healthy knots up to a size of 15 mm, substantially below 1/5 of the cross-sectional area of the boards. This visual preselection and assessment of boards allowed us to grade the sample population into the S10 and S13 classes [20], and therefrom, to assign the C24 and C30 strength grading classes for the selected boards [21]. Most of the boards were initially visually graded into the top S13 class. However, some of the boards (n = 5; 11.1% of the samples), due to growth rings of widths greater than 6 mm, and therefore, lower wood density, were graded into the lower S10 grading class.

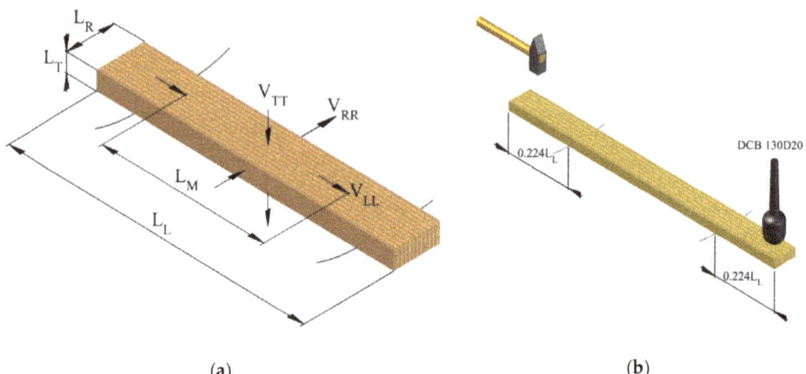



**Figure 1.** (**a**) The experimental setup for determination of velocity of ultrasound in longitudinal- ($v_{LL}$), radial- ($v_{RR}$) and tangential ($v_{TT}$) wood direction; (**b**) principle of the analysis of flexural vibration response of structural timber specimens.

The initial weight and dimensions of the selected boards were determined afterwards for the ranging of timber into 9 density classes. We made nine density classes by ranking boards from the smallest to the highest density of wood. Five boards were placed successively in each density class. Small cut-off specimens (L = 25 mm; 300 mm from the board end) were made afterwards from each board to gravimetrically determine wood equilibrium moisture content.

*2.2. Methods*

2.2.1. Heat Treatment

Industrial dry vacuum heat treatment of wood was carried out by the Silvaprodukt company (Ljubljana, SI) according to patented SilvaproTM industrial vacuum procedure with a pre-drying phase (T = 105 °C; t = 24 h), stepwise heating phase ($\Delta$T = +15 °C/h), heating at maximum temperature for 3 hours, followed by cooling ($\Delta$T = $-$15 °C/h) and conditioning in normal climate (20 °C, RH = 65%). One board per wood density class was taken for this purpose of the control group and treated at 4 heat treatment intensities (9 boards per treatment), having the maximum temperature of 170 °C, 190 °C, 210 °C, and 230 °C. A one-month conditioning period (20 °C, 65%) was used prior to determining the final weight of the boards and their equilibrium moisture content (EMC) and wood density ($\rho$). The board weight loss ($WT_{loss}$) after heat treatment was calculated on the dry mass basis.

2.2.2. Determination of Wood Color

Standard color measurement (CIELAB) was performed on every board (3 measurements per sample) at the initial and heat treated state by X-Rite OptotronikTM SP62 (XRITE Inc., Rapids, MI, USA) spectrophotometer. The total color difference of wood before and after thermal modification was determined by the $\Delta$E* colorimetric parameter (Equation (1)).

$$\Delta E^* = \sqrt{\Delta L^{*2} + \Delta a^{*2} + \Delta b^{*2}}, \tag{1}$$

where $\Delta$L* is a difference in color lightness, $\Delta$a* is difference in green-red axis and $\Delta$b* is difference in blue-yellow color axis.

2.2.3. Determination of Elastomechanical Properties of Structural Timber with Ultrasound

The velocity of ultrasound has been added to the measurement 3-times per board in the radial ($v_{RR}$, $L_R$ = 120 mm), tangential ($v_{TT}$, $L_T$ = 45 mm) and longitudinal ($v_{LL}$, $L_M$ = 1500 mm) board direction by Proceq Pundit PL-200PE (Proceq Inc., Scharzenbach, Switzerland) pulse ultrasonic device, equipped with 54 kHz exponential transducers (Figure 1a). The velocity of ultrasound ($v_{ii}$) and wood density ($\rho$) were used to determine the moduli of elasticity ($E_i$) in longitudinal- ($E_L$), radial- ($E_R$) and tangential ($E_T$) direction of the boards (Equation (2)). Acoustic anisotropy was determined by ratios of ultrasound velocity in all three wood anatomical directions ($v_{LL}/v_{RR}$, $v_{LL}/v_{TT}$, $v_{RT}/v_{TT}$; two-letter index: the first letter represents the direction of the ultrasonic wave, and the second represents its polarization).

$$E_i = \rho \cdot v_{ii}^2, \tag{2}$$

2.2.4. Analysis of Flexural Vibration of Structural Timber Boards and Strength Grading

For free-free flexural vibration, the test specimens were placed on soft thin rubber supports from their nodes of the 1st vibration mode (0.224 L) and excited using a steel hammer (mass 100 g) from a free end. The sound was recorded by unidirectional condenser microphone (PCB-130D20; PCB Piezotronics Inc, Depew, NY, USA) on the other free end of the board, and acquired by NI-9234 DAQ-module (National Instruments Inc, Austin, TX, USA) in 24-bit resolution with 51.2 kHz sampling frequency (Figure 1b). Euler-Bernoulli's moduli of elasticity ($E_B$) were determined based on each of the five initial modes ($1 \leq n \leq 5$) of flexural vibration (Equation (3)):

$$E_B = \frac{4 \cdot \pi^2 \cdot L_L^4 \cdot \rho \cdot f_n^2 \cdot A}{I \cdot k_n^4}, \tag{3}$$

where $L_L$ is the length of a board, $\rho$ is the mean density of a board, A is a board's cross section, I is the moment of inertia and $k_n$ is a constant depending on vibration mode number n ($k_n$ = ((2 n + 1)

$\pi)/2$)). The analysis of the theoretical linear decreasing slope of the evaluated moduli of elasticity with increasing vibration mode number was accomplished by calculating the difference between sequential moduli ($\Delta E_{Bi} = \Delta E_{Bi} - \Delta E_{B(i-1)}$; $2 \leq i \leq 5$), and finally by calculating the coefficient of variation of the moduli difference between vibration modes (q) (Equations (4)–(6)):

$$\overline{\Delta E_B} = \frac{1}{n} \sum_{i=1}^{n} \Delta E_{Bi}, \tag{4}$$

$$SD = \sqrt{\frac{\sum_{i=1}^{n} \left(\Delta E_{Bi} - \overline{\Delta E_B}\right)^2}{n}}, \tag{5}$$

$$q = \frac{SD}{\left|\overline{\Delta E_B}\right|} \cdot 100\%, \tag{6}$$

where $\overline{\Delta E_B}$ is mean sequential moduli difference and SD is standard deviation of the sequential moduli difference.

We used the Euler-Bernoulli's modulus of elasticity in 1st vibration mode ($E_{B1}$) and wood density for the strength grading of boards according to standards EN 14081 [22] and EN 338 [23]. As a criterion for classification in a particular strength class, we took into account the achievement of the characteristic value of the wood density ($\rho_c$) and 95% of the average modulus of elasticity ($E_m$). The ANOVA (Analysis of variance) statistical tool and Duncan's multiple range test at the 95% level of significance were used for all the tested properties, to analyze the difference among group means in the sample of boards.

## 3. Results

### 3.1. Impact of Heat Treatment on Wood Density, Weight Loss, Hygroscopicity and Color of Fir Structural Timber

The rising of the heat treatment temperature induced a significant increase in the weight loss of fir wood (*Abies alba* Mill.; ANOVA, $p = 1.42 \times 10^{-12}$). This caused a drop in the mean density of wood after the heat treatment of 2.5% at a temperature of 170 °C ($\rho_{170} = 415$ kg/m$^3$) and up to 10.3% at the heat treatment temperature of 230 °C ($\rho_{230} = 392$ kg/m$^3$; ANOVA, $p = 0.07$). The hygroscopic nature of wood was significantly improved by the heat treatment (EMC; ANOVA, $p = 1.11 \times 10^{-16}$). Even after the lightest thermal modification (T = 170 °C), the equilibrium moisture content of the wood in the normal climate dropped to 8.0%. Only slightly lower values, i.e., between 6.9% and 7.6%, we recorded in stronger heat-treated wood (Table 1, Figure 2a).

**Table 1.** Average wood density ($\rho$), weight loss (WT$_{loss}$), equilibrium moisture content (EMC) and color parameters (L*—lightness, a*, b*—chromaticity on green-red and blue-yellow axis; $\Delta$E—total color difference) of wood after heat treatment (2nd row present Coef. of variation (%)).

| Heat Treatment (°C) | $\rho$ (kg/m$^3$) | WT$_{loss}$ (%) | EMC (%) | L* | a* | b* | $\Delta$E* |
|---|---|---|---|---|---|---|---|
| Control | 425 | | 12.4 | 74.9 | 6.1 | 25.1 | |
| | (6.4) | | (4.1) | (2.8) | (10.5) | (5.7) | |
| 170 | 415 | 2.5 | 8.0 | 57.8 | 13.0 | 31.5 | 19.6 |
| | (6.7) | (36.6) | (12.7) | (6.2) | (9.2) | (4.8) | (24.0) |
| 190 | 415 | 3.1 | 7.5 | 47.6 | 11.7 | 24.8 | 28.0 |
| | (7.3) | (26.8) | (11.9) | (7.1) | (6.5) | (10.9) | (13.9) |
| 210 | 397 | 5.9 | 6.9 | 44.2 | 11.6 | 23.5 | 31.3 |
| | (6.5) | (24.9) | (8.1) | (6.0) | (5.8) | (4.9) | (10.8) |
| 230 | 392 | 10.3 | 7.6 | 37.6 | 10.3 | 18.8 | 38.2 |
| | (6.7) | (15.9) | (13.9) | (4.4) | (8.5) | (9.9) | (8.6) |

By increasing the intensity of the heat treatment, the color lightness of the wood was significantly reduced (ANOVA, $p = 1.1 \times 10^{-16}$). The mean color lightness was the highest in the control samples

(L* = 74.9), and the lowest in the samples after the heat treatment at 230 °C (L* = 37.6) (Table 1). Changes in the color parameters a* and b* were not as large, and were insignificant with respect to the intensity of the treatment (ANOVA, $p = 0.13$). Otherwise, the values of the two parameters under mild treatment conditions ($\leq$190 °C) increased slightly, while for the more intensively heat-treated wood they dropped again. The total color difference in wood ($\Delta$E*), compared to the color of the test specimens, was largely due to the change in color lightness. With the intensity of the heat treatment, the total color difference $\Delta$E* was significantly increased (ANOVA, $p = 1.47 \times 10^{-9}$). It has also been shown that there is a positive correlation of $\Delta$E* with the weight loss of test specimens (Table 1, Figure 2b).

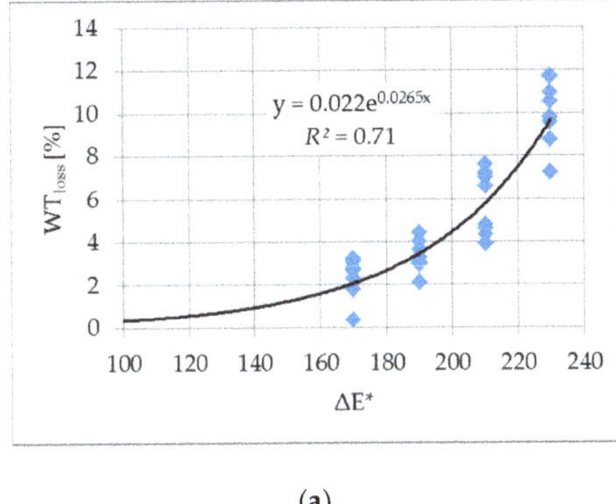


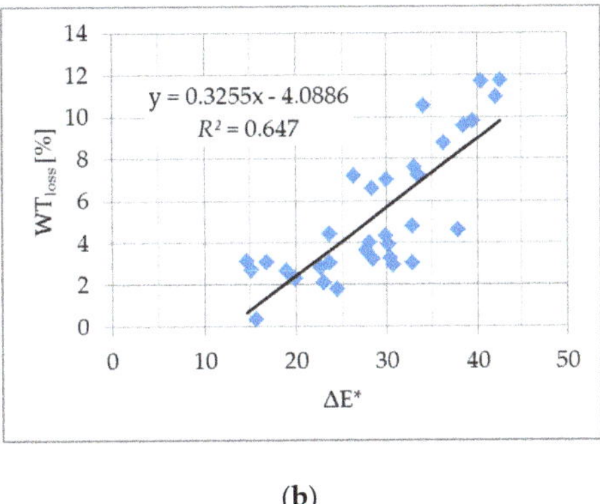


(a)         (b)

**Figure 2.** (**a**) The relationship between the individual weight loss (WT$_{loss}$) of wood and the intensity of the heat treatment; (**b**) the relationship between the individual weight loss (WT$_{loss}$) of wood and its total color difference ($\Delta$E*).

### 3.2. Elastomechanical Properties and Anisotropy of Heat-Treated Structural Timber

The ultrasound velocity was significantly improved in the longitudinal ($v_{LL}$) and radial direction ($v_{RR}$) of the heat-treated structural timber, up to a treatment temperature of 210 °C (ANOVA, $p = 0.049$). The velocity of ultrasound in these two anatomical directions was again slightly lower only in the most intense heat-treated structural timber (T = 230 °C). In the tangential anatomical wood direction ($v_{TT}$), the velocity of ultrasound didn't significantly change with the intensity of the heat treatment (Table 2; ANOVA, $p = 0.82$).

A somewhat smaller increase than in ultrasound velocity was recorded in the longitudinal- ($E_L$; ANOVA, $p = 0.014$) and radial direction of wood ($E_R$; ANOVA, $p = 0.046$) with the intensity of heat treatment of structural timber. This difference in trends in ultrasound velocity and stiffness of wood is attributed to the simultaneous decrease in the density of structural timber by increasing the intensity of the treatment. The latter also causes a reduction, however statistically insignificant (ANOVA, $p = 0.15$), in the modulus of elasticity in the tangential direction of the wood ($E_T$) by increasing the intensity of the thermal process (Table 2).

The elastomechanical anisotropy of the structural timber changed slightly but insignificantly with the intensity of the thermal process (ANOVA, $p = 0.21$). The largest anisotropy was determined in the longitudinal-tangential plane (4.7 to 5.4) and somewhat smaller in the longitudinal-radial plane (3.5 to 3.9). As expected, elastomechanical anisotropy was the smallest in the radial-tangential plane (1.2 to 1.5) of structural timber (Table 2).

**Table 2.** Mean velocity of ultrasound in longitudinal- ($v_{LL}$), radial- ($v_{RR}$) and tangential wood direction ($v_{TT}$), elastomechanical anisotropy ($v_{LL}/v_{RR}$, $v_{LL}/v_{TT}$, $v_{RR}/v_{TT}$) and mean moduli of elasticity ($E_L$—longitudinal, $E_R$—radial, $E_T$—tangential) of heat-treated structural timber (2nd row present Coef. of variation (%)).

| Heat Treatment (°C) | $v_{LL}$ (m/s) | $v_{RR}$ (m/s) | $v_{TT}$ (m/s) | $v_{LL}/v_{RR}$ | $v_{LL}/v_{TT}$ | $v_{RR}/v_{TT}$ | $E_L$ (GPa) | $E_R$ (GPa) | $E_T$ (GPa) |
|---|---|---|---|---|---|---|---|---|---|
| Control | 4991 | 1399 | 1079 | 3.7 | 4.7 | 1.3 | 10.6 | 0.88 | 0.51 |
| | (6.8) | (22.8) | (14.8) | (25.0) | (16.4) | (13.0) | (15.5) | (50.3) | (35.8) |
| 170 | 5424 | 1610 | 1090 | 3.5 | 5.0 | 1.5 | 12.2 | 1.13 | 0.50 |
| | (6.3) | (21.1) | (10.7) | (21.6) | (14.6) | (15.0) | (10.5) | (44.3) | (22.8) |
| 190 | 5365 | 1473 | 1045 | 3.8 | 5.4 | 1.5 | 12.0 | 0.93 | 0.39 |
| | (4.1) | (21.7) | (7.7) | (24.6) | (14.1) | (21.7) | (11.7) | (40.1) | (20.3) |
| 210 | 5520 | 1587 | 1071 | 3.5 | 5.0 | 1.5 | 11.7 | 0.99 | 0.49 |
| | (4.2) | (17.4) | (9.7) | (28.6) | (15.6) | (18.2) | (10.4) | (31.0) | (26.3) |
| 230 | 5161 | 1225 | 1028 | 3.9 | 4.7 | 1.2 | 10.1 | 0.60 | 0.42 |
| | (6.5) | (13.4) | (8.5) | (24.8) | (17.6) | (15.4) | (13.5) | (30.1) | (21.3) |

*3.3. Vibration Response of Heat-Treated Structural Timber*

The modulus of elasticity in the test specimens increased initially with the intensity of heat treatment ($\leq$190 °C); however, at higher temperatures, i.e., particularly at 230 °C, it was significantly reduced compared to control samples (ANOVA, $p = 0.021$). This trend was present at the modulus of the elasticity of the specimens in all five vibration modes ($E_{B1}$ to $E_{B5}$) (Table 3).

**Table 3.** Mean moduli of elasticity of heat-treated structural timber determined by flexural vibration at individual vibration mode ($1 \leq n \leq 5$) and its mean bending strength according to EN 338 (2nd row present Coef. of variation (%)).

| Heat Treatment (°C) | $E_{B1}$ (GPa) | $E_{B2}$ (GPa) | $E_{B3}$ (GPa) | $E_{B4}$ (GPa) | $E_{B5}$ (GPa) | Bending Strength Grade (MPa) |
|---|---|---|---|---|---|---|
| Control | 12.74 | 13.01 | 12.38 | 11.83 | 11.45 | 31.6 |
| | (12.2) | (12.6) | (12.9) | (12.7) | (12.4) | (21.2) |
| 170 | 13.58 | 13.48 | 12.96 | 12.47 | 12.06 | 34.1 |
| | (9.6) | (11.1) | (10.7) | (10.2) | (10.3) | (17.6) |
| 190 | 13.59 | 13.73 | 12.96 | 12.50 | 12.12 | 34.9 |
| | (12.5) | (10.6) | (10.3) | (10.2) | (10.1) | (20.3) |
| 210 | 12.94 | 12.81 | 12.12 | 11.71 | 11.40 | 29.4 |
| | (14.5) | (10.8) | (11.7) | (11.8) | (11.4) | (20.9) |
| 230 | 11.58 | 12.45 | 11.69 | 11.21 | 10.92 | 26.8 |
| | (23.4) | (17.8) | (18.9) | (18.2) | (18.3) | (36.6) |

The uniform, close to the linear decreasing slope of the flexural moduli of elasticity with increasing vibration mode number was confirmed only for the control structural timber (Table 3; Figure 3a). The sequential moduli difference between 1st and 2nd vibration modes was initially significantly changed for the structural timber already after the low intense heat treatment ($\leq$190 °C). Major changes between the sequencing moduli with regard to vibration mode, especially for higher modal numbers, occurred at greater heat treatment temperatures ($\geq$210 °C). The variation of the modulus of elasticity (q-coefficient) of heat-treated structural timber at 210 °C significantly increased (ANOVA, $p = 0.045$) compared to the rest of the tested population (Figure 3b).

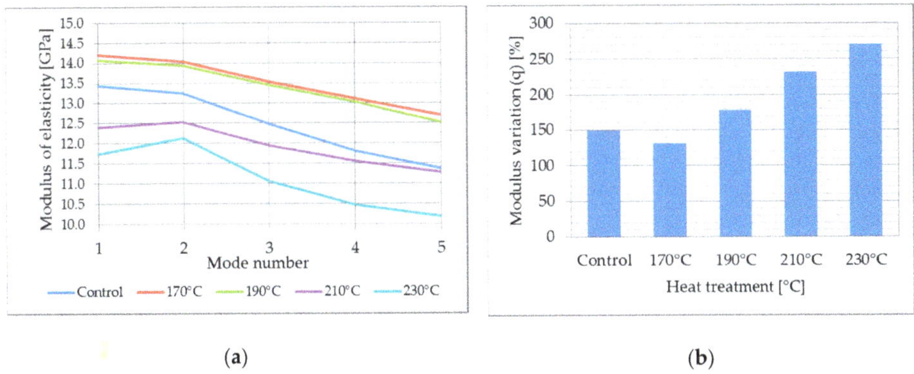

(**a**)  (**b**)

**Figure 3.** (**a**) Modal evaluated values of modulus of elasticity of heat-treated structural timber; (**b**) the coefficient of variation of the moduli difference in-between vibration modes (q-coefficient) of heat-treated wood.

Strength Grading of Heat-Treated Structural Timber

The small reduction in wood density and significant increase in the modulus of elasticity in the moderate heat-treated construction wood ($\leq$190 °C) cause improved classification, i.e., into the EN 338 strength classes. In the case of a moderate heat-treated structural timber ($\leq$190 °C), the resulting mean bending strength was greater (Table 3, Figure 4a).

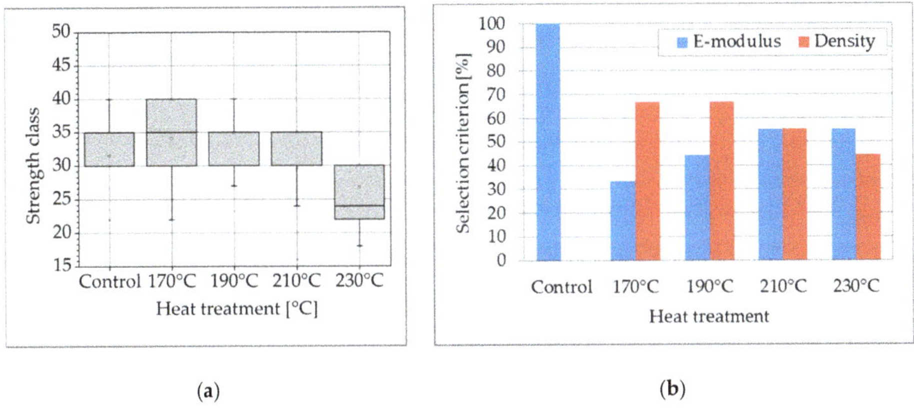

(**a**)  (**b**)

**Figure 4.** (**a**) Strength class distribution of heat-treated structural timber; (**b**) the decision making selection criteria for strength grading of heat-treated structural timber.

Classification into lower strength grades was required for heat-treated structural timber with a temperature of 210 °C or more. The estimated mean bending strength of the wood in this case is typically reduced below the values of the control specimens (Figure 4a), having also asymmetric distribution. In control specimens, the determining classification criterion was exclusively the individual modulus of elasticity. For heat-treated structural timber, the wood density was more often used for strength grading criteria (Figure 4b).

## 4. Discussion

The weight loss of various wood species and, consequently, the decrease in the wood density, was confirmed in both dry heat treatments, as well as in hydrothermal heat treatment processes [24–28].

Studies confirmed smaller degradation of heat-treated wood in a vacuum than under nitrogen or water vapor under the same conditions [26]. This is explained by the effect of vacuum allowing removal the of volatile degradation products limiting the acidic degradation of polysaccharides due to formation of acetic acid and the recondensation of volatile degradation products within the wood structure. Therefore, in the used dry vacuum heat treatment, the weight loss of wood is attributed mainly to the degradation of hemicelluloses, i.e., the most reactive wood components that hydrolyze into oligomeric and monomeric structures [29,30]. However, it is also suggested that other phenomena, such as structural modifications and chemical changes of lignin, also play an important part [31].

In comparable heat treatment in conifers, a similar increase in the weight loss of the wood was observed with increasing treatment temperature, and was dominant over the impact of the treatment time. As in previous studies, we also confirmed a rising decrease in mass correlated with EMC reduction [14,25,27,32]. This is explained by less moisture-accessible hydroxyl groups of the heat-treated specimens compared with untreated controls [33]. However, the EMC remained stable beyond the limit value of approximately 3% decrease in weight reached at 190 °C treatment temperature. This achieved limit value is lower compared to the data from related studies, ranging between 6% and 9%; however, it indicates the completion of decomposition of moisture-accessible hydroxyl groups by the heat treatment [25,32]. In these heat treatment processes, we achieved up to 40% reduction in EMC of wood, while research indicates that the EMC can be reduced in thermally-treated timber even up to 60% [28,34].

Color changes in wood during the heat treatment were found to be related to the process intensity. The excellent but non-linear negative relationship was observed between the lightness of wood (L*) and the used temperature of heat treatment, which was confirmed also in some previous studies [25,35–37]. Researchers even proposed more reliable means of measuring the intensity of a thermal modification process by combining parameter L*+b* and by milling of wood, to prevent scattering of color parameters on real wood surfaces [35]. Additionally, the relation between chemical composition and lightness decrease for heat-treated wood is reported [38]. However, the total color difference ΔE* is most often used, and was found also in this study to have the same exponential relationship with the increase of the treatment temperature, determined even in a case of wood weight loss (Figure 2a,b).

The increase in the ultrasound velocity, significantly in the longitudinal and partially in the radial direction of fir wood, coincided with its weight loss and overall color difference ΔE*, but just up to the treatment temperature of 210 °C. The effect of heat treatment on the longitudinal and radial sound velocity may vary, since it also increases greatly, i.e., up to 0.8% with a decrease of 1% of EMC, in the range of 5 to 30% equilibrium moisture content of wood [39,40]. Otherwise, the ultrasound velocity in the tangential direction of wood ($v_{TT}$) remained unchanged at these conditions, which induced the increase of elastomechanical anisotropy of heat-treated fir wood with respect to its tangential plane ($v_{LL}/v_{TT}$, $v_{RR}/v_{TT}$). In the heat treatment of wood up to approx. 200 °C, the increase in the ultrasound velocity, especially in the longitudinal direction of the wood, which is equivalent to a specific modulus of elasticity ($E/\rho$), is also indicated by other studies [41,42]. Elastomechanical anisotropy of heat-treated wood has not been widely studied so far. However, some researchers report the increase in the mechanical anisotropy of wood, but already in the area of plastic deformations, where they determined the increase in the ratio of compression strength of wood along- and transverse to the grain [43,44].

The positive correlation of the modulus of elasticity with the treatment temperature up to 190 °C, and then its decrease at higher treatment temperatures ($\geq$210 °C), were equivalent regardless of the method used, i.e., at the ultrasound velocity and flexural vibration. A similar trend in heat-treated wood is reported by some related studies [45]. Otherwise, we measured on average a 14% lower modulus of elasticity in in the 1st flexural vibration mode compared to the ultrasound velocity measurements. Lower values of modulus of elasticity determined by ultrasound can be a consequence of a small distance between sensors ($L_M$ = 1500 mm). The surface wave propagation may have affected the compression wave passage lengthwise, and skewed the final passage time of the wave through material, as reported elsewhere [41].

The control specimens showed a steady and semi-uniform decrease in the evaluated moduli of elasticity from flexural vibration related to the increasing mode number, which suggests an approximation to properties of homogeneous axial isotropic (orthotropic) material [46,47]. After the heat treatment, this steady decrease line showed some breakages (Figure 3a), determined also by the increase of the q-coefficient, as the measure of modulus variation in-between vibration modes (Figure 3b). The latter finding indicates the possible presence and increase of structural inhomogeneity in the heat-treated wood, intensified also by increasing of the heat treatment temperature. The principle of slope breakage of modally evaluated moduli of elasticity was, in the past, already successfully used to recognize the severity of artificially made defects in wooden beams [48], or for the determination of density inhomogeneity along the boards of various wood species [49]. The same methodology has been proposed for detecting concentrated mass due to knottiness in solid wood [50], as well as at the determination of surface- and end-cracks in kiln dried wood [51]. This research suggests that the evaluation of moduli of elasticity from flexural vibration can be successfully used to determine inhomogeneous structural changes in full size heat-treated wood, which are likely present in material at increased treatment temperature. Depending on treatment parameters such as treatment temperature, the heating rate, the holding time at the maximum temperature, or the gas humidity, cracks can appear and the cell structure can be partially degraded as well [13,24,28,52].

The q-coefficient was found to be in a negative relationship, however, not significant, with the strength grading class of heat-treated wood after the treatment. It is important to note that the increase in q-coefficient by increasing the treatment temperature is also likely to be due to the change in the density of the heat-treated wood. Wood density, however, was a common decision criterion for strength grading of the heat-treated wood. The findings indicate the potential of both, i.e., density and q-coefficient, together with the modal evaluation of the modulus of elasticity, for use in strength grading of the heat-treated structural timber.

## 5. Conclusions

The weight loss and the total CIELAB color difference $\Delta E^*$ of structural fir timber positively correlate with the increase of the heat treatment temperature, in the range between 170 °C and 230 °C. The maximum 40% reduction of hygroscopicity of fir wood is already reached at 210 °C treatment temperature.

The ultrasound velocity, and consequently, modulus of elasticity, increases initially in the longitudinal and partially in the radial direction of fir structural timber, up to the treatment temperature of 210 °C, and decreases under more intensive heat treatment conditions. Due to the constant ultrasound velocity in the tangential direction ($v_{TT}$) of heat-treated wood, the increase of its elastomechanical anisotropy with respect to the tangential plane is confirmed.

As with the ultrasonic method, an initial positive correlation exists of the modulus of elasticity from the flexural vibration of boards in all vibration modes with heat treatment temperatures up to 190 °C, which then decreases at higher treatment temperatures.

The Euler-Bernoulli model, used in free-free flexural vibration, was found to be valid only in the 1st vibration mode at structural timber. This discovery has potential for use in timber strength grading. The visually highly-graded preselected structural timber, having minimal structural anomalies, shows a steady decrease in the evaluated moduli of elasticity related to the increasing mode number. After the heat treatment, this steady decrease line has some breakages, with increased modulus variation between vibration modes.

**Author Contributions:** Conceptualization and experiments design, A.S., G.F. and B.G.; A.S. and B.G. performed the experiments; A.S. and B.G. analyzed the data; validation and formal analysis, G.F.; A.S., writing—original draft preparation; G.F. and B.G., writing—review and editing.

**Funding:** This work was supported by Ministry of Education, Science and Sport of the Republic of Slovenia within the framework of the Program P4-0015 and Program P2-0182.

**Acknowledgments:** Thanks are to the Breza Commerce d.o.o. Company (Todraž, Slovenia) for providing the timber for this study and to the Silvaprodukt d.o.o. Company (Ljubljana, Slovenia) for the heat treatment of wood.

**Conflicts of Interest:** The authors declare no conflict of interest. The funders had no role in the design of the study; in the collection, analyses, or interpretation of data; in the writing of the manuscript, or in the decision to publish the results.

## References


1. Stamm, A.J. Themal degradation of wood and cellulose. *Ind. Eng. Chem.* **1956**, *48*, 413–417. [CrossRef]
2. Kollmann, F.; Schneider, A. Über das Sorptionsverhaltnis wärmebehandelter Hölzer. *Holz Roh- Werkst.* **1963**, *21*, 77–85. (In German) [CrossRef]
3. Tjeerdsma, B.F.; Boonstra, M.; Pizzi, A.; Tekely, P.; Militz, H. Characterisation of thermally modified wood: Molecular reasons for wood performance improvement. *Holz Roh- Werkst.* **1998**, *56*, 149–153. [CrossRef]
4. Metsä-Kortelainen, S.; Antikainen, T.; Viitaniemi, P. The water absorption of sapwood and heartwood of Scots pine and Norway spruce heat-treated at 170 °C., 190 °C., 210 °C. and 230 °C. *Holz Roh- Werkst.* **2006**, *64*, 192–197. [CrossRef]
5. Hakkou, M.; Pétrissans, M.; Zoulalian, A.; Gérardin, P. Investigation of wood wettability changes during heat treatment on the basis of chemical analysis. *Polym. Dégrad. Stab.* **2005**, *89*, 1–5. [CrossRef]
6. Burmeister, A. Einfluss einer Wärme-Druck-Behandlung halbtrockenen Holzes auf eine seine Formbeständigkeit. *Holz Roh- Werkst.* **1973**, *31*, 237–243. (In German) [CrossRef]
7. Cetera, P.; Negro, F.; Cremonini, C.; Todaro, L. Physico-Mechanical Properties of Thermally Treated Poplar OSB. *Forests* **2018**, *9*, 345. [CrossRef]
8. Santos, J.A. Mecanical behaviour of Eucalyptus wood modified by heat. *Wood Sci. Technol.* **2000**, *34*, 39–43. [CrossRef]
9. Shi, J.L.; Kocaefe, D.; Zhang, J. Mechanical behaviour of Quebec wood species heat-treated using ThermoWood process. *Holz Roh- Werkst.* **2007**, *65*, 255–259. [CrossRef]
10. Bengtsson, C.; Jermer, J.; Brem, F. Bending strength of heat-treated spruce and pine timber. In Proceedings of the IRG/WP 02-40242, Jackson Lake, WJ, USA, 20–24 May 2007.
11. Giebeler, E. Dimensionstabilisierung von Holz durch eine Feuchte/Wärme/Druck- Behandlung. *Holz Roh- Werkst.* **1983**, *41*, 87–94. (In German) [CrossRef]
12. Kubojima, Y.; Okano, T.; Ohta, M. Bending strength and toughness of heat-treated wood. *J. Wood Sci.* **2000**, *46*, 8–15. [CrossRef]
13. Poncsak, S.; Kocaefe, D.; Bouazara, M.; Pichette, A. Effect of high temperature treatment on the mechanical properties of birch (*Betula papyrifera*). *Wood Sci. Technol.* **2006**, *40*, 647–663. [CrossRef]
14. Borůvka, V.; Zeidler, A.; Holeček, T.; Dudik, R. Elastic and Strength Properties of Heat-Treated Beech and Birch Wood. *Forests* **2018**, *9*, 197. [CrossRef]
15. Zeidler, A.; Borůvka, V.; Schönfelder, O. Comparison of Wood Quality of Douglas Fir and Spruce from Afforested Agricultural Land and Permanent Forest Land in the Czech Republic. *Forests* **2017**, *9*, 13. [CrossRef]
16. Rusche, H. Festigkeitseigenschaften von trockenem Holz nach thermischer Behandlung. *Holz Roh- Werkst.* **1973**, *31*, 273–281. (In German) [CrossRef]
17. Widman, R.; Fernandez-Cabo, J.L.; Steiger, R. Mechanical properties of thermally modified beech timber for structural purposes. *Eur. J. Wood Wood Prod.* **2012**, *70*, 775–784. [CrossRef]
18. Newton, P.F. Acoustic-Based Non-Destructive Estimation of Wood Quality Attributes within Standing Red Pine Trees. *Forests* **2017**, *8*, 380. [CrossRef]
19. Škorpik, P.; Konrad, H.; Geburek, T.; Schuh, M.; Vasold, D.; Eberhardt, M.; Schuler, S. Solid Wood Properties Assessed by Non-Destructive Measurements of Standing European Larch (*Larix decidua* Mill.): Environmental Effects on Variation within and among Trees and Forest Stands. *Forests* **2018**, *9*, 276. [CrossRef]
20. German Institute for Standardisation. *Strength Grading of Wood—Part 1: Coniferous Sawn Timber*; German Institute for Standardisation: Berlin, Germany, 2012; Volume DIN 4071-1, p. 8.

21. European Committee for Standardization. *Structural timber—Strength classes—Assignment of visual grades and species*; CEN: Brussels, Belgium, 2007; Volume EN 1912: 2004, p. 16.

22. European Committee for Standardization. *Timber Structures—Strength Graded Structural Timber with Rectangular Cross Section—Part 1: General Requirements*; CEN: Brussels, Belgium, 2010; Volume EN 14081-1: 2005, p. 12.

23. European Committee for Standardization. *Structural Timber—Stress Classes*; CEN: Brussels, Belgium, 2009; Volume EN 338: 2009, p. 10.

24. Boonstra, M.; van Acker, J.; Tjeerdsma, B.F.; Kegel, E.V. Strength properties of thermally modified softwoods and its relation to polymeric structural wood constituents. *Ann. For. Sci.* **2007**, *64*, 679–690. [CrossRef]

25. Welzbacher, C.R.; Brischke, C.; Rapp, A.O. Influence of treatment temperature and duration on selected biological, mechanical, physical and optical properties of thermally modified timber. *Wood Mater. Sci. Eng.* **2007**, *2*, 66–76. [CrossRef]

26. Candelier, K.; Dumarcay, S.; Pétrassans, A.; Desharnais, L.; Gérardin, P. Comparison of chemical composition and decay durability of heat treated wood cured under different inert atmospheres: Nitrogen or vacuum. *Polym. Dégrad. Stab.* **2013**, *98*, 677–681. [CrossRef]

27. Alén, R.; Kotilainen, R.; Zaman, A. Thermochemical behaviour of Norway spruce (*Picea abies*) at 180 to 225 °C. *Wood Sci. Technol.* **2002**, *36*, 163–171. [CrossRef]

28. Esteves, B.M.; Pereira, H. Wood modification by heat treatment: A. review. *BioResources* **2009**, *4*, 370–404.

29. Bobleter, O.; Binder, H. Dynamischer hydrothermaler Abbau von Holz. *Holzforschung* **1980**, *34*, 48–51. [CrossRef]

30. Tjeerdsma, B.F.; Militz, H. Chemical changes in hydrothermal treated wood: FTIR analysis of combined hydrothermal and dry heat-treated wood. *Holz Roh- Werkst.* **2005**, *63*, 102–111. [CrossRef]

31. Repellin, V.; Guyonnet, R. Evaluation of heat treated wood swelling by differential scanning calotrimetry in relation with chemical composition. *Holzforschung* **2005**, *59*, 28–34. [CrossRef]

32. Paul, W.; Ohlmeyer, M.; Leithoff, H.; Boonstra, M. Optimising the properties of OSB by one-step heat pre-treatment process. *Holz Roh- Werkst.* **2006**, *64*, 227–234. [CrossRef]

33. Boonstra, M.; Tjeerdsma, B.F. Chemical analysis of heat treated softwoods. *Holz Roh- Werkst.* **2006**, *64*, 204–211. (In German) [CrossRef]

34. Esteves, B.; Marques, A.V.; Domingos, I.; Pereira, H. Influence of steam heating on the properties of pine (*Pinus pinaster*) and eucalypt (*Eucalyptus globulus*) wood. *Wood Sci. Technol.* **2006**, *41*, 193–207. [CrossRef]

35. Brischke, C.; Welzbacher, C.R.; Brandt, K.; Rapp, A.O. Quality control of thermally modified timber: Interrelationship between heat treatment intensities and CIEL*a*b* color data on homogenized wood samples. *Holzforschung* **2007**, *61*, 19–22. [CrossRef]

36. Bekhta, P.; Niemz, P. Effect of high temperature on the change in color, dimensional stability and mechanical properties of spruce wood. *Holzforschung* **2003**, *57*, 539–546. [CrossRef]

37. Wei, Y.; Zhang, P.; Liu, Y.; Chen, Y.; Gao, J.; Fan, Y. Kinetic Analysis of the Color of Larch Sapwood and Heartwood during Heat Treatment. *Forests* **2018**, *9*, 289. [CrossRef]

38. Esteves, B.; Marques, A.V.; Domingos, I.; Pereira, H. Heat-induced colour changes of pine (*Pinus pinaster*) and eucalypt (*Eucalyptus globulus*) wood. *Wood Sci. Technol.* **2008**, *42*, 369–384. [CrossRef]

39. Sandoz, J.L. Grading of construction timber by ultrasound. *Wood Sci. Technol.* **1989**, *23*, 95–108. [CrossRef]

40. Llana, D.F.; Iniguez-Gonzales, G.; Arriaga, F.; Niemz, P. Influence of Temperature and Moisture Content in Non-destructive values of Scots pine (*Pinus sylvestris* L.). In Proceedings of the 18th International Nondestructive Testing and Evaluation of Wood Symposium, Madison, WI, USA, 24–27 September 2013; pp. 451–458.

41. Holeček, T.; Gašparik, M.; Lagaňa, R.; Borůvka, V.; Oberhofnerová, E. Measuring the Modulus of Elasticity of Thermally Treated Spruce Wood using the Ultrasound and Resonance Methods. *BioResources* **2017**, *12*, 819–838. [CrossRef]

42. Del Menezzi, C.H.S.; Amorim, M.R.S.; Costa, M.A.; Garcez, L.R.O. Evaluation of Thermally Modified Wood by Means of Stress Wave and Ultrasound Nondestructive Methods. *Mater. Sci.* **2014**, *20*, 61–66. [CrossRef]

43. Heräjärvi, H. Effect of Drying Technology on Aspen Wood Properties. *Silva Fennica* **2009**, *43*, 433–445. [CrossRef]

44. Straže, A.; Fajdiga, G.; Pervan, S.; Gorišek, Ž. Hygro-mechanical behaviour of thermally treated beech subjected to compression loads. *Constr. Build. Mater.* **2016**, *113*, 28–33. [CrossRef]

45. Kubojima, Y.; Okano, T.; Ohta, M. Vibrational properties of Sitka spruce heat-treated in nitrogen gas. *J. Wood Sci.* **1998**, *44*, 73–77. [CrossRef]

46. Bodig, J.; Jayne, B. *Mechanics of Wood and Wood Composites*; Krieger Publishing Co.: Malabar, India, 1993; p. 712.

47. Harris, C.M.; Paez, T.L. *Harris' Shock and Vibration Handbook*, 6th ed.; McGraw-Hill: New York, NY, USA, 2009; p. 1168.

48. Roohnia, M.; Tajdini, A. Identification of the Severity and Position of a Single Defect in a Wooden Beam. *BioResources* **2014**, *9*, 3428–3438. [CrossRef]

49. Kubojima, Y.; Tonosaki, M.; Yoshihara, H. Young's modulus obtained by flexural vibration test of a wooden beam with inhomogeneity of density. *J. Wood Sci.* **2006**, *52*, 20–24. [CrossRef]

50. Kubojima, Y.; Suzuki, S.; Tonosaki, M. Effect of Additional Mass on the Apparent Young's Modulus of a Wooden Bar by Longitudinal Vibration. *BioResources* **2014**, *9*, 5088–5098. [CrossRef]

51. Roohnia, M.; Yavari, A.; Tajdini, A. Elastic parameters of poplar wood with end-cracks. *Ann. For. Sci.* **2010**, *67*, 409. [CrossRef]

52. Johansson, D. Influence of drying on internal checking of spruce (*Picea abies* L.) heat-treated at 212°C. *Holzforschung* **2006**, *60*, 558–560. [CrossRef]

*Article*

# Effect of Selected Factors on the Bending Deflection at the Limit of Proportionality and at the Modulus of Rupture in Laminated Veneer Lumber

**Adam Sikora, Tomáš Svoboda *, Vladimír Záborský and Zuzana Gaffová**

Department of Wood Processing and Biomaterials, Faculty of Forestry and Wood Sciences,
Czech University of Life Sciences in Prague, Kamýcká 1176, Prague–Suchdol 16521, Czech Republic;
sikoraadam@seznam.cz (A.S.); zaborsky@fld.czu.cz (V.Z.); gaffova@fld.czu.cz (Z.G.)
* Correspondence: tomassvo@seznam.cz; Tel.: +420-608-321-941

Received: 18 March 2019; Accepted: 5 May 2019; Published: 9 May 2019

**Abstract:** The deflection of a test material occurs under bending stress that is caused by force. In terms of plasticity and elasticity, the deflection can be quantified at two main areas, which are the limit of proportionality and the modulus of rupture. Both of these deflections are of great importance in terms of the scientific and practical use. These characteristics are particularly important when designing structural elements that are exposed to bending stress in terms of the size of the deflection in their practical application. This study analyzed the effect on the size of the deflection at the limit of proportionality and at the modulus of rupture. Wood species (*Fagus sylvatica* L. and *Populus tremula* L.), material thickness (6 mm, 10 mm, and 18 mm), non-wood component (glass and carbon fiber), position of the non-wood component in the layered material (up and down side with respect to the loading direction), and adhesive used to join the individual layers (polyurethane and polyvinyl acetate) were the observed factors. Glass fiber reinforcement proved to be a better option; however, the effect of correctly selected glue for individual wood species was also apparent. For the aspen laminated materials, polyurethane adhesive (PUR) adhesive was shown to be a more effective adhesive and PVAc adhesive was better for the beech-laminated materials. These results are of great importance for the production of new wood-based materials and materials were based on non-wood components, with specific properties for their intended use.

**Keywords:** technological and product innovations; cyclic loading; laminated wood; deflection at the limit of proportionality; deflection at the modulus of rupture; wood-processing industry performance

---

## 1. Introduction

The composition design and production of Laminated Veneer Lumber materials is primarily focused on the intended use of the material. Therefore, it is necessary to know the proposed composition under the influence of various factors in order to create such a material.

The basic component of LVL materials is wood component, however non-wood component can be also used. The single layers of LVL material could be modified or unmodified. The modification can be carried out in a variety of ways, such as high temperature, chemical, impregnation, pressure, microwave irradiation, lamination, etc. The possibilities of increasing the strength and the stiffness of wood lie in its combination with non-wood components at various material bases [1]. These are steel, glass, and carbon fibers in building materials [2–4]. Authors [5] conducted research with the effect of embedded carbon fiber fabric in the bonded joint of five-layer plywood to increase the flexural strength.

Flexibility is a technical property of wood that expresses its ability to bend. It is a material property that may be classified as either a positive or negative characteristic, depending on the purpose. In the case of materials that are intended for bending, this property is desirable [6–8], whereas it is undesirable in materials that are intended for construction [9,10].

Bending is defined as a torque that acts on a material perpendicular to the cross section, which results in normal and tangential stress that causes stress through bending or twisting [11]. Each change in a beam subjected to bending is the result of work that is directly dependent on the force that is used and the resulting deformation [12,13]. The range of external forces doing the work changes in bending within the maximum boundary. The internal forces that are caused by deformation are also displaced [14]. Potential energy accumulates in an elastically deformed object, and then the energy is converted into work that is consumed after release to enable the object to return to its original shape [15–17].

The deflection at the limit of proportionality of wood is characterized as the boundary after which deformation becomes elastic over time and plastic. It is necessary to use the force at the limit of proportionality to achieve deflection at the limit of proportionality [18]. Up to this point, the wood is loaded with a force that only causes elastic deformation and it is therefore only flexibly loaded [19–21]. This deflection can be achieved without any apparent permanent deformation of its size or shape. The deflection at the modulus of rupture is characterized as the deflection when the material breaks. It is necessary to use force at the modulus of rupture to achieve this deflection (Figure 1) [20,22,23].

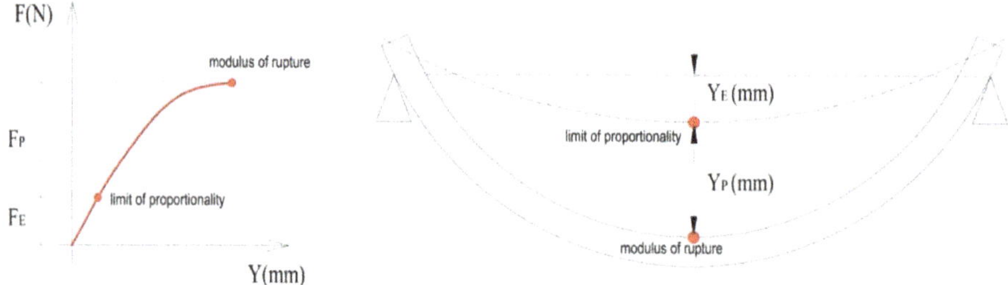


**Figure 1.** Force-deflection diagram of the bending stress and determining the limit of proportionality and the modulus of rupture [24].

The bending characteristics of the wood can be improved by reinforcing it with non-wood components on a different basis [1,25]. Carbon fibers, aramid fibers, basalt fibers, fiberglass, and polyvinyl alcohol (PVA), and others, are suitable non-wood components that can be used for wood reinforcement [26–28]. In addition to synthetic fiber reinforcement, natural fiber reinforcement may also be used, e.g., in the form of a nonwoven fabric. Experimental research has shown that correctly locating the reinforcement component has the potential to increase the bending characteristics of laminated wood in the case of application on the stressed tensile zone [21,29,30].

The literature points to the influence of surface structure on the properties of the bonded surface joint [31,32]. In view of this, the selection of a suitable adhesive for joining wood-based materials and non-wood components is very critical [33]. Wood itself is a hygroscopic material and, as such, it continually absorbs and releases moisture, which, among other things, causes a change in size. These changes can cause the adhesive to separate from the wood component. This effect can be avoided by appropriately selecting an adhesive for a specific bonded element [34].

The objective of this study was to determine the effect of the composition of laminated wood (wood species, type of non-wood component, position of the component in the structure, thickness of the material, and the type of adhesive used for bonding individual layers) on the deflection characteristics of the force-deflection diagram (deflection at the limit of proportionality and at the modulus of rupture).

## 2. Materials and Methods

### 2.1. Material

For the experiment, layered beech (*Fagus sylvatica* L.) and aspen (*Populus tremula* L.) wood (Polana, Slovakia), with a non-wood reinforcing component (glass and carbon fibers), were used. The test specimens were made with three thicknesses of 6 mm, 10 mm, and 18 mm, with a lamella width of 35 mm and length of 600 mm. The test specimens were divided into 48 test groups, with respect to the thickness and species of wood component, the type of non-wood component, and the position of the non-wood component relative to the load direction. Figure 2 shows the categorization of the test specimens. The individual wood and non-wood components were glued while using two types of adhesive, which were single-component waterproof polyvinyl acetate adhesive (PVAc) (AG-COLL 8761/L D3, EOC, Oudenaarde, Belgium) and single-component polyurethane adhesive (PUR) (NEOPUR 2238R, NEOFLEX, Madrid, Spain). Table 1 lists the detailed parameters of these adhesives. The test specimens were climatized at a 12% equilibrium moisture content (EMC) in a climatic chamber (Binder ED, APT Line II, Tuttlingen, Germany) at set relative humidity (65%) and temperature values (20 °C).

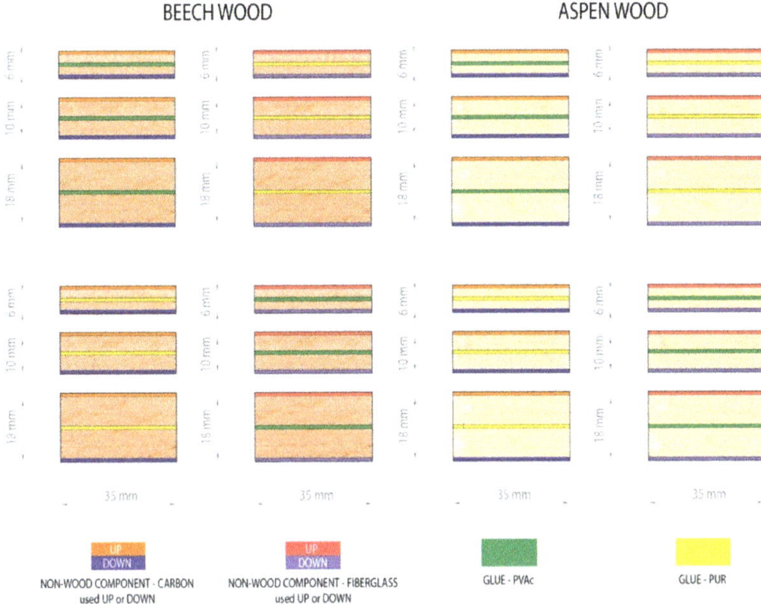


**Figure 2.** Categorization of the test specimens.

**Table 1.** Parameters of the polyvinyl acetate adhesive (PVAc) and polyurethane adhesive (PUR) Adhesives.

| Technical Data | AG-COLL 8761/L D3 | NEOPUR 2238R |
|---|---|---|
| Viscosity (mPa) | 5000 to 7000 by 23 °C | 2000 to 4500 by 25 °C |
| Working Time (min) | 15 to 20 | 60 |
| Density (g/cm$^3$) | 0.9 to 1.1 by 23 °C | approx. 1.13 |
| NCO Content (%) | - | approx. 15.5 to 16.5 |
| Color | White, milk | Brown |
| Open Time (min) | 15 | approx. 20 to 25 |
| Dry Matter Content | 49 to 51 | 100 |
| pH | 3.8 to 4.5 | - |

### 2.2. Methods

#### 2.2.1. Determining Selected Characteristics

The three-point bending test was performed according to the predetermined conditions. The lower support span was set to 20 times the thickness of the test specimen (this span varied, depending on the thickness of the test specimens). The bending tests were performed on a universal testing machine (FPZ 100, TIRA, Schalkau, Germany), according to [35–37]. To adhere to the duration limit for the test (2 min.), the top support feed rate was set to 3 mm/min. The deflections at the limit of proportionality and at the modulus of rupture were measured using an ALMEMO 2690-8 datalogger (Ahlborn GmbH, Braunschweig, Germany).

A force-deflection diagram was used to determine the limit of proportionality and the modulus of rupture. To identify all of the necessary characteristics, a program was developed for identifying data that can be obtained from the force-deflection diagram.

#### 2.2.2. Evaluation and Calculation

To determine the influence of the sample type on the bending characteristics, an analysis of variance (ANOVA) and a Fischer's F-test were performed using Statistica 12 software (Statsoft Inc., Tulsa, OK, USA).

For the determination of deflection at the limit of proportionality and at the modulus of rupture, it was necessary to calculate the limit of proportionality and the modulus of rupture.

We calculate the limit of proportionality "*LOP*" in three point bending according to EN 310 [35] and Equation (1):

$$LOP = \frac{3F_E l_0}{2bh^2} \tag{1}$$

where *LOP* is the limit of proportionality of material (MPa), $F_E$ is force at the limit of proportionality (N), $l_0$ is the distance between supporting span (mm), $b$ is width of test samples (mm), and $h$ is the height (thickness) of the sample (mm).

The bending strength "modulus of rupture (*MOR*)" in three point bending was calculated in accordance with ISO 13061-3 [38] and Equation (2):

$$MOR = \frac{3F_{max} l_0}{2bh^2} \tag{2}$$

where *MOR* is (bending strength) of material (MPa), $F_{max}$ is maximum (breaking) force (N), $l_0$ is the distance between supporting span (mm), $b$ is width of test samples (mm), and $h$ is the height (thickness) of the sample (mm).

### 3. Results and Discussion

Tables 2 and 3 show the average value and the coefficients of variation for the deflection at the limit of proportionality ($Y_E$) and deflection at the modulus of rupture ($Y_P$) in the laminated aspen and beech materials reinforced with a non-wood component (glass and carbon fiber) that is glued with PUR and PVAc adhesives. For the aspen wood (Table 2), the highest deflection at the limit of proportionality (4.89 mm) was measured on the 18-mm-thick test specimens that were glued with PVAc adhesive and reinforced with carbon fibers on the underside of the loaded specimen with respect to the loading direction. The greatest deflection at the modulus of rupture (16.70 mm) was measured on the 18-mm-thick laminated material that was glued with PUR adhesive and reinforced with glass fibers placed on the underside with respect to the loading direction. A comparison with the results of Sikora et al. verified the effect of the use of a non-wood material reinforcement to improve the deflection characteristics [39], and it was concluded that the use of a non-wood component significantly affected the deflection characteristics.

**Table 2.** Mean Values of the Deflection at the Limit of Proportionality and the Deflection at the Modulus of Rupture, and Coefficient of Variance of the Aspen Wood.

| WS | NWC | NWC Location | Glue | $T$ (mm) | Code | $Y_E$ (mm) | $Y_P$ (mm) |
|---|---|---|---|---|---|---|---|
| A | CA | U | PUR | 6 | A-CA-U-PUR-6 | 1.59 (14.0) | 3.00 (18.1) |
| A | CA | U | PUR | 10 | A-CA-U-PUR-10 | 2.06 (16.8) | 8.01 (16.6) |
| A | CA | U | PUR | 18 | A-CA-U-PUR-18 | 2.45 (13.0) | 9.32 (13.2) |
| A | CA | U | PVAc | 6 | A-CA-U-PVAc-6 | 2.10 (9.0) | 7.97 (3.3) |
| A | CA | U | PVAc | 10 | A-CA-U-PVAc-10 | 2.81 (17.7) | 11.32 (5.5) |
| A | CA | U | PVAc | 18 | A-CA-U-PVAc-18 | 3.29 (21.0) | 12.24 (16.8) |
| A | LA | U | PUR | 6 | A-LA-U-PUR-6 | 1.98 (17.9) | 7.46 (14.4) |
| A | LA | U | PUR | 10 | A-LA-U-PUR-10 | 2.73 (15.4) | 7.74 (15.9) |
| A | LA | U | PUR | 18 | A-LA-U-PUR-18 | 3.81 (8.0) | 8.74 (17.1) |
| A | LA | U | PVAc | 6 | A-LA-U-PVAc-6 | 2.28 (5.7) | 9.07 (13.6) |
| A | LA | U | PVAc | 10 | A-LA-U-PVAc-10 | 2.70 (7.7) | 9.68 (19.5) |
| A | LA | U | PVAc | 18 | A-LA-U-PVAc-18 | 3.32 (16.5) | 9.90 (11.0) |
| A | CA | D | PUR | 6 | A-CA-D-PUR-6 | 1.54 (20.2) | 5.11 (19.4) |
| A | CA | D | PUR | 10 | A-CA-D-PUR-10 | 2.97 (7.4) | 9.07 (19.9) |
| A | CA | D | PUR | 18 | A-CA-D-PUR-18 | 3.22 (16.4) | 13.30 (16.9) |
| A | CA | D | PVAc | 6 | A-CA-D-PVAc-6 | 3.82 (15.0) | 8.46 (20.4) |
| A | CA | D | PVAc | 10 | A-CA-D-PVAc-10 | 4.09 (20.6) | 12.95 (6.9) |
| A | CA | D | PVAc | 18 | A-CA-D-PVAc-18 | 4.89 (19.1) | 14.48 (17.9) |
| A | LA | D | PUR | 6 | A-LA-D-PUR-6 | 3.21 (18.5) | 8.23 (19.3) |
| A | LA | D | PUR | 10 | A-LA-D-PUR-10 | 3.56 (7.7) | 10.33 (16.4) |
| A | LA | D | PUR | 18 | A-LA-D-PUR-18 | 4.93 (8.8) | 16.70 (18.7) |
| A | LA | D | PVAc | 6 | A-LA-D-PVAc-6 | 3.48 (9.5) | 9.21 (6.2) |
| A | LA | D | PVAc | 10 | A-LA-D-PVAc-10 | 4.05 (12.9) | 10.51 (16.3) |
| A | LA | D | PVAc | 18 | A-LA-D-PVAc-18 | 4.22 (14.3) | 13.62 (21.1) |

WS—wood species; NWC—non-wood component, $T$—thickness, $Y_E$—deflection at the limit of proportionality, $Y_P$—Deflection at the modulus of rupture, A—aspen, CA—carbon, LA—fiberglass, U—up, and D—down

**Table 3.** Mean Values of the Deflection at the Limit of Proportionality and the Deflection at the Modulus of Rupture, and Coefficient of Variance of the Beech Wood.

| WS | NWC | NWC Location | Glue | $T$ (mm) | Code | $Y_E$ (mm) | $Y_P$ (mm) |
|---|---|---|---|---|---|---|---|
| B | CA | U | PUR | 6 | B-CA-U-PUR-6 | 2.11 (20.1) | 5.13 (17.6) |
| B | CA | U | PUR | 10 | B-CA-U-PUR-10 | 2.22 (14.4) | 6.81 (19.8) |
| B | CA | U | PUR | 18 | B-CA-U-PUR-18 | 2.45 (8.7) | 7.38 (8.8) |
| B | CA | U | PVAc | 6 | B-CA-U-PVAc-6 | 2.59 (19.8) | 7.85 (9.3) |
| B | CA | U | PVAc | 10 | B-CA-U-PVAc-10 | 2.87 (8.1) | 9.48 (9.7) |
| B | CA | U | PVAc | 18 | B-CA-U-PVAc-18 | 3.48 (14.4) | 10.39 (9.5) |
| B | LA | U | PUR | 6 | B-LA-U-PUR-6 | 1.85 (12.6) | 5.42 (10.3) |
| B | LA | U | PUR | 10 | B-LA-U-PUR-10 | 2.85 (18.5) | 7.80 (17.0) |
| B | LA | U | PUR | 18 | B-LA-U-PUR-18 | 4.39 (18.7) | 9.25 (16.8) |
| B | LA | U | PVAc | 6 | B-LA-U-PVAc-6 | 2.37 (20.2) | 7.14 (4.8) |
| B | LA | U | PVAc | 10 | B-LA-U-PVAc-10 | 3.20 (17.6) | 10.25 (11.2) |
| B | LA | U | PVAc | 18 | B-LA-U-PVAc-18 | 4.03 (7.6) | 11.69 (19.3) |
| B | CA | D | PUR | 6 | B-CA-D-PUR-6 | 2.83 (19.5) | 7.26 (8.4) |
| B | CA | D | PUR | 10 | B-CA-D-PUR-10 | 3.04 (19.6) | 7.72 (14.9) |
| B | CA | D | PUR | 18 | B-CA-D-PUR-18 | 7.07 (20.1) | 15.12 (15.7) |
| B | CA | D | PVAc | 6 | B-CA-D-PVAc-6 | 3.00 (19.0) | 8.99 (12.9) |
| B | CA | D | PVAc | 10 | B-CA-D-PVAc-10 | 3.71 (16.1) | 10.07 (18.9) |
| B | CA | D | PVAc | 18 | B-CA-D-PVAc-18 | 6.69 (19.0) | 12.70 (16.1) |
| B | LA | D | PUR | 6 | B-LA-D-PUR-6 | 4.36 (17.9) | 11.38 (7.4) |
| B | LA | D | PUR | 10 | B-LA-D-PUR-10 | 5.33 (20.5) | 12.45 (17.4) |
| B | LA | D | PUR | 18 | B-LA-D-PUR-18 | 5.63 (12.4) | 15.33 (15.6) |
| B | LA | D | PVAc | 6 | B-LA-D-PVAc-6 | 2.87 (20.7) | 7.52 (8.1) |
| B | LA | D | PVAc | 10 | B-LA-D-PVAc-10 | 4.07 (19.3) | 11.84 (17.9) |
| B | LA | D | PVAc | 18 | B-LA-D-PVAc-18 | 6.77 (19.7) | 20.01 (18.9) |

WS—wood species, NWC—non-wood component, $T$—thickness, $Y_E$—deflection at the proportionality limit, $Y_P$—deflection at the modulus of rupture, B—beech, CA—carbon, LA—fiberglass, U—up, and D—down.

Table 3 shows the deflection characteristics ($Y_E$ and $Y_P$) of the layered beech materials that were reinforced with non-wood components. The greatest deflection at the limit of proportionality (7.07 mm) was measured on the 18-mm-thick test specimens that were bonded with PUR adhesive and reinforced with carbon fibers placed on the underside of the loaded test specimen relative to the loading direction. In the case of deflection at the modulus of rupture, the highest values (20.01 mm) were measured on the 18-mm-thick test specimens that were glued with PVAc adhesive and reinforced with glass fibers placed on the underside with respect to the loading direction. The effect of the adhesive was verified while using the results of Gáborík et al. [40], who presented results of the deflection of beech lamellas that were glued using PUR and PVAc adhesives. Regarding the effect of using a non-wood component to reinforce the layered material, the results of the present study were compared with the results of Sikora et al. [39], with the same conclusion as for the layered aspen materials.

The measured data were statistically evaluated using one-factor analyses (ANOVA), where the factor was the type of test specimen. The code of test samples includes the wood species that are used in the laminated material, type of non-wood component and its position in the structure relative to the loading direction, type of adhesive, and total thickness of the test sample. The evaluation was based on the $p$ significance level, which was $p = 0.005$. Table 4 shows the results of the statistical evaluation of the effect of the aspen and beech test sample type on the deflection at the limit of proportionality and deflection at the modulus of rupture of the laminated materials with a non-wood component on the underside and topside. It was clear from these results that the type of test sample had a significant effect on the deflection at the limit of proportionality and on the deflection at the modulus of rupture.

From Figure 3 and Table 4, it was clear that it was more beneficial to apply the reinforcing material to the down side of the laminated aspen material for the deflection at the limit of proportionality. Higher values were achieved in all of the cases, except one, where the values were almost equal. The values of the reinforced material glued on the down side were approximately 41.32% higher than the average values when it was glued on the top. When evaluating the material that was used for reinforcing solid wood, the average values with glass fiber were 15.62% higher than those measured with carbon fiber. The last important factor was the adhesive used to connect the individual elements together. The average values measured using PVAc adhesive were 20.56% higher than those measured with PUR adhesive.

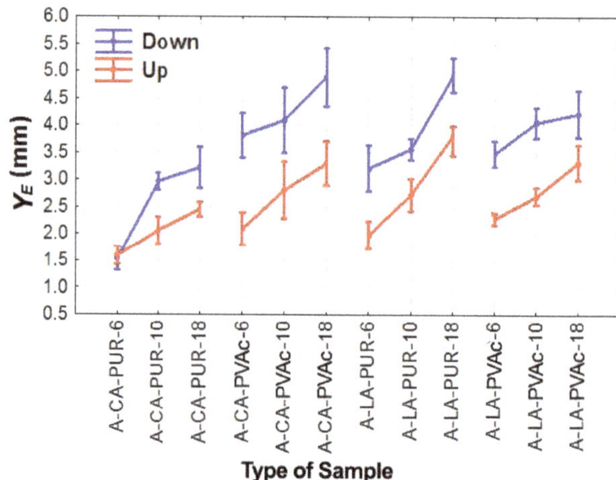


**Figure 3.** Effect of the aspen wood species on the deflection at the limit of proportionality.

The ANOVA test results in Figure 4 and Table 4 confirmed the same trend in the deflection at the limit of proportionality in the reinforced laminated beech material, as in the deflection at the limit of proportionality in the laminated aspen material. Higher values were achieved when the reinforcing material was glued on the down side. The average values in this case were up to 60.91% higher. The reinforcing materials were evaluated in the same way. The glass fiber had 13.46% higher deflection values than the carbon fiber. As for the adhesive used, there were no major differences. Expressed as a percentage difference of the mean values, the difference between PVAc and PUR was approximately 3.5%, where PVAc had higher deflection values at the limit of proportionality.

**Table 4.** Statistical Evaluation of the Effect of Type of Samples on the Deflection at the Limit of Proportionality and Deflection at the Modulus of rupture.

| Deflection at the Limit of Proportionality of Aspen and NWC Down | | | | | |
|---|---|---|---|---|---|
| **Monitored Factor** | **Sum of Squares** | **Degree of Freedom** | **Variance** | **Fisher's F-test** | **Significance Level** *p* |
| Intercept | 1469.556 | 1 | 1469.556 | 5410.635 | *** |
| (1) Type of Sample | 76.586 | 11 | 6.962 | 25.634 | *** |
| Error | 29.333 | 108 | 0.272 | | - |
| **Deflection at the Limit of Proportionality of Aspen and NWC UP** | | | | | |
| **Monitored Factor** | **Sum of Squares** | **Degree of Freedom** | **Variance** | **Fisher's F-test** | **Significance Level** *p* |
| Intercept | 469.9151 | 1 | 469.9151 | 5033.475 | *** |
| (1) Type of Sample | 19.7016 | 11 | 1.7911 | 19.185 | *** |
| Error | 9.4292 | 101 | 0.0934 | | - |
| **Deflection at the Modulus of Rupture of Aspen and NWC Down** | | | | | |
| **Monitored Factor** | **Sum of Squares** | **Degree of Freedom** | **Variance** | **Fisher's F-test** | **Significance Level** *p* |
| Intercept | 2373.329 | 1 | 2373.329 | 2497.682 | *** |
| (1) Type of Sample | 298.265 | 11 | 27.115 | 28.536 | *** |
| Error | 102.623 | 108 | 0.950 | | - |
| **Deflection at the Modulus of Rupture of Aspen and NWC Up** | | | | | |
| **Monitored Factor** | **Sum of Squares** | **Degree of Freedom** | **Variance** | **Fisher's F-test** | **Significance Level** *p* |
| Intercept | 5186.339 | 1 | 5186.339 | 4235.214 | *** |
| (1) Type of Sample | 288.155 | 11 | 26.196 | 21.392 | *** |
| Error | 139.602 | 114 | 1.225 | | - |
| **Deflection at the Limit of Proportionality of Beech and NWC Down** | | | | | |
| **Monitored Factor** | **Sum of Squares** | **Degree of Freedom** | **Variance** | **Fisher's F-test** | **Significance Level** *p* |
| Intercept | 13,863.51 | 1 | 13,863.51 | 3716.494 | *** |
| (1) Type of Sample | 1281.94 | 11 | 116.54 | 31.242 | *** |
| Error | 402.87 | 108 | 3.73 | | - |
| **Deflection at the Limit of Proportionality of Beech and NWC Up** | | | | | |
| **Monitored Factor** | **Sum of Squares** | **Degree of Freedom** | **Variance** | **Fisher's F-test** | **Significance Level** *P* |
| Intercept | 6405.696 | 1 | 6405.696 | 4427.627 | *** |
| (1) Type of Sample | 373.911 | 11 | 33.992 | 23.495 | *** |
| Error | 146.122 | 101 | 1.447 | | - |
| **Deflection at the Modulus of Rupture of Beech and NWC Down** | | | | | |
| **Monitored Factor** | **Sum of Squares** | **Degree of Freedom** | **Variance** | **Fisher's F-test** | **Significance Level** *P* |
| Intercept | 15,277.64 | 1 | 15,277.64 | 4198.613 | *** |
| (1) Type of Sample | 1739.17 | 11 | 158.11 | 43.451 | *** |
| Error | 392.98 | 108 | 3.64 | | - |
| **Deflection at the Modulus of Rupture of Beech and NWC Up** | | | | | |
| **Monitored Factor** | **Sum of Squares** | **Degree of Freedom** | **Variance** | **Fisher's F-test** | **Significance Level** *P* |
| Intercept | 5186.339 | 1 | 5186.339 | 4235.214 | *** |
| (1) Type of Sample | 288.155 | 11 | 26.196 | 21.392 | *** |
| Error | 139.602 | 114 | 1.225 | | - |

NS—not significant, ***—significant at $p < 0.005$.

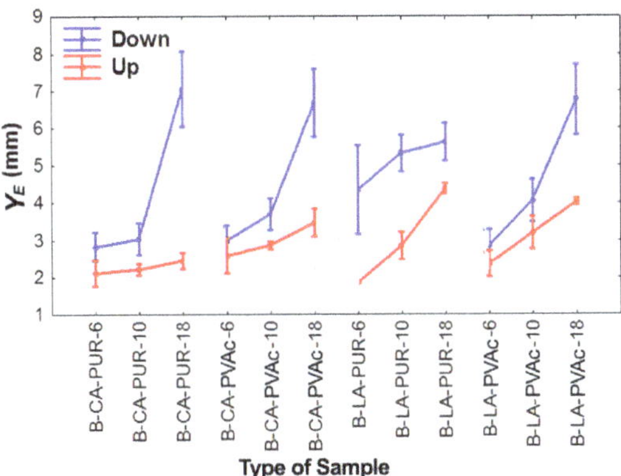


**Figure 4.** Effect of the beech wood species on the deflection at the limit of proportionality.

Figure 5 graphically depicts a comparison of the factors that were focused on the deflection at the modulus of rupture in the aspen wood. The dependencies were similar to those of the deflection at the limit of proportionality, as seen in Table 4. A greater deflection was observed when the reinforcing material was glued opposite to the loaded side (by 26.35% in this case). The average deflection values were 5.17% higher when glass fiber. The adhesive used was the last important factor. With PVAc glue, the deflection at the modulus of rupture was 20.93% higher.

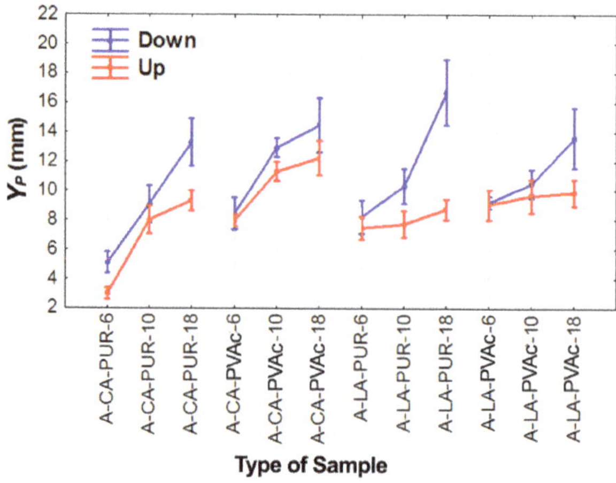

**Figure 5.** Effect of the aspen wood species on the deflection at the modulus of rupture.

As for all of the previous cases, higher deflection at the modulus of rupture values was achieved in the reinforced laminated beech materials by gluing the reinforcing material to the down side relative to the loading direction (Figure 6). These values were approximately 36% higher than these measured in the materials with the non-wood component glued on the top (Table 4). With glass fibers, the deflection at the modulus of rupture values increased by approximately 20% when compared with carbon fibers. A further increase in the deflection values can be achieved by using the correct adhesive; in this study, the deflection values for the samples with PVAc were 15.2% higher.

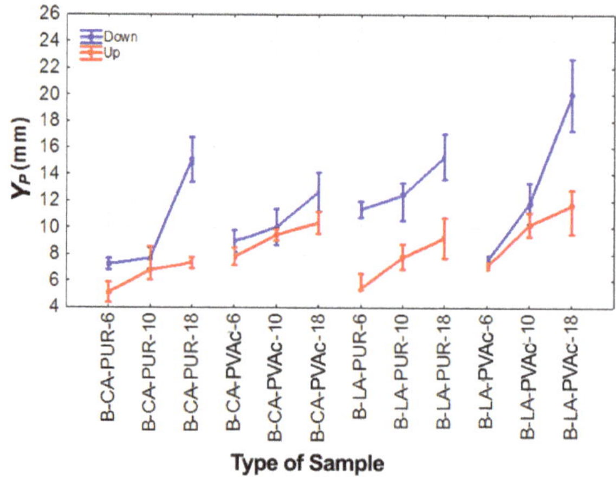

**Figure 6.** Effect of the beech wood species on the deflection at the modulus of rupture.

## 4. Conclusions

1. For the laminated aspen and beech materials that were reinforced with non-wood components, the lowest deflection values were measured in the test specimens with the non-wood component on the up side with respect to the loading direction, and the highest deflection values were measured in the test specimens with the non-wood component on the down side with respect to the loading direction.

2. Glass fiber was proven to be better non-wood component for both the laminated beech and aspen materials with regard to the deflection at the limit of proportionality and at the modulus of rupture in terms of the non-wood component itself.

3. In the case of the adhesive used, it cannot be clearly stated which adhesive was the most effective. However, higher average values of deflection at the limit of proportionality and at the modulus of rupture were measured with the PVAc adhesive.

4. As was expected, the thickness of the material proved to be an important factor that affected the deflection at both the limit of proportionality and at the modulus of rupture.

5. In the case of aspen composition for better bendability, we suggest gluing with PUR adhesive and reinforcement with glass fiber on the down side with respect to the loading direction. In addition, in the case of beech composition, we suggest gluing with PVAc adhesive and reinforcement with glass fiber on the down side with respect to the loading direction.

**Author Contributions:** Resources, A.S., T.S., V.Z. and Z.G.; Supervision, A.S.; Visualization, T.S.; Writing—original draft, A.S. and T.S.; Writing—review & editing, A.S. and T.S.

**Funding:** The authors are grateful for the support of the Advanced Research Supporting the Forestry and Wood-processing Sector's Adaptation to Global Change and the 4th Industrial Revolution (Project No. CZ.02.1.01/0.0/0.0/16_019/0000803), financed by OP RDE. The authors are also grateful for the support of the Internal Grant Agency (IGA) of the Faculty of Forestry and Wood Sciences (Project No. B 06/17).

**Conflicts of Interest:** The authors declare no conflict of interest.

## References

1. Blomberg, J.; Persson, B. Swelling pressure of semi-isostatically densified wood under different mechanical restraints. *Wood Sci. Technol.* **2017**, *41*, 401–415. [CrossRef]
2. Dai, L. *Carbon Nanotechnology*; Elsevier B.V.: Oxford, UK, 2006; p. 395s.
3. Greason, W.D. Analysis of the Charge Transfer of Models for Electrostatic Discharge (EDS) and Semiconductor Devices. *IEEE Trans. Ind. Appl.* **1996**, *32*, 726–734. [CrossRef]
4. Parvez, A.; Ansell, M.P.; Smedley, D. Mechanical repair of timber beams fractured in flexure using bonded-in reinforcements. *Compos. Part B* **2009**, *40*, 95–106.
5. Brezovič, M.; Jambrekovič, V.; Pervan, S. Bending properties of carbon fibres in forced plywood. *Wood Res.* **2003**, *48*, 1336–4561.
6. Bond, I.P.; Ansell, M.P. Fatigue properties of joined wood composites: Part I—Statistical analysis, fatigue master curves and constant life diagrams. *J. Mater. Sci.* **1998**, *33*, 2751–2762. [CrossRef]
7. Kurjatko, S.; Čunderlík, I.; Dananajová, J.; Dibdiaková, J.; Dudas, J.; Gáborík, J.; Gaff, M.; Hrčka, R.; Hudec, J.; Kačík, F.; et al. *Parametre Kvality Dreva Určujúce Jeho Finálne Použitie [Wood Quality Parameters Determining Its End Use]*; Technical University in Zvolen: Zvolen, Slovakia, 2010.
8. Bal, B.C. Flexural properties, bonding performance and splitting strength of LVL reinforced with woven glass fiber. *Constr. Build. Mater.* **2014**, *51*, 9–14. [CrossRef]
9. Bao, F.; Fu, F.; Choong, E.T.; Hse, C.Y. Contribution factor of wood properties of three poplar clones to strength of laminated veneer lumber. *Wood Fiber Sci.* **2001**, *33*, 345–352.
10. Guntekin, E.; Ozkan, S.; Yilmaz, T. Prediction of bending properties for beech lumber using stress wave method. *Maderas. Cienc. Tecnol.* **2014**, *16*, 93–98. [CrossRef]
11. Požgaj, A.; Chovanec, D.; Kurjatko, S.; Babiak, M. *Štruktúra a Vlastnosti Dreva [Structure and Properties of Wood]*; Príroda, A.S.: Bratislava, Slovakia, 1997.

12. Wagenführ, A.; Buchelt, B.; Pfriem, A. Material behaviour of veneer during multidimensional moulding. *HolzRoh. Werkst.* **2006**, *64*, 83–89. [CrossRef]

13. Gaff, M.; Gašparík, M. Effect of cyclic loading on modulus of elasticity of aspen wood. *BioResources* **2015**, *10*, 290–298. [CrossRef]

14. Kamke, F.A. Densified radiata pine for structural composites. *Maderas. Cienc. Technol.* **2006**, *8*, 83–92. [CrossRef]

15. Sandberg, D.; Navi, P. *Introduction to Thermo-Hydro-Mechanical (THM) Wood Processing*; Växjö University: Växjö, Sweden, 2007.

16. Igaz, R.; Ružiak, I.; Krišťák, L.; Réh, R.; Iždinský, J.; Šiagiová, P. Optimization of pressing parameters of crosswise bonded timber formwork sheets. *Acta Fac. Xylologiae* **2015**, *57*, 83–88.

17. Igaz, R.; Krišťák, L.; Ružiak, I.; Réh, R.; Danihelová, Z. Heat transfer during pressing of 3D moulded veneer plywood composite materials. *Key Eng. Mat.* **2016**, *688*, 131–137. [CrossRef]

18. Mackes, K.H.; Lynch, D.L. The effect of aspen wood characteristics and properties on utilization. In *Sustaining Aspen in Western Landscapes: Symposium Proceedings*; Rocky Mountain Research Station: Grand Junction, CO, USA, 2001; pp. 429–440.

19. Gaff, M.; Babiak, M.; Vokatý, V.; Ruman, D. Bending characteristics of hardwood lamellae in the elastic region. *Compos. Part B Eng.* **2017**, *116*, 61–65. [CrossRef]

20. Gaff, M.; Babiak, M. Methods for determining the plastic work in bending and impact of selected factors on its value. *Compos. Struct.* **2018**, *202*, 66–76. [CrossRef]

21. Hýsek, Š.; Gaff, M.; Sikora, A.; Babiak, M. New composite material based on winter rapeseed and his elasticity properties as a function of selected factors. *Compos. Part B Eng.* **2018**, *153*, 108–116. [CrossRef]

22. Gaff, M.; Babiak, M. Tangent modulus as a function of selected factors. *Compos. Struct.* **2018**, *202*, 436–446. [CrossRef]

23. Sikora, A.; Gaff, M.; Hysek, Š.; Babiak, M. The plasticity of composite material based on winter rapeseed as a function of selected factor. *Compos. Struct.* **2018**, *202*, 783–792. [CrossRef]

24. Svoboda, T.; Gaffová, Z.; Rajnoha, R.; Šatanová, A.; Kminiak, R. Bending forces at the proportionality limit and the maximum—Technological innovations for better performance in wood processing companies. *BioResources* **2017**, *12*, 4146–4165. [CrossRef]

25. Corigliano, P.; Crupi, V.; Epasto, G.; Guglielmino, E.; Maugeri, N.; Marinò, A. Experimental and theoretical analyses of Iroko wood laminates. *Compos. Part B Eng.* **2013**, *112*, 251–264. [CrossRef]

26. Plevris, N.; Triantafillou, T.C. Creep behavior of FRP-reinforced wood members. *J. Struct. Eng.* **1995**, *121*, 174–186. [CrossRef]

27. Redon, C.; Li, V.C.; Wu, C.; Hoshiro, H.; Saito, T.; Ogawa, A. Measuring and modifying interface properties of PVA fibres in ECC matrix. *J. Mater. Civil. Eng.* **2001**, *13*, 399–406. [CrossRef]

28. Sviták, M.; Ruman, D. Tensile-shear strength of layered wood reinforced by carbon materials. *Wood Res. Slovakia* **2017**, *62*, 243–252.

29. Raftery, G.M.; Harte, A.M. Nonlinear numerical modelling of FRP reinforced glued laminated timber. *Compos. Part B Eng.* **2013**, *52*, 40–50. [CrossRef]

30. Mosallam, A.S. Structural evaluation and design procedure for wood beams repaired and retrofitted with FRP laminates and honeycomb sandwich panels. *Compos. Part B Eng.* **2015**, *87*, 196–213. [CrossRef]

31. Sandoz, J.L. *Industrial Ultrasonic Grinding for Multi-glued Laminated Timber*; Études& Constructions Bois, Concept Bois Technologie: Paris, France, 2009.

32. Steiger, R.; Gehri, E. Quality control of glulam: Improved method for shear testing of glue lines. In Proceedings of the Final Conference of COST Action E53: The Future of Quality Control for Wood & Wood Products, Edinburgh, Scotland, 4–7 May 2010.

33. Lorenzo, A. Numerical Tools for Moisture-stress and Fracture Analysis of Timber Structures. Ph.D. Dissertation, University of Calabria, Mendicino, Italy, 2010.

34. Saracoglu, E. Finite-element Simulations of the Influence of Cracks on the Strength of Glulam Beams. Master's Thesis, Blekinge Institute of Technology, Karlskrona, Sweden, 2011.

35. EN 310. *Wood-based panels: Determination of modulus of elasticity in bending and of bending strength*; European Committee for Standardization: Brussels, Belgium, 1993.

36. ISO 13061-1. *Physical and Mechanical Properties of Wood—Test Methods for Small Clear Wood—Part 1: Determination of Moisture Content for Physical and Mechanical Tests*; International Organization for Standardization: Geneva, Switzerland, 2014.

37. ISO 13061-2. *Physical and Mechanical Properties of Wood—Test Methods for Small Slear Wood Specimens—Part 2: Determination of Density for Physical and Mechanical Tests*; International Organization for Standardization: Geneva, Switzerland, 2014.

38. ISO 13061-3. *Physical and Mechanical Properties of Wood—Test Methods for Small Clear Wood Specimens—Part 3: Determination of Density for Physical and Mechanical Tests*; International Organization for Standardization: Geneva, Switzerland, 2014.

39. Sikora, A.; Gaffová, Z.; Rajnoha, R.; Šatanová, A.; Kminiak, R. Deflection of densified beech and aspen woods as a function of selected factors. *Bio Resour.* **2017**, *12*, 3192–3210. [CrossRef]

40. Gáborík, J.; Gaff, M.; Ruman, D.; Záborský, V.; Kašíčková, V.; Sikora, A. Adhesive as a factor affecting the properties of laminated wood. *Bio Resour.* **2016**, *11*, 10565–10574. [CrossRef]

*Article*

# Laminated Veneer Lumber with Non-Wood Components and the Effects of Selected Factors on Its Bendability

**Tomáš Svoboda *, Adam Sikora, Vladimír Záborský and Zuzana Gaffová**

Department of Wood Processing and Biomaterials, Faculty of Forestry and Wood Sciences,
Czech University of Life Sciences in Prague, Kamýcká 1176, 16521 Prague-Suchdol, Czech Republic;
sikoraa@fld.czu.cz (A.S.); zaborsky@fld.czu.cz (V.Z.); gaffova@fld.czu.cz (Z.G.)
* Correspondence: tomassvo@seznam.cz; Tel.: +420-608-321-941

Received: 23 April 2019; Accepted: 27 May 2019; Published: 30 May 2019

**Abstract:** Knowledge of the coefficients of wood bendability ($K_{bendC}$ and $K_{bendB}$) and of the effects of selected factors on the listed characteristics in bending stress has both scientific and practical significance. It forms a foundation for designing tools for bending and determines the stress that products and their parts can be exposed to during use. This study analyzes the effects of selected factors on the selected characteristics, such as the coefficients of wood bendability ($K_{bendC}$ and $K_{bendB}$). The selected factors of this study were wood species (WS) (*Fagus sylvatica* L. and *Populus tremula* L.), non-wood component (carbon fiber and glass fiber), position of the non-wood component in the laminated material (top and bottom), material thickness ($T$) (6 mm, 10 mm, and 18 mm), and adhesive (polyvinyl acetate and polyurethane), as well as their combined interaction on the monitored characteristics described above. The results contribute to the advancement of knowledge necessary for the study and development of new materials with specific properties for their intended use. The measured values of laminated structures can be compared with the values measured on the samples from the wood. The results can improve the innovative potential of wood processing companies and increase their performance and competitiveness in the market.

**Keywords:** coefficient of wood bendability; laminated wood; technological and product innovations; minimal curve radius

---

## 1. Introduction

The effective use of wood and its by-products has gained increased attention in recent years due to limited natural resources [1,2]. It is in society's general interest to efficiently utilize our limited forest resources and improve recycling [2].

Because composite material production uses materials of varying characteristics, it is necessary to verify their quality to ensure good product performance and market competitiveness. Composite production is a complex process [3]; it requires immediate consideration of various parameters (cutting geometry, production volume, matrix types, machine requirements, market economy, etc.). One of the main aspects limiting the structural use of high-strength composites is their weak interlaminar resistance [4]. Several strategies for enhancing the resistance of composites have been proposed [4–6], such as using a harder resin for hybrid composites, harder adhesive layers, and others.

Material stratification is very important in industrial practice, both in construction and in the manufacturing industry [7,8]. In the woodworking industry, homogeneity leads to better performance, thereby reducing the possible negative properties of wood that could lead to material failures. In environmental modification of wood, a number of studies have focused on thermal modification [9–15]. Densification of wood is also one of the ways to modify the basic properties of wood. It is a process whereby wood is pressed, for example, by rolling or by the action of various presses, thereby reducing its volume and increasing density. Such wood is then harder, firmer, and darker to look at. Densification of wood reduces the porosity and moisture content of the wood [16]. Gaff et al. [17] examined the effect of densification on bond strength. They found that the effect of the densification on bond strength is statistically very significant.

Another way to modify the properties of wood elements is through the use of non-wood reinforcing materials to form wood-based composite materials [18]. Such reinforcing materials include carbon, aramid, basalt, and glass fibers [19–21]. The application of non-wood components in a wood-based laminated veneer lumber material is usually intended to strengthen the material, increase its resistance to stress, and reduce bending values [22,23]. Such materials are characterized by different specific properties for their intended uses [24–27]. The intended use is a determining factor of the desired characteristics in a given material [22,28–30]. In some cases, emphasis is placed on materials with high strength values, while other cases see the creation of materials with high elasticity values [23] or high bendability values [22].

Bendability is a characteristic that has recently attracted great interest. The effect of the placement of a non-wood component in such a material has not yet been given much attention, and the interactions of different types of materials with other factors influencing this characteristic have also not been studied. A mathematical interpretation of the bending coefficient [17] was only recently established for the correct description of bendability.

The bending coefficient ($K_{bend}$) is a quantitative characteristic that is defined as the ratio of the thickness ($h$) of the bent material to the minimum bend radius ($R$) (Figure 1). For most types of wood, the limit ratio is $h{:}R$ = 1:35 to 1:45. The critical area for bending wood is the tensile zone. The maximum tensile deformation of wood in its original unmodified state is 0.75% to 1%. This can be increased with plasticization to 1.5% to 2%. By contrast, the compressibility of wood is greater at optimum humidity and temperature; if its porosity allows it, it reaches up to 40% [31].

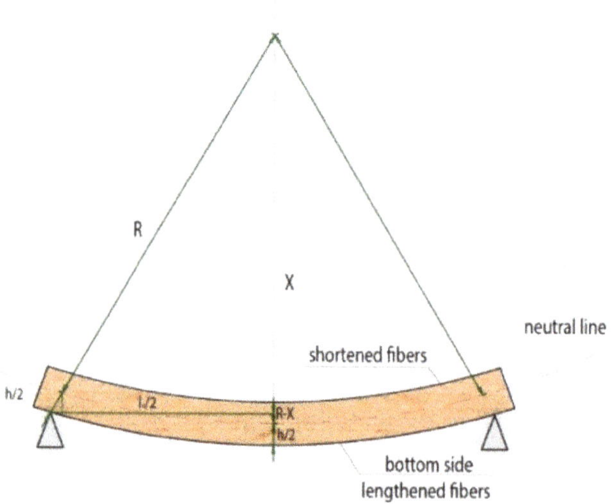


**Figure 1.** Bending geometry.

There is very little scientific knowledge about the minimum bend radius and bending coefficient. The aim of our work is therefore to deepen the knowledge of the bending coefficient of wood ($K_{bendC}$ and $K_{bendB}$), namely beech and aspen, under three-point bending. Gaff et al. [32] showed that as the material thickness increases, the value of the bending coefficient decreases, and the force needed for bending increases. Gaff et al. [17] created a model for analyzing bendability, with which it is possible to define the correct relations for determining the minimum bend radius ($R_{min}$), which can then be used to calculate the bending coefficient ($K_{bend}$).

## 2. Materials and Methods

### 2.1. Material

The wooden lamellas used in this experiment were made of beech wood (*Fagus sylvatica* L.) and aspen wood (*Populus tremula* L.) with thicknesses of 3 mm, 5 mm, and 9 mm, widths of 35 mm, and lengths of 600 mm. The beech and aspen wood came from Polana, Slovakia. Polyvinyl acetate (PVAc) and polyurethane (PUR) adhesives were used to produce laminated wood using the above lamellas. Carbon fibers (SikaWrap-150 C/30, 155 g/m$^2$ ± 5 g/m$^2$) and glass fibers (Kittfort, 355 g/m$^2$) were used as the reinforcing materials, which were first glued on the convex sides and then on the concave sides with respect to the direction of loading. Categorization of the test specimens is shown in Figure 2.

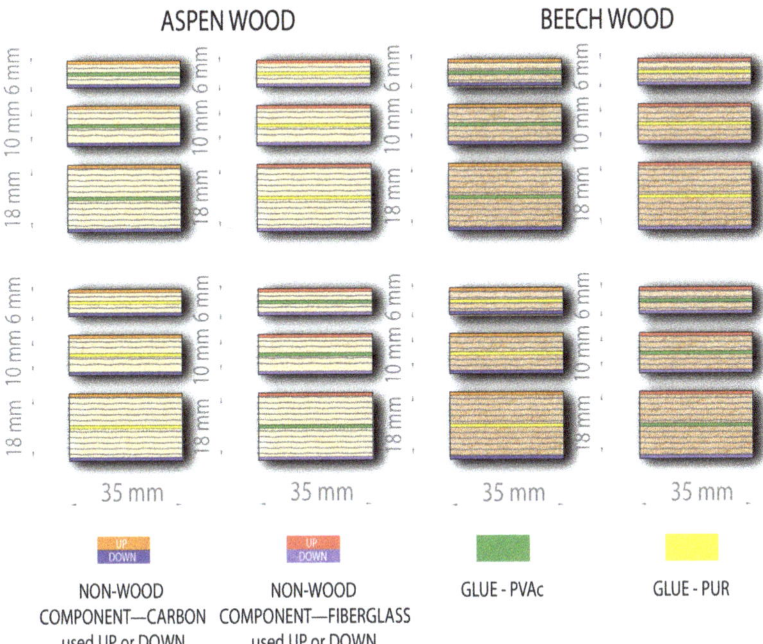


**Figure 2.** Categorization of test specimens.

After all test specimens were created, they were climatized in a climatic chamber (ED, APT Line II, Binder, Tuttlingen, Germany) to 12% moisture content at 65% relative humidity and 20 °C.

### 2.2. Methods

#### 2.2.1. Determining Selected Characteristics

Testing was performed with three-point bending (Figure 3), with the bottom support span set to 20 times the total thickness of the test specimen. The top support crossbeam was set in a center position relative to the distance of the bottom support crossbeam. Testing was performed according to EN 310 (1993) [33] using a universal testing machine (FPZ 100, TIRA, Schalkau, Germany). Testing took place in the tangential direction relative to the fiber direction. The feed rate of the top support was set to 3 mm/min due to the duration of the test. An ALMEMO 2690-8 datalogger (AhlbornGmbH, Braunschweig, Germany) was used to record all the forces during the test.

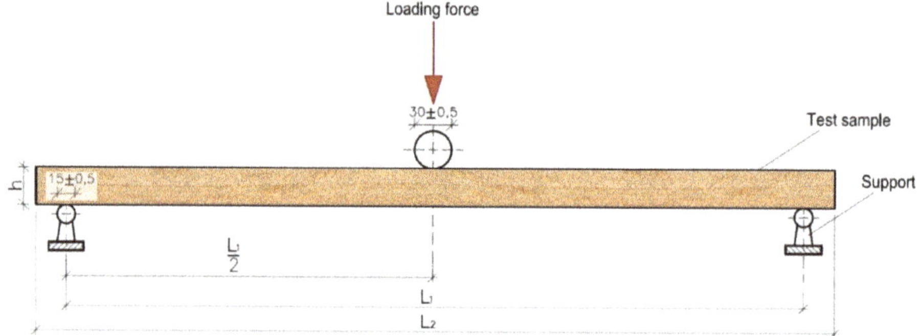


**Figure 3.** Principle of the three-point bending test [33].

Before the measurement was performed, the densities and humidities of the used wooden components were measured according to ISO 13061-1 (2014) [34] and ISO 13061-2 (2014) [35]. After the tests were completed, all test specimens were dried to 0% moisture content, as necessary to calculate the moisture content at the time of the test.

2.2.2. Evaluation and Calculation of $K_{bend}$ and $R_{min}$

Based on data obtained from the stress–strain diagram, exact identification of the boundary points between the linear and non-linear parts of the diagram was used to determine forces at the limit of proportionality ($F_E$) and at the yield point ($F_P$), along with the deflections at the limit of proportionality ($Y_E$) and at the yield point ($Y_P$). These characteristics were identified using the MATESS program, which is currently being developed by the Czech University of Life Sciences in Prague.

In the next step, the bendability of the tested material was evaluated based on the minimum bend radius ($R_{minB}$ and $R_{minC}$) and the bending coefficient ($K_{bendB}$ and $K_{bendC}$). Two approaches were used to evaluate the bendability. The first approach was based on bending geometry (Equations (1) and (2)), while the second approach was based on the simple bending equations (Equations (3) and (4)), which were used in the work of Gaff et al. [17]:

$$R_{minB} = \frac{l_0^2}{8\,Y_P} + \frac{Y_P}{2} - \frac{h}{2} \tag{1}$$

$$K_{bendB} = \frac{h}{R_{minB}} = \frac{h}{\frac{l_0^2}{8\,Y_P} + \frac{Y_P}{2} - \frac{h}{2}} \tag{2}$$

$$R_{minC} = \frac{l_0^2}{12\,Y_P} \tag{3}$$

$$K_{bendC} = \frac{h}{R_{minC}} = \frac{h}{\frac{l_0^2}{12\,Y_P}} \tag{4}$$

where $R_{minB}$ is the minimum bend radius based on bending geometry (mm), $R_{minC}$ is the minimum bend radius based on the simple bending equations (mm), $K_{bendB}$ is the bending coefficient based on bending geometry, $K_{bendC}$ is the bending coefficient based on the simple bending equations, $Y_P$ is the deflection at the yield point (mm), $l_0$ is the bottom support span (mm), and $h$ is the total material thickness (mm).

The results were statistically evaluated with an analysis of variance (ANOVA), specifically Fisher's F-test, with STATISTICA 12 software (Statsoft Inc., Tulsa, OK, USA). The results were evaluated using a 95% confidence interval, which represents a significance level of 0.05 ($p < 0.05$). Duncan's test was also used for deeper analysis to compare all sets of test specimens.

## 3. Results and Discussion

Tables 1 and 2 show the average values and coefficients of variation (in parentheses) of the monitored characteristics, the average density values measured in individual sets of test specimens, and the corresponding coefficients of variation. Table 1 shows the average values of $K_{bendC}$, $K_{bendB}$, $R_{minC}$, and $R_{minB}$ measured in the aspen test specimens.

**Table 1.** Values of bending characteristics and the coefficients of variance of layered aspen material.

| WS | NWC | Location | Glue | $T$ (mm) | Code of Test Sample | $K_{bendC}$ | $K_{bendB}$ | $R_{minC}$ (mm) | $R_{minB}$ (mm) |
|---|---|---|---|---|---|---|---|---|---|
| A | CA | U | PUR | 6 | A-CA-U-PUR-6 | 0.014 (17.0) | 0.012 (13.5) | 519.28 (9.4) | 717.76 (11.9) |
| A | CA | U | PUR | 10 | A-CA-U-PUR-10 | 0.024 (13.3) | 0.015 (16.8) | 526.38 (14.9) | 693.13 (20.4) |
| A | CA | U | PUR | 18 | A-CA-U-PUR-18 | 0.012 (8.4) | 0.007 (7.8) | 1616.03 (14.7) | 2448.28 (13.5) |
| A | CA | U | PVAc | 6 | A-CA-U-PVAc-6 | 0.039 (4.1) | 0.026 (4.1) | 150.72 (3.3) | 227.08 (3.3) |
| A | CA | U | PVAc | 10 | A-CA-U-PVAc-10 | 0.034 (5.0) | 0.023 (5.1) | 280.19 (12.0) | 406.81 (17.0) |
| A | CA | U | PVAc | 18 | A-CA-U-PVAc-18 | 0.013 (14.0) | 0.008 (13.6) | 1279.40 (15.5) | 1651.62 (11.4) |
| A | LA | U | PUR | 6 | A-LA-U-PUR-6 | 0.031 (11.3) | 0.021 (17.5) | 159.95 (14.3) | 238.00 (18.0) |
| A | LA | U | PUR | 10 | A-LA-U-PUR-10 | 0.024 (16.0) | 0.015 (14.4) | 511.83 (18.5) | 671.01 (16.6) |
| A | LA | U | PUR | 18 | A-LA-U-PUR-18 | 0.015 (17.2) | 0.010 (17.2) | 1233.71 (14.0) | 1848.16 (14.0) |
| A | LA | U | PVAc | 6 | A-LA-U-PVAc-6 | 0.043 (14.7) | 0.028 (14.1) | 134.11 (11.4) | 202.81 (11.0) |
| A | LA | U | PVAc | 10 | A-LA-U-PVAc-10 | 0.022 (11.3) | 0.015 (17.6) | 401.48 (10.8) | 628.74 (9.6) |
| A | LA | U | PVAc | 18 | A-LA-U-PVAc-18 | 0.016 (10.3) | 0.011 (8.9) | 1220.28 (10.8) | 1688.32 (13.4) |
| A | CA | D | PUR | 6 | A-CA-D-PUR-6 | 0.026 (19.4) | 0.019 (17.9) | 248.51 (19.1) | 355.31 (16.0) |
| A | CA | D | PUR | 10 | A-CA-D-PUR-10 | 0.022 (17.9) | 0.014 (15.4) | 379.73 (18.4) | 570.20 (18.3) |
| A | CA | D | PUR | 18 | A-CA-D-PUR-18 | 0.020 (19.6) | 0.013 (20.4) | 878.98 (13.2) | 1365.45 (18.2) |
| A | CA | D | PVAc | 6 | A-CA-D-PVAc-6 | 0.036 (19.2) | 0.026 (21.0) | 165.21 (11.8) | 253.26 (7.4) |
| A | CA | D | PVAc | 10 | A-CA-D-PVAc-10 | 0.037 (15.7) | 0.024 (15.3) | 280.82 (16.4) | 427.67 (15.8) |
| A | CA | D | PVAc | 18 | A-CA-D-PVAc-18 | 0.024 (17.2) | 0.014 (18.3) | 884.80 (12.0) | 1454.81 (11.9) |
| A | LA | D | PUR | 6 | A-LA-D-PUR-6 | 0.031 (15.6) | 0.024 (20.4) | 144.14 (15.0) | 218.42 (14.7) |
| A | LA | D | PUR | 10 | A-LA-D-PUR-10 | 0.032 (16.5) | 0.021 (16.3) | 330.58 (16.3) | 495.74 (16.2) |
| A | LA | D | PUR | 18 | A-LA-D-PUR-18 | 0.028 (18.7) | 0.019 (18.5) | 670.60 (21.4) | 1004.97 (21.3) |
| A | LA | D | PVAc | 6 | A-LA-D-PVAc-6 | 0.036 (7.2) | 0.027 (12.1) | 143.00 (14.7) | 225.48 (15.4) |
| A | LA | D | PVAc | 10 | A-LA-D-PVAc-10 | 0.027 (16.6) | 0.018 (16.5) | 402.30 (18.6) | 602.38 (18.5) |
| A | LA | D | PVAc | 18 | A-LA-D-PVAc-18 | 0.023 (19.2) | 0.014 (19.9) | 886.44 (17.9) | 1322.03 (18.0) |

WS—wood species; NWC—non-wood component; $T$—thickness; $K_{bendC}$—bending coefficient based on simple bending equations; $K_{bendB}$—bending coefficient based on bending geometry; $R_{minC}$—minimum bend radius based on simple bending equations; $R_{minB}$—minimum bend radius based on bending geometry; A—aspen; CA—carbon; LA—glass fiber; U—top; D—bottom.

**Table 2.** Average values of bending characteristics and the coefficients of variance of layered beech material.

| WS | NWC | Location | Glue | $T$ (mm) | Code of Test Sample | $K_{bendC}$ | $K_{bendB}$ | $R_{minC}$ (mm) | $R_{minB}$ (mm) |
|---|---|---|---|---|---|---|---|---|---|
| B | CA | U | PUR | 6 | B-CA-U-PUR-6 | 0.024 (17.6) | 0.016 (17.6) | 231.20 (12.9) | 340.25 (8.4) |
| B | CA | U | PUR | 10 | B-CA-U-PUR-10 | 0.026 (15.4) | 0.017 (18.6) | 459.30 (20.7) | 663.96 (17.4) |
| B | CA | U | PUR | 18 | B-CA-U-PUR-18 | 0.011 (6.0) | 0.006 (19.4) | 2262.89 (18.2) | 3288.95 (16.4) |
| B | CA | U | PVAc | 6 | B-CA-U-PVAc-6 | 0.037 (9.1) | 0.024 (8.8) | 153.98 (8.9) | 232.03 (8.7) |
| B | CA | U | PVAc | 10 | B-CA-U-PVAc-10 | 0.031 (9.6) | 0.021 (9.6) | 354.65 (9.5) | 531.16 (9.5) |
| B | CA | U | PVAc | 18 | B-CA-U-PVAc-18 | 0.014 (16.9) | 0.009 (9.7) | 1416.71 (12.4) | 1804.00 (21.0) |
| B | LA | U | PUR | 6 | B-LA-U-PUR-6 | 0.044 (5.1) | 0.030 (0.8) | 134.43 (11.2) | 169.19 (8.5) |
| B | LA | U | PUR | 10 | B-LA-U-PUR-10 | 0.025 (16.4) | 0.016 (16.3) | 439.62 (18.5) | 657.95 (18.4) |
| B | LA | U | PUR | 18 | B-LA-U-PUR-18 | 0.008 (19.3) | 0.005 (17.2) | 2442.86 (20.3) | 3391.53 (14.3) |
| B | LA | U | PVAc | 6 | B-LA-U-PVAc-6 | 0.033 (5.6) | 0.022 (5.5) | 168.33 (4.7) | 253.25 (4.6) |
| B | LA | U | PVAc | 10 | B-LA-U-PVAc-10 | 0.032 (11.8) | 0.022 (10.8) | 317.76 (16.7) | 476.68 (16.4) |
| B | LA | U | PVAc | 18 | B-LA-U-PVAc-18 | 0.013 (19.2) | 0.008 (19.2) | 1451.32 (18.4) | 2171.66 (18.4) |
| B | CA | D | PUR | 6 | B-CA-D-PUR-6 | 0.034 (7.7) | 0.023 (7.6) | 166.27 (8.6) | 250.18 (8.4) |
| B | CA | D | PUR | 10 | B-CA-D-PUR-10 | 0.025 (17.9) | 0.015 (17.9) | 427.22 (18.8) | 644.19 (18.5) |
| B | CA | D | PUR | 18 | B-CA-D-PUR-18 | 0.015 (12.3) | 0.015 (18.5) | 1232.78 (15.1) | 1488.03 (17.2) |
| B | CA | D | PVAc | 6 | B-CA-D-PVAc-6 | 0.042 (14.0) | 0.028 (13.6) | 135.40 (12.5) | 204.75 (12.2) |
| B | CA | D | PVAc | 10 | B-CA-D-PVAc-10 | 0.030 (14.9) | 0.021 (19.5) | 296.37 (16.9) | 489.47 (19.4) |
| B | CA | D | PVAc | 18 | B-CA-D-PVAc-18 | 0.021 (18.9) | 0.014 (18.8) | 793.45 (18.4) | 1248.23 (11.0) |
| B | LA | D | PUR | 6 | B-LA-D-PUR-6 | 0.046 (7.4) | 0.030 (7.0) | 105.94 (7.5) | 162.18 (7.0) |
| B | LA | D | PUR | 10 | B-LA-D-PUR-10 | 0.024 (13.0) | 0.015 (17.9) | 426.66 (13.0) | 683.36 (19.0) |
| B | LA | D | PUR | 18 | B-LA-D-PUR-18 | 0.026 (18.4) | 0.016 (16.7) | 663.54 (18.5) | 1209.10 (14.0) |
| B | LA | D | PVAc | 6 | B-LA-D-PVAc-6 | 0.035 (8.5) | 0.023 (8.3) | 160.41 (7.9) | 241.58 (7.7) |
| B | LA | D | PVAc | 10 | B-LA-D-PVAc-10 | 0.034 (19.3) | 0.024 (13.5) | 296.34 (19.6) | 459.85 (13.9) |
| B | LA | D | PVAc | 18 | B-LA-D-PVAc-18 | 0.033 (17.3) | 0.022 (17.0) | 526.61 (16.3) | 795.58 (16.4) |

WS—wood species; NWC—non-wood component; $T$—thickness; $K_{bendC}$—bending coefficient based on simple bending equations; $K_{bendB}$—bending coefficient based on bending geometry; $R_{minC}$—minimum bend radius based on simple bending equations; $R_{minB}$—minimum bend radius based on bending geometry; B—beech; CA—carbon; LA—glass fiber; U—top; D—bottom.

The highest average values of $R_{minC}$ (1616 mm) and $R_{minB}$ (2448 mm) were measured in the material with a thickness of 18 mm glued with PUR adhesive and reinforced with carbon fibers placed on the top side with respect to the direction of loading. The lowest average values of $R_{minC}$ (134 mm) and $R_{minB}$ (202 mm) were measured in the material with a thickness of 6 mm glued with PVAc adhesive and reinforced with glass fibers placed on the top side with respect to the direction of loading.

Higher average values for the bending coefficient were obtained in calculations based on the simple bending equation $K_{bendC}$ (0.01 to 0.04) than in calculations based on bending geometry $K_{bendB}$ (0.01 to 0.03), which corresponds with the results reported in the work of Gaff et al. [17], who also studied the bending coefficient of unmodified aspen wood.

Table 2 shows the average values of $K_{bendC}$, $K_{bendB}$, $R_{minC}$, and $R_{minB}$ calculated in beech test specimens. The layered beech materials showed the same tendency of the bending coefficient as the laminated aspen materials. In the laminated beech materials, $K_{bendC}$ values (0.01 to 0.05) were greater than $K_{bendB}$ values (0.01 to 0.03). Comparing these results with those of Gaff et al. [17] confirms the trend of greater $K_{bendC}$ values.

The greatest average value of $R_{minB}$ (3391 mm) was measured in the material with a thickness of 18 mm glued with PUR adhesive and reinforced with glass fibers on the top side with respect to the direction of loading. The lowest value of $R_{minB}$ (162 mm) was measured in the material with a total thickness of 6 mm bonded with PUR adhesive and reinforced with glass fiber on the bottom side of the test specimen relative to the direction of loading. The greatest (2442 mm) and lowest (105 mm) average values of $R_{minC}$ were measured in the same materials as the greatest and lowest values of $R_{minB}$.

All the measured data were statistically evaluated using a single-factor analysis in which the test specimen type was chosen as the default factor. The evaluation was based on the significance level $p$, which was less than 0.005. Tables 3–6 show the statistical evaluation of the effect of the test specimen type on the bending coefficient based on the simple bending equations ($K_{bendC}$) in laminated aspen and beech materials with the non-wood component placed on the top or bottom side with respect to the direction of loading.

**Table 3.** Statistical evaluation of the effect of the factors and their interaction on the coefficient of wood bendability ($K_{bendC}$) for aspen and non-wood component (NWC) on the bottom.

| Monitored Factor | Sum of Squares | Degrees of Freedom | Variance | Fisher's F-Test | Significance Level |
|---|---|---|---|---|---|
| Intercept | 0.101015 | 1 | 0.101015 | 4182.484 | *** |
| 1) Type of Sample | 0.003778 | 11 | 0.000343 | 14.220 | *** |
| Error | 0.002608 | 108 | 0.000024 | | |

NS—not significant, ***—significant at $p < 0.005$.

**Table 4.** Statistical evaluation of the effect of the factors and their interaction on the coefficient of wood bendability ($K_{bendC}$) for aspen and NWC on top.

| Monitored Factor | Sum of Squares | Degrees of Freedom | Variance | Fisher's F-Test | Significance Level |
|---|---|---|---|---|---|
| Intercept | 0.058817 | 1 | 0.058817 | 5814.366 | *** |
| 1) Type of Sample | 0.010364 | 11 | 0.000942 | 93.141 | *** |
| Error | 0.001022 | 101 | 0.000010 | | |

NS—not significant, ***—significant at $p < 0.005$.

**Table 5.** Statistical evaluation of the effect of the factors and their interaction on the coefficient of wood bendability ($K_{bendC}$) for beech and NWC on the bottom.

| Monitored Factor | Sum of Squares | Degrees of Freedom | Variance | Fisher's F-Test | Significance Level |
|---|---|---|---|---|---|
| Intercept | 0.114036 | 1 | 0.114036 | 5587.265 | *** |
| 1) Type of Sample | 0.008469 | 11 | 0.000770 | 37.724 | *** |
| Error | 0.002204 | 108 | 0.000020 | | |

NS—not significant, ***—significant at $p < 0.005$.

**Table 6.** Statistical evaluation of the effect of the factors and their interaction on the coefficient of wood bendability ($K_{bendC}$) for beech and NWC on top.

| Monitored Factor | Sum of Squares | Degrees of Freedom | Variance | Fisher's F-Test | Significance Level |
|---|---|---|---|---|---|
| Intercept | 0.055837 | 1 | 0.055837 | 5566.564 | *** |
| 1) Type of Sample | 0.011935 | 11 | 0.001085 | 108.163 | *** |
| Error | 0.001144 | 114 | 0.000010 | | |

NS—not significant, ***—significant at $p < 0.005$.

Tables 7–10 show the statistical evaluation of the effect of the test specimen type on the bending coefficient based on bending geometry ($K_{bendB}$) in laminated aspen and beech materials with the non-wood component placed on the top or bottom with respect to the direction of loading.

**Table 7.** Statistical evaluation of the effect of the factors and their interaction on the coefficient of wood bendability ($K_{bendB}$) for aspen and NWC on the bottom.

| Monitored Factor | Sum of Squares | Degrees of Freedom | Variance | Fisher's F-Test | Significance Level |
|---|---|---|---|---|---|
| Intercept | 0.046953 | 1 | 0.046953 | 3654.379 | *** |
| 1) Type of Sample | 0.002775 | 11 | 0.000252 | 19.633 | *** |
| Error | 0.001388 | 108 | 0.000013 | | |

NS—not significant, ***—significant at $p < 0.005$.

**Table 8.** Statistical evaluation of the effect of the factors and their interaction on the coefficient of wood bendability ($K_{bendB}$) for aspen and NWC on top.

| Monitored Factor | Sum of Squares | Degrees of Freedom | Variance | Fisher's F-Test | Significance Level |
|---|---|---|---|---|---|
| Intercept | 0.026766 | 1 | 0.026766 | 5122.907 | *** |
| 1) Type of Sample | 0.004369 | 11 | 0.000397 | 76.012 | *** |
| Error | 0.000528 | 101 | 0.000005 | | |

NS—not significant, ***—significant at $p < 0.005$.

**Table 9.** Statistical evaluation of the effect of the factors and their interaction on the coefficient of wood bendability ($K_{bendB}$) for beech and NWC on the bottom.

| Monitored Factor | Sum of Squares | Degrees of Freedom | Variance | Fisher's F-Test | Significance Level |
|---|---|---|---|---|---|
| Intercept | 0.052061 | 1 | 0.052061 | 5814.871 | *** |
| 1) Type of Sample | 0.003206 | 11 | 0.000291 | 32.553 | *** |
| Error | 0.000967 | 108 | 0.000009 | | |

NS—not significant, ***—significant at $p < 0.005$.

**Table 10.** Statistical evaluation of the effect of the factors and their interaction on the coefficient of wood bendability ($K_{bendB}$) for beech and NWC on top.

| Monitored Factor | Sum of Squares | Degrees of Freedom | Variance | Fisher's F-Test | Significance Level |
|---|---|---|---|---|---|
| Intercept | 0.024553 | 1 | 0.024553 | 5158.266 | *** |
| 1) Type of Sample | 0.005388 | 11 | 0.000490 | 102.915 | *** |
| Error | 0.000543 | 114 | 0.000005 | | |

NS—not significant, ***—significant at $p < 0.005$.

Tables 11–14 show the statistical evaluation of the effect of the test specimen type on the minimum bend radius based on the simple bending equations ($R_{minC}$) in laminated aspen and beech materials with the non-wood component placed on the top or bottom with respect to the direction of loading.

**Table 11.** Statistical evaluation of the effect of the factors and their interaction on the minimum bend radius at the yield point ($R_{minC}$) for aspen and NWC on the bottom.

| Monitored Factor | Sum of Squares | Degrees of Freedom | Variance | Fisher's F-Test | Significance Level |
|---|---|---|---|---|---|
| Intercept | 24436158 | 1 | 24436158 | 3264.754 | *** |
| 1) Type of Sample | 9717956 | 11 | 883451 | 118.032 | *** |
| Error | 808363 | 108 | 7485 | | |

NS—not significant, ***—significant at $p < 0.005$.

**Table 12.** Statistical evaluation of the effect of the factors and their interaction on the minimum bend radius at the yield point ($R_{minC}$) for aspen and NWC on top.

| Monitored Factor | Sum of Squares | Degrees of Freedom | Variance | Fisher's F-Test | Significance Level |
|---|---|---|---|---|---|
| Intercept | 44421882 | 1 | 44421882 | 2710.337 | *** |
| 1) Type of Sample | 28160468 | 11 | 2560043 | 156.197 | *** |
| Error | 1655370 | 101 | 16390 | | |

NS—not significant, ***—significant at $p < 0.005$.

**Table 13.** Statistical evaluation of the effect of the factors and their interaction on the minimum bend radius at the yield point ($R_{minC}$) for beech and NWC on the bottom.

| Monitored Factor | Sum of Squares | Degrees of Freedom | Variance | Fisher's F-Test | Significance Level |
|---|---|---|---|---|---|
| Intercept | 22802737 | 1 | 22802737 | 2892.018 | *** |
| 1) Type of Sample | 12097814 | 11 | 1099801 | 139.485 | *** |
| Error | 851549 | 108 | 7885 | | |

NS—not significant, ***—significant at $p < 0.005$.

**Table 14.** Statistical evaluation of the effect of the factors and their interaction on the minimum bend radius at the yield point ($R_{minC}$) for beech and NWC on top.

| Monitored Factor | Sum of Squares | Degrees of Freedom | Variance | Fisher's F-Test | Significance Level |
|---|---|---|---|---|---|
| Intercept | 58491893 | 1 | 58491893 | 1178.097 | *** |
| 1) Type of Sample | 81191097 | 11 | 7381009 | 148.662 | *** |
| Error | 5660040 | 114 | 49649 | | |

NS—not significant, ***—significant at $p < 0.005$.

Tables 15–18 show the statistical evaluation of the effect of the test specimen type on the minimum bend radius based on bending geometry ($R_{minB}$) in laminated aspen and beech materials with the non-wood component placed on the top or bottom with respect to the direction of loading.

**Table 15.** Statistical evaluation of the effect of the factors and their interaction on the minimum bend radius at the yield point ($R_{minB}$) for aspen and NWC on the bottom.

| Monitored Factor | Sum of Squares | Degrees of Freedom | Variance | Fisher's F-Test | Significance Level |
|---|---|---|---|---|---|
| Intercept | 57349324 | 1 | 57349324 | 2933.356 | *** |
| 1) Type of Sample | 24093243 | 11 | 2190295 | 112.031 | *** |
| Error | 2111481 | 108 | 19551 | | |

NS—not significant, ***—significant at $p < 0.005$.

**Table 16.** Statistical evaluation of the effect of the factors and their interaction on the minimum bend radius at the yield point ($R_{minB}$) for aspen and NWC on top.

| Monitored Factor | Sum of Squares | Degrees of Freedom | Variance | Fisher's F-Test | Significance Level |
|---|---|---|---|---|---|
| Intercept | 89797468 | 1 | 89797468 | 2744.915 | *** |
| 1) Type of Sample | 61286334 | 11 | 5571485 | 170.308 | *** |
| Error | 3304126 | 101 | 32714 | | |

NS—not significant, ***—significant at $p < 0.005$.

**Table 17.** Statistical evaluation of the effect of the factors and their interaction on the minimum bend radius at the yield point ($R_{minB}$) for beech and NWC on the bottom.

| Monitored Factor | Sum of Squares | Degrees of Freedom | Variance | Fisher's F-Test | Significance Level |
|---|---|---|---|---|---|
| Intercept | 51699287 | 1 | 51699287 | 2759.401 | *** |
| 1) Type of Sample | 22194201 | 11 | 2017655 | 107.690 | *** |
| Error | 2023455 | 108 | 18736 | | |

NS—not significant, ***—significant at $p < 0.005$.

**Table 18.** Statistical evaluation of the effect of the factors and their interaction on the minimum bend radius at the yield point ($R_{minB}$) for beech and NWC on top.

| Monitored Factor | Sum of Squares | Degrees of Freedom | Variance | Fisher's F-Test | Significance Level |
|---|---|---|---|---|---|
| Intercept | 118241938 | 1 | 118241938 | 1681.398 | *** |
| 1) Type of Sample | 159214450 | 11 | 14474041 | 205.821 | *** |
| Error | 8016889 | 114 | 70324 | | |

NS—not significant, ***—significant at $p < 0.005$.

Duncan's test was performed for a detailed comparison of the differences in the bending coefficients ($K_{bendC}$ and $K_{bendB}$) among individual types of laminated aspen and beech materials, and the results are shown in Tables 19–26.

**Table 19.** Comparison of the effects of individual factors using Duncan's test on the coefficient of bendability ($K_{bendC}$) for aspen and NWC on the bottom.

| No. | Type of Sample | (1) 0.026 | (2) 0.022 | (3) 0.020 | (4) 0.036 | (5) 0.037 | (6) 0.024 | (7) 0.031 | (8) 00.32 | (9) 0.028 | (10) 0.036 | (11) 0.027 | (12) 0.023 |
|---|---|---|---|---|---|---|---|---|---|---|---|---|---|
| 1. | A-CA-D-PUR-6 | | | | | | | | | | | | |
| 2. | A-CA-D-PUR-10 | 0.048 | | | | | | | | | | | |
| 3. | A-CA-D-PUR-18 | 0.007 | 0.414 | | | | | | | | | | |
| 4. | A-CA-D-PVAc-6 | 0.000 | 0.000 | 0.000 | | | | | | | | | |
| 5. | A-CA-D-PVAc-10 | 0.000 | 0.000 | 0.000 | 0.973 | | | | | | | | |
| 6. | A-CA-D-PVAc-18 | 0.261 | 0.331 | 0.092 | 0.000 | 0.000 | | | | | | | |
| 7. | A-LA-D-PUR-6 | 0.049 | 0.000 | 0.000 | 0.032 | 0.035 | 0.003 | | | | | | |
| 8. | A-LA-D-PUR-10 | 0.016 | 0.000 | 0.000 | 0.085 | 0.096 | 0.001 | 0.595 | | | | | |
| 9. | A-LA-D-PUR-18 | 0.445 | 0.008 | 0.001 | 0.001 | 0.001 | 0.077 | 0.182 | 0.078 | | | | |
| 10. | A-LA-D-PVAc-6 | 0.000 | 0.000 | 0.000 | 0.943 | 0.965 | 0.000 | 0.036 | 0.100 | 0.001 | | | |
| 11. | A-LA-D-PVAc-10 | 0.895 | 0.040 | 0.005 | 0.000 | 0.000 | 0.238 | 0.057 | 0.019 | 0.494 | 0.000 | | |
| 12. | A-LA-D-PVAc-18 | 0.213 | 0.403 | 0.120 | 0.000 | 0.000 | 0.840 | 0.002 | 0.000 | 0.056 | 0.000 | 0.187 | |

**Table 20.** Comparison of the effects of individual factors using Duncan's test on the coefficient of bendability ($K_{bendC}$) for aspen and NWC on top.

| No. | Type of Sample | (1) 0.014 | (2) 0.024 | (3) 0.012 | (4) 0.039 | (5) 0.034 | (6) 0.013 | (7) 0.031 | (8) 0.024 | (9) 0.015 | (10) 0.043 | (11) 0.022 | (12) 0.016 |
|---|---|---|---|---|---|---|---|---|---|---|---|---|---|
| 1. | A-CA-U-PUR-6 | | | | | | | | | | | | |
| 2. | A-CA-U-PUR-10 | 0.000 | | | | | | | | | | | |
| 3. | A-CA-U-PUR-18 | 0.198 | 0.000 | | | | | | | | | | |
| 4. | A-CA-U-PVAc-6 | 0.000 | 0.000 | 0.000 | | | | | | | | | |
| 5. | A-CA-U-PVAc-10 | 0.000 | 0.000 | 0.000 | 0.002 | | | | | | | | |
| 6. | A-CA-U-PVAc-18 | 0.886 | 0.000 | 0.221 | 0.000 | 0.000 | | | | | | | |
| 7. | A-LA-U-PUR-6 | 0.000 | 0.000 | 0.000 | 0.000 | 0.075 | 0.000 | | | | | | |
| 8. | A-LA-U-PUR-10 | 0.000 | 0.935 | 0.000 | 0.000 | 0.000 | 0.000 | 0.000 | | | | | |
| 9. | A-LA-U-PUR-18 | 0.603 | 0.000 | 0.086 | 0.000 | 0.000 | 0.535 | 0.000 | 0.000 | | | | |
| 10. | A-LA-U-PVAc-6 | 0.000 | 0.000 | 0.000 | 0.011 | 0.000 | 0.000 | 0.000 | 0.000 | 0.000 | | | |
| 11. | A-LA-U-PVAc-10 | 0.000 | 0.242 | 0.000 | 0.000 | 0.000 | 0.000 | 0.000 | 0.240 | 0.000 | 0.000 | | |
| 12. | A-LA-U-PVAc-18 | 0.264 | 0.000 | 0.022 | 0.000 | 0.000 | 0.228 | 0.000 | 0.000 | 0.502 | 0.000 | 0.000 | |

**Table 21.** Comparison of the effects of individual factors using Duncan's test on the coefficient of bendability ($K_{bendC}$) for beech and NWC on the bottom.

| No. | Type of Sample | (1) 0.034 | (2) 0.025 | (3) 0.015 | (4) 0.042 | (5) 0.030 | (6) 0.021 | (7) 0.046 | (8) 0.024 | (9) 0.026 | (10) 0.035 | (11) 0.034 | (12) 0.033 |
|---|---|---|---|---|---|---|---|---|---|---|---|---|---|
| 1. | B-CA-D-PUR-6 | | | | | | | | | | | | |
| 2. | B-CA-D-PUR-10 | 0.000 | | | | | | | | | | | |
| 3. | B-CA-D-PUR-18 | 0.000 | 0.000 | | | | | | | | | | |
| 4. | B-CA-D-PVAc-6 | 0.000 | 0.000 | 0.000 | | | | | | | | | |
| 5. | B-CA-D-PVAc-10 | 0.029 | 0.032 | 0.000 | 0.000 | | | | | | | | |
| 6. | B-CA-D-PVAc-18 | 0.000 | 0.042 | 0.009 | 0.000 | 0.000 | | | | | | | |
| 7. | B-LA-D-PUR-6 | 0.000 | 0.000 | 0.000 | 0.107 | 0.000 | 0.000 | | | | | | |
| 8. | B-LA-D-PUR-10 | 0.000 | 0.655 | 0.000 | 0.000 | 0.012 | 0.089 | 0.000 | | | | | |
| 9. | B-LA-D-PUR-18 | 0.000 | 0.702 | 0.000 | 0.000 | 0.061 | 0.020 | 0.000 | 0.438 | | | | |
| 10. | B-LA-D-PVAc-6 | 0.861 | 0.000 | 0.000 | 0.000 | 0.025 | 0.000 | 0.000 | 0.000 | 0.000 | | | |
| 11. | B-LA-D-PVAc-10 | 0.910 | 0.000 | 0.000 | 0.000 | 0.027 | 0.000 | 0.000 | 0.000 | 0.000 | 0.941 | | |
| 12. | B-LA-D-PVAc-18 | 0.592 | 0.000 | 0.000 | 0.000 | 0.078 | 0.000 | 0.000 | 0.000 | 0.001 | 0.516 | 0.544 | |

**Table 22.** Comparison of the effects of individual factors using Duncan's test on the coefficient of bendability ($K_{bendC}$) for beech and NWC on top.

| No. | Type of Sample | (1) 0.024 | (2) 0.026 | (3) 0.011 | (4) 0.037 | (5) 0.031 | (6) 0.014 | (7) 0.044 | (8) 0.025 | (9) 0.008 | (10) 0.033 | (11) 0.032 | (12) 0.013 |
|---|---|---|---|---|---|---|---|---|---|---|---|---|---|
| 1. | B-CA-U-PUR-6 | | | | | | | | | | | | |
| 2. | B-CA-U-PUR-10 | 0.174 | | | | | | | | | | | |
| 3. | B-CA-U-PUR-18 | 0.000 | 0.000 | | | | | | | | | | |
| 4. | B-CA-U-PVAc-6 | 0.000 | 0.000 | 0.000 | | | | | | | | | |
| 5. | B-CA-U-PVAc-10 | 0.000 | 0.006 | 0.000 | 0.001 | | | | | | | | |
| 6. | B-CA-U-PVAc-18 | 0.000 | 0.000 | 0.062 | 0.000 | 0.000 | | | | | | | |
| 7. | B-LA-U-PUR-6 | 0.000 | 0.000 | 0.000 | 0.000 | 0.000 | 0.000 | | | | | | |
| 8. | B-LA-U-PUR-10 | 0.689 | 0.296 | 0.000 | 0.000 | 0.000 | 0.000 | 0.000 | | | | | |
| 9. | B-LA-U-PUR-18 | 0.000 | 0.000 | 0.060 | 0.000 | 0.000 | 0.000 | 0.000 | 0.000 | | | | |
| 10. | B-LA-U-PVAc-6 | 0.000 | 0.000 | 0.000 | 0.017 | 0.304 | 0.000 | 0.000 | 0.000 | 0.000 | | | |
| 11. | B-LA-U-PVAc-10 | 0.000 | 0.001 | 0.000 | 0.005 | 0.581 | 0.000 | 0.000 | 0.000 | 0.000 | 0.585 | | |
| 12. | B-LA-U-PVAc-18 | 0.000 | 0.000 | 0.313 | 0.000 | 0.000 | 0.333 | 0.000 | 0.000 | 0.006 | 0.000 | 0.000 | |

**Table 23.** Comparison of the effects of individual factors using Duncan's test on the coefficient of bendability ($K_{bendB}$) for aspen and NWC on the bottom.

| No. | Type of Sample | (1) 0.019 | (2) 0.014 | (3) 0.013 | (4) 0.026 | (5) 0.024 | (6) 0.014 | (7) 0.024 | (8) 0.021 | (9) 0.019 | (10) 0.027 | (11) 0.018 | (12) 0.014 |
|---|---|---|---|---|---|---|---|---|---|---|---|---|---|
| 1. | A-CA-D-PUR-6 | | | | | | | | | | | | |
| 2. | A-CA-D-PUR-10 | 0.004 | | | | | | | | | | | |
| 3. | A-CA-D-PUR-18 | 0.003 | 0.936 | | | | | | | | | | |
| 4. | A-CA-D-PVAc-6 | 0.000 | 0.000 | 0.000 | | | | | | | | | |
| 5. | A-CA-D-PVAc-10 | 0.004 | 0.000 | 0.000 | 0.124 | | | | | | | | |
| 6. | A-CA-D-PVAc-18 | 0.003 | 0.996 | 0.937 | 0.000 | 0.000 | | | | | | | |
| 7. | A-LA-D-PUR-6 | 0.005 | 0.000 | 0.000 | 0.105 | 0.859 | 0.000 | | | | | | |
| 8. | A-LA-D-PUR-10 | 0.112 | 0.000 | 0.000 | 0.005 | 0.153 | 0.000 | 0.181 | | | | | |
| 9. | A-LA-D-PUR-18 | 0.918 | 0.005 | 0.004 | 0.000 | 0.003 | 0.004 | 0.005 | 0.110 | | | | |
| 10. | A-LA-D-PVAc-6 | 0.000 | 0.000 | 0.000 | 0.808 | 0.093 | 0.000 | 0.074 | 0.003 | 0.000 | | | |
| 11. | A-LA-D-PVAc-10 | 0.503 | 0.021 | 0.020 | 0.000 | 0.001 | 0.018 | 0.001 | 0.034 | 0.540 | 0.000 | | |
| 12. | A-LA-D-PVAc-18 | 0.005 | 0.844 | 0.795 | 0.000 | 0.000 | 0.837 | 0.000 | 0.000 | 0.006 | 0.000 | 0.022 | |

**Table 24.** Comparison of the effects of individual factors using Duncan's test on the coefficient of bendability ($K_{bendB}$) for aspen and NWC on top.

| No. | Type of Sample | (1) 0.012 | (2) 0.015 | (3) 0.007 | (4) 0.026 | (5) 0.023 | (6) 0.008 | (7) 0.021 | (8) 0.015 | (9) 0.010 | (10) 0.028 | (11) 0.015 | (12) 0.011 |
|---|---|---|---|---|---|---|---|---|---|---|---|---|---|
| 1. | A-CA-U-PUR-6 | | | | | | | | | | | | |
| 2. | A-CA-U-PUR-10 | 0.002 | | | | | | | | | | | |
| 3. | A-CA-U-PUR-18 | 0.000 | 0.000 | | | | | | | | | | |
| 4. | A-CA-U-PVAc-6 | 0.000 | 0.000 | 0.000 | | | | | | | | | |
| 5. | A-CA-U-PVAc-10 | 0.000 | 0.000 | 0.000 | 0.005 | | | | | | | | |
| 6. | A-CA-U-PVAc-18 | 0.004 | 0.000 | 0.393 | 0.000 | 0.000 | | | | | | | |
| 7. | A-LA-U-PUR-6 | 0.000 | 0.000 | 0.000 | 0.000 | 0.153 | 0.000 | | | | | | |
| 8. | A-LA-U-PUR-10 | 0.003 | 0.820 | 0.000 | 0.000 | 0.000 | 0.000 | 0.000 | | | | | |
| 9. | A-LA-U-PUR-18 | 0.071 | 0.000 | 0.056 | 0.000 | 0.000 | 0.242 | 0.000 | 0.000 | | | | |
| 10. | A-LA-U-PVAc-6 | 0.000 | 0.000 | 0.000 | 0.024 | 0.000 | 0.000 | 0.000 | 0.000 | 0.000 | | | |
| 11. | A-LA-U-PVAc-10 | 0.002 | 0.997 | 0.000 | 0.000 | 0.000 | 0.000 | 0.000 | 0.810 | 0.000 | 0.000 | | |
| 12. | A-LA-U-PVAc-18 | 0.480 | 0.000 | 0.003 | 0.000 | 0.000 | 0.024 | 0.000 | 0.000 | 0.226 | 0.000 | 0.000 | |

**Table 25.** Comparison of the effects of individual factors using Duncan's test on the coefficient of bendability ($K_{bendB}$) for beech and NWC on the bottom.

| No. | Type of Sample | (1) 0.023 | (2) 0.015 | (3) 0.015 | (4) 0.028 | (5) 0.021 | (6) 0.014 | (7) 0.030 | (8) 0.015 | (9) 0.016 | (10) 0.023 | (11) 0.024 | (12) 0.022 |
|---|---|---|---|---|---|---|---|---|---|---|---|---|---|
| 1. | B-CA-D-PUR-6 | | | | | | | | | | | | |
| 2. | B-CA-D-PUR-10 | 0.000 | | | | | | | | | | | |
| 3. | B-CA-D-PUR-18 | 0.000 | 0.973 | | | | | | | | | | |
| 4. | B-CA-D-PVAc-6 | 0.000 | 0.000 | 0.000 | | | | | | | | | |
| 5. | B-CA-D-PVAc-10 | 0.455 | 0.000 | 0.000 | 0.000 | | | | | | | | |
| 6. | B-CA-D-PVAc-18 | 0.000 | 0.428 | 0.416 | 0.000 | 0.000 | | | | | | | |
| 7. | B-LA-D-PUR-6 | 0.000 | 0.000 | 0.000 | 0.179 | 0.000 | 0.000 | | | | | | |
| 8. | B-LA-D-PUR-10 | 0.000 | 0.977 | 0.954 | 0.000 | 0.000 | 0.431 | 0.000 | | | | | |
| 9. | B-LA-D-PUR-18 | 0.000 | 0.524 | 0.521 | 0.000 | 0.000 | 0.178 | 0.000 | 0.514 | | | | |
| 10. | B-LA-D-PVAc-6 | 0.863 | 0.000 | 0.000 | 0.001 | 0.382 | 0.000 | 0.000 | 0.000 | 0.000 | | | |
| 11. | B-LA-D-PVAc-10 | 0.291 | 0.000 | 0.000 | 0.007 | 0.087 | 0.000 | 0.000 | 0.000 | 0.000 | 0.340 | | |
| 12. | B-LA-D-PVAc-18 | 0.473 | 0.000 | 0.000 | 0.000 | 0.935 | 0.000 | 0.000 | 0.000 | 0.000 | 0.405 | 0.094 | |

**Table 26.** Comparison of the effects of individual factors using Duncan's test on the coefficient of bendability ($K_{bendB}$) for beech and NWC on top.

| No. | Type of Sample | (1)<br>0.016 | (2)<br>0.017 | (3)<br>0.006 | (4)<br>0.024 | (5)<br>0.021 | (6)<br>0.009 | (7)<br>0.030 | (8)<br>0.016 | (9)<br>0.005 | (10)<br>0.022 | (11)<br>0.022 | (12)<br>0.008 |
|---|---|---|---|---|---|---|---|---|---|---|---|---|---|
| 1. | B-CA-U-PUR-6 | | | | | | | | | | | | |
| 2. | B-CA-U-PUR-10 | 0.398 | | | | | | | | | | | |
| 3. | B-CA-U-PUR-18 | 0.000 | 0.000 | | | | | | | | | | |
| 4. | B-CA-U-PVAc-6 | 0.000 | 0.000 | 0.000 | | | | | | | | | |
| 5. | B-CA-U-PVAc-10 | 0.000 | 0.002 | 0.000 | 0.003 | | | | | | | | |
| 6. | B-CA-U-PVAc-18 | 0.000 | 0.000 | 0.014 | 0.000 | 0.000 | | | | | | | |
| 7. | B-LA-U-PUR-6 | 0.000 | 0.000 | 0.000 | 0.000 | 0.000 | 0.000 | | | | | | |
| 8. | B-LA-U-PUR-10 | 0.682 | 0.621 | 0.000 | 0.000 | 0.000 | 0.000 | 0.000 | | | | | |
| 9. | B-LA-U-PUR-18 | 0.000 | 0.000 | 0.690 | 0.000 | 0.000 | 0.006 | 0.000 | 0.000 | | | | |
| 10. | B-LA-U-PVAc-6 | 0.000 | 0.000 | 0.000 | 0.031 | 0.329 | 0.000 | 0.000 | 0.000 | 0.000 | | | |
| 11. | B-LA-U-PVAc-10 | 0.000 | 0.000 | 0.000 | 0.032 | 0.303 | 0.000 | 0.000 | 0.000 | 0.000 | 0.902 | | |
| 12. | B-LA-U-PVAc-18 | 0.000 | 0.000 | 0.038 | 0.000 | 0.000 | 0.615 | 0.000 | 0.000 | 0.018 | 0.000 | 0.000 | |

As shown in Figure 4, the greatest $K_{bendC}$ values in the aspen samples were found in the samples with thicknesses of 6 mm and 10 mm with a carbon fiber non-wood component placed on the bottom of the laminated material. These lamellas were bonded with PVAc adhesive. By contrast, the lowest $K_{bendC}$ values were found in 18 mm thick lamellas, where the non-wood component was placed on the top of the laminated material. In the samples with a carbon fiber non-wood component, the effect of the adhesive used was also significant. The 6 mm thick and 10 mm thick samples with carbon fibers on the top of the material glued with PUR adhesive had significantly lower values (by more than 50%) than all the other samples.

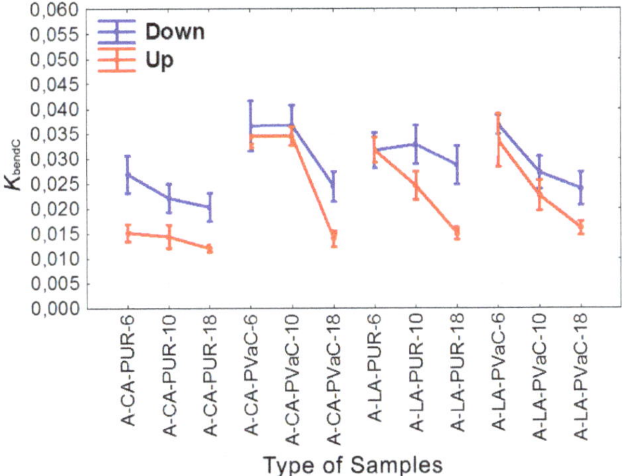


**Figure 4.** $K_{bendC}$ in case of aspen.

The situation was similar with beech lamellas (Figure 5). The greatest $K_{bendC}$ values were found in the beech samples with a 6 mm thickness, but unlike the aspen lamellas, no significant differences were found when different types of non-wood components were used. As with aspen lamellas, the lowest $K_{bendC}$ values were measured in lamellas with a thickness of 18 mm. The lamella thickness affected the bending coefficient. In beech samples, there was one extreme in the case of lamellas with glass fibers bonded with PVAc adhesive. The values measured on these 6 mm thick and 18 mm thick samples were no different from those measured on the 10 mm thick samples.

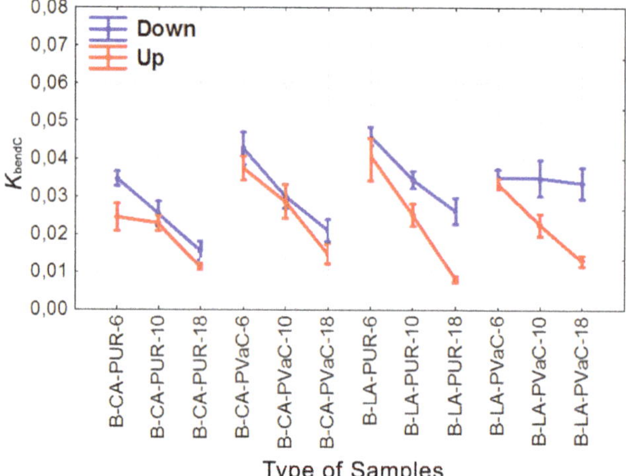

**Figure 5.** $K_{bendC}$ in case of beech.

Figure 6 shows that the $K_{bendB}$ values were affected most by the material thickness. Materials with lower thicknesses had greater $K_{bendB}$ values. The greatest $K_{bendB}$ values in the aspen lamellas were reached in samples with a non-wood component on the bottom of the laminated material. These lamellas were bonded with PVAc adhesive, and the non-wood component had no influence on these values. By contrast, the lowest values were measured in the aspen lamellas with carbon fibers on the top side of the material and bonded with PUR adhesive. The lowest $K_{bendB}$ values were also found in the samples with a carbon fiber non-wood component placed on the top of the material. In the 6 mm thick and 10 mm thick samples bonded with PUR adhesive with a non-wood carbon fiber component, the values were more than 50% lower than in the samples bonded with PVAc adhesive.

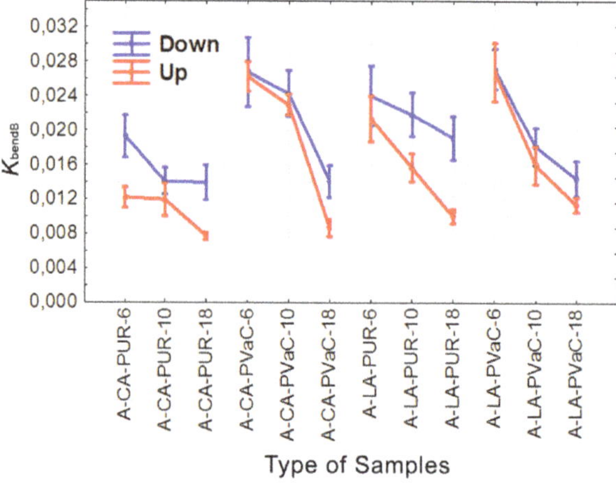

**Figure 6.** $K_{bendB}$ in case of aspen.

The lowest $K_{bendB}$ values in the beech samples were measured in the 18 mm thick samples bonded with PUR adhesive (Figure 7). In similar samples bonded with PVAc adhesive, the samples reached about 30% greater values. The highest $K_{bendB}$ values were measured in the 6 mm thick beech samples, but unlike aspen lamellas, no significant differences were found with the use of different types of non-wood components and adhesives. As in aspen lamellas, the lowest $K_{bendB}$ values were measured in the 18 mm thick lamellas. The lamella thickness affected the bending coefficient.

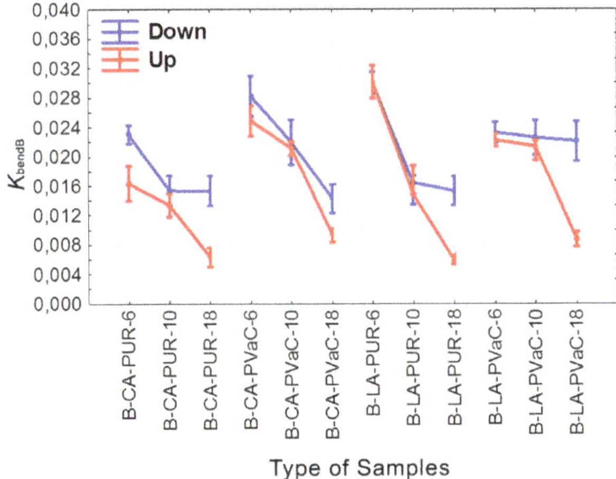


**Figure 7.** $K_{bendB}$ in case of beech.

## 4. Conclusions

1.  The type and position of the non-wood component used in the laminated materials had a significant effect on all the observed characteristics: $K_{bendC}$, $K_{bendB}$, $R_{minC}$, and $R_{minB}$.
2.  Bending coefficient values based on the simple bending equation ($K_{bendC}$) tended to be greater than bending coefficient values based on bending geometry ($K_{bendB}$).
3.  The greatest values of the bending coefficient based on the simple bending equation ($K_{bendC}$) and the bending coefficient based on bending geometry ($K_{bendB}$) were generally found in materials of lower thickness.
4.  No rule was observed for the high or low measured values of the observed characteristics ($K_{bendC}$, $K_{bendB}$, $R_{minC}$, and $R_{minB}$) in relation to the wood species used.

**Author Contributions:** Data curation, A.S., V.Z. and Z.G.; Formal analysis, T.S.; Resources, A.S.; Supervision, T.S.; Visualization, T.S.; Writing—original draft, T.S.; Writing—review & editing, A.S.

**Funding:** The authors are grateful for the support of the Advanced Research Supporting the Forestry and Wood-processing Sector's Adaptation to Global Change and the 4th Industrial Revolution (Project No. CZ.02.1.01/0.0/0.0/16_019/0000803), financed by OP RDE. The authors are also grateful for the support of the Internal Grant Agency (IGA) of the Faculty of Forestry and Wood Sciences (Project No. B 06/17).

**Conflicts of Interest:** The authors declare no conflict of interest.

## References


1.  Pokharel, R.; Grala, R.K.; Grebner, D.L. Woody residue utilization for bioenergy by primary forest products manufacturers: An exploratory analysis. *For. Policy Econ.* **2017**, *85*, 161–171. [CrossRef]
2.  Long, Z.; Wu, J.; Xu, W.; Lin, W. Study of the coordination mechanism of a wood processing residue-based reverse supply chain. *BioResources* **2018**, *13*, 2562–2577. [CrossRef]
3.  Bhoominathan, R.; Divyabarathi, P.; Manimegalai, R.; Nithya, T.; Shanmugapriya, S. Infra-red thermography based inspection of hybrid composite laminates under flexure loading. *IJVSS* **2018**, *10*, 6–9. [CrossRef]
4.  Silva, F.G.A.; de Moura, M.F.S.F.; Magalhães, A.G. Lowvelo city impact behaviour of a hybrid carbon-epoxy/corklaminate. *Strain* **2017**, *53*. [CrossRef]
5.  Abrate, S. Impact on laminated composite materials. *Appl. Mech. Rev.* **1991**, *44*, 155–190. [CrossRef]

6.  Bigg, D.M. The impact behavior of thermoplastic sheet composites. *J Reinf. Plast. Comp.* **1994**, *13*, 339–354. [CrossRef]

7.  Glos, P.; Denzler, J.K.; Linsenmann, P. Strength and stiffness behaviour of beech laminations for high strength glulam. In Proceedings of the Meeting 37 CIB Working Commission W18-Timber Structures, Edinburgh, Scotland, UK, August 2004.

8.  Frese, M.; Blaß, H.J. Characteristic bending strength of beech glulam. *Mater. Struct.* **2007**, *40*, 3–13. [CrossRef]

9.  Hill, C.A.S. *Wood Modification: Chemical, Thermal and Other Processes*; John Wiley & Sons: Chichester, UK, 2006. [CrossRef]

10. Kubovský, I.; Babiak, M. Color changes induced by $CO_2$ laser irradiation of wood surface. *Wood Res.* **2009**, *54*, 61–66.

11. Hrčka, R.; Babiak, M. Some non-traditional factors influencing thermal properties of wood. *Wood Res.* **2012**, *57*, 367–374.

12. Gašparík, M.; Barcík, Š. Impact of plasticization by microwave heating on the total deformation of beech wood. *BioResources* **2013**, *8*, 6297–6308. [CrossRef]

13. Gašparík, M.; Barcík, Š. Effect of plasticizing by microwave heating on bending characteristics of beech wood. *BioResources* **2014**, *9*, 4808–4820. [CrossRef]

14. Svoboda, T.; Ruman, D.; Gaff, M.; Gašparík, M.; Miftieva, E.; Dundek, L. Bending characteristics of multilayered soft and hardwood materials. *BioResources* **2015**, *10*, 8461–8473. [CrossRef]

15. Miftieva, E.; Gaff, M.; Svoboda, T.; Babiak, M.; Gašparík, M.; Ruman, D.; Suchopár, M. Effects of selected factors on bending characteristics of beech wood. *BioResources* **2016**, *11*, 599–611. [CrossRef]

16. Fang, C.-H.; Mariotti, N.; Cloutier, A.; Koubaa, A.; Blanchet, P. Densification of wood veneers by compression combined with heat and steam. *Eur. J. Wood Prod.* **2012**, *70*, 155–163. [CrossRef]

17. Gaff, M.; Vokatý, V.; Babiak, M.; Bal, B.C. Coefficient of wood bendability as a function of selected factors. *Constr. Build Mater.* **2016**, *126*, 632–640. [CrossRef]

18. Blomberg, J.; Persson, B. Swelling pressure of semi-isostaticallydensified wood under different mechanical resraints. *Wood Sci. Technol.* **2007**, *41*, 401–415. [CrossRef]

19. Plevris, N.; Triantafillou, T.C. CreepbehaviorofFRP-reinforcedwoodmembers. *J. Struct. Eng.* **1995**, *121*, 174–186. [CrossRef]

20. Redon, C.; Li, V.C.; Wu, C.; Hoshiro, H.; Saito, T.; Ogawa, A. Measuring and modifying interface propertiesof PVA fibers in ECC matrix. *J Mater. Civil Eng.* **2001**, *13*, 399–406. [CrossRef]

21. Sviták, M.; Ruman, D. Tensile-shear strength of layered wood reinforced by carbon materials. *Wood Res.* **2017**, *62*, 243–252.

22. Gaff, M.; Babiak, M.; Vokatý, V.; Gašparík, M.; Ruman, D. Bending characteristics of hardwood lamellae in the elastic region. *Compos. Part B-Eng.* **2017**, *116*, 61–75. [CrossRef]

23. Babiak, M.; Gaff, M.; Sikora, A.; Hysek, Š. Modulus of elasticity in three- and four-point bending of wood. *Compos. Struct.* **2018**, *204*, 454–465. [CrossRef]

24. Gaff, M.; Gašparík, M.; Babiak, M.; Vokatý, V. Bendability characteristics of wood lamellae in plastic region. *Compos. Struct.* **2017**, *163*, 410–422. [CrossRef]

25. Gaff, M.; Babiak, M. Methods for determining the plastic work in bending and impact of selected factors on its value. *Compos. Struct.* **2018**, *202*, 66–76. [CrossRef]

26. Sikora, A.; Gaff, M.; Hysek, Š.; Babiak, M. The plasticity of composite material based on winter rapeseed as a function of selected factors. *Compos. Struct.* **2018**, *202*, 783–792. [CrossRef]

27. Gaff, M.; Babiak, M. Tangent modulus as a function of selected factors. *Compos. Struct.* **2018**, *202*, 436–446. [CrossRef]

28. Saracoglu, E. Finite-element Simulations of the Influence of Cracks on the Strength of Glulam Beams. Master's Thesis, Blekinge Institute of Technology, Karlskrona, Sweden, 2011.

29. Khorasan, S.R. Finite-element Simulations of Glulam Beams with Natural Cracks. Master's Thesis, Blekinge Institute of Technology, Karlskrona, Sweden, 2012.

30. Hýsek, Š.; Gaff, M.; Sikora, A.; Babiak, M. New composite material based on winter rapeseed and his elasticity properties as a function of selected factors. *Compos. Part B-Eng.* **2018**, *153*, 108–116. [CrossRef]

31. Požgaj, A.; Chovanec, D.; Kurjatko, S.; Babiak, M. Štruktúra a Vlastnosti Dreva. In *Structure and Properties of Wood*; Príroda: Bratislava, Slovakia, 1997.

32. Gaff, M.; Gašparík, M.; Borůvka, V.; Haviarová, E. Stress simulation in layered wood-based materials under mechanical loading. *Mater. Des.* **2015**, *87*, 1065–1071. [CrossRef]
33. European Committee for Standardization, Wood-Based Panels. *EN 310: Determination of Modulus of Elasticity in Bending and of Bending Strength*; European Committee for Standardization: Brussels, Belgium, 1993.
34. International Organization for Standardization. *ISO 13061-1: Physical and Mechanical Properties of Wood—Test Methods for Small Clear Wood Specimens—Part 1: Determination of Moisture Content for Physical and Mechanical Tests*; International Organization for Standardization: Geneva, Switzerland, 2014.
35. International Organization for Standardization. *ISO 13061-2: Physical and Mechanical Properties of Wood—Test Methods for Small Clear Wood Specimens—Part 2: Determination of Density for Physical and Mechanical Tests*; International Organization for Standardization: Geneva, Switzerland, 2014.

 *forests*

*Article*

# Influence of Natural and Artificial Weathering on the Colour Change of Different Wood and Wood-Based Materials

Davor Kržišnik, Boštjan Lesar, Nejc Thaler and Miha Humar *

Department of Wood Science and Technology, Biotechnical Faculty, University of Ljubljana,
Jamnikarjeva 101, SI-1000 Ljubljana, Slovenia; davor.krzisnik@bf.uni-lj.si (D.K.); bostjan.lesar@bf.uni-lj.si (B.L.);
nejc.thaler@bf.uni-lj.si (N.T.)
* Correspondence: miha.humar@bf.uni-lj.si; Tel.: +386-1320-3638

Received: 1 July 2018; Accepted: 8 August 2018; Published: 10 August 2018

**Abstract:** The importance of the aesthetic performance of wood is increasing and the colour is one of the most important parameters of aesthetics, hence the colour stability of twelve different wood-based materials was evaluated by several in-service and laboratory tests. The wood used for wooden façades and decking belongs to a group of severely exposed surfaces. Discolouration of wood in such applications is a long-known phenomenon, which is a result of different biotic and abiotic causes. The ongoing in-service trial started in October 2013, whilst a laboratory test mimicking seasonal exposure was performed in parallel. Samples were exposed to blue stain fungi (*Aureobasidium pullulans* and *Dothichiza pithyophila*) in a laboratory test according to the EN 152 procedure. Afterwards, the same samples were artificially weathered and re-exposed to the same blue stain fungi for the second time. The purpose of this experiment was to investigate the synergistic effect of weathering and staining. The broader aim of the study was to determine the correlation factors between artificial and natural weathering and to compare laboratory and field test data of fungal disfigurement of various bio-based materials. During the four years of exposure, the most prominent colour changes were determined on decking. Respective changes on the façade elements were significantly less prominent, being the lest evident on the south and east façade. The results showed that there are positive correlations between natural weathering and the combination of artificial weathering and blue staining. Hence, the artificial weathering of wood-based materials in the laboratory should consist of two steps, blue staining and artificial weathering, in order to simulate colour changes.

**Keywords:** artificial weathering; blue staining fungi; colour change; natural weathering; wood

## 1. Introduction

During their service life, buildings and building components are exposed to a wide variety of environmental conditions. For wood-based materials, moisture stress and biological factors like mould, blue-stain, and decay fungi are often critical, especially for cladding and decking applications in exterior use conditions, representing two commodities where wood is frequently used [1,2]. Service life prediction, service life cost analysis, and the aesthetic performance of newly available bio-based building materials are essential for their promotion and increased use in the construction sector. The appearance of bio-based building materials usually changes during their service life. Therefore, the aesthetic service life is often a decisive criterion for these applications [3,4].

The service life of different building products and commodities is determined by very different criteria, e.g., colour stability of coated or uncoated surfaces; cracking and checking; the occurrence of moulds, stain, or fungal decay; damage by insects or marine borers; resistance to abrasion and wear, etc. In addition, the effect of other factors, like solar radiation, surface erosion, and mechanical impact, has a role in the service life of wood, which makes it a huge challenge to take into account the many different factors having a potential impact on the service life of wooden components [1,5,6]. In general, the service life of wood can be categorised into a group of functional, technical, or aesthetic service lives. While in the past, the majority of research efforts were focused on the functional service life, nowadays, aesthetic service life is gaining importance.

Wooden façades are a group of severely exposed wooden surfaces and as such are very susceptible to discolouration by different biotic and abiotic causes. The most important biotic factors for discolouration are fungi and bacteria due to the production of pigments, e.g., melanin by blue stain fungi [7,8]. The wood-discolouring moulds and staining fungi live on nutrients in the parenchyma cells of sapwood [9,10]. Conifers and hardwoods, round wood, lumber, finished wood, and wood products can all be colonized if the moisture content is high enough. Discolouring fungi do not cause any or only very little cell wall attack [11,12]. Tertiary blue stain fungi, which develop on the surface of the wood in use, are frequently *Aureobasidium pullulans* and *Dothichiza pithyophila* [13]. These two fungal species develop on timber that has been converted into products and was painted, and re-imbibe moisture while in service, like wooden façades, window frames, garage doors, and garden furniture [14]. Frequently, moulds are recognizable by their fast growth on the surface of substrates, on which conidia can develop rapidly. Due to the species-specific colour of the conidia, wood colonized by several mould species can result in a multi-coloured impression, or be dominated by a single colour, e.g., black due to *Aspergillus niger*, or green due to *Penicillium* spp. or *Trichoderma* spp. In parallel to biotic factors, wood is exposed to solar radiation, which initially leads to rapid colour change due to the absorption of sunlight, and in the further stages, to large chemical modifications and breakdown of the wood surface layer [15,16]. This complex degradation process in combination with exposure to precipitation is often described as weathering [17]. The UV spectrum of solar radiation is one of the most effective parameters amongst all environmental factors that contribute to the weathering process of wood [15,18,19]. Although the UV spectrum only represents 5% of energy in sunlight, its strong effect on wood degradation is well documented [19,20]. The photo-degraded wood surface is a good substrate for bacteria and fungi, including staining fungi; the light colour of the photo-degraded wood surface and the black colour of fungal hyphae make up the majority of the grey surface of outdoor weathered wood [17,19,21]. Besides UV and fungi, there are some additional agents contributing to weathering, namely; water (leaching of extractives, stress in the material leading to fractures), the atmosphere (oxygen, free radicals, pollutants), and wind (wind-driven rain, hail, particles like sand) [22,23]. Leaching of extractives is predominantly important for wood species with a high extractive content, like oak [24]. However, in the first year of exposure in a temperate zone, oak wood usually exhibits rather low colour changes, while spruce wood is rather prone to discolouration [25–27]. Even higher colour changes are usually reported for dark thermally-modified wood. One of the wood species that is frequently used for cladding applications is larch. Larch can perform very well in applications with a lot of UV radiation and a low relative humidity. In these climates, a grey colour develops. On the other hand, the aesthetic performance of larch in a moderate climate with a higher relative humidity and less sunny days is not meeting users' expectations due to the development of mould fungi on the surface, which leads to several users' complains [28].

As already stated, aesthetic service life is becoming more and more important. End users want to see the final visual appearance of their wooden facades in the planning phase. Therefore, ageing and weathering models have to be included in BIM (Building Information Modelling) software [29]. In order to model this parameter, we have to prove the factors contributing to this phenomena and provide a methodology for the characterization of new materials The aim of this research was to study the correlations between natural and artificial weathering and their effect on the colour change of twelve different bio-based wooden materials. Although there have been several reports on the colour and structural changes after natural and artificial weathering [17,30–35], these correlations were usually less significant, as the artificial weathering did not include the step which would enable mould growth. Hence, there was no fungal staining considered. To our knowledge, this study is the first report where significant correlation factors between laboratory tests (combination of blue staining and artificial weathering) and natural weathering have been determined for various wood-based materials commercially used in outdoor applications in a temperate climate. In the respective study, the aesthetic performance of twelve different wood-based materials used for cladding applications was investigated.

## 2. Material and Methods

The group of selected materials consists of four untreated wood species (Norway spruce, European larch, European beech, and English oak) and eight materials that are treated or modified in different ways (Table 1).

**Table 1.** Twelve different investigated wood species and wood-based products.

| | Wood Species | | | | Treatment | | | |
|---|---|---|---|---|---|---|---|---|
| | Norway Spruce (*Picea abies*) | European Larch (*Larix decidua*) | European Beech (*Fagus sylvatica*) | English Oak (*Quercus robur*) | Thermal Modification | Impregnation with Suspension of NATURAL Wax | Copper-Ethanolamine Impregnation | Water Borne Acrylic Surface Coating |
| Abbreviation | PA | LD | FS | Q | TM | NW | CE | AC |
| PA | × | | | | | | | |
| PA-NW | × | | | | | × | | |
| PA-AC | × | | | | | | | × |
| PA-CE | × | | | | | | × | |
| PA-CE-NW | × | | | | | × | × | |
| PA-TM | × | | | | × | | | |
| PA-TM-NW | × | | | | × | × | | |
| PA-TM-CE | × | | | | × | | × | |
| LD | | × | | | | | | |
| LD-TM | | × | | | × | | | |
| FS | | | × | | | | | |
| Q | | | | × | | | | |

Thermal modification (TM) was performed according to the commercial process Silvapro® (Silvaprodukt, Ljubljana, Slovenia) with an initial vacuum in the first step of the treatment [36,37]. The modification was performed for three hours at the target temperature (ranging between 210 °C and 230 °C, depending on the wood species). Impregnation was performed with a commercial copper-ethanolamine solution Silvanolin® (Silvaprodukt, Ljubljana, Slovenia), which consists of copper, ethanolamine, boric acid, and quaternary ammonium compounds [38]. The concentration of active ingredients and consequent retention meet the use of class 3 requirements [39]. Impregnation was performed according to the full cell process in a laboratory impregnation setup. It consisted of 30 min vacuum (80 kPa), 180 min pressure (1 MPa), and 20 min vacuum (80 kPa), respectively. The same procedure was used for the impregnation of wood with 5% commercially available natural wax dispersion with a solid content of up to 50% by weight (Montax 50, Romonta, Seegebiet Mansfelder Land, Germany) [40]. The acrylic surface coating Silvanol® Lazura B (Silvaprodukt, Ljubljana, Slovenia) was manually applied to wood by brushing in two layers, with a 24-h drying time between each layer.

### 2.1. In-Service Testing

Figure 1 shows the wooden model house unit at the Department of Wood Science and Technology in Ljubljana, Slovenia (46°02′55.7″ N 14°28′47.3″ E, elevation above sea level 293 m), where the in-service performance of the façade and decking cladding elements was tested. The test specimens with a cross-section of 2.5 × 5.0 cm were exposed horizontally on the walls of the model house facing north, south, east, and west. At least seven specimens of the same material were exposed on each wall and deck. The in-service testing started in October 2013 and the prime objective was to monitor aesthetical properties (aesthetic service life), the presence of decay (functional service life), and moisture performance.

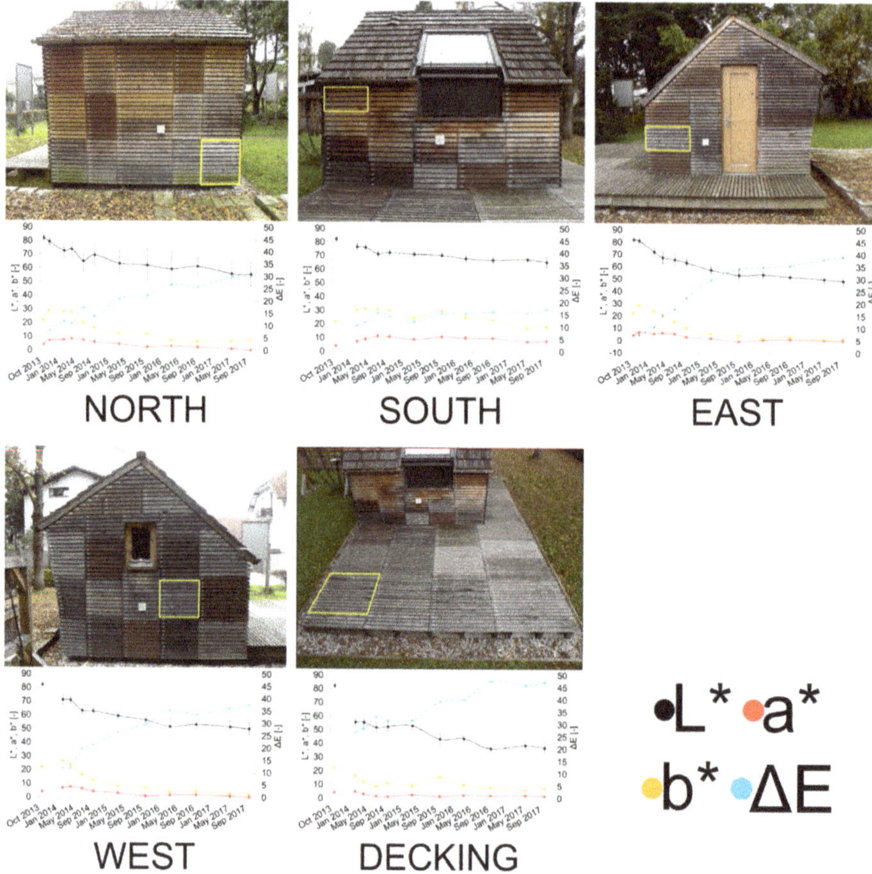



**Figure 1.** Exposure sites on the wooden model house unit in October 2017, after four years of weathering. In yellow frames, the samples of Norway spruce (PA) are shown and under the pictures, graphs showing the measured values of L*, a*, and b* plotted against exposure time and calculated ΔE are presented.

To determine the colour development due to weathering of the outdoor exposed samples, the colour of the test specimens was recorded in the CIE Lab system during the year with a portable colourimeter (EasyCo 566, Erichsen, Hemer, Germany). Colour was determined on the exposed surfaces only. To limit the influence of moisture content, measurements were performed at least two days after the last rain event. There were five to seven measurements performed on every respective element. We have tried to avoid fasteners, knots, and resin pockets, if present. Also, according to modified Johansson et al. [41], the evaluation of surface blue staining was performed using a rating scale from 0 to 4 as follows:

0 = not blue stained;

1 = weakly blue stained: few spots of blue stain on the surface;

2 = slightly blue stained: up to 1.5 mm in width and 4 mm in length;

3 = moderately blue stained: up to one-third of the surface;

4 = severely blue stained.

The evaluation scale from Johansson [41] was adopted, in order to meet our needs and to enable an easier comparison between laboratory and field trials.

## 2.2. Artificial Weathering Test

In parallel to the in-service test, we prepared a set of five samples (110 mm × 40 mm × 10 mm) made of the same materials as those listed in Table 1. All obtained specimens were exposed to blue stain fungi in a laboratory test according to the EN 152 [42] procedure. Afterwards, samples were exposed to artificial weathering and then re-exposed to blue stain fungi for the second time. With this setup, we somehow mimicked the natural processes. We have decided on this sequence, as we have noticed that the occurrence of the blue staining on wood is a rather fast process and that the first discolourations on wood are the consequence of blue staining and not weathering. The purpose of this experiment was to determine the semi-synergistic effect of weathering and staining.

The laboratory test was conducted with the blue-stain fungi *Aureobasidium pullulans* (de Bary) G. Arnaud strain ZIM L060 and *Dothichiza pithyophila* (Corda) Petr. strain ZIM L070. Both strains were obtained from the Collection of wood decay fungi from the Department of Wood Science and Technology (Biotechnical Faculty, Ljubljana, Slovenia) [43]. Prior to the inoculation, the wood specimens were sterilized in an autoclave with hot steam at 121 °C and 150 kPa for 15 min. Later on, the sterilized specimens were dipped into a spore suspension and placed horizontally in a Kolle flask, which was also inoculated with 15 mL of spore suspension. Afterwards, the flasks were placed in an incubation chamber at 25 °C and 80% RH for six weeks. After that time, the samples were visually evaluated and ranked according to the ranking system prescribed in the EN 152 [31] from 0 to 3 (rank 0 = not blue stained; 1 = small spots less than 2 mm; 2 = blue stained up to one third of surface; 3 = strongly blue stained). Only the uppermost side was evaluated and scanned for colour measurements.

An artificial weathering test (AW) between two EN 152 [42] tests was performed in a solar simulation chamber (Suntest XXL+, Atlas Material Testing Technology, Linsengericht, Germany). The climatic test chamber for solar simulation is equipped with three 1700 W xenon lamps emitting light with wavelengths between 300 and 420 nm. The standard EN 16474-1test [44] was followed to reproduce the artificial radiation in outdoor conditions. The exposure experiment was performed at a black standard temperature (BST) of 63 °C and an irradiance of approximately 0.35 W m$^{-2}$ (340 nm) with daylight filters, and with a total cycle time of 2 h, composed of a 102 min dry cycle at 40%–60% relative humidity and an 18 min wet cycle with water spray. The colour of the specimens was measured every 100 h for the total duration of 500 h of artificial weathering. Every period, the uppermost side was visually evaluated and scanned.

After the artificial weathering, the specimens were re-exposed to blue-stain fungi for the second time, according to the standard EN 152 procedure [42] described above.

## 2.3. Colour Analysis

Colour was determined on semi-radial surfaces of the samples. There were no measurements performed on axial surfaces. Prior measurement samples were conditioned under laboratory conditions (23 °C; 65%) to limit the influence of wood moisture content on colour. The colour measurements were performed on samples from the artificial weathering test according to the CIE Lab system, a method created by the Commission International de l'Eclairage. The CIE Lab system is characterized by three parameters: L*, a*, and b*. The L* axis represents the lightness, which varies from a hundred (white) to zero (black), representing the achromatic axis of greys, whereas a* and b* are the chromaticity coordinates. A positive value of a* denotes a redder colour on a green-red scale, whereas a positive value of b* denotes a more yellow colour on a blue-yellow scale. Together, those three components form a three-dimensional colour space [45].

Colour measurements of in-service testing were performed several times a year with a portable Colour Measuring Device (EasyCo 566, Erichsen, Hemer, Germany) and expressed in the CIE L*a*b* system. This device enables contact-free precise colour measurement. The diameter of the measurement spot is 20 mm. However, laboratory test specimens were scanned and processed with Corel Photo-Paint 8 software. Corel Photo paint was used as colour analysis as this technique provides the colour of the whole surface and not of individual spots. This technique provides reliable measurements, as indicated by the comparison of both techniques in our laboratory.

Total colour difference $\Delta E$ (Equation (1)) from a reference colour (L*0, a*0, b*0) to a target colour (L*1, a*1, b*1) in the CIE Lab space is calculated by determining the Euclidean distance between two colours given by:

$$\Delta E = \sqrt{(\Delta L^*)^2 + (\Delta a^*)^2 + (\Delta b^*)^2} \tag{1}$$

Summing up $\Delta E$ ($\Sigma \Delta E$) for all analysed points in time, the course of the colour changes can be taken into account:

$$\sum_{i-1,1}^{n} \Delta E_{i,n} = \Delta E_{0,1} + \Delta E_{1,2} + \cdots + \Delta E_{n-1,n} \tag{2}$$

*2.4. Data Processing*

The total colour difference ($\Delta E$ values) for different directions from the in-service and laboratory trial was used to calculate the Pearson product moment correlation coefficient ($r$) using Statgraphics Centurion XVII software, version 17.2.05 (Statpoint Technologies Inc., The Plains Virginia, USA). The Multiple-Variable Analysis (Correlations) procedure was used to calculate the correlation coefficients to measure the strength of the linear relationship between all the variables. $p$-values (P) below 0.05 were considered to be significant.

## 3. Results and Discussion

*3.1. Colour Changes of In-Service Testing*

As already stated in the introduction, in-service testing provided the most accurate information regarding the colour changes during weathering. Figure 1 shows the visual appearance of the model house and plotted colour values (L*, a*, and b*) against exposure time for Norway spruce (PA) only. Norway spruce was studied in more detail, as it is one of the most frequently used wood species in central Europe. Lightness values (L*) for south exposure exhibited the lowest change, being the most light-coloured after four years of exposure, and decking had the highest rate of change; moreover, the latter samples were the darkest of the group at the end of the exposure. One of the reasons for this observed difference originated in the construction details. Samples exposed on the south façade were the most protected against rainfall events, while decking was the most exposed. In addition, the samples on the north façade and decking showed seasonal fluctuations in lightness, e.g., after the summer months, the L* values were higher, and after the winter, the L* values were lower.

The most important reason for seasonal fluctuations was related to fungal melanin; with its formation on one hand being countered by bleaching on the other. During the wet autumn and winter months, blue stain fungi grew on the surface (Table 2). In the summer months, melanin was at least partially bleached [31], which resulted in fluctuations of lightness (in the autumn and winter the wood gets darker and in summer it turns back to lighter colours) (Figure 1). The extent of discolouration depended on microclimate conditions, the susceptibility of the material, UV radiation, etc. The rate of change of the two chromatic components, a* and b* values, decreased over time for north, east, and west façade exposure and was close to zero, meaning that samples were close to being achromatic, i.e., grey, which was also clearly evident from their visual appearances. Values of parameter a* on Norway spruce samples (PA) exposed on the east façade dropped beneath 0 in October 2015, which meant that the chromatic value changed from "red" to "green". However, after October 2015, values were positive again.

**Table 2.** Visual evaluation of surface mould growth on the façade and decking of the model house in Ljubljana with the standard deviations.

| Material | Exposure Direction | 1st Evaluation | 2nd Evaluation | 3rd Evaluation | 4th Evaluation | 5th Evaluation | 6th Evaluation |
|---|---|---|---|---|---|---|---|
| | | 29 November 2013 | 7 January 2014 | 18 March 2014 | 5 June 2014 | 7 October 2014 | 3 July 2015 |
| PA | north | 0.18 ± 0.60 | 1.09 ± 1.22 | 2.73 ± 1.10 | 3.18 ± 0.98 | 3.14 ± 1.46 | 3.00 ± 0.89 |
| | south | 0.00 ± 0.00 | 0.00 ± 0.00 | 0.00 ± 0.00 | 0.20 ± 0.45 | 0.20 ± 0.45 | 0.20 ± 0.45 |
| | east | 0.00 ± 0.00 | 1.57 ± 0.79 | 1.86 ± 1.07 | 2.14 ± 1.07 | 2.29 ± 0.95 | - |
| | west | 0.00 ± 0.00 | 1.67 ± 0.71 | 2.44 ± 0.73 | 2.67 ± 0.71 | 3.50 ± 0.58 | 2.25 ± 0.46 |
| | decking | 1.79 ± 1.31 | 2.50 ± 1.16 | - | - | - | - |
| PA-NW | north | 0.36 ± 0.81 | 0.73 ± 0.79 | 1.00 ± 1.10 | 1.36 ± 1.43 | 1.73 ± 1.35 | 2.73 ± 1.01 |
| | south | 0.80 ± 1.10 | 2.00 ± 0.71 | 3.20 ± 0.45 | 3.60 ± 0.55 | 3.40 ± 0.55 | 3.40 ± 0.55 |
| | east | 0.00 ± 0.00 | 1.75 ± 0.46 | 2.00 ± 0.00 | 2.00 ± 0.00 | 2.00 ± 0.00 | 1.75 ± 0.46 |
| | west | 0.73 ± 0.79 | 1.91 ± 0.83 | 3.36 ± 0.50 | 3.64 ± 0.50 | 2.33 ± 2.08 | - |
| | decking | 1.36 ± 1.60 | 2.14 ± 1.03 | - | - | - | - |
| PA-AC | north | 0.00 ± 0.00 | 0.00 ± 0.00 | 0.00 ± 0.00 | 0.00 ± 0.00 | 0.00 ± 0.00 | - |
| | south | 0.00 ± 0.00 | 0.40 ± 0.89 | 0.40 ± 0.89 | 0.40 ± 0.89 | 0.00 ± 0.00 | - |
| | east | 0.00 ± 0.00 | 0.71 ± 0.49 | 0.71 ± 0.49 | 0.29 ± 0.49 | 0.14 ± 0.38 | - |
| | west | 0.00 ± 0.00 | 0.73 ± 1.10 | 1.27 ± 1.62 | 1.27 ± 1.62 | 1.27 ± 1.62 | 1.00 ± 1.18 |
| | decking | 0.00 ± 0.00 | 0.43 ± 0.65 | 0.36 ± 0.50 | 0.50 ± 0.65 | 0.00 ± 0.00 | 0.00 ± 0.00 |
| PA-CE | north | 0.00 ± 0.00 | 0.00 ± 0.00 | 0.00 ± 0.00 | 0.00 ± 0.00 | 0.00 ± 0.00 | 0.83 ± 0.98 |
| | south | 0.00 ± 0.00 | 0.00 ± 0.00 | 0.00 ± 0.00 | 0.00 ± 0.00 | 0.00 ± 0.00 | 0.00 ± 0.00 |
| | east | 0.00 ± 0.00 | 0.00 ± 0.00 | 0.00 ± 0.00 | 0.00 ± 0.00 | 0.00 ± 0.00 | 0.00 ± 0.00 |
| | west | 0.00 ± 0.00 | 0.00 ± 0.00 | 0.00 ± 0.00 | 0.00 ± 0.00 | 0.00 ± 0.00 | 0.78 ± 0.83 |
| | decking | 0.00 ± 0.00 | 0.00 ± 0.00 | - | - | - | - |
| PA-CE-NW | north | 0.00 ± 0.00 | 0.00 ± 0.00 | 0.00 ± 0.00 | 0.00 ± 0.00 | 0.00 ± 0.00 | 0.00 ± 0.00 |
| | south | 0.00 ± 0.00 | 0.00 ± 0.00 | 0.20 ± 0.45 | 0.20 ± 0.45 | 1.00 ± 1.00 | 3.00 ± 0.00 |
| | east | 0.00 ± 0.00 | 0.00 ± 0.00 | 0.56 ± 0.88 | 0.56 ± 0.88 | 0.00 ± 0.00 | 0.00 ± 0.00 |
| | west | 0.00 ± 0.00 | 0.08 ± 0.28 | 0.23 ± 0.60 | 0.00 ± 0.00 | 0.00 ± 0.00 | 2.31 ± 0.48 |
| | decking | 0.00 ± 0.00 | 0.00 ± 0.00 | - | - | - | - |
| PA-TM | north | 0.00 ± 0.00 | 0.00 ± 0.00 | 0.00 ± 0.00 | 1.67 ± 0.82 | 2.80 ± 0.45 | 3.00 ± 0.00 |
| | south | 0.00 ± 0.00 | 0.40 ± 0.55 | 1.60 ± 1.14 | 1.80 ± 0.84 | 2.00 ± 0.71 | 2.00 ± 0.71 |
| | east | 0.00 ± 0.00 | 0.00 ± 0.00 | 0.00 ± 0.00 | 0.83 ± 0.39 | 1.50 ± 0.52 | 2.18 ± 0.40 |
| | west | 0.00 ± 0.00 | 0.36 ± 0.67 | 0.64 ± 0.81 | 0.91 ± 0.94 | 1.40 ± 0.84 | 2.67 ± 0.50 |
| | decking | 0.00 ± 0.00 | 0.00 ± 0.00 | 0.00 ± 0.00 | 0.00 ± 0.00 | - | - |
| PA-TM-NW | north | 0.00 ± 0.00 | 0.83 ± 0.75 | 1.50 ± 0.84 | 1.50 ± 0.84 | 1.50 ± 0.84 | 0.50 ± 0.84 |
| | south | 0.00 ± 0.00 | 0.40 ± 0.55 | 0.80 ± 0.45 | 1.00 ± 0.00 | - | - |
| | east | 0.00 ± 0.00 | 0.00 ± 0.00 | 0.83 ± 0.41 | 1.00 ± 0.00 | 1.17 ± 0.41 | 0.33 ± 0.52 |
| | west | 0.00 ± 0.00 | 1.00 ± 0.00 | 1.00 ± 0.00 | 1.30 ± 0.48 | - | - |
| | decking | 0.14 ± 0.53 | 0.14 ± 0.53 | - | - | - | - |
| PA-TM-CE | north | 0.00 ± 0.00 | 0.00 ± 0.00 | 0.00 ± 0.00 | 0.00 ± 0.00 | 0.00 ± 0.00 | 0.00 ± 0.00 |
| | south | 0.00 ± 0.00 | 0.00 ± 0.00 | 0.00 ± 0.00 | 0.00 ± 0.00 | 0.00 ± 0.00 | 0.00 ± 0.00 |
| | east | 0.00 ± 0.00 | 0.00 ± 0.00 | 0.00 ± 0.00 | 0.00 ± 0.00 | 0.00 ± 0.00 | 0.00 ± 0.00 |
| | west | 0.00 ± 0.00 | 0.00 ± 0.00 | 0.00 ± 0.00 | 0.00 ± 0.00 | 0.00 ± 0.00 | 0.40 ± 0.84 |
| | decking | 0.00 ± 0.00 | 0.00 ± 0.00 | 0.00 ± 0.00 | 0.00 ± 0.00 | 0.00 ± 0.00 | - |
| LD | north | 0.00 ± 0.00 | 0.36 ± 0.92 | 1.00 ± 1.10 | 2.82 ± 0.40 | 3.27 ± 0.65 | 3.67 ± 0.82 |
| | south | 0.00 ± 0.00 | 1.00 ± 0.71 | 1.00 ± 0.71 | 0.80 ± 0.84 | - | - |
| | east | 0.00 ± 0.00 | 1.29 ± 0.49 | 2.00 ± 0.00 | 2.00 ± 0.00 | 2.86 ± 0.38 | 3.00 ± 0.58 |
| | west | 0.00 ± 0.00 | 0.33 ± 0.59 | 2.00 ± 0.69 | 2.56 ± 0.70 | 3.22 ± 0.81 | 4.00 ± 0.00 |
| | decking | 0.70 ± 0.82 | 1.00 ± 0.00 | - | - | - | - |
| LD-TM | north | 0.00 ± 0.00 | 0.00 ± 0.00 | 0.00 ± 0.00 | 0.00 ± 0.00 | 0.00 ± 0.00 | 1.27 ± 0.79 |
| | south | 0.00 ± 0.00 | 0.00 ± 0.00 | 0.00 ± 0.00 | 0.00 ± 0.00 | 0.00 ± 0.00 | 0.00 ± 0.00 |
| | east | 0.00 ± 0.00 | 0.14 ± 0.38 | 0.57 ± 0.53 | 0.71 ± 0.76 | 0.86 ± 0.69 | 2.00 ± 1.00 |
| | west | 0.00 ± 0.00 | 0.00 ± 0.00 | 0.45 ± 0.52 | 0.90 ± 0.32 | 0.80 ± 0.45 | 2.17 ± 0.41 |
| | decking | 0.00 ± 0.00 | 0.00 ± 0.00 | - | - | - | - |
| FS | north | 0.00 ± 0.00 | 0.00 ± 0.00 | 0.00 ± 0.00 | 0.09 ± 0.30 | 0.45 ± 0.52 | 1.80 ± 0.92 |
| | south | 0.80 ± 1.10 | 2.20 ± 0.45 | 3.20 ± 0.45 | 3.20 ± 0.45 | - | - |
| | east | 0.00 ± 0.00 | 2.57 ± 0.79 | 2.86 ± 0.38 | 3.71 ± 0.49 | 3.71 ± 0.49 | 3.00 ± 0.00 |
| | west | 0.00 ± 0.00 | 2.64 ± 0.50 | 3.64 ± 0.50 | 4.00 ± 0.00 | 4.00 ± 0.00 | 4.00 ± 0.00 |
| | decking | 1.00 ± 0.00 | 1.29 ± 0.47 | - | - | - | - |
| Q | north | 0.00 ± 0.00 | 0.00 ± 0.00 | 0.20 ± 0.45 | 0.80 ± 0.45 | 1.00 ± 0.71 | 3.40 ± 0.55 |
| | south | 0.00 ± 0.00 | 1.00 ± 0.00 | 2.00 ± 0.00 | 2.40 ± 0.55 | 2.33 ± 0.58 | - |
| | east | 0.00 ± 0.00 | 0.00 ± 0.00 | 3.00 ± 0.82 | 3.14 ± 0.90 | 0.00 ± 0.00 | - |
| | decking | 0.00 ± 0.00 | 3.38 ± 0.81 | - | - | - | - |

The total colour difference ($\Delta E$ from Equation (1)) after four years of weathering was the highest on decking and lowest on the south facing façade. Colour changes on the other three exposures (north, east, and west) were comparable. Furthermore, seasonal fluctuation in $\Delta E$ could be observed on all façades and decking, but mostly on the north facing façade and on the deck. The horizontal position is the most exposed to sun radiation, as well as to precipitation, which assured conditions for colour changes in abiotic and biotic ways, in particular the formation of a grey colour, as well as the sequential formation of UV-induced free radicals, the degradation of lignin into quinones, the leaching of quinones [20], and the development of staining fungi [38,46].

Figures 2–6 show the averaged colour for each material and the calculated $\Delta E$ value (see Equation (1)), defined as the colour difference between the initial and the final stage. In general, the biggest colour difference was observed on the decking, with one exception, the thermally modified European larch (LD-TM), for which the $\Delta E$ was lowest on decking compared to other exposures. In addition, colour changes of thermally modified Norway spruce (PA-TM) on decking were not the most prominent; on the east and west façade, the colour difference was more notable. Generally, there was no correlation between which orientation of façade resulted in lower or higher colour changes. The prime reason for this was that the micro-positions of each material on respective façades differed, hence introducing an additional level of uncertainty. If individual materials were compared, it can be concluded that the lowest $\Delta E$ was observed on PA-TM-CE (copper-treated thermally-modified Norway spruce), followed by LD-TM (thermally-modified European larch) and PA-CE (copper-treated Norway spruce), and on the other side, the highest $\Delta E$ was measured on Norway spruce (PA), followed by wax treated Norway spruce (PA-NW) and European larch (LD). The modifications in wood structure occurring at high temperature were accompanied by several favourable changes in physical structure, i.e., enhanced weather resistance and a decorative, dark colour [47,48]. Values for $\Sigma\Delta E$ (see Equation (2)) indicated that the course of the colour changes was fairly similar for all the exposure directions. One of the reasons for these observed differences originated in the initial colour. Norway spruce (PA), wax-treated Norway spruce (PA-NW), and European larch (LD) were lightly-coloured materials, so the development of the dark pigmented blue stain fungi was the most notable. On the other hand, the presence of fungicides (copper-based wood preservatives) prevented the development of the blue stain fungi (Table 2), whilst the presence of copper slowed down UV-induced degradation to a certain extent [49], which resulted in less noticeable colour differences (PA-CE and PA-TM-CE in Figures 2–6).

| | Oct 2013 | Dec 2013 | Mar 2014 | May 2014 | Aug 2014 | Oct 2014 | Apr 2015 | Oct 2015 | Apr 2016 | Oct 2016 | May 2017 | Oct 2017 |
|---|---|---|---|---|---|---|---|---|---|---|---|---|
| PA | | 8.05 | 11.82 | 11.05 | 17.50 | 14.03 | 21.15 | 22.32 | 26.96 | 25.94 | 30.06 | 34.30 |
| PA-NW | | 5.04 | 6.27 | 7.48 | 11.21 | 7.78 | 10.28 | 9.46 | 14.40 | 14.59 | 17.43 | 25.18 |
| PA-AC | | 3.96 | 1.99 | 2.76 | 2.85 | 2.99 | 6.40 | 2.39 | 5.34 | 5.38 | 6.94 | 8.04 |
| PA-CE | | 4.03 | 3.26 | 5.65 | 7.60 | 9.06 | 5.88 | 12.69 | 8.59 | 7.92 | 7.29 | 15.64 |
| PA-CE-NW | | 3.96 | 3.78 | 5.59 | 4.86 | 6.87 | 3.12 | 9.99 | 3.18 | 4.86 | 2.71 | 15.34 |
| PA-TM | | 6.97 | 8.60 | 7.40 | 2.38 | 6.50 | 8.42 | 7.87 | 8.88 | 8.62 | 9.70 | 17.67 |
| PA-TM-NW | | 4.23 | 3.41 | 6.45 | 6.20 | 7.00 | 8.55 | 9.91 | 6.10 | 7.71 | 6.76 | 13.39 |
| PA-TM-CE | | 2.79 | 3.16 | 4.82 | 4.64 | 5.62 | 5.29 | 9.54 | 5.16 | 7.01 | 5.53 | 7.73 |
| LD | | 5.35 | 3.68 | 4.36 | 8.39 | 6.78 | 14.85 | 15.74 | 21.19 | 21.47 | 24.12 | 29.55 |
| LD-TM | | 10.97 | 11.89 | 13.68 | 13.23 | 14.75 | 12.46 | 13.11 | 10.63 | 11.51 | 11.00 | 13.21 |
| FS | | 4.85 | 4.04 | 5.71 | 7.31 | 7.64 | 4.50 | 8.93 | 12.03 | 9.26 | 13.57 | 18.83 |
| Q | | 9.00 | 9.90 | 10.01 | 8.60 | 7.52 | 2.85 | 2.95 | 3.73 | 6.00 | 11.53 | 18.16 |

**Figure 2.** North façade of the model house. Colour representations are averaged colour measurements. Presented values are $\Delta E$.

| | Oct 2013 | Dec 2013 | Mar 2014 | May 2014 | Aug 2014 | Oct 2014 | Apr 2015 | Oct 2015 | Apr 2016 | Oct 2016 | May 2017 | Oct 2017 |
|---|---|---|---|---|---|---|---|---|---|---|---|---|
| PA | | | 10.74 | 12.23 | 15.79 | 14.41 | 12.35 | 15.53 | 15.59 | 16.32 | 15.83 | 17.33 |
| PA-NW | | | 5.37 | 7.01 | 12.01 | 11.83 | 19.95 | 18.73 | 23.18 | 24.48 | 28.30 | 27.78 |
| PA-AC | | | 5.84 | 4.44 | 9.91 | 12.76 | 15.03 | 18.81 | 20.10 | 21.09 | 23.13 | 23.61 |
| PA-CE | | | 5.56 | 8.34 | 9.73 | 10.92 | 8.28 | 12.74 | 9.52 | 9.28 | 9.75 | 9.94 |
| PA-CE-NW | | | 6.72 | 8.23 | 2.43 | 5.72 | 10.78 | 13.34 | 16.25 | 17.76 | 21.38 | 21.49 |
| PA-TM | | | 13.90 | 15.48 | 13.18 | 16.64 | 14.40 | 13.78 | 13.69 | 11.23 | 6.62 | 3.57 |
| PA-TM-NW | | | 8.01 | 8.88 | 3.76 | 8.62 | 11.05 | 10.98 | 14.50 | 15.01 | 17.86 | 18.80 |
| PA-TM-CE | | | 9.89 | 11.93 | 11.88 | 12.38 | 10.61 | 8.37 | 8.04 | 5.19 | 5.69 | 6.51 |
| LD | | | 3.50 | 8.82 | 24.77 | 26.95 | 30.30 | 33.87 | 35.06 | 34.20 | 36.80 | 35.20 |
| LD-TM | | | 11.61 | 11.65 | 11.94 | 12.65 | 11.27 | 12.08 | 9.47 | 8.56 | 6.88 | 2.76 |
| FS | | | 7.57 | 8.44 | 18.15 | 20.62 | 25.83 | 26.66 | 28.38 | 22.99 | 25.13 | 26.37 |
| Q | | | 5.72 | 5.72 | 2.14 | 5.30 | 5.43 | 2.69 | 8.33 | 9.37 | 11.48 | 11.70 |

**Figure 3.** South façade of the model house. Colour representations are averaged colour measurements. Presented values are ΔE.

| | Oct 2013 | Dec 2013 | Mar 2014 | May 2014 | Aug 2014 | Oct 2014 | Apr 2015 | Oct 2015 | Apr 2016 | Oct 2016 | May 2017 | Oct 2017 |
|---|---|---|---|---|---|---|---|---|---|---|---|---|
| PA | | 7.26 | 10.48 | 15.04 | 17.80 | 22.27 | 29.76 | 34.24 | 34.83 | 35.55 | 38.16 | 39.07 |
| PA-NW | | 5.69 | 11.24 | 10.77 | 12.96 | 16.20 | 24.05 | 24.53 | 28.76 | 27.22 | 33.30 | 33.28 |
| PA-AC | | 3.87 | 7.67 | 2.92 | 3.85 | 5.23 | 11.40 | 13.70 | 13.81 | 16.53 | 19.27 | 20.24 |
| PA-CE | | 5.03 | 2.27 | 5.60 | 3.74 | 2.89 | 5.49 | 10.08 | 13.77 | 15.65 | 17.50 | 18.25 |
| PA-CE-NW | | 5.57 | 6.53 | 10.12 | 10.45 | 10.44 | 6.74 | 7.73 | 9.51 | 8.58 | 10.72 | 12.62 |
| PA-TM | | 8.90 | 10.82 | 12.88 | 15.12 | 18.01 | 18.50 | 16.02 | 15.19 | 13.97 | 15.46 | 16.40 |
| PA-TM-NW | | 2.87 | 4.27 | 4.48 | 4.36 | 8.05 | 9.17 | 9.36 | 12.51 | 10.10 | 12.93 | 13.84 |
| PA-TM-CE | | 2.84 | 7.61 | 11.83 | 11.14 | 8.97 | 9.54 | 6.92 | 10.28 | 9.20 | 10.83 | 12.07 |
| LD | | 6.25 | 0.62 | 1.27 | 5.74 | 8.94 | 18.97 | 22.34 | 25.07 | 26.96 | 30.93 | 32.60 |
| LD-TM | | 9.40 | 6.40 | 6.26 | 8.65 | 9.04 | 10.63 | 10.93 | 11.12 | 9.55 | 10.18 | 10.08 |
| FS | | 4.17 | 7.64 | 5.82 | 7.48 | 11.17 | 18.36 | 21.03 | 22.96 | 17.11 | 22.08 | 23.05 |
| Q | | 9.61 | 3.45 | 4.78 | 10.39 | 11.94 | 16.48 | 18.87 | 21.91 | 23.11 | 24.82 | 25.27 |

**Figure 4.** East façade of the model house. Colour representations are averaged colour measurements. Presented values are ΔE.

| | Oct 2013 | Dec 2013 | Mar 2014 | May 2014 | Aug 2014 | Oct 2014 | Apr 2015 | Oct 2015 | Apr 2016 | Oct 2016 | May 2017 | May 2017 |
|---|---|---|---|---|---|---|---|---|---|---|---|---|
| PA | | | 12.10 | 12.06 | 19.69 | 21.42 | 26.60 | 29.76 | 35.36 | 33.78 | 36.20 | 37.64 |
| PA-NW | | | 9.61 | 12.16 | 16.98 | 19.09 | 24.52 | 25.33 | 30.88 | 31.38 | 33.92 | 33.80 |
| PA-AC | | | 3.33 | 3.97 | 6.83 | 5.19 | 9.52 | 9.10 | 14.65 | 14.51 | 16.31 | 16.62 |
| PA-CE | | | 4.00 | 7.52 | 6.40 | 6.43 | 3.64 | 4.42 | 8.05 | 10.75 | 13.75 | 15.31 |
| PA-CE-NW | | | 9.79 | 10.72 | 8.65 | 10.01 | 7.39 | 7.90 | 9.41 | 8.03 | 10.28 | 12.00 |
| PA-TM | | | 13.70 | 16.71 | 15.05 | 16.51 | 16.90 | 16.02 | 16.40 | 14.98 | 16.54 | 16.54 |
| PA-TM-NW | | | 8.54 | 14.44 | 10.86 | 14.51 | 15.32 | 12.30 | 13.67 | 11.05 | 12.49 | 14.16 |
| PA-TM-CE | | | 7.78 | 9.68 | 5.44 | 7.42 | 7.01 | 5.06 | 5.39 | 7.88 | 9.30 | 10.27 |
| LD | | | 4.89 | 12.86 | 6.25 | 8.58 | 14.08 | 17.61 | 23.07 | 26.74 | 30.73 | 31.94 |
| LD-TM | | | 10.35 | 11.36 | 5.21 | 5.32 | 6.26 | 6.01 | 6.65 | 8.14 | 9.30 | 9.92 |
| FS | | | 4.88 | 6.09 | 11.37 | 14.66 | 19.87 | 21.28 | 24.49 | 18.83 | 24.04 | 23.96 |

**Figure 5.** West façade of the model house. Colour representations are averaged colour measurements. Presented values are ΔE.

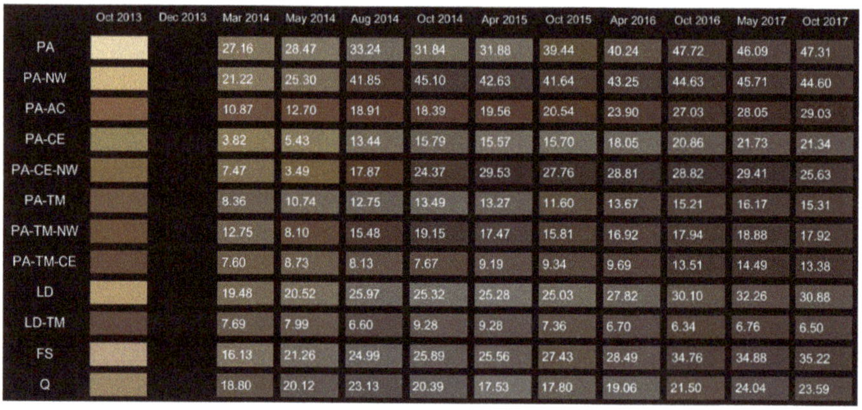

| | Oct 2013 | Dec 2013 | Mar 2014 | May 2014 | Aug 2014 | Oct 2014 | Apr 2015 | Oct 2015 | Apr 2016 | Oct 2016 | May 2017 | Oct 2017 |
|---|---|---|---|---|---|---|---|---|---|---|---|---|
| PA | | | 27.16 | 28.47 | 33.24 | 31.84 | 31.88 | 39.44 | 40.24 | 47.72 | 46.09 | 47.31 |
| PA-NW | | | 21.22 | 25.30 | 41.85 | 45.10 | 42.63 | 41.64 | 43.25 | 44.63 | 45.71 | 44.60 |
| PA-AC | | | 10.87 | 12.70 | 18.91 | 18.39 | 19.56 | 20.54 | 23.90 | 27.03 | 28.05 | 29.03 |
| PA-CE | | | 3.82 | 5.43 | 13.44 | 15.79 | 15.57 | 15.70 | 18.05 | 20.86 | 21.73 | 21.34 |
| PA-CE-NW | | | 7.47 | 3.49 | 17.87 | 24.37 | 29.53 | 27.76 | 28.81 | 28.82 | 29.41 | 25.63 |
| PA-TM | | | 8.36 | 10.74 | 12.75 | 13.49 | 13.27 | 11.60 | 13.67 | 15.21 | 16.17 | 15.31 |
| PA-TM-NW | | | 12.75 | 8.10 | 15.48 | 19.15 | 17.47 | 15.81 | 16.92 | 17.94 | 18.88 | 17.92 |
| PA-TM-CE | | | 7.60 | 8.73 | 8.13 | 7.67 | 9.19 | 9.34 | 9.69 | 13.51 | 14.49 | 13.38 |
| LD | | | 19.48 | 20.52 | 25.97 | 25.32 | 25.28 | 25.03 | 27.82 | 30.10 | 32.26 | 30.88 |
| LD-TM | | | 7.69 | 7.99 | 6.60 | 9.28 | 9.28 | 7.36 | 6.70 | 6.34 | 6.76 | 6.50 |
| FS | | | 16.13 | 21.26 | 24.99 | 25.89 | 25.56 | 27.43 | 28.49 | 34.76 | 34.88 | 35.22 |
| Q | | | 18.80 | 20.12 | 23.13 | 20.39 | 17.53 | 17.80 | 19.06 | 21.50 | 24.04 | 23.59 |

**Figure 6.** Decking of the model house. Colour representations are averaged colour measurements. Presented values are ΔE.

The L*, a*, and b* values for all twelve materials in all exposure directions are shown in Figure 7. Grey and yellow rectangles mark the winter and summer months, respectively, to distinguish the predominant effect of each part of the year, e.g., darkening due to moulding in winter months and lightening due to UV irradiation in summer months (Table 2). On the north façade, the highest absolute ΔL* value after four years of natural weathering was calculated for Norway spruce (PA; 26.50), followed by European larch (LD; 17.29), wax-treated Norway spruce (PA-NW; 16.45), and European beech (FS; 14.07). These wood species are respectively classified as moderately, slightly, or not durable to attack by decay fungi according to EN 350 [50] and are susceptible if not treated to fungal disfigurement. All copper (II) ethanolamine impregnated samples, i.e., copper- and wax-treated Norway spruce (PA-CE-NW; 0.67), copper-treated Norway spruce (PA-CE; 3.25), and copper-treated thermally-modified Norway spruce (PA-TM-CE; 4.66), were among the materials with the lowest absolute ΔL*. The reason for the observed phenomena was mentioned previously and is assigned to the presence of copper-based active ingredients with a clearly proven fungicidal effect [51]. Acrylic-coated Norway spruce (PA-AC) stood out from all other materials with the highest a* values, but the Δa* value after four years of weathering was the lowest (0.92), because of the pigments in the acrylic coating. This proved that the pigments remained in the acrylic coating and effectively influenced the aesthetic performance of coated wood. On the other hand, the least prominent change of Δa* was determined to be copper-treated thermally-modified Norway spruce (PA-TM-CE; 1.05), whilst European beech (FS; 1.66) also had the slightest change. The main reason for the quite insignificant absolute Δa* change of beech wood specimens was the constant exposure of fresh beech wood due to wasps, which would constantly remove any degraded wood surface. It is a known fact that wasps construct paper covers for their nests using weathered wood [52]. The highest absolute Δa* was measured on thermally-modified Norway spruce (PA-TM; 7.59), followed by European larch (LD; 6.83) and copper-treated Norway spruce (PA-CE; 6.35). In general, a higher change was noted on the yellow-blue axis, i.e., the b* axis. The lowest difference in absolute Δb* was determined for acrylic-coated Norway spruce (PA-AC; 6.04), followed by copper-treated thermally-modified Norway spruce (PA-TM-CE; 6.07), thermally-modified European larch (LD-TM; 7.21), and thermally-modified wax-treated Norway spruce (PA-TM-NW; 11.99). With the exception of acrylic-coated wood, the other three materials belong to the group of thermally-modified materials, which clearly indicated their better colour stability on this blue–yellow chromatic axis. The better colour stability of thermally-modified wood is in accordance with previously published results [49]. In the respective article, a considerably lower colour change of thermally-modified wood during UV light exposure was determined when compared to that of non-modified wood. Both thermally-modified wood and non-treated wood showed similar, but less pronounced, changes visible from IR spectra, indicating that there is still some degradation of lignin and non-cellulosic polysaccharides, and formation of non-conjugated carbonyl groups, that ultimately resulted in some degree of colour change to light irradiated wood. A similar trend can be observed in all other exposure directions, but the ranking was also influenced by the precise position on the façade or decking, because of the additional protection by the roof overhang or shade from close-by trees, distance from the ground, and other factors that influence the weathering.

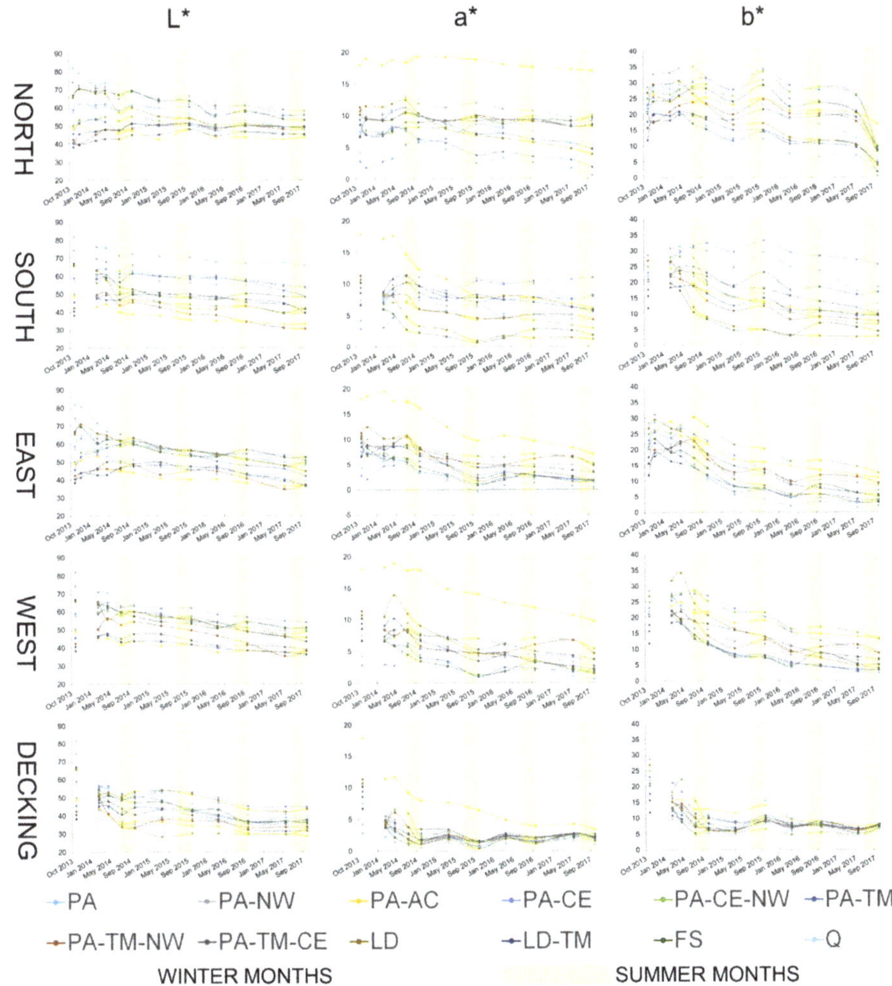


**Figure 7.** Twelve materials' colour values (L*, a*, and b*) for outdoor exposure for all exposure directions.

Blue staining assessment of the façade and decking elements of the model house was performed periodically (Table 2). As the samples were mounted, only the visible surface was assessed with the grades 0–4. Table 2 represents the changes after the first 20 months of exposure on the model house. Later assessments were not possible, as the grey colour of the fungal pigments was hard to distinguish from the grey surface of the weathered wood. During the study, it could be seen that the progression of blue staining processes was rather fast. The first materials were stained after the first month of exposure. Staining predominantly appeared (or was more visible) on light-coloured materials. Blue staining even continued through the following winter months, which was a result of a rather mild winter. After three months of exposure, only copper-treated materials were not affected by blue staining, i.e., copper-treated Norway spruce (PA-CE), copper- and wax-treated Norway spruce (PA-CE-NW), and thermally-modified copper treated Norway spruce (PA-TM-CE). After one year, only PA-CE and PA-TM-CE remained unaffected by staining fungi; however, after 18 months of exposure, at least some fungal disfigurement was evident on all tested materials. Because of the issues related to recognising fungal staining on weathered surfaces, visual evaluation was determined after 20 months of natural weathering.

### 3.2. Influence of Blue Staining and Artificial Weathering Test on Colour Changes

In the next part of the study, we decided to simulate weathering in the laboratory. Therefore, two types of tests were combined, namely exposure to blue stain fungi according to the EN 152 test [42] and artificial weathering. The EN 152 [42] standard test is designed for testing the efficacy of materials against blue staining organisms. The standard prescribes that the blue stained surface is evaluated with marks between 0 and 3 (Figure 8). After the first exposure of the samples to the suspension of fungal spores, only three out of twelve materials developed surface stains. Beechwood (FS) had the highest visual grade of blue stain discolouration (3.00 ± 0.00), followed by Norway spruce wood (PA; 2.60 ± 0.55) and wax-treated Norway spruce (PA-NW; 1.40 ± 0.55). As this type of evaluation is deterministic by its definition, we decided to evaluate blue staining with colour change measurements. Therefore, samples were scanned and their colour determined. Figure 9 shows colour changes of wood after the first blue staining, artificial weathering (AW), and the second blue staining. After the first exposure to blue stain fungi, the majority of the samples became considerably darker as a result of pigments from the micro-organisms, e.g., melanin, which is a dark, high-molecular-weight pigment within hyphal walls [7]. Only two materials, acrylic-coated Norway spruce (PA-AC) and thermally-modified wax-treated Norway spruce (PA-TM-NW), retained their original colours, predominately due to the dark colour of the thermally-modified wood (PA-TM-NW) and the dark colour of pigments in the acrylic coating (PA-AC). Considering all the colour changes expressed as a ΔE value in Figure 9, it can be clearly seen that the most prominent colour change was determined for light-coloured wood species; Norway spruce (PA; 24.0 ± 17.0) and European beech (FS; 18.0 ± 2.1). The least evident colour changes, where ΔE was lower than 3, were determined on thermally-modified wood. This is somehow expected, due to the dark colour of the thermally-modified wood. The respective difference is lower than the threshold (ΔE = 5) for the obvious difference in colour, as perceived by casual observations [53]. This result indicated that the original dark colour can make fungal disfigurement less notable.

In the second part of this study, an artificial weathering test (AW) was performed. Samples were exposed to water, UV, and IR light according to the prescribed regime described in Section 2.2. Specimens were visually assessed and scanned every 100 h of AW. As can be seen from Figure 8, it is notable that the staining became less visible. For example, European beech wood (FS) after 100 h of artificial weathering was downgraded from 3.00 ± 0.00 to 2.00 ± 0.00. A similar influence of artificial weathering was determined for Norway spruce (PA; 2.00 ± 1.41) and wax-treated Norway spruce (PA-NW; 0.50 ± 0.58). After the first 100 h of AW, all materials became lighter, with the exception of Norway spruce (PA), acrylic-coated Norway spruce (PA-AC), and European larch (LD). After 200 h of AW, fungal stains became even less visible; for example, wax-treated Norway spruce had no visible fungal stains (PA-NW; 0.00 ± 0.00). It can be deduced that wax-treated Norway spruce (PA-NW), acrylic-coated Norway spruce (PA-AC), and copper- and wax-treated Norway spruce (PA-CE-NW), after 100 h of AW, almost established the original colour prior to exposure. The rapid colour change observed during the first 200 h of irradiation is in line with other literature findings [49,54]. Srinivas and Pandey [30] reported that thermally-modified wood had a dark brown colour that lightened upon UV light exposure in contrast to the unmodified wood, which became darker. A similar trend was observed in this study, but only for Norway spruce (PA) and European larch (LD), not for European beech (FS) nor English oak (Q), and only after the first 100 h of UV irradiation. After 200 h, the lightness parameter (L*) increased on all twelve materials. The negative values of Δa* revealed a tendency for the samples to become greenish or to lose red pigmentation. In general (but with some exceptions), samples exhibited a lowering of the Δa* values throughout the experiment. It was also observed that more or less all the values showed negative values for Δb*, indicating a loss of yellow pigmentation and an increase of blue tones of weathered wood surfaces. Regarding the summed ΔE colour change (Equation (2)), throughout the whole test, acrylic-coated Norway spruce (PA-AC) was the most colour-stable material (ΣΔE = 11.90), followed by copper- and wax-treated Norway spruce (PA-CE-NW; ΣΔE = 28.84), copper-treated Norway spruce (PA-CE; ΣΔE = 30.83), and copper-treated thermally-modified Norway spruce (PA-TM-CE; ΣΔE = 37.32). On the other hand, European beech (FS; ΣΔE = 114.47), Norway spruce (PA; ΣΔE = 91.75), and English oak (Q; ΣΔE = 81.25) expressed the most prominent colour changes. The observed colour stability of coated Norway spruce can be associated with pigments used in coatings because they can absorb UV light and allow good colour retention by minimizing the degradation of the wood by UV light [55]. The value of summed ΔE of copper-treated Norway spruce (PA-CE) after 500 h of AW was 41.7 % less than the value of Norway spruce (PA), similar to the results presented in other studies [49,54]. Both studies explain better colour stability with the formation of phenolate and other complexes due to the interaction of copper ethanolamine with phenolic groups of lignin that inhibit the formation of free radicals. After the last 100 h of AW, the presence of fungal pigments became less and less visible and almost disappeared; in other words, Norway spruce (PA) was graded 1.50 ± 1.29 and European beech (FS) was 0.25 ± 0.50. *A. pullulans*, one of the most successful ascomycete fungi at colonizing weathered wood, has the ability to increase its production of melanin when exposed to UV radiation [8,56,57]. However, the conditions (UV exposure, heat) in the AW test were too harsh for fungi to survive, and consequently, the exposure to artificial weathering caused the degradation of polyphenolics (lignin) and depolymerisation of polysaccharides (cellulose, hemicelluloses) [17,58]. Wet cycles with water spray during AW washed out soluble degradation products from the wood surface and reduced the visibility of the discolouration previously caused by blue-stain fungi.

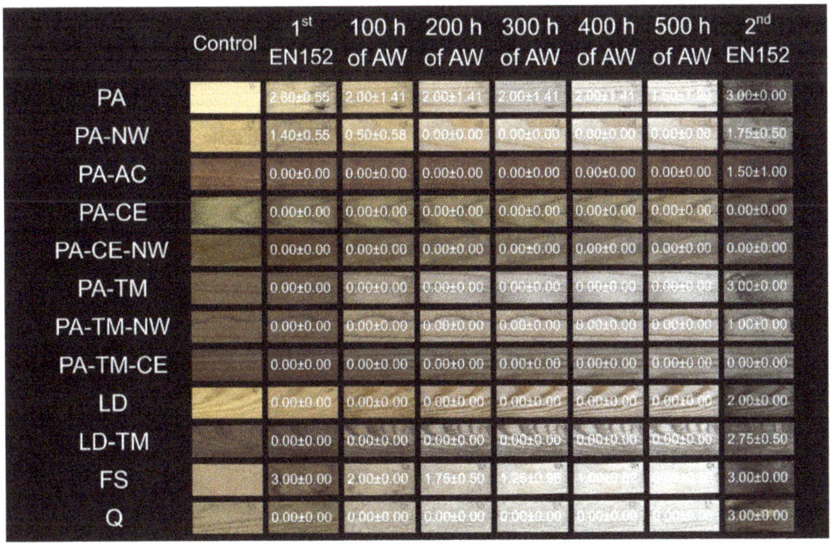

| | Control | 1st EN152 | 100 h of AW | 200 h of AW | 300 h of AW | 400 h of AW | 500 h of AW | 2nd EN152 |
|---|---|---|---|---|---|---|---|---|
| PA | | 2.60±0.55 | 2.00±1.41 | 2.00±1.41 | 2.00±1.41 | 2.00±1.41 | 1.50±1.29 | 3.00±0.00 |
| PA-NW | | 1.40±0.55 | 0.50±0.58 | 0.00±0.00 | 0.00±0.00 | 0.00±0.00 | 0.00±0.00 | 1.75±0.50 |
| PA-AC | | 0.00±0.00 | 0.00±0.00 | 0.00±0.00 | 0.00±0.00 | 0.00±0.00 | 0.00±0.00 | 1.50±1.00 |
| PA-CE | | 0.00±0.00 | 0.00±0.00 | 0.00±0.00 | 0.00±0.00 | 0.00±0.00 | 0.00±0.00 | 0.00±0.00 |
| PA-CE-NW | | 0.00±0.00 | 0.00±0.00 | 0.00±0.00 | 0.00±0.00 | 0.00±0.00 | 0.00±0.00 | 0.00±0.00 |
| PA-TM | | 0.00±0.00 | 0.00±0.00 | 0.00±0.00 | 0.00±0.00 | 0.00±0.00 | 0.00±0.00 | 3.00±0.00 |
| PA-TM-NW | | 0.00±0.00 | 0.00±0.00 | 0.00±0.00 | 0.00±0.00 | 0.00±0.00 | 0.00±0.00 | 1.00±0.00 |
| PA-TM-CE | | 0.00±0.00 | 0.00±0.00 | 0.00±0.00 | 0.00±0.00 | 0.00±0.00 | 0.00±0.00 | 0.00±0.00 |
| LD | | 0.00±0.00 | 0.00±0.00 | 0.00±0.00 | 0.00±0.00 | 0.00±0.00 | 0.00±0.00 | 2.00±0.00 |
| LD-TM | | 0.00±0.00 | 0.00±0.00 | 0.00±0.00 | 0.00±0.00 | 0.00±0.00 | 0.00±0.00 | 2.75±0.50 |
| FS | | 3.00±0.00 | 2.00±0.00 | 1.75±0.50 | 1.25±0.96 | 1.00±0.82 | [illegible] | 3.00±0.00 |
| Q | | 0.00±0.00 | 0.00±0.00 | 0.00±0.00 | 0.00±0.00 | 0.00±0.00 | 0.00±0.00 | 3.00±0.00 |

**Figure 8.** Scanned samples of surface blue staining and the artificial weathering test. Presented values are visual grades with standard deviation.

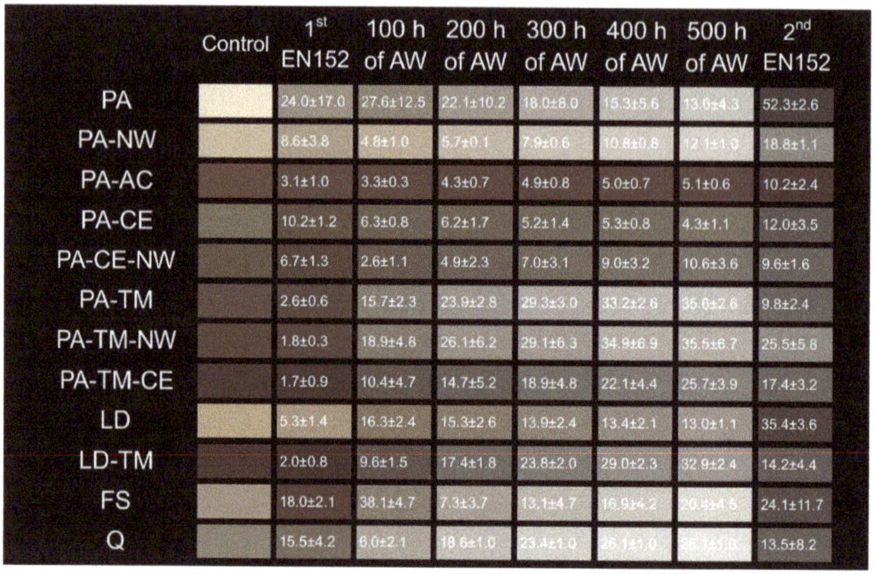

| | Control | 1st EN152 | 100 h of AW | 200 h of AW | 300 h of AW | 400 h of AW | 500 h of AW | 2nd EN152 |
|---|---|---|---|---|---|---|---|---|
| PA | | 24.0±17.0 | 27.6±12.5 | 22.1±10.2 | 18.0±8.0 | 15.3±5.6 | 13.6±4.3 | 52.3±2.6 |
| PA-NW | | 8.6±3.8 | 4.8±1.0 | 5.7±0.1 | 7.9±0.6 | 10.8±0.8 | 12.1±1.0 | 18.8±1.1 |
| PA-AC | | 3.1±1.0 | 3.3±0.3 | 4.3±0.7 | 4.9±0.8 | 5.0±0.7 | 5.1±0.6 | 10.2±2.4 |
| PA-CE | | 10.2±1.2 | 6.3±0.8 | 6.2±1.7 | 5.2±1.4 | 5.3±0.8 | 4.3±1.1 | 12.0±3.5 |
| PA-CE-NW | | 6.7±1.3 | 2.6±1.1 | 4.9±2.3 | 7.0±3.1 | 9.0±3.2 | 10.6±3.6 | 9.6±1.6 |
| PA-TM | | 2.6±0.6 | 15.7±2.3 | 23.9±2.8 | 29.3±3.0 | 33.2±2.6 | 35.6±2.6 | 9.8±2.4 |
| PA-TM-NW | | 1.8±0.3 | 18.9±4.8 | 26.1±6.2 | 29.1±6.3 | 34.9±6.9 | 35.5±6.7 | 25.5±5.8 |
| PA-TM-CE | | 1.7±0.9 | 10.4±4.7 | 14.7±5.2 | 18.9±4.8 | 22.1±4.4 | 25.7±3.9 | 17.4±3.2 |
| LD | | 5.3±1.4 | 16.3±2.4 | 15.3±2.6 | 13.9±2.4 | 13.4±2.1 | 13.0±1.1 | 35.4±3.6 |
| LD-TM | | 2.0±0.8 | 9.6±1.5 | 17.4±1.8 | 23.8±2.0 | 29.0±2.3 | 32.9±2.4 | 14.2±4.4 |
| FS | | 18.0±2.1 | 38.1±4.7 | 7.3±3.7 | 13.1±4.7 | 16.9±4.2 | 20.4±4.5 | 24.1±11.7 |
| Q | | 15.5±4.2 | 6.0±2.1 | 18.6±1.0 | 23.4±1.0 | 26.1±1.0 | 26.1±1.0 | 13.5±8.2 |

**Figure 9.** Colour changes of twelve materials under two types of tests; exposure to blue stain fungi according to EN 152 and artificial weathering. Colour representations are averaged colour measurements. Presented values are ΔE with standard deviation.

In the third part of the laboratory trial, EN 152 [42] was performed again on weathered wood samples and more prominent colour changes were noted for all exposed specimens, as shown in Figure 9. The greatest colour change was measured for Norway spruce (PA; 52.3 ± 2.6) and European larch (LD; 35.4 ± 3.6). It is obvious that artificial weathering makes the surface more susceptible to fungal disfigurement. There are several explanations for this phenomenon. Firstly, UV radiation and spraying degrade and leach coloured secondary metabolites from the surface. In addition, partial degradation of lignin and cellulose occurred, which made them more susceptible to fungal infestation. Average ΔE among all specimens after first exposure to blue-staining fungi was 13.3, while average ΔE after second exposure was considerably higher, at 20.2. This can be explained by the ability of *A. pullulans* to utilise products of lignocellulosic photo-degradation, due to the ability of the fungus to use lignin-breakdown products as a sole carbon and energy source [59,60]. In addition, grades of visual assessment also increased significantly. After the second exposure, three materials remained non-discoloured, namely: copper-treated Norway spruce (PA-CE), copper- and wax-treated Norway spruce (PA-CE-NW), and thermally-modified copper-treated Norway spruce (PA-TM-CE), all of which were impregnated with a copper-based preservative. This observation is in line with the in-service testing presented in Section 3.1. This indicates that only biocides can limit fungal discolouration after severe weathering.

*3.3. Pearson's Correlations Factors for Colour Change between In-Service and Laboratory Testing*

One of the prime objectives of the respective manuscript was to determine the most predictable laboratory method for the simulation of colour changes on outdoor exposed wood. Table 3 gives the Pearson's correlation factors for colour change (ΔE) for all twelve materials between the in-service results, i.e., natural weathering on the model house, and laboratory testing, i.e., the combination of EN 152 and artificial weathering. In general, the correlations were low or even negative for all exposure directions from one year of natural weathering (October 2014) or with the laboratory testing from 200 h of artificial weathering. South and west exposure presented good correlations with March and April

2014 and 200 h to 500 h of AW. In addition, the north exposure also exhibited quite high correlations with December 2013 to April 2014 and 200 h to 500 h of AW, but probably as a result of better protection against wetting and a lower moulding rate (Table 2). Decking gave the best correlations with the first and the second exposure to blue-stain fungi for almost all periods of in-service testing. This indicated that the staining was the predominant cause of colour changes in the respective application. East and west exposure sites also displayed good correlations with the first and the second exposure to blue-stain fungi, and north exposure only with the second exposure to the EN 152 test.

As shown in Table 4, a significant correlation between east and west exposure was observed for almost all the materials, with the highest values for Norway spruce (PA; $r = 0.9812$, $p < 0.0001$) and wax-treated Norway spruce (PA-NW; $r = 0.9798$, $p < 0.0001$). This can be explained, since the exposure locations on both façades are comparable, with the absence of roof overhang, i.e., the influence of UV radiation. Only one material had a negative correlation between the east and west direction: thermally-modified European larch (LD-TM; $r = -0.5800$, $P = 0.0788$). In addition, decking correlated quite well with all exposure sites for Norway spruce, acrylic-coated Norway spruce, European larch, and European beech (PA, PA-AC, LD, and FS), with a statistically significant correlation.

However, the most important objective of the respective research was to show how the in-service test correlated with the laboratory trial. The best correlations with the laboratory trial and in-service test (for each exposure site separately) were determined for wax-treated Norway spruce (PA-NW) with a south ($r = 0.8643$, $P = 0.0264$), east ($r = 0.7957$, $P = 0.0324$), and west ($r = 0.9158$, $P = 0.0103$) direction. A strong positive correlation was also found for acrylic-coated Norway spruce (PA-AC) between the south and laboratory trial ($r = 0.8160$, $P = 0.0477$), and for thermally-modified Norway spruce (PA-TM) between the east and laboratory trial ($r = 0.7702$, $P = 0.0428$).

**Table 3.** Pearson correlation for colour change (ΔE) between the in-service test and artificial weathering test for five different exposures. Asterisks represent the significance of Pearson correlation (**** $p < 0.0001$, *** $p < 0.001$, ** $p < 0.01$, * $p < 0.05$).

| | | December 2013 | March 2014 | May 2014 | August 2014 | October 2014 | April 2015 | October 2015 | October 2016 | October 2016 | May 2017 | October 2017 |
|---|---|---|---|---|---|---|---|---|---|---|---|---|
| North | 1st EN152 | 0.2532 | 0.3795 | 0.3051 | 0.5954 * | 0.4108 | 0.2575 | 0.3622 | 0.5125 | 0.4533 | 0.5986 * | 0.6380 * |
| | 100 h of AW | 0.0838 | 0.1350 | 0.1040 | 0.2742 | 0.2710 | 0.3420 | 0.4173 | 0.5228 | 0.4278 | 0.4829 | 0.4105 |
| | 200 h of AW | 0.4715 | 0.5256 | 0.4837 | 0.2042 | 0.3276 | 0.4210 | 0.2751 | 0.2082 | 0.2813 | 0.2566 | 0.1982 |
| | 300 h of AW | 0.5005 | 0.5124 | 0.5061 | 0.0507 | 0.2503 | 0.1669 | 0.0308 | −0.0357 | 0.0241 | 0.0288 | −0.0299 |
| | 400 h of AW | 0.4499 | 0.4402 | 0.4747 | −0.0303 | 0.1948 | 0.0340 | −0.0809 | −0.1663 | −0.1129 | −0.1085 | −0.1532 |
| | 500 h of AW | 0.4377 | 0.4193 | 0.4641 | −0.0697 | 0.1709 | −0.0456 | −0.1422 | −0.2292 | −0.1820 | −0.1724 | −0.2220 |
| | 2nd EN152 | 0.1468 | 0.2683 | 0.2106 | 0.6746 * | 0.4468 | 0.8077 ** | 0.7820 ** | 0.8716 *** | 0.8840 *** | 0.8622 *** | 0.7754 ** |
| South | 1st EN152 | | −0.1013 | −0.1073 | 0.1510 | 0.0445 | 0.0439 | 0.0772 | 0.1256 | 0.1228 | 0.1306 | 0.2097 |
| | 100 h of AW | | 0.2786 | 0.3490 | 0.5295 | 0.5555 | 0.4627 | 0.4225 | 0.3873 | 0.2317 | 0.1564 | 0.1929 |
| | 200 h of AW | | 0.5735 | 0.5999 * | −0.0073 | 0.0354 | −0.2184 | −0.2998 | −0.2910 | −0.3103 | −0.3887 | −0.4064 |
| | 300 h of AW | | 0.6487 * | 0.5906 * | −0.1047 | −0.0331 | −0.2384 | −0.3832 | −0.3564 | −0.4190 | −0.4867 | −0.5196 |
| | 400 h of AW | | 0.6298 * | 0.5395 | −0.1745 | −0.0882 | −0.2370 | −0.4036 | −0.3652 | −0.4380 | −0.4883 | −0.5261 |
| | 500 h of AW | | 0.6380 * | 0.5296 | −0.1764 | −0.0916 | −0.2198 | −0.4034 | −0.3659 | −0.4557 | −0.5006 | −0.5423 |
| | 2nd EN152 | | 0.0098 | 0.2104 | 0.5323 | 0.4462 | 0.3585 | 0.3909 | 0.3806 | 0.3858 | 0.3425 | 0.3813 |
| East | 1st EN152 | 0.2084 | 0.1446 | 0.2478 | 0.3850 | 0.4875 | 0.6086 * | 0.7382 ** | 0.7495 ** | 0.6927 * | 0.6772 * | 0.6768 * |
| | 100 h of AW | −0.1115 | 0.1531 | 0.1129 | 0.1671 | 0.3753 | 0.4782 | 0.4860 | 0.4502 | 0.2683 | 0.2952 | 0.3039 |
| | 200 h of AW | 0.3524 | 0.0036 | 0.2001 | 0.3383 | 0.4179 | 0.2532 | 0.1461 | 0.0963 | 0.0719 | 0.0009 | −0.0056 |
| | 300 h of AW | 0.3784 | 0.0574 | 0.1631 | 0.2989 | 0.3400 | 0.1196 | −0.0361 | −0.0889 | −0.1648 | −0.2186 | −0.2331 |
| | 400 h of AW | 0.3202 | 0.0647 | 0.1190 | 0.2242 | 0.2541 | 0.0146 | −0.1564 | −0.1986 | −0.3007 | −0.3362 | −0.3545 |
| | 500 h of AW | 0.3105 | 0.0938 | 0.1219 | 0.2204 | 0.2247 | −0.0255 | −0.2096 | −0.2513 | −0.3697 | −0.3977 | −0.4172 |
| | 2nd EN152 | −0.0328 | 0.0421 | 0.1845 | 0.3122 | 0.4960 | 0.6795 * | 0.7399 ** | 0.7440 ** | 0.7077 * | 0.6954 * | 0.7081 * |
| West | 1st EN152 | | 0.0362 | −0.2075 | 0.5939 | 0.5767 | 0.6030 * | 0.6915 * | 0.7063 * | 0.6259 * | 0.6614 * | 0.6732 * |
| | 100 h of AW | | 0.0559 | 0.0595 | 0.3951 | 0.5087 | 0.5758 | 0.5787 | 0.5088 | 0.3588 | 0.4068 | 0.4027 |
| | 200 h of AW | | 0.5974 | 0.7549 ** | 0.2914 | 0.3615 | 0.2721 | 0.1887 | 0.0770 | 0.0542 | 0.0030 | 0.0179 |
| | 300 h of AW | | 0.6130 * | 0.7152 * | 0.1806 | 0.2708 | 0.1692 | 0.0518 | −0.0928 | −0.1549 | −0.1975 | −0.2008 |
| | 400 h of AW | | 0.5754 | 0.6723 * | 0.1106 | 0.2141 | 0.1030 | −0.0326 | −0.1841 | −0.2628 | −0.2999 | −0.3098 |
| | 500 h of AW | | 0.5557 | 0.6147 * | 0.0556 | 0.1630 | 0.0576 | −0.0792 | −0.2359 | −0.3209 | −0.3525 | −0.3685 |
| | 2nd EN152 | | 0.1333 | 0.2183 | 0.4793 | 0.5378 | 0.6494 * | 0.7102 * | 0.7152 * | 0.7255* | 0.7227 * | 0.7533 ** |
| Decking | 1st EN152 | | 0.6605 * | 0.6882 * | 0.5994 * | 0.5102 | 0.4812 | 0.6223 * | 0.6000 * | 0.6865 * | 0.6686 * | 0.6978 * |
| | 100 h of AW | | 0.4234 | 0.4539 | 0.1997 | 0.1391 | 0.1052 | 0.2182 | 0.1943 | 0.3051 | 0.2716 | 0.3101 |
| | 200 h of AW | | 0.2149 | 0.0812 | −0.2044 | −0.2799 | −0.3676 | −0.3100 | −0.3345 | −0.3031 | −0.3180 | −0.2944 |
| | 300 h of AW | | 0.0447 | −0.0419 | −0.3443 | −0.3986 | −0.4915 | −0.4749 | −0.5086 | −0.4884 | −0.4987 | −0.4800 |
| | 400 h of AW | | −0.0568 | −0.1314 | −0.3935 | −0.4171 | −0.5091 | −0.5248 | −0.5618 | −0.5569 | −0.5644 | −0.5525 |
| | 500 h of AW | | −0.1110 | −0.1624 | −0.4270 | −0.4432 | −0.5274 | −0.5538 | −0.5950 * | −0.5917 * | −0.5986 * | −0.5897 * |
| | 2nd EN152 | | 0.7716 ** | 0.6711 * | 0.5101 | 0.4251 | 0.4002 | 0.5374 | 0.5155 | 0.5883 * | 0.5632 | 0.5897 * |

**Table 4.** Pearson correlation between the in-service test and artificial weathering test for five different exposures for each tested material regarding the colour change for the full test span. Asterisks represent the significance of Pearson correlation (**** $p < 0.0001$, *** $p < 0.001$, ** $p < 0.01$, * $p < 0.05$).

| PA | North | South | East | West | Decking | AW | PA-TM-NW | North | South | East | West | Decking | AW |
|---|---|---|---|---|---|---|---|---|---|---|---|---|---|
| North | 1.0000 | | | | | | North | 1.0000 | | | | | |
| South | 0.7852 ** | 1.0000 | | | | | South | 0.5413 | 1.0000 | | | | |
| East | 0.9474 **** | 0.7579 * | 1.0000 | | | | East | 0.6711 * | 0.9185 *** | 1.0000 | | | |
| West | 0.9621 **** | 0.8020 ** | 0.9812 **** | 1.0000 | | | West | 0.5053 | 0.2718 | 0.3918 | 1.0000 | | |
| Decking | 0.9246 *** | 0.8600 ** | 0.9035 *** | 0.9234 *** | 1.0000 | | Decking | 0.3724 | 0.4759 | 0.7005 * | 0.1609 | 1.0000 | |
| AW | 0.3143 | 0.2760 | 0.3523 | 0.3495 | 0.6308 | 1.0000 | AW | 0.6277 | 0.1450 | 0.6998 | 0.8033 | 0.6470 | 1.0000 |

| PA-NW | North | South | East | West | Decking | AW | PA-TM-CE | North | South | East | West | Decking | AW |
|---|---|---|---|---|---|---|---|---|---|---|---|---|---|
| North | 1.0000 | | | | | | North | 1.0000 | | | | | |
| South | 0.8407 ** | 1.0000 | | | | | South | −0.4673 | 1.0000 | | | | |
| East | 0.8482 *** | 0.9844 **** | 1.0000 | | | | East | 0.2749 | −0.0188 | 1.0000 | | | |
| West | 0.8201 ** | 0.9928 **** | 0.9798 **** | 1.0000 | | | West | −0.0889 | −0.2488 | 0.5296 | 1.0000 | | |
| Decking | 0.5652 | 0.7771 ** | 0.7121 | 0.8043 ** | 1.0000 | | Decking | 0.4215 | −0.8985 *** | 0.3512 | 0.5073 | 1.0000 | |
| AW | 0.4204 | 0.8643 * | 0.7957 * | 0.9158 * | 0.6876 | 1.0000 | AW | 0.5139 | 0.2543 | 0.6927 | −0.2958 | 0.3548 | 1.0000 |

| PA-AC | North | South | East | West | Decking | AW | LD | North | South | East | West | Decking | AW |
|---|---|---|---|---|---|---|---|---|---|---|---|---|---|
| North | 1.0000 | | | | | | North | 1.0000 | | | | | |
| South | 0.7681 ** | 1.0000 | | | | | South | 0.8302 ** | 1.0000 | | | | |
| East | 0.7699 ** | 0.9249 *** | 1.0000 | | | | East | 0.9782 **** | 0.8900 *** | 1.0000 | | | |
| West | 0.8656 ** | 0.9435 **** | 0.9196 *** | 1.0000 | | | West | 0.9465 **** | 0.7199 * | 0.9278 *** | 1.0000 | | |
| Decking | 0.8193 ** | 0.9583 **** | 0.8645 ** | 0.9603 **** | 1.0000 | | Decking | 0.8970 *** | 0.8885 *** | 0.8945 *** | 0.8453 ** | 1.0000 | |
| AW | −0.1464 | 0.8160 * | 0.7329 | 0.6814 | 0.6653 | 1.0000 | AW | 0.5908 | 0.3699 | 0.5754 | 0.6334 | 0.1141 | 1.0000 |

| PA-CE | North | South | East | West | Decking | AW | LD-TM | North | South | East | West | Decking | AW |
|---|---|---|---|---|---|---|---|---|---|---|---|---|---|
| North | 1.0000 | | | | | | North | 1.0000 | | | | | |
| South | 0.7422 * | 1.0000 | | | | | South | 0.3221 | 1.0000 | | | | |
| East | 0.5899 | 0.3193 | 1.0000 | | | | East | −0.2922 | −0.3275 | 1.0000 | | | |
| West | 0.4870 | 0.1706 | 0.8410 ** | 1.0000 | | | West | −0.1847 | −0.4317 | −0.5800 | 1.0000 | | |
| Decking | 0.6192 | 0.5841 | 0.7878 ** | 0.6476 * | 1.0000 | | Decking | 0.5429 | 0.5844 | −0.2029 | −0.2184 | 1.0000 | |
| AW | 0.3691 | 0.5695 | 0.7043 | −0.2040 | 0.1235 | 1.0000 | AW | 0.6665 | 0.0745 | 0.2824 | −0.6144 | 0.5873 | 1.0000 |

| PA-CE-NW | North | South | East | West | Decking | AW | FS | North | South | East | West | Decking | AW |
|---|---|---|---|---|---|---|---|---|---|---|---|---|---|
| North | 1.0000 | | | | | | North | 1.0000 | | | | | |
| South | 0.2818 | 1.0000 | | | | | South | 0.6038 | 1.0000 | | | | |
| East | 0.4946 | 0.3070 | 1.0000 | | | | East | 0.7403 ** | 0.9003 *** | 1.0000 | | | |
| West | 0.4491 | 0.2511 | 0.7293* | 1.0000 | | | West | 0.7271 * | 0.9673 **** | 0.9678 **** | 1.0000 | | |
| Decking | 0.0895 | 0.5894 | 0.0820 | −0.3457 | 1.0000 | | Decking | 0.8100 ** | 0.7687** | 0.7618* | 0.8357 ** | 1.0000 | |
| AW | 0.2101 | −0.0241 | 0.6212 | −0.3288 | 0.6704 | 1.0000 | AW | −0.2983 | −0.0857 | 0.2286 | −0.0450 | −0.4772 | 1.0000 |

Table 4. *Cont.*

| PA-TM | North | South | East | West | Decking | AW | Q | North | South | East | West | Decking | AW |
|---|---|---|---|---|---|---|---|---|---|---|---|---|---|
| North | 1.0000 | | | | | | North | 1.0000 | | | | | |
| South | −0.7545 * | 1.0000 | | | | | South | 0.5416 | 1.0000 | | | | |
| East | 0.1603 | −0.0571 | 1.0000 | | | | East | 0.0448 | 0.6575 * | 1.0000 | | | |
| West | 0.2688 | −0.1126 | 0.7073 * | 1.0000 | | | West | | | | 1.0000 | | |
| Decking | 0.3617 | −0.6223 | 0.5705 | 0.4952 | 1.0000 | | Decking | 0.7403 * | 0.5233 | 0.3485 | | 1.0000 | |
| AW | −0.2530 | 0.4051 | 0.7702* | 0.4695 | 0.7050 | 1.0000 | AW | −0.3479 | −0.0395 | 0.4405 | | 0.2566 | 1.0000 |

## 4. Conclusions

The average summed ΔE for all materials exposed in the in-service experiment on a model house was the lowest for the south (40.82) and east (40.95) exposure directions. The highest average summed ΔE was determined on decking elements (48.64). In addition, material that exhibited the lowest summed ΔE for all exposure directions was acrylic-coated Norway spruce (PA-AC; 33.70), followed by thermally-modified and copper-treated Norway spruce (PA-TM-CE; 35.63) and European larch (LD-TM; 35.65). On the contrary, Norway spruce, PA, displayed the highest summed ΔE (55.94).

To some extent, colour measurements show the seasonal fluctuating pattern, especially for lightness (L* value). This phenomenon was tested with the laboratory test, where samples were exposed to artificial weathering (AW) for 500 h after performing the EN 152 test. At the end of AW, EN 152 was performed again, to see the effect of weathering on the growth of blue staining fungi. It was figured out that only three materials, Norway spruce (PA), wax-treated Norway spruce (PA-NW), and European beech (FS), were stained after the first EN 152 trial and that after 500 h of UV irradiation, almost all stains were bleached out. After the second EN 152 test, all but copper-ethanolamine treated samples were stained and to a significantly greater extent than after the first blue-staining test. This indicates that only biocides can limit fungal discolouration after severe weathering.

The aim of the respective work was to determine the correlations between the in-service test and the laboratory test. The results showed that there are some high correlations with statistical significance, but it would be impossible to precisely predict the level of colour change using only the described laboratory test. The best correlations between the laboratory trial and in-service test (for each exposure site separately) were determined for wax-treated Norway spruce (PA-NW) with south ($r = 0.8643$, $P = 0.0264$), east ($r = 0.7957$, $P = 0.0324$), and west ($r = 0.9158$, $P = 0.0103$) directions. A strong positive correlation was also found for acrylic-coated Norway spruce (PA-AC) between the south direction and laboratory trial ($r = 0.8160$, $P = 0.0477$), and for thermally-modified Norway spruce (PA-TM) between the east direction and laboratory trial ($r = 0.7702$, $P = 0.0428$).

Further to the work herein, the influence of the weather, e.g., relative humidity, surface temperature, and solar irradiation as described in Charisi et al. [61], on the colour change of different wood-based materials and the possibility of modelling and predicting the colour changes in an outdoor environment will be investigated.

**Author Contributions:** M.H. provided the overall idea for this research, provided founding, and designed the experiment. D.K. performed the majority of the laboratory experiments and data analysis and wrote a major part of the manuscript. M.H., B.L., D.K. and N.T. set up the model house and performed outdoor monitoring. All of the authors contributed to the discussion, commenting on, and editing of the paper.

**Funding:** The research was founded through the support of the Slovenian Research Agency within the framework of projects L4-5517, L4-7547, program P4-0015, and the infrastructural centre (IC LES PST 0481-09). Part of the research was supported by the program: Sustainable and innovative construction of smart buildings—TIGR4smart (C3330-16-529003).

**Acknowledgments:** Technical support of Samo Grbec and Andreja Žagar is acknowledged.

**Conflicts of Interest:** The authors declare no conflict of interest.

## References

1. Viitanen, H.; Toratti, T.; Makkonen, L.; Thelandersson, S.; Isaksson, T.; Früwald, E.; Jermer, J.; Englund, F.; Suttie, E. Modelling of service life and durability of wooden structures. In Proceedings of the 9th Nordic Symposium on Building Physics—NSB 2011, Tampere, Finland, 29 May–2 June 2011; Volume 2, pp. 925–932.

2. Isaksson, T.; Thelandersson, S.; Brischke, C.; Jermer, J. *Service Life of Wood in Outdoor above Ground Applications—Engineering Design Guidline*; Report TVBK-3060; Division of Structural Engineering, Lund University: Lund, Sweden, 2011. Available online: http://www.kstr.lth.se/fileadmin/kstr/pdf_files/Guideline/TVBK3067.pdf (accessed on 29 November 2017).

3. Burud, I.; Smeland, K.A.; Thiis, T.K.; Gobakken, L.R.; Sandak, A.; Sandak, J.; Liland, K.H. Modeling weather degradation of wooden facades using NIR hyperspectral imaging on thin wood samples. In Proceedings of the World Conference on Timber Engineering, Vienna, Austria, 22–25 August 2016; p. 8.

4. Sandak, A.; Sandak, J. Prediction of service life—Does aesthetic matter? In Proceedings of the IRG Annual Meeting 2017, Ghent, Belgium, 4–8 June 2017.

5. George, B.; Suttie, E.; Merlin, A.; Deglise, X. Photodegradation and photostabilisation of wood—The state of the art. *Polym. Degrad. Stab.* **2005**, *88*, 268–274. [CrossRef]

6. Isaksson, T.; Brischke, C.; Thelandersson, S. Development of decay performance models for outdoor timber structures. *Mater. Struct.* **2013**, *46*, 1209–1225. [CrossRef]

7. Zink, P.; Fengel, D. Studies on the Colouring Matter of Blue-stain Fungi. Part 1. General Characterization and the Associated Compounds. *Holzforschung* **1988**, *42*, 217–220. [CrossRef]

8. Hernández, V.A.; Evans, P.D. Effects of UV radiation on melanization and growth of fungi isolated from weathered wood surfaces. In Proceedings of the IRG Annual Meeting 2015, Viña del Mar, Chile, 10–14 May 2015; pp. 1–12.

9. Walther, T.; Reinsch, H.; Grosse, A.; Ostermann, K.; Deutsch, A.; Bley, T. Mathematical modeling of regulatory mechanisms in yeast colony development. *J. Theor. Biol.* **2004**, *229*, 327–338. [CrossRef] [PubMed]

10. Van Den Bulcke, J.; Van Acker, J.; Stevens, M. Laboratory testing and computer simulation of blue stain growth on and in wood coatings. *Int. Biodeterior. Biodegrad.* **2007**, *59*, 137–147. [CrossRef]

11. Humar, M.; Vek, V.; Bučar, B. Properties of blue-stained wood. *Drv. Ind.* **2008**, *59*, 75–79.

12. Fojutowski, A. The influence of fungi causing blue—stain on absorptiveness of Scotch pine wood. In Proceedings of the IRG Annual Meeting 2005, Viña del Mar, Chile, 10–14 May 2005; pp. 1–5.

13. Sharpe, P.R.; Dickinson, D.J. Blue stain in service on wood surface coatings. Part 1: The nutritional requirements of Aureobasidium pullulans. In Proceedings of the IRG Annual Meeting 1992, Harrogate, UK, 10–15 May 1992.

14. Schmidt, O. *Wood and Tree Fungi: Biology, Damage, Protection, and Use;* Springer: Berlin/Heidelberg, Germany, 2006.

15. Ayadi, N.; Lejeune, F.; Charrier, F.; Charrier, B.; Merlin, A. Colour stability of heat-treated wood during artificial weathering. *Holz als Roh- und Werkstoff* **2003**, *61*, 221–226. [CrossRef]

16. Hayoz, P.; Peter, W.; Rogez, D. A new innovative stabilization method for the protection of natural wood. *Prog. Org. Coat.* **2003**, *48*, 297–309. [CrossRef]

17. Feist, W.C. *Outdoor Wood Weathering and Protection;* Rowell, R.M., Barbour, R.J., Eds.; American Chemical Society: Washington, DC, USA, 1990; pp. 263–298. [CrossRef]

18. Kalnins, M.A. *Surface Characteristics of Wood as They Affect Durability of Finishes;* United States Department of Agriculture, Forest Products Laboratory: Madison, WI, USA, 1966; 60p.

19. Feist, W.C.; Hon, D.N.-S. Chemistry of weathering and protection. *Chem. Solid Wood* **1984**, *207*, 401–451.

20. Hon, D.N.-S. Weathering and photochemistry of wood. *Wood Cell. Chem.* **2001**, *2*, 512–546.

21. Kielmann, B.C.; Mai, C. Natural weathering performance and the effect of light stabilizers in water-based coating formulations on resin-modified and dye-stained beech-wood. *J. Coat. Technol. Res.* **2016**, *13*, 1065–1074. [CrossRef]

22. Cogulet, A.; Blanchet, P.; Landry, V. The Multifactorial Aspect of Wood Weathering: A Review Based on a Holistic Approach of Wood Degradation Protected by Clear Coating. *BioResources* **2018**, *13*, 2116–2138. [CrossRef]

23. Williams, R.S. Weathering of Wood. In *Handbook of Wood Chemistry and Wood Composites;* CRC Press: Boca Raton, FL, USA, 2005.

24. Pánek, M.; Oberhofnerová, E.; Zeidler, A.; Šedivka, P. Efficacy of Hydrophobic Coatings in Protecting Oak Wood Surfaces during Accelerated Weathering. *Coatings* **2017**, *7*, 172. [CrossRef]

25. Herrera, R.; Sandak, J.; Robles, E.; Krystofiak, T.; Labidi, J. Weathering Resistance of Thermally Modified Wood Finished with Coatings of Diverse Formulations. *Prog. Org. Coat.* **2018**, *119*, 145–154. [CrossRef]

26. Pandey, K.K. A note on the influence of extractives on the photo-discoloration and photodegradation of wood. *Polym. Degrad. Stab.* **2005**, *87*, 375–379. [CrossRef]

27. Oberhofnerová, E.; Pánek, M.; García-Cimarras, A. The Effect of Natural Weathering on Untreated Wood Surface. *Maderas Cienc. Tecnol.* **2017**, *19*, 173–184. [CrossRef]

28. Ugovšek, A.; Šubic, B.; Starman, J.; Rep, G.; Humar, M.; Lesar, B.; Thaler, N.; Brischke, C.; Meyer-Veltrup, L.; Jones, D.; et al. Short-Term Performance of Wooden Windows and Facade Elements Made of Thermally Modified and Non-Modified Norway Spruce in Different Natural Environments. *Wood Mater. Sci. Eng.* **2018**, 1–6. [CrossRef]

29. Forsthuber, B.; Grüll, G. Prediction of Wood Surface Discoloration for Applications in the Field of Architecture. *Wood Sci. Technol.* **2018**, *52*, 1093–1111. [CrossRef]

30. Srinivas, K.; Pandey, K.K. Photodegradation of thermally modified wood. *J. Photochem. Photobiol. B Biol.* **2012**, *117*, 140–145. [CrossRef] [PubMed]

31. Rüther, P.; Jelle, B.P. Colour changes of wood and wood-based materials due to natural and artificial weathering. *Wood Mater. Sci. Eng.* **2013**, *8*, 13–25. [CrossRef]

32. Pánek, M.; Reinprecht, L. Effect of vegetable oils on the colour stability of four tropical woods during natural and artificial weathering. *J. Wood Sci.* **2016**, *62*, 74–84. [CrossRef]

33. Teacă, C.-A.; Roşu, D.; Bodîrlău, R.; Roşu, L. Structural Changes in Wood under Artificial UV Light Irradiation Determined by FTIR Spectroscopy and Color Measurements—A Brief Review. *BioResources* **2013**, *8*, 1478–1507. [CrossRef]

34. Agresti, G.; Bonifazi, G.; Calienno, L.; Capobianco, G.; Lo Monaco, A.; Pelosi, C.; Picchio, R.; Serranti, S. Surface Investigation of Photo-Degraded Wood by Colour Monitoring, Infrared Spectroscopy, and Hyperspectral Imaging. *J. Spectrosc.* **2013**, *2013*, 380536. [CrossRef]

35. Pandey, K.K. Study of the Effect of Photo-Irradiation on the Surface Chemistry of Wood. *Polym. Degrad. Stab.* **2005**, *90*, 9–20. [CrossRef]

36. Rep, G.; Pohleven, F. Wood modification—A promising method for wood preservation. *Drv. Ind.* **2001**, *52*, 71–76.

37. Rep, G.; Pohleven, F.; Bučar, B. Characteristics of thermally modified wood in vacuum. In Proceedings of the IRG Annual Meeting 2004, Ljubljana, Slovenia, 6–10 June 2004; p. 9.

38. Humar, M.; Pohleven, F. Solution for Wood Preservation. European Patent EP 1791682 (B1), 3 September 2008.

39. Thelandersson, S.; Isaksson, T.; Suttie, E.; Frühwald, E.; Toratti, T.; Grüll, G.; Viitanen, H.; Jermer, J. Quantitative design guideline for wood outdoors above ground applications. In Proceedings of the IRG Annual Meeting 2011, Queenstown, New Zealand, 8–12 May 2011; International research group on wood protection: Stockholm, Sweden, 2011; p. 19. Available online: http://lup.lub.lu.se/search/ws/files/5572865/2154681.pdf (accessed on 26 May 2018).

40. Humar, M.; Kržišnik, D.; Lesar, B.; Thaler, N.; Ugovšek, A.; Zupančič, K.; Žlahtič, M. Thermal modification of wax-impregnated wood to enhance its physical, mechanical, and biological properties. *Holzforschung* **2017**, *71*, 57–64. [CrossRef]

41. Johansson, P.; Ekstrand-Tobin, A.; Svensson, T.; Bok, G. Laboratory study to determine the critical moisture level for mould growth on building materials. *Int. Biodeterior. Biodegrad.* **2012**, *73*, 23–32. [CrossRef]

42. CEN. European Standard EN 152. Wood Preservatives. Determination of the protective effectiveness of a preservative treatment against blue stain in wood in service. Laboratory method. In *Proceedings of the B/515, Brussels, Belgium, 2012*; BSI: London, UK, 2012. Available online: https://shop.bsigroup.com/ProductDetail/?pid=000000000030212121 (accessed on 12 October 2014).

43. Raspor, P.; Smole-Možina, S.; Podjavoršek, J.; Pohleven, F.; Gogala, N.; V Nekrep, F.; Rogelj, I.; Hacin, J. *ZIM: Zbirka Industrijskih Mikroorganizmov. Katalog Biokultur*; Biotehniška fakulteta, Katedra za Biotehnologijo: Ljubljana, Slovenia, 1995.

44. CEN. Paints and varnishes—Methods of exposure to laboratory light sources—Part 1: General guidance. In *Proceedings of ISO/TC 35/SC 9 General Test Methods for Paints and Varnishes, Brussels, Belgium, 2013*; BSI: London, UK, 2013. Available online: https://www.iso.org/obp/ui/#iso:std:iso:16474:-1:ed-1:v1:en (accessed on 12 October 2014).

45. CIE, Colourimetry—Part 4: CIE 1976 L*a*b* Colour Space. 2007. Available online: http://www.cie.co.at/publications/colorimetry-part-4-cie-1976-lab-colour-space (accessed on 12 October 2014).

46. Humar, M.; Kržišnik, D.; Lesar, B.; Ugovšek, A.; Rep, G.; Šubic, B.; Thaler, N.; Žlahtič, M. Performance of window, door, decking and façade elements made of thermally modified spruce wood on model house in Ljubljana. In Proceedings of the 27th International Conference on Wood Science and Technology (ICWST) Implementation of Wood Science in Woodworking Sector, Zagreb, Croatia, 13–14 October 2016; pp. 75–82.

47. Yildiz, S.; Gümüşkaya, E. The effects of thermal modification on crystalline structure of cellulose in soft and hardwood. *Build. Environ.* **2007**, *42*, 62–67. [CrossRef]

48. Yildiz, S.; Tomak, E.D.; Yildiz, U.C.; Ustaomer, D. Effect of artificial weathering on the properties of heat treated wood. *Polym. Degrad. Stab.* **2013**, *98*, 1419–1427. [CrossRef]

49. Deka, M.; Humar, M.; Rep, G.; Kričej, B.; Šentjurc, M.; Petrič, M. Effects of UV light irradiation on colour stability of thermally modified, copper ethanolamine treated and non-modified wood: EPR and DRIFT spectroscopic studies. *Wood Sci. Technol.* **2008**, *42*, 5–20. [CrossRef]

50. CEN. EN 350–Durability of wood and wood-based products. Testing and classification of the durability to biological agents of wood and wood-based materials. In *Proceedings of the B/515, October, 2016*; BSI: London, UK, 2016.

51. Thaler, N.; Lesar, B.; Humar, M. Performance of Copper-Ethanolamine-impregnated Scots Pine Wood during Exposure to Terrestrial Microorganisms. *Bioresources* **2013**, *8*, 3299–3308. [CrossRef]

52. Schmolz, E.; Brüders, N.; Daum, R.; Lamprecht, I. Thermoanalytical investigations on paper covers of social wasps. *Thermochim. Acta* **2000**, *361*, 121–129. Available online: https://ac.els-cdn.com/S00406031000005530/ 1-s2.0-S00406031000005530-main.pdf?_tid=ad8fc58b-6cd3-4df1-b1a0-34deaf292e7c&acdnat=1520343132_ d977f0f4bb497addb1bfa7223354e1e9 (accessed on 6 March 2018). [CrossRef]

53. Mokrzycki, W.S.; Tatol, M. Colour difference Delta E—A survey. *Mach. Graph. Vis.* **2011**, *20*, 383–411.

54. Kamdem, D.P.; Grelier, S. Surface Roughness and Colour Change of Copper Amine and UV Absorber-Treated Red Maple (Acer rubrum) Exposed to Artificial Ultraviolet Ligh. *Holzforschung* **2002**, *56*, 473–478. Available online: https://www.degruyter.com/downloadpdf/j/hfsg.2002.56.issue-5/hf.2002.073/hf.2002.073.pdf (accessed on 22 November 2017). [CrossRef]

55. Landry, V.; Blanchet, P. Weathering resistance of opaque PVDF-acrylic coatings applied on wood substrates. *Prog. Org. Coat.* **2012**, *75*, 494–501. [CrossRef]

56. Hernández, V.A.; Evans, P.D. Technical Note: Melanization of the wood-staining fungus Aureobasidium pullulans in response to UV radiation. *Wood Fiber Sci.* **2015**, *47*, 120–124.

57. Hernández, V.A. Role of Non-Decay Fungi on the Weathering of Wood. Ph.D. Thesis, The University of British Columbia, Vancouver, BC, Canada, September 2012.

58. Evans, P.D.; Thay, P.D.; Schmalzl, K.J. Degradation of wood surfaces during natural weathering. Effects on lignin and cellulose and on the adhesion of acrylic latex primers. *Wood Sci. Technol.* **1996**, *30*, 411–422. [CrossRef]

59. Schoeman, M.; Dickinson, D.J. Growth of Aureobasidium pullulans on lignin breakdown products at weathered wood surfaces. *Mycologist* **1997**, *11*, 168–172. [CrossRef]

60. Evans, P.D. Wood Products: Weathering. *Encycl. Mater. Sci. Technol.* **2001**, 9716–9721. [CrossRef]

61. Charisi, S.; Thiis, T.K.; Stefansson, P.; Burud, I. Prediction model of microclimatic surface conditions on building façades. *Build. Environ.* **2018**, *128*, 46–54. [CrossRef]

 *forests*

*Article*

# Surface Changes of Selected Hardwoods Due to Weather Conditions

Ivan Kubovský [1,*], Eliška Oberhofnerová [2], František Kačík [1,2] and Miloš Pánek [2]

[1]   Faculty of Wood Sciences and Technology, Technical University in Zvolen, T.G. Masaryka 24,
      960 53 Zvolen, Slovakia; kacik@tuzvo.sk
[2]   Faculty of Forestry and Wood Sciences, Czech University of Life Sciences in Prague, Kamýcká 129,
      16521 Praha 6-Suchdol, Czech Republic; eliska.oberhofnerova@seznam.cz (E.O.);
      panekmilos@fld.czu.cz (M.P.)
*    Correspondence: kubovsky@tuzvo.sk; Tel.: +421-045-520-6462

Received: 14 August 2018; Accepted: 7 September 2018; Published: 11 September 2018

**Abstract:** The study is focused on the surface changes of five hardwoods (oak, black locust, poplar, alder and maple) that were exposed to natural weathering for 24 months in the climatic conditions of Central Europe. Colour, roughness, visual and chemical changes of exposed surface structures were examined. The lowest total colour changes ($\Delta E^*$) were found for oak (23.77), the highest being recorded for maple (34.19). Roughness differences after 24-month exposure ($\Delta R_a$) showed minimal changes in poplar wood (9.41); the highest changes in roughness were found on the surface of alder (22.18). The presence of mould and blue stains was found on the surface of maple, alder and poplar. Chemical changes were characterized by lignin and hemicelluloses degradation. Decreases of both methoxy and carbonyl groups, cleavage of bonds in lignin and hemicelluloses, oxidation reaction and formation of new chromophores were observed. In the initial phases of the degradation process, the discoloration was related to chemical changes; in the longer period, the greying due to settling of dust particles and action of mould influenced the wood colour. The data were confirmed by confocal laser scanning microscopy. The obtained results revealed degradation processes of tested hardwood surfaces exposed to external environmental factors.

**Keywords:** chemical changes; colour changes; infrared spectroscopy; hardwoods; roughness

## 1. Introduction

One of the major contributions to securing sustainable development is the use of renewable natural materials, which undoubtedly include wood. In an effort to reduce the ecological burden, the surface of wooden structures is often left untreated. In addition to the traditional interior design elements, the use of non-treated wood is expanding even further into the exterior. Contemporary trends even prefer untreated wood that turns grey after exposure to weather under aboveground conditions over non-durable wood with applied surface coatings [1], since wood with higher natural durability [2] allows outdoor utilization without any deterioration risk [3]. However, wood is susceptible to environmental degradation similar to other natural materials [4]—it changes surface properties due to weather and can even be attacked by biotic agents if the basic design principles are not respected. Weathering is defined as the slow decomposition of materials subjected to weather factors [5], causing mostly unwanted and premature product failures [6]. The main weathering factors are solar radiation and water (moisture) acting synergistically with temperature, wind, airborne pollutants, biological agents, and acid rain, etc. [6–8].

The main organic components in wood, lignin, hemicelluloses and cellulose, react to weathering. Lignin is the wood component most susceptible to photodegradation caused mainly by ultraviolet (UV) light [9–11]. This aromatic biopolymer strongly absorbs UV radiation [12]. The initial decrease in lignin content is accompanied by generation of carbonyl groups, whereas degradation of cellulose is indicated by a loss of weight and reduction in the degree of polymerization. Extractives, such as terpenes, terpenoids, phenols, lignans, tannins, flavonoids, etc. [10,13], responsible for wood colour, odour and natural durability against biodegradation [4], are also affected by weathering. Photodegraded products are leached out of the wood surface by water [1]. The changes caused by weathering develop rapidly and cannot be fully avoided, as presented in previous work focused on weathering of untreated wood [14,15]. The rate of degradation is usually related to wood species [12], intensity of light and light wavelengths [16], period of irradiation [12,17,18] and climatic factors occurring during exposure [19].

Weathering affects only the surface layers of wood [20]. Atmospheric degradation is then manifested by the change of colour reflecting the chemical changes [21,22], followed by the formation of cracks and increased roughness of the samples [23,24]. Untreated wood specimens exhibit higher colour changes in a shorter time than treated specimens [25]. Some wood species turn yellow or brown; eventually, they turn grey due to growth of fungi and moulds [26] and by the dust particles in air which penetrate the porous structure of wood [19]. These changes occur to a depth of only 0.05–2.5 mm [11,23]. The depth has a relatively high correlation with the total colour difference value [15]. The next parameter used to analyse the weathering effects is surface roughness. During natural weathering, increasing roughness has been reported [15,27–29]. The increase is caused by unequal erosion of the surface in the late wood and early wood with thin walled cells and lower density [30]. As the result, the early wood is degraded faster [31], mainly in the case of softwoods with lower density [15]. The wood also reacts to moisture and water from ambient atmosphere [32]. Its surface wettability significantly increases during weathering [27]. Some species reached full wettability after 1 year of exposure [33]. Chemical changes in wood during weathering are often investigated by Fourier Transform Infrared Spectroscopy (FTIR) analysis [10,14,34,35]. According to Pandey [35], the total colour difference was related to the lignin degradation and increase of carbonyl groups. Hardwoods underwent a faster degradation than softwoods because the syringyl structure in hardwoods degraded faster than guaiacyl structures in softwoods. Liu et al. [36] and Reinprecht et al. [37] studied surface changes of 10 tropical woods during weathering. According to the results, colour changes might be caused by leaching of extractives, oxidation and degradation of lignin; however, the degree of change was different for every wood species. In the study of Gupta et al. [14], FTIR was used for prediction of wooden cladding service life. There are efforts to minimize the use of harmful chemicals [38]; therefore, it is important to understand the degradation processes in untreated wood during weathering. The service life determination of wooden components is a crucial research constraint that needs a thorough investigation of material properties before the prediction is done [14,39].

The objective of this paper is to characterize hardwood' overall performance under specific exterior conditions of Central Europe (surface, visual and chemical parameters) using different scientific methods, considering the increasing application of untreated wood outdoors. The results can lead to deepening knowledge in this area, to comparison of naturally durable and less durable species and to proper and more frequent use of untreated hardwoods in applications directly exposed to weathering. During weathering, we assume that gradual lignin decomposition and changes in the hemicellulose complex are manifested by pronounced colour changes and increased surface roughness of the wood surface. These findings are supported by microscopic analysis. This study follows and expands the findings of the previously published study of Oberhofnerová et al. [15], which presents the results of colour and roughness changes after one year of natural weathering.

## 2. Materials and Methods

### 2.1. Material Preparation

The experiment was conducted using three samples for each weathering period, all of them with dimensions 375 × 78 × 20 mm (Longitudinal × Tangential × Radial), from five hardwood species (black locust, oak, maple, alder and poplar wood). A total of 75 samples were used for the experiment.

Only the heartwood was tested. Before exposure, the test samples were sanded with sandpaper grit of 120 (P120). The specification of used wood species is given in Table 1.

**Table 1.** Specification of used hardwoods. Natural durability against fungi is given in the range of 1–5, where 1 signifies the best durability. Wood density was determined at 12% wood moisture according to ČSN 49 0108 [40].

| Hardwood | Natural Durability Against Fungi (EN 350) | Average Density (kg/m$^3$) | Initial Appearance |
|---|---|---|---|
| Oak (*Quercus robur* L.) | 2 | 710 | |
| Black locust (*Robinia pseudoacacia* L.) | 1–2 | 827 | |
| Poplar (*Populus* sp.) | 5 | 413 | |
| Alder (*Alnus glutinosa* (L.) Gaertn) | 5 | 534 | |
| Maple (*Acer pseudoplatanus* L.) | 5 | 599 | |

## 2.2. Natural Weathering

The natural weathering test was performed at Suchdol, Prague, Czech Republic (50°07′49.68″ N, 14°22′13.87″ E, 285 m altitude) and lasted 24 months (from 15 December 2014 to 20 December 2016 for four periods—3, 6, 12 and 24 months). The overview of weather conditions is given in Table 2. The samples were exposed outdoors, at 45° inclination angle, facing south and placed approximately 1 m above the ground according to [41]. The specimens were stabilized at $20 \pm 2\ °C$ and 65% relative humidity (RH) before the measurements for 7 days. The colour, roughness, FTIR, and confocal microscopy were measured first on the surface before exposure (measurements were repeated after each weathering period).

**Table 2.** Weather conditions during the natural weathering test. Values are based on data from a weather station [42].

| Period (Months) | Average Temperature (°C) | | Relative Humidity (%) | | Total Precipitation (mm) | | Average Total Solar Radiation (kJ/m$^2$) | |
|---|---|---|---|---|---|---|---|---|
| | 2015 | 2016 | 2015 | 2016 | 2015 | 2016 | 2015 | 2016 |
| 0–2 | 2.0 | 0.58 | 77.3 | 82.58 | 23.3 | 66.2 | 10,678 | 11,977 |
| 2–4 | 5.2 | 4.53 | 67.1 | 74.98 | 58.8 | 44.2 | 18,722 | 18,067 |
| 4–6 | 14.1 | 18.02 | 61.5 | 66.15 | 71 | 178.8 | 20,488 | 20,164 |
| 6–8 | 21.2 | 19.09 | 55.2 | 64.39 | 91.3 | 134 | 17,432 | 16,496 |
| 8–10 | 14.4 | 15.86 | 67.8 | 67.63 | 72.3 | 81.1 | 8496 | 5082 |
| 10–12 | 6.7 | 1.05 | 80.6 | 75.16 | 53.7 | 48 | 2549 | 2271 |

## 2.3. Colour Measurement

The colour of the wood surface was measured using a spectrophotometer CM-600d (Konica Minolta, Osaka, Japan). A standard illuminant D65, viewing angle of 10° and SCI (Specular Component Included), was used. The colour values were measured at 8 given positions of each sample (Figure 1) and expressed in the Lab system [43]. This colorimetric system comprises three independent axes perpendicular to each other. The $L^*$ axis determines the lightness (0-black, 100-white), the $a^*$ axis determines the ratio between red (positive) and green (negative) and finally the $b^*$ axis represents the ratio between yellow (positive) and blue (negative). Total colour change $\Delta E^*$ was calculated using the Equation (1) [44]:

$$\Delta E^* = \sqrt{\Delta L^{*2} + \Delta a^{*2} + \Delta b^{*2}} \tag{1}$$

where $\Delta L^*$, $\Delta a^*$, $\Delta b^*$ are differences in individual axes (difference between the value measured after and before exposure).

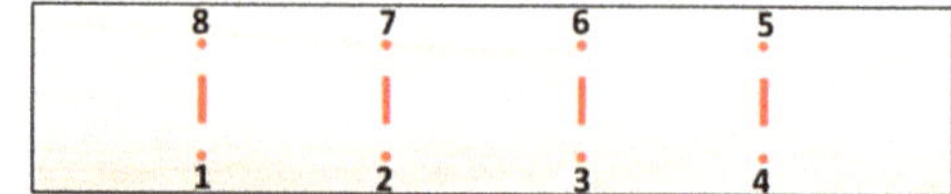

**Figure 1.** Points on the samples for measuring colour (dots) and roughness (lines).

*2.4. Roughness Measurement*

Surface roughness was assessed with a stylus profilometer Talysurf Form Intra (Taylor-Hobson, Leicester, UK). The measurements were made according to standards [45,46] on the surface of each sample at the same positions, Figure 1. The $R_a$ (arithmetical mean profile deviation) was measured in the perpendicular direction to the fibres. These measurements were considered representative surface-roughness indicators. The tracing length was 15 mm and cut-off lengths were 2.5 mm.

*2.5. ATR-FTIR Analysis*

FTIR spectra of weathered wood surfaces were recorded on a Nicolet iS10 FT-IR spectrometer (Thermo Fisher Scientific, Waltham, MA, USA), equipped with Smart iTR using an attenuated total reflectance (ATR) sampling accessory attached to a diamond crystal. The spectra were acquired by accumulating 64 scans at a spectral resolution of 4 cm$^{-1}$ in an absorbance mode from 4000 to 400 cm$^{-1}$ standardised using the baseline method and analysed using OMNIC 8.0 software (Thermo Fisher Scientific, Waltham, MA, USA). Measurements were performed on four replicates per sample. The spectral range 1800–800 cm$^{-1}$ was evaluated. Each specimen was measured on the tangential face.

*2.6. Confocal Laser Scanning Microscopy*

To evaluate the surface degradation of hardwoods, microscopic structural changes of wood surfaces were studied using a confocal laser scanning microscope Lext Ols 4100 (Olympus, Tokyo, Japan) before and after each exposure of samples to natural weathering.

*2.7. Statistical Evaluation*

The statistical evaluation was done in Statistica 12 software (Statsoft, Palo Alto, CA, USA) and MS Excel 2013 (Microsoft, Redmond, WA, USA) using mean values, standard deviations, analysis of variance (ANOVA), and Tukey´s N HSD multiple comparison test at $\alpha = 0.05$ significance level.

## 3. Results

The effect of wood species and period of exposure and their interactions on the measured properties (colour parameters and roughness) after weathering were evaluated as statistically significant ($p < 0.05$)—see Table 3. For further evaluation, Tukey's HSD test was used to compare specific differences between wood species.

**Table 3.** Results of multifactorial ANOVA test (* denotes $p < 0.05$).

| Factors | Response Variables | | | | |
|---|---|---|---|---|---|
| | $L^*$ | $a^*$ | $b^*$ | $\Delta E^*$ | $R_a{}^*$ |
| Wood Species (WS) | * | * | * | * | * |
| Period of Exposure (PE) | * | * | * | * | * |
| WS × PE | * | * | * | * | * |

*3.1. Colour Changes*

The colour parameters ($L^*$, $a^*$, $b^*$) expressed as colour differences ($\Delta L^*$, $\Delta a^*$, $\Delta b^*$) measured on the surface of investigated hardwoods (Tables 4–6) and corresponding colour differences ($\Delta E^*$) are presented (Table 7). Total colour difference $\Delta E^*$ increased over the duration of external exposure for all the hardwoods [25,28,47]. The highest colour difference after two years of weathering was observed for maple ($\Delta E^* = 34.2$) and the lowest for oak wood ($\Delta E^* = 23.8$).

**Table 4.** Exposure time-dependence of $L^*$ value (lightness differences $\Delta L^*$ are in brackets). The data represent mean values.

| Exposure Time | Oak | | Black Locust | | Poplar | | Alder | | Maple | |
|---|---|---|---|---|---|---|---|---|---|---|
| 0 m | 66.02 | (0) | 71.47 | (0) | 82.38 | (0) | 72.03 | (0) | 79.23 | (0) |
| 3 m | 68.06 | (2.04) | 61.77 | (−9.70) | 75.73 | (−6.65) | 66.54 | (−5.49) | 77.44 | (−1.79) |
| 6 m | 70.15 | (4.13) | 64.96 | (−6.51) | 65.68 | (−16.70) | 58.70 | (−13.33) | 74.29 | (−4.94) |
| 12 m | 50.57 | (−15.45) | 49.29 | (−22.18) | 51.68 | (−30.70) | 53.19 | (−18.84) | 54.48 | (−24.75) |
| 24 m | 49.29 | (−16.73) | 51.82 | (−19.65) | 52.35 | (−30.03) | 54.00 | (−18.03) | 47.81 | (−31.42) |

**Table 5.** Exposure time-dependence of $a^*$ value (chromaticity differences $\Delta a^*$ are in brackets). The data represent mean values.

| Exposure Time | Oak | | Black Locust | | Poplar | | Alder | | Maple | |
|---|---|---|---|---|---|---|---|---|---|---|
| 0 m | 6.34 | (0) | 2.71 | (0) | 4.16 | (0) | 11.11 | (0) | 5.20 | (0) |
| 3 m | 4.49 | (−1.85) | 9.95 | (7.24) | 5.18 | (1.02) | 7.10 | (−4.01) | 4.85 | (−0.35) |
| 6 m | 1.73 | (−4.61) | 4.76 | (2.05) | 1.49 | (−2.67) | 1.98 | (−9.13) | 1.40 | (−3.8) |
| 12 m | 0.77 | (−5.57) | 1.44 | (−1.27) | 0.76 | (−3.4) | 1.24 | (−9.87) | 0.51 | (−4.69) |
| 24 m | 1.15 | (−5.19) | 1.45 | (−1.26) | 1.39 | (−2.77) | 1.50 | (−9.61) | 1.78 | (−3.42) |

**Table 6.** Exposure time-dependence of $b^*$ value (chromaticity differences $\Delta b^*$ are in brackets). The data represent mean values.

| Exposure Time | Oak | | Black Locust | | Poplar | | Alder | | Maple | |
|---|---|---|---|---|---|---|---|---|---|---|
| 0 m | 20.49 | (0) | 26.77 | (0) | 16.69 | (0) | 21.78 | (0) | 16.98 | (0) |
| 3 m | 16.64 | (−3.85) | 20.79 | (−5.98) | 20.24 | (3.55) | 21.36 | (−0.42) | 19.06 | (2.08) |
| 6 m | 9.18 | (−11.31) | 12.16 | (−14.61) | 8.50 | (−8.19) | 8.52 | (−13.26) | 7.19 | (−9.79) |
| 12 m | 4.87 | (−15.62) | 6.64 | (−20.13) | 3.84 | (−12.85) | 4.84 | (−16.94) | 3.12 | (−13.86) |
| 24 m | 4.43 | (−16.06) | 4.41 | (−22.36) | 4.43 | (−12.26) | 4.37 | (−17.41) | 3.96 | (−13.02) |

**Table 7.** Exposure time-dependence of $\Delta E^*$ value. The data represent mean values.

| Exposure Time | Oak | Black Locust | Poplar | Alder | Maple |
|---|---|---|---|---|---|
| 0 m | 0 | 0 | 0 | 0 | 0 |
| 3 m | 4.74 | 13.50 | 7.61 | 6.90 | 2.77 |
| 6 m | 12.90 | 16.12 | 18.79 | 20.91 | 11.61 |
| 12 m | 22.67 | 29.97 | 33.46 | 27.19 | 28.76 |
| 24 m | 23.77 | 29.79 | 32.56 | 26.84 | 34.19 |

*3.2. Roughness Changes*

The average roughness values ($R_a$) of unweathered (control) and weathered samples expressed as roughness differences after and before each exposure period ($\Delta R_a$) are showed in Table 8.

**Table 8.** Exposure time-dependence of arithmetical mean profile deviation $R_a$ (roughness differences $\Delta R_a$ are in parentheses).

| Exposure Time | Oak | | Black Locust | | Poplar | | Alder | | Maple | |
|---|---|---|---|---|---|---|---|---|---|---|
| 0 m | 7.08 | (0) | 4.48 | (0) | 6.27 | (0) | 5.25 | (0) | 4.10 | (0) |
| 3 m | 8.75 | (1.67) | 6.57 | (2.09) | 9.54 | (3.27) | 8.62 | (3.37) | 6.11 | (2.01) |
| 6 m | 15.76 | (8.68) | 13.32 | (8.85) | 10.80 | (4.53) | 8.76 | (3.51) | 6.17 | (2.07) |
| 12 m | 14.77 | (7.69) | 8.36 | (3.88) | 14.46 | (8.19) | 7.70 | (2.44) | 8.07 | (3.97) |
| 24 m | 21.93 | (14.85) | 19.67 | (15.19) | 15.68 | (9.41) | 27.43 | (22.18) | 23.62 | (19.52) |

The following figures (Figures 2–6) show the 3D microstructure of the surface of all examined wood samples after 2 years of weathering obtained by confocal laser scanning microscopy. The degree of surface erosion, determining the wood roughness change, is represented by a colour range ranging from purple to red.

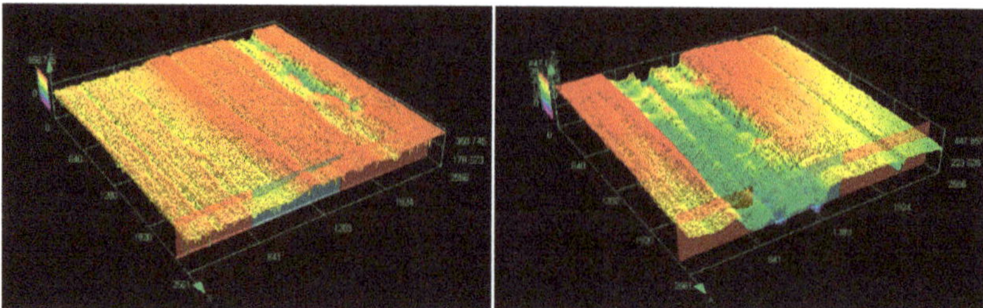

**Figure 2.** 3D image of oak wood surface before (**left**) and after 2 years of weathering (**right**)—confocal laser scanning microscopy (area of 800 × 800 μm).

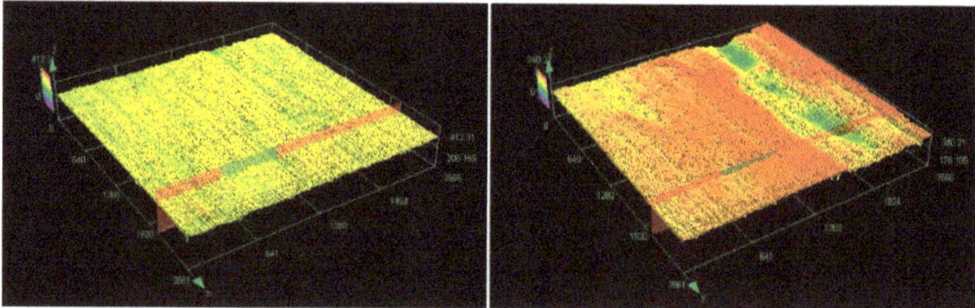

**Figure 3.** 3D image of black locust wood surface before (**left**) and after 2 years of weathering (**right**)—confocal laser scanning microscopy (area of 800 × 800 μm).

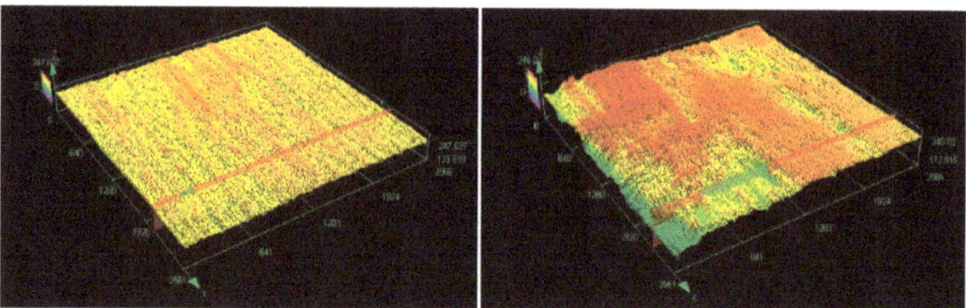

**Figure 4.** 3D image of poplar wood surface before (**left**) and after 2 years of weathering (**right**)—confocal laser scanning microscopy (area of 800 × 800 μm).

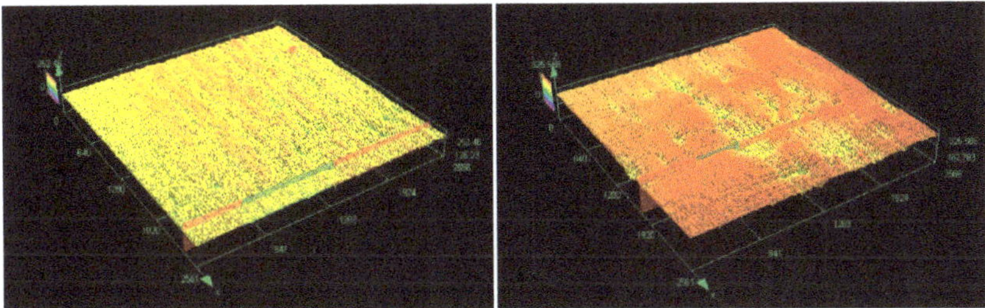

**Figure 5.** 3D image of alder wood surface before (**left**) and after 2 years of weathering (**right**)—confocal laser scanning microscopy (area of 800 × 800 μm).

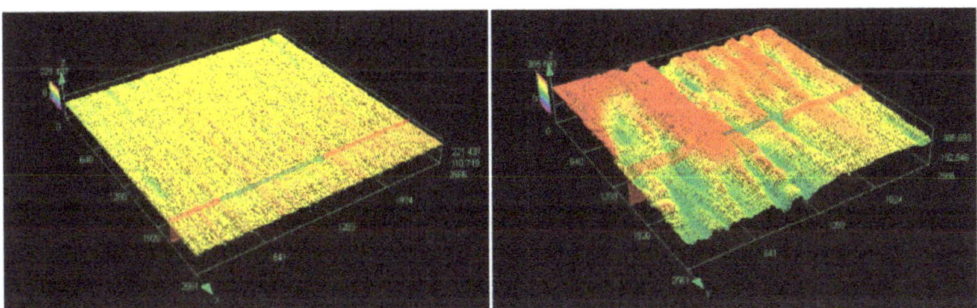

**Figure 6.** 3D image of maple wood surface before (**left**) and after 2 years of weathering (**right**)—confocal laser scanning microscopy (area of 800 × 800 μm).

### 3.3. Visual Evaluation

Visual evaluation of samples was performed using confocal laser scanning microscopy of the wood sample surface before and during exposure (Figure 7). The detailed analysis of surface changes during weathering of wood species was performed by confocal laser microscope (Figures 8–12).

**Figure 7.** Confocal microscopy of samples during exposure to natural weathering (each scan corresponds to an area of 2.5 × 2.5 mm).

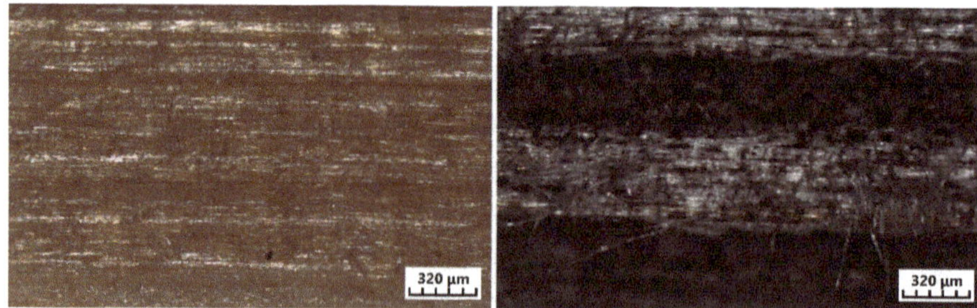

**Figure 8.** Confocal laser scanning microscopy of oak wood before and after 2 years of weathering (detail).

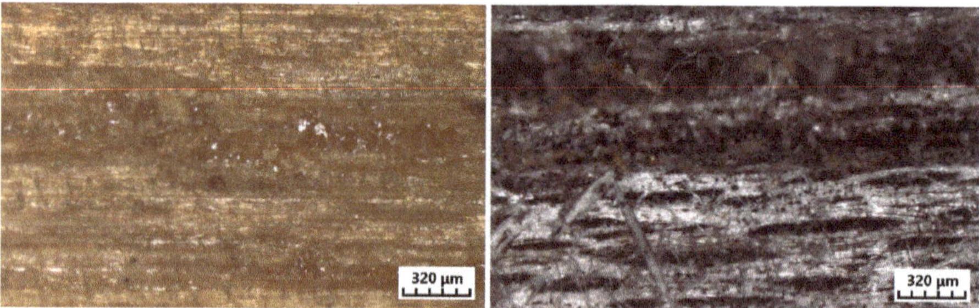

**Figure 9.** Confocal laser scanning microscopy of black locust wood before and after 2 years of weathering (detail).

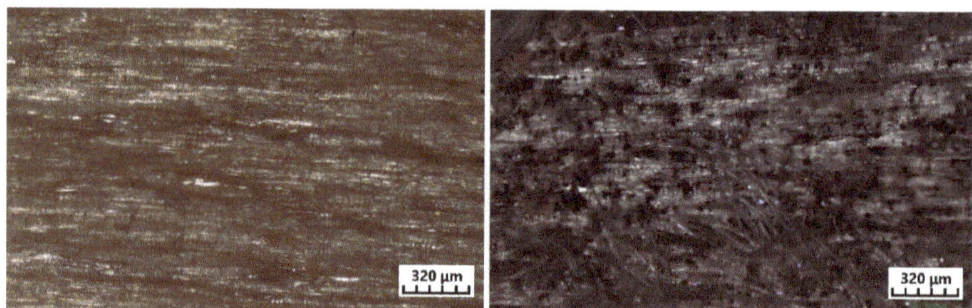

**Figure 10.** Confocal laser scanning microscopy of poplar wood before and after 2 years of weathering (detail).

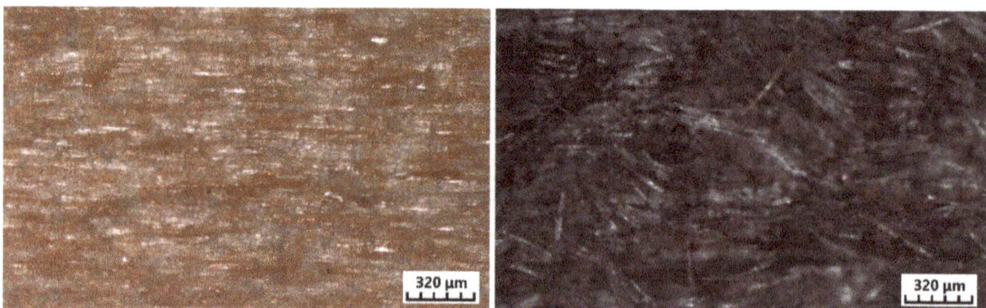

**Figure 11.** Confocal laser scanning microscopy of alder wood before and after 2 years of weathering (detail).

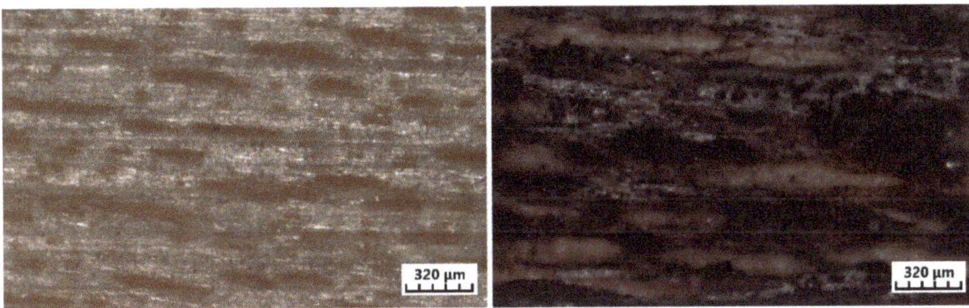

**Figure 12.** Confocal laser scanning microscopy of maple wood before and after 2 years of weathering (detail).

### 3.4. Changes in the FTIR Spectra

Chemical changes of wood surface components were evaluated by ATR-FTIR analysis. They are expressed as the differential FTIR spectra (difference in absorbance after and before of each exposure period, denoted as AD3 (Absorbances Difference after 3 month of weathering), AD6—after 6 months and so on, Figures 13–17). The positive values refer to the increase, negative values to the decrease in the absorbance. The absorbance of bands at 1730 cm$^{-1}$ (C=O vibration of non-conjugated carbonyl groups in hemicelluloses), 1600 cm$^{-1}$ and 1505 cm$^{-1}$ (C=C skeletal vibrations in aromatic structure of lignin), 1460 cm$^{-1}$ (C$-$H bending in CH$_3$ group in lignin), 1235 cm$^{-1}$ (C$-$O stretching in lignin and xylan) and 1267 cm$^{-1}$ (C$-$O stretching in lignin structure) decreased during natural weathering. These results indicate that the structure of the lignin polymer and hemicelluloses was degraded to a significant extent.

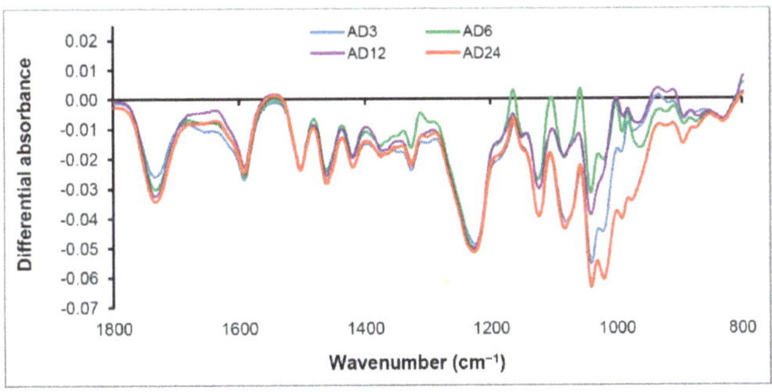

**Figure 13.** Differential Fourier Transform Infrared Spectroscopy (FTIR) spectra of oak wood samples.

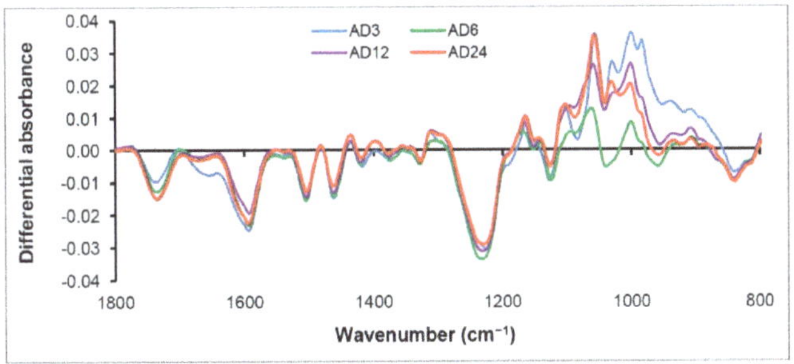

**Figure 14.** Differential FTIR spectra of black locust samples.

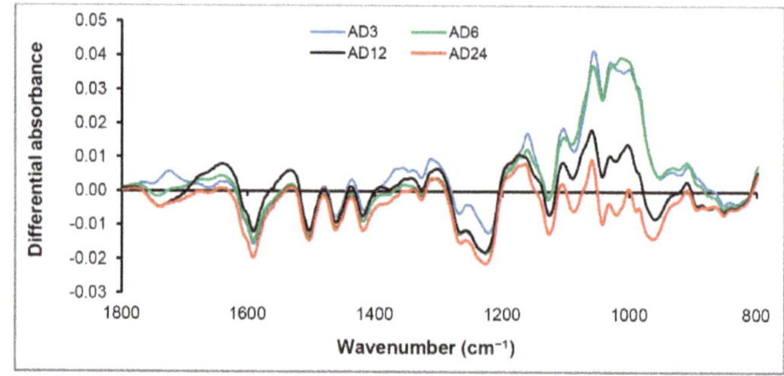

**Figure 15.** Differential FTIR spectra of poplar wood samples.

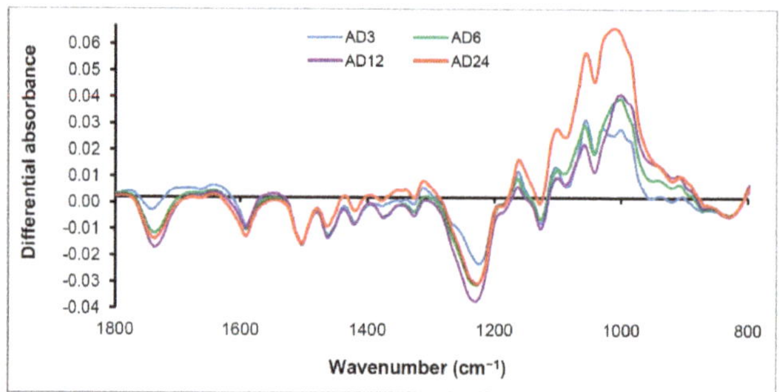

**Figure 16.** Differential FTIR spectra of alder wood samples.

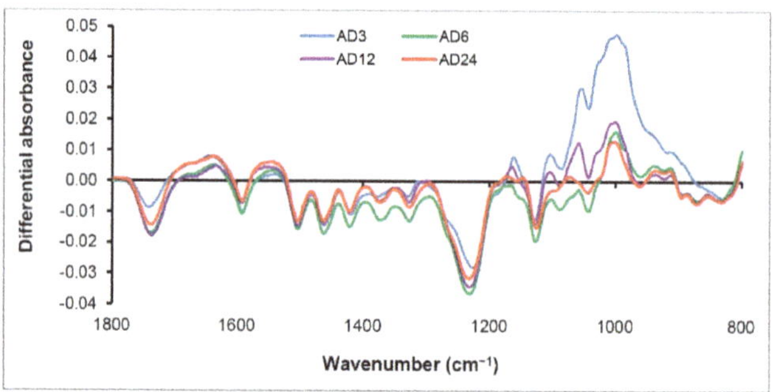

**Figure 17.** Differential FTIR spectra of maple wood samples.

## 4. Discussion

During the weathering process, colour parameter changes showed practically identical trends for all analysed wood samples. A decreased $L*$ (Table 4, negative values of $\Delta L*$) value was observed, similar to Turkoglu et al. [25]. Darkening of the wood surface was more pronounced after twelve months of natural weathering. The value $a*$ gradually decreased (negative trend in $\Delta a*$), mainly after 12 months. The lowest values were found for oak and alder (Table 5). The decrease in the chromaticity coordinate $b*$ (negative trend in $\Delta b*$) was similar (Table 6). A sharp drop in the $b*$ value (increase in $b*$ difference) was observed in the first half of the 24-month cycle, with the remainder of the exposure declining only slightly. The reduction is caused by leaching of decomposed lignin and extractives from the surface [25,48]. At the end of the two-year exposure, values $a*$ and $b*$ stabilized at a level which correspond to the grey colour. Undoubtedly, untreated wood exhibits higher colour changes than treated wood during natural weathering [1,25,27]. The highest changes in colour parameters were observed in initial phases of weathering as in the studies of Lesar et al. [1] and Sharrat et al. [49]. The more pronounced changes were noted during the first 12 months; then, the other changes were negligible due to the formation of the grey degraded layer which acts as a protective barrier and slows the weathering process [50]. The greying caused by the action of weather and dust settling into the porous and degraded surface can be visibly observed after 6 months, especially for less durable wood species such as maple and alder (Figure 7). After 12 months of weathering, the greying was obvious for all the tested hardwoods. The tearing of wood fibers was observed after 24 months of weathering (Figures 8–12). The presence of slight fungal growth was observed after a few months of weathering on poplar, alder and maple (Figures 10–12) as in the study of Mohebby and Saei [27].

As shown by the measured values, the surface roughness value of native wood samples increased with weathering time (Table 8 and Figures 2–6), as presented in other studies [27,28]. However, the surface roughness of some wood samples (black locust, oak and alder) partly decreased with weathering between the 6th and 12th month period. Similar trends have also been observed by Turkoglu et al. [51]. Increasing roughness is caused by lignin decomposition [26]. Another effect can be the absorption of the thermal component of solar radiation. Turkulin et al. [52] reported that this radiation mostly degraded the middle lamella, which is between two cell walls and holds the cells together. This degradation increases the roughness of the wood surface [31]. Kerber et al. [53] also reported that in addition to the leaching of lignin degraded by natural weathering reactions, the increase in the roughness of the wood is also related to sudden changes in humidity (absorption and desorption of the humidity) causing the presence of superficial cracks. Also, rain water helps to remove loosened fibres and particles produced during irradiation and to move leached lignin fragments to the top layer that is mainly composed of cellulose (which causes the grey colour of the wood and a rough texture [12]). Elevated temperature [54], high moisture content [52] and diluted acid [55] can increase the photodegradation rate. Additional factors contributing to weathering are superficial wetting and drying, generating surface stresses that can cause checking [7,56]. The changes in surface structure after 2 years of weathering were apparent in laser scanning microscopy as well (Figures 2–6), where tearing of the wood fibres (Figures 8–12) and surface erosion, more obvious in less dense earlywood than in latewood [30], caused by abiotic factors, were observed (Figures 8–12).

The dominant factor responsible for wood degradation is lignin, which absorbs ultraviolet and visible radiation due to its chromophoric groups. However, the effect of visible light to surface degradation at later stages of weathering was observed [57–59]. According to Norrstrom [60], lignin is responsible for absorbing 80% to 95% of the total UV light absorbed by wood, carbohydrates 5 to 20% and extractives about 2%. FTIR spectra of weathered wood (Figures 13–17) showed that the absorbance at 1730 cm$^{-1}$ (C=O vibration of the non-conjugated carbonyl groups, stretching in xylan) decreased at higher exposure times. This is in accordance with findings of natural and artificial weathering of seven tropical woods [37]. The decreasing trend of this band was probably caused by the condensation reactions of lignin as well as by hemicellulose deacetylation [61]. Hemicelluloses are the least resistant to thermal treatment, and carboxylic acids (predominantly acetic acid) are formed by their decomposition. Carboxylic acids cause the depolymerization of cellulose and cleavage of bonds in lignin [62,63].

The band occurring at 1600 cm$^{-1}$ is characteristic of aromatic compounds and is attributed to aromatic skeleton vibrations. In our case, a sustained decrease in absorbance was recorded. The absorbance at 1505 cm$^{-1}$ (and C=C skeletal stretching vibrations in aromatic rings in lignin) showed a slight decrease. The samples did not show remarkable changes in this regard. A similar effect in the intensity was also observed after UV treatment of wood [8,64]. Decreased absorption at that band is interpreted as lignin decay combined with the formation of new carbonyl groups, evidencing photo-induced oxidation of the wood surface. During the weathering test, the decrease of the band at 1462 cm$^{-1}$ (C–H asymmetric bending in CH$_3$ group in lignin) revealed the loss of lignin [65]. The peaks at 1235 cm$^{-1}$ (C–O stretching in lignin and xylan) and 1267 cm$^{-1}$ (C–O stretching and breathing of guaiacyl ring) are assigned to lignin [66,67]. Described phenomena may indicate changes in lignin structures leading to a decrease of methoxy groups in lignin.

The degradation mechanism is complex with different paths leading to water soluble products and finally to chromophoric groups like carboxylic acids, quinines or hydroperoxides [68]. These water-soluble compounds can be extracted from wood by rain during weathering processes, resulting in the decrease of carbonyl groups and the increase in surface roughness.

## 5. Conclusions

The study revealed the overall degradation process of hardwoods caused by abiotic factors in the exterior using different experimental methods. Five hardwoods from the temperate climate zone with different natural durability (oak, black locust, poplar, alder and maple wood) were exposed to natural weathering for 24 months. Significant colour changes were noticeable after 3 months of weathering. During the first year of weathering, all the hardwoods were characterized by high colour changes. The decrease in lightness $L^*$ indicated gradual darkening of the samples, decreasing $a^*$ and $b^*$ values showed a more pronounced shift to the grey colour. During the second year of weathering, the colour changed only slightly. The highest colour difference after 2 years of weathering was recorded for maple wood, the lowest for oak. The roughness change $R_a$ had a similarly increasing trend (with the exception of poplar), but as opposed to the colour changes, the largest increase was observed during the second year of weathering, specifically for alder. ATR-FTIR analysis confirmed the assumption of degradation, in particular, in the structure of lignin and hemicelluloses of hardwoods. Decreases of both methoxy and carbonyl groups, cleavage of bonds in lignin and hemicelluloses, oxidation reaction and formation of new chromophores were observed during weathering. Based on the presented results, the use of more durable wood species (oak, black locust) in the exterior can be recommended (the lowest discolouration, relatively low roughness changes, no formation of moulds or fungi). Alder was also characterized by relatively good values of colour changes, but mould and degraded wood fibers were formed on its surface.

The significant impact of weathering factors on the quality and wood colour and roughness was demonstrated. The obtained findings regarding surface parameters (colour, roughness) were supported by visual performance, microscopic and chemical analysis. Presented results give some useful information about the surface degradation and related chemical changes of different hardwoods during exposure to weathering. The obtained results can increase the possible use of untreated hardwoods in applications directly exposed to weathering.

**Author Contributions:** E.O. and M.P. conceived and designed the experiments; E.O. and M.P. measured colour and roughness; I.K. analyzed the colour data; I.K. and F.K. measured and analyzed the FTIR data; I.K., E.O., F.K. and M.P. wrote the paper.

**Funding:** This research was funded by the Slovak Research and Development Agency under the contract No. APVV-16-0326 (30%), by the VEGA agency of the Ministry of Education, Science, Research and Sport of the Slovak Republic No. 1/0387/18 (30%) and by the Internal Grant Agency of the Faculty of Forestry and Wood Sciences, project IGA No. A07/18 (40%).

**Conflicts of Interest:** The authors declare no conflict of interest. The funders had no role in the design of the study; in the collection, analyses, or interpretation of data; in the writing of the manuscript, and in the decision to publish the results.

## References

1. Lesar, B.; Humar, M.; Kržišnik, D.; Thaler, N.; Žlahtič, M. Performance of façade elements made of five different thermally modified wood species on model house in Ljubljana. In Proceedings of the World Conference on Timber Engineering, Vienna, Austria, 22–25 August 2016; ISBN 978-390303900-1. Available online: http://wcte2016.conf.tuwien.ac.at/ (accessed on 3 September 2016).

2. EN 350: 2016. *Durability of Wood and Wood-Based Products. Testing and Classification of the Durability to Biological Agents of Wood and Wood-Based Materials*; European Committee for Standardization: Brussels, Belgium, 2016.

3. Ganne-Chédeville, C.; Volkmer, T.; Letsch, B.; Lehmann, M. Measures for the maintenance of untreated wood facades. In Proceedings of the 12th World Conference on Timber Engineering, Auckland, New Zealand, 15–19 July 2012; Quenneville, P., Ed.; Poster presentation. pp. 504–509.

4. Feist, W.C.; Hon, D.N.S. Chemistry of weathering and protection. In *The Chemistry of Solid Wood*; Rowell, R.M., Ed.; American Chemical Society: Washington, DC, USA, 1984; Chapter 11; pp. 401–451. ISBN 0-8412-0796-8.

5. Sandak, J.; Sandak, A.; Riggio, M. Characterization and monitoring of surface weathering on exposed timber structures with a multi-sensor approach. *Int. J. Archit. Herit.* **2015**, *9*, 674–688. [CrossRef]

6. Kržišnik, D.; Lesar, B.; Thaler, N.; Humar, M. Influence of Natural and Artificial Weathering on the Colour Change of Different Wood and Wood-Based Materials. *Forests* **2018**, *9*, 488. [CrossRef]

7. Evans, P.D. Weathering and photoprotection of wood. In *Development of Commercial Wood Preservatives: Efficacy, Environmental, and Health Issues*; Schultz, T.P., Militz, H., Freeman, M.H., Goodell, B., Nicholas, D.D., Eds.; ACS Symposium Series 982; American Chemical Society: Washington, DC, USA, 2008; Chapter 5; pp. 69–117. ISBN 978-0-8412-3951-7.

8. Müller, U.; Rätzsch, M.; Schwanninger, M.; Steiner, M.; Zöbl, H. Yellowing and IR- changes of spruce wood as result of UV-irradiation. *J. Photochem. Photobiol. B* **2003**, *69*, 97–105. [CrossRef]

9. Kubovský, I.; Kačík, F.; Reinprecht, L. The impact of UV radiation on the change of colour and composition of the surface of lime wood treated with a $CO_2$ laser. *J. Photochem. Photobiol. A* **2016**, *322*, 60–66. [CrossRef]

10. Pandey, K.K. A note on the influence of extractives on the photo-discoloration and photo-degradation of wood. *Polym. Degrad. Stab.* **2005**, *87*, 375–379. [CrossRef]

11. Reinprecht, L. *Wood Deterioration, Protection and Maintenance*; John Wiley & Sons, Ltd.: Chichester, UK, 2016; p. 376. ISBN 978-1-119-10653-1.

12. Hon, D.N.S.; Minemura, N. Color and discoloration. In *Wood and Cellulosic Chemistry*; Hon, D.N.S., Shirashi, N., Eds.; Marcel Dekker: New York, NY, USA, 2001; pp. 385–442.

13. Brocco, V.F.; Paes, J.B.; da Costa, L.G.; Brazolin, S.; Arantes, M.D.C. Potential of teak heartwood extracts as a natural wood preservative. *J. Clean. Prod.* **2017**, *142*, 2093–2099. [CrossRef]

14. Gupta, B.S.; Jelle, B.P.; Hovde, P.J.; Rüther, P. Studies of wooden cladding materials degradation by spectroscopy. *Proc. Inst. Civ. Eng. Constr. Mater.* **2011**, *164*, 329–340. [CrossRef]

15. Oberhofnerová, E.; Pánek, M.; García-Cimarras, A. The effect of natural weathering on untreated wood surface. *Maderas. Ciencia y Tecnología* **2017**, *19*, 173–184. [CrossRef]

16. Tolvaj, L.; Varga, D. Photodegradation of timber of three hardwood species caused by different light sources. *Acta Silv. Lign. Hung.* **2012**, *8*, 145–155. [CrossRef]

17. Kačík, F.; Kubovský, I. Chemical changes of beech wood due to $CO_2$ laser irradiation. *J. Photochem. Photobiol. A* **2011**, *222*, 105–110. [CrossRef]

18. Kúdela, J.; Kubovský, I. Accelerated-ageing-induced photo-degradation of beech wood surface treated with selected coating materials. *Acta Facultatis Xylologiae Zvolen* **2016**, *58*, 27–36. [CrossRef]

19. Hon, D.N.S.; Chang, S.T. Surface degradation of wood by ultraviolet light. *J. Polym. Sci. Polym. Chem. Ed.* **1984**, *22*, 2227–2241. [CrossRef]

20. Kataoka, Y.; Kiguchi, M. Depth profiling of photo-induced degradation in wood by FT-IR microspectroscopy. *J. Wood Sci.* **2001**, *47*, 325–327. [CrossRef]

21. Tolvaj, L.; Papp, G. Outdoor weathering of impregnated and steamed black locust. In Proceedings of the ICWSF'99 Conference, Missenden Abbey, UK, 14–16 July 1999; pp. 112–115.

22. Williams, R.S.; Feist, W.C. *Water Repellents and Water-Repellent Preservatives for Wood*; General Technical Report FPL-GTR-109; USDA, Forest Products Laboratory: Madison, WI, USA, 1999; pp. 1–12.

23. Feist, W.C. Outdoor wood weathering and protection. In *Archaeological Wood: Properties, Chemistry, and Preservation*; Advances in Chemistry Series 225; Rowell, R.M., Barbour, R.J., Eds.; American Chemical Society: Washington, DC, USA, 1989; pp. 263–298.

24. Ozgenc, O.; Hiziroglu, S.; Yildiz, U.C. Weathering properties of wood species treated with different coating applications. *BioResources* **2012**, *7*, 4875–4888. [CrossRef]

25. Turkoglu, T.; Baysal, E.; Toker, H. The effects of natural weathering on color stability of impregnated and varnished wood materials. *Adv. Mater. Sci. Eng.* **2015**, 1–9. [CrossRef]

26. Feist, W.C. Weathering of wood in structural uses. In *Structural Uses of Wood in Adverse Environments*; Van Nostrand Reinhold Company: New York, NY, USA, 1982; pp. 156–178.

27. Mohebby, B.; Saei, A.M. Effects of geographical directions and climatological parameters on natural weathering of fir wood. *Constr. Build. Mater.* **2015**, *94*, 684–690. [CrossRef]
28. Nzokou, P.; Kamdem, D.P.; Temiz, A. Effect of accelerated weathering on discoloration and roughness of finished ash wood surfaces in comparison with red oak and hard maple. *Progr. Org. Coat.* **2011**, *71*, 350–354. [CrossRef]
29. Xie, Y.; Krause, A.; Militz, H.; Mai, C. Weathering of uncoated and coated wood treated with methylated 1,3-dimethylol-4,5-dihydroxyethyleneurea (mDMDHEU). *Holz Roh. Werkst.* **2008**, *66*, 455–464. [CrossRef]
30. Williams, R.S.; Knaebe, M.T.; Feist, W.C. Erosion rates of wood during natural weathering. Part II. Earlywood and latewood erosion rates. *Wood Fiber Sci.* **2001**, *33*, 43–49.
31. Tolvaj, L.; Molnar, Z.; Magoss, E. Measurement of photodegradation-caused roughness of wood using a new optical method. *J. Photochem. Photobiol. B* **2014**, *134*, 23–26. [CrossRef] [PubMed]
32. Feist, W.C. Natural weathering of wood and its control by water-repellent preservatives. *Am. Paint. Contr.* **1992**, *69*, 18–25.
33. Oberhofnerová, E.; Pánek, M. Surface wetting of selected wood species by water during initial stages of weathering. *Wood Res.* **2016**, *61*, 545–552.
34. Žlahtič, M.; Humar, M. Influence of artificial and natural weathering on the hydrophobicity and surface properties of wood. *BioResources* **2016**, *11*, 4964–4989. [CrossRef]
35. Pandey, K.K. Study of the effect of photo-irradiation on the surface chemistry of wood. *Polym. Degrad. Stab.* **2005**, *90*, 9–20. [CrossRef]
36. Liu, R.; Pang, X.; Yang, Z. Measurement of three wood materials against weathering during long natural sunlight exposure. *Measurement* **2017**, *102*, 179–185. [CrossRef]
37. Reinprecht, L.; Mamoňová, M.; Pánek, M.; Kačík, F. The impact of natural and artificial weathering on the visual, colour and structural changes of seven tropical woods. *Eur. J. Wood Prod.* **2018**, *76*, 175–190. [CrossRef]
38. Singh, T.; Singh, A.P. A review on natural products as wood protectant. *Wood Sci. Technol.* **2012**, *46*, 851–870. [CrossRef]
39. Brischke, C.; Rapp, A.O. Dose–response relationships between wood moisture content, wood temperature and fungal decay determined for 23 European field test sites. *Wood Sci. Technol.* **2008**, *42*, 507–518. [CrossRef]
40. ČSN 49 0108. *Drevo. Zist'ovanie Hustoty [Wood. Determination of the Density]*; Český Normalizační Institut: Prague, Czech Republic, 1993.
41. EN 927-3: 2006. *Paints and Varnishes—Coating Materials and Coating Systems for Exterior Wood—Part 3: Natural Weathering Test*; CEN: Brussels, Belgium, 2006.
42. Meteostanice: 2017. Meteorological Station of the Czech University of Life Sciences in Prague, Faculty of Agrobiology, Food and Natural Resources. Available online: http://meteostanice.agrobiologie.cz (accessed on 4 December 2017).
43. ISO 11664-4. *Colorimetry—Part 4: CIE 1976 L*a*b* Colour Space*; International Organization for Standardization: Geneva, Switzerland, 2008.
44. ISO 11664-6. *Colorimetry—Part 6: CIEDE2000 Colour-Difference Formula*; International Organization for Standardization: Geneva, Switzerland, 2013.
45. EN ISO 4287. *Geometrical Product Specifications (GPS)—Surface Texture: Profile Method-Terms, Definitions, and Surface Texture Parameters*; International Organization for Standardization: Geneva, Switzerland, 1997.
46. EN ISO 4288. *Geometrical Product Specifications (GPS)—Surface Texture: Profile Method—Rules and Procedures for the Assessment of Surface Texture*; International Organization for Standardization: Geneva, Switzerland, 1996.
47. Kropf, F.W.; Sell, J.; Feist, W.C. Comparative weathering tests of North American and European exterior wood finishes. *For. Prod. J.* **1994**, *44*, 33–41.
48. Pastore, T.C.; Santos, K.O.; Rubim, J.C. A spectrocolorimetric study on the effect of ultraviolet irradiation of four tropical hardwoods. *Bioresour. Technol.* **2004**, *93*, 37–42. [CrossRef] [PubMed]
49. Sharratt, V.; Hill, C.A.S.; Kint, D.P.R. A study of early colour change due to simulated accelerated sunlight exposure in Scots pine (*Pinus sylvestris*). *Polym. Degrad. Stabil.* **2009**, *94*, 1589–1594. [CrossRef]
50. Browne, F.L.; Simonson, H.C. The penetration of light into wood. *For. Prod. J.* **1957**, *7*, 308–314.
51. Turkoglu, T.; Kabasakal, Y.; Baysal, E.; Gunduz, A.; Kucuktuvek, M.; Bayraktar, D.K.; Toker, H.; Peker, H. Surface characteristics of heated and varnished oriental beech after accelerated weathering. *Wood Res.* **2017**, *62*, 961–972.

52. Turkulin, H.; Derbyshire, H.; Miller, E.R. Investigations into the photodegradation of wood using microtensile testing—Part 5: The influence of moisture on photodegradation rates. *Holz als Roh-und Werkstoff* **2004**, *62*, 307–312. [CrossRef]

53. Kerber, P.R.; Stangerlin, D.M.; Pariz, E.; de Melo, R.R.; de Souza, A.P.; Calegari, L. Colorimetry and surface roughness of three amazonian woods submitted to natural weathering. *Nativa* **2016**, *4*, 303–307. [CrossRef]

54. Derbyshire, H.; Miller, E.R.; Turkulin, H. Investigations into the photodegradation of wood using microtensile testing. Part 3: The influence of temperature on photodegradation rates. *Holz als Roh-und Werkstoff* **1997**, *55*, 287–291. [CrossRef]

55. Hon, D.N.S. Degradative effects of ultraviolet light and acid rain on wood surface quality. *Wood Fiber Sci.* **1994**, *26*, 185–191.

56. Williams, R.S. Weathering of wood. In *Handbook of Wood Chemistry and Wood Composites*; Rowell, R.M., Ed.; CRC Press: Boca Raton, FL, USA, 2005; pp. 139–185.

57. Kataoka, Y.; Kiguchi, M.; Williams, R.; Evans, D. Violet light causes photodegradation of wood beyond the zone affected by ultraviolet radiation. *Holzforschung* **2007**, *61*, 23–27. [CrossRef]

58. Živković, V.; Arnold, M.; Radmanović, K.; Richter, K.; Turkulin, H. Spectral sensitivity in the photodegradation of fir wood (*Abies alba* Mill.) surfaces: Colour changes in natural weathering. *Wood Sci. Technol.* **2014**, *48*, 239–252. [CrossRef]

59. Živković, V.; Arnold, M.; Pandey, K.K.; Richter, K.; Turkulin, H. Spectral sensitivity in the photodegradation of fir wood (*Abies alba* Mill.) surfaces: Correspondence of physical and chemical changes in natural weathering. *Wood Sci. Technol.* **2016**, *50*, 989–1002. [CrossRef]

60. Norrstrom, H. Light absorbing properties of pulp and paper components. *Sven. Papperstidn.* **1969**, *72*, 25–38.

61. Li, R.; Xu, W.; Wang, X.; Wang, C. Modeling and predicting of the color changes of wood surface during $CO_2$ laser modification. *J. Clean. Prod.* **2018**, *183*, 818–823. [CrossRef]

62. Kačík, F.; Podzimek, S.; Vizárová, K.; Kačíková, D.; Čabalová, I. Characterization of cellulose degradation during accelerated ageing by SEC-MALS, SEC-DAD, and A4F-MALS methods. *Cellulose* **2016**, *23*, 357–366. [CrossRef]

63. Sun, Z.; Fridrich, B.; de Santi, A.; Elangovan, S.; Barta, K. Bright Side of Lignin Depolymerization: Toward New Platform Chemicals. *Chem. Rev.* **2018**, *118*, 614–678. [CrossRef] [PubMed]

64. Pandey, K.K.; Vuorinen, T. Comparative study of photodegradation of wood by a UV laser and a Xenon light source. *Polym. Degrad. Stabil.* **2008**, *93*, 2138–2146. [CrossRef]

65. Shen, H.; Zhang, S.; Cao, J.; Jiang, J.; Wang, W. Improving anti-weathering performance of thermally modified wood by $TiO_2$ sol or/and paraffin emulsion. *Constr. Build. Mater.* **2018**, *169*, 372–378. [CrossRef]

66. Faix, O. Fourier transform infrared spectroscopy. In *Methods in Lignin Chemistry*; Lin, S.Y., Dence, C.W., Eds.; Springer: Berlin, Germany, 1992; pp. 83–109.

67. Müller, G.; Schöpper, C.; Vos, H.; Kharazipour, A.; Polle, A. FTIR-ATR spectroscopic analysis of changes in wood properties during particle and fibreboard production of hard and softwood trees. *BioResources* **2009**, *4*, 49–71. [CrossRef]

68. Wang, X.; Ren, H. Comparative study of the photo-discoloration of moso bamboo (*Phyllostachys pubescens* Mazel) and two wood species. *Appl. Surf. Sci.* **2008**, *254*, 7029–7034. [CrossRef]

*Article*

# The Colour of Tropical Woods Influenced by Brown Rot

**Zuzana Vidholdová * and Ladislav Reinprecht**

Technical University in Zvolen, Faculty of Wood Sciences and Technology, T. G. Masaryka 24, Zvolen, SK 96001, Slovakia; reinprecht@tuzvo.sk
* Correspondence: zuzana.vidholdova@tuzvo.sk; Tel.: +421-45-520-6389

Received: 26 February 2019; Accepted: 2 April 2019; Published: 10 April 2019

**Abstract:** Interesting aesthetic properties of tropical woods, like surface texture and colour, are rarely impaired due to weathering, rotting and other degradation processes. This study analyses the colour of 21 tropical woods before and after six weeks of intentional attack by the brown-rot fungus *Coniophora puteana*. The CIEL$^*$a$^*$b$^*$ colour system was applied for measuring the lightness, redness and yellowness, and from these parameters the hue tone angle and colour saturation were calculated. Lighter tropical woods tended to appear a less red and a more yellow, and had a greater hue tone angle. However, for the original woods was not found dependence between the lightness and colour saturation. Tropical woods at attack by *C. puteana* lost a weight from 0.08% to 6.48%. The lightest and moderately light species—like okoumé, iroko, ovengol and sapelli—significantly darkened, while the darkest species—wengé and ipé—significantly lightened. The majority of tropical woods obtained a brighter shade of yellow, typically wengé, okoumé and blue gum, while some of them also a brighter shade of green, typically sapelli, padouk and macaranduba. *C. puteana* specifically affected the hue tone angle and colour saturation of tested tropical woods, but without an apparent changing the tendency of these colour parameters to lightness. The total colour difference of tested tropical woods significantly increased in connection with changes of their lightness ($\Delta E^*_{ab} = 5.92 - 0.50 \cdot \Delta L^*$; $R^2 = 0.37$), but it was not influenced by the red and yellow tint changes, and weight losses.

**Keywords:** tropical woods; brown rot; *Coniophora puteana*; colour; CIEL$^*$a$^*$b$^*$ system

---

## 1. Introduction

The surface appearance of wood is often evaluated by examining its texture, roughness, and colour [1–3]. The colour of an individual wood species is predetermined by the type and amount of extractives, by the surface roughness and moisture, and by the direction of light irradiation [4].

The CIE 1976 L$^*$a$^*$b$^*$ colour system classifies the temperate and tropical wood species into the positive octant with the lightness ($L^*$) from 20 to 90, the redness index ($+ a^*$) from 0 to 20, and the yellowness index ($+ b^*$) from 10 to 30 [1,5]. The CIE 1976 L$^*$a$^*$b$^*$ colour system also allows visualisation of the cylindrical parameters of wood, the colour saturation—chromaticity ($C^*_{ab}$) and the hue tone angle ($h_{ab}$) [6]. The tropical wood species occupy a much greater portion of the colour space in comparison with temperate (for example European) species [7,8].

Within a defined wood species, the colour variations can be influenced by more factors, mainly by its chemical and anatomical structure [9] and specific genetic parameters [10], and also by environmental conditions at growth [11,12], atmospheric effects at exposure in exteriors or interiors [13,14], and biodeterioration processes [15–17].

Wooden products having a higher moisture content—usually above 20%–30%—are no rarely damaged by biodeterioration processes in presence of wood decaying fungi, staining fungi, moulds, or bacteria.

Generally, brown-rot fungi cause firstly yellowing and gradually browning of woods in a connection with decomposition of whiter hemicelluloses and cellulose, while darker lignin is less evidently damaged [18,19]. Wood extractives, which give to different wood species a characteristic colour, are specifically resistant to individual species of brown-rot fungi [20–22]. Therefore, the accidental or deliberate exposures of wood products to brown-rot fungi can lead to typical changes in their colour and aesthetic parameters.

The rotting of damp wooden products is common not only outdoors, but also inside of buildings. In interiors, the rotting of wood is first of all caused by brown-rot fungi [23–25]. The genus *Coniophora* comprises about 20 species frequently occurring in buildings. Gabriel and Švec [26] listed the species abundance of seven indoor wood decay basidiomycetes reported in Europe, when the brown-rot fungi *Serpula lacrymans* and *Coniophora puteana (C. puteana)* were most frequent. *C. puteana* causes—already in the early period of wood attack—a disruption of linkages between hemicelluloses and lignin, decomposition of polysaccharides, while lignin is oxidative modified and partly damaged [27]. Rot of wood with *C. puteana* is not usually homogenous and it's specific parts can be more damaged [15].

In the literature, it is also noted that wood-inhabiting fungi are able add colour to wood due to a pigment residues left by fungi in wood (called as spalting) [17,28–34]. Spalting occurs in growing tree in form of zone lines formation or pigmentation. Also bleaching is marked as kinds of wood spalting [17,29,31]. Bleaching is caused by the breakdown of coloured lignin from the wood cell wall, generally by fungi classified as white-rotting, which results in a lightening of the natural wood colour and structural integrity of the wood. A lightening in colour can also be due to a build-up of white mycelium [18,20,35]. Wood with zone lines (thin and winding lines of dark melanin) are formed due to inter- or intra- fungal antagonism for example a pairing of the white-rot fungi *Trametes versicolor*/*Bjerkadera adusta* or by solitary isolates of *Xylaria polymorpha* ascomycete that causes soft rot [31]. Also, the brown-rot fungus *Fistulina hepatica* stains oaks and some other woods from light gold, yellow brown to reddish brown shades [36,37]. Pigment-type spalting fungi are a select group of soft-rotting ascomycetes with extracellular pigments production into wood, for example fungi from genera *Chlorocibolia, Ceratocystis, Ophiostoma, Scytalidium* and others [32,33]. Pigment penetration into wood depends on moisture content, the digestive capabilities of the fungus, the permeability of the wood structure, differences between heartwood and sapwood, and the type of pigment produced [29]. Mold fungi, such as *Trichoderma* spp., are not considered to be spalting fungi, as their hyphae do not colonize the wood internally and they do not produce the enzymes necessary to digest the wood cell wall components.

From several tropical woods are manufactured various products for interiors, for example, furniture, flooring, stairs, windows, doors, claddings or structural elements—using usually massive, glued timbers, veneers or plywood [38]. For more of these products, both strength and aesthetic are important. The interesting aesthetic and specific colours of the individual tropical wood species, existing in the natural state as well as in the primarily fungal-pigmented state, can be changed or even worsened at additional rotting processes. In interiors, the degree and range of rotting is influenced mainly by: (1) the natural durability of wood to decaying fungi (e.g., [39] classifies woods into five classes of durability), and (2) the enough moisture of wood above 20%, depending on presence of condensed, capillary, plumbing or rain water.

The aim of this study was to determine the colour changes of 21 tropical woods when exposed to the brown-rot fungus *C. puteana*, which can cause an important deterioration of wood products in the interiors of buildings.

## 2. Materials and Methods

### 2.1. Woods and Specimens

Twenty-one tropical woods, in a form of naturally dried and conditioned boards with a moisture content of 13 ± 2.5%, was bought from the trading company JAF Holz, Ltd., Slovakia (Table 1).

**Table 1.** Tropical wood species used in the experiment.

| Family | Species Common Name [1] | Species Scientific Name | Density at MC 12% (kg·m$^{-3}$) | |
|---|---|---|---|---|
| | | | "Literature" [40] | "Experiment" |
| *Bignoniaceae* | Ipé | *Handroanthus serratifolius* (Vahl) S.O.Grose [3] [4] | 960–1100 | 968 *(26)* |
| *Burseraceae* | Okoumé | *Aucoumea klaineana* Pierre [2] | 370–560 | 566 *(27)* |
| *Cunoniaceae* | Tineo | *Weinmannia trichosperma* Cav. | 570–650 | 646 *(31)* |
| *Dipterocarpaceae* | Dark red meranti | *Shorea curtisii* Dyer ex King [2] | 590–890 | 592 *(10)* |
| | Yellow balau [5] | *Shorea laevis* Ridl. [2] | 900–1100 | 925 *(55)* |
| *Ebenaceae* | Macassar ebony | *Diospyros celebica* Bakh. [2] | 1100–1200 | 1 013 *(82)* |
| *Fabaceae* | Doussié | *Afzelia bipindensis* Harms [2] | 750–950 | 889 *(18)* |
| | Cerejeira | *Amburana cearensis* A. C. Sm. [2] | 550–650 | 651 *(10)* |
| | Bubinga | *Guibourtia demeusei* J. Léonard | 830–950 | 830 *(13)* |
| | Ovengol | *Guibourtia ehie* J. Léonard [2] | 700–910 | 755 *(27)* |
| | Merbau | *Intsia bijuga* O. Ktze. [2] | 830–900 | 837 *(50)* |
| | Santos rosewood | *Machaerium scleroxylon* Tul. [2] | 900–1000 | 904 *(6)* |
| | Zebrano | *Microberlinia brazzavillensis* Chev. [2] | 700–850 | 718 *(23)* |
| | Wengé | *Millettia laurentii* De Wild. [2] | 810–950 | 823 *(37)* |
| | Padouk | *Pterocarpus soyauxii* Taub. | 650–850 | 647 *(37)* |
| *Meliaceae* | Sapelli | *Entandrophragma cylindricum* Sprague [2] | 510–750 | 631 *(38)* |
| *Moraceae* | Iroko | *Milicia excelsa* C. C. Berg [2] | 550–850 | 551 *(16)* |
| *Myrtaceae* | Karri | *Eucalyptus diversicolor* F. Muell. | 800–870 | 804 *(30)* |
| | Blue gum | *Eucalyptus globulus* Labill. | 720–770 | 760 *(59)* |
| *Sapotaceae* | Maçaranduba [6] | *Manilkara bidenta* A. Chev. | 900–1000 | 916 *(19)* |
| | Makoré | *Tieghemella heckelii* Pierre [2] | 530–720 | 570 *(25)* |

Notes: [1] by Association Technique Internationale des Bois Tropicaux (ATIBT) in France; [2] registered in The IUCN Red List of Threatened Species™ [41]; [3] name by EN 350 [39]; [4] previous name by [40] was *Tabebuia serratifolia*; [5] its other name is Bangkirai; [6] its other name is Massaranduba; Mean values of experimental densities are from 16 specimens (4 of those were used in this experiment) and standard deviations are in italic and parentheses.

From the heart-zone of each wood species were prepared and tested four specimens 25 mm × 25 mm × 3 mm (longitudinal × radial × tangential)—without biological damage, knots or other defects. Before the fungal attack, the top surfaces of specimens were sanded along fibres using 240-grit sandpaper. Subsequently, the specimens were conditioned on a moisture content of 12 ± 1%, weighted with an accuracy of 0.001 g, sterilized with 30 W germicidal lamp (Chirana, Slovakia) at a temperature of 22 ± 2 °C per 20 min for each side, and finally their top surfaces submitted in sterilized room to colour analyses (point 2.3).

After fungal attack (point 2.2), the specimens were carefully cleaned from surface fungal mycelia, slowly dried in a laboratory, conditioned on a moisture content of 12 ± 1%, weighted with an accuracy of 0.001 g, the top surfaces sanded along fibres with 240-grit sandpaper, and finally the top surfaces again submitted to colour analyses (point 2.3).

At sanding of the top surfaces of tested specimens, performed before and after fungal attack, in both cases the thickness of specimens declined about approximately 10 micrometers.

### 2.2. Fungal Attack of Woods

The specimens of tropical woods were exposed to the brown-rot fungus *Coniophora puteana* (Schumacher ex Freist) Karsten, strain BAM Ebw. 15 (Bundesanttalt für Materialforshung und—prüfung, Berlin) in glass Petri dishes with a diameter of 100 mm. Two replicates of the same wood species were placed into one dish on plastic mats under which a fungal mycelium was already grown on an autoclave sterilized and solidified 3–4 mm thick layer of 4.5 wt.% malt agar medium (HiMedia, Ltd., India). Fungal attacks lasted 6 weeks at a temperature of $24 \pm 2\ ^{\circ}\mathrm{C}$ and a relative humidity of $90 \pm 5\%$.

The weight loss $\Delta m$ (%) of specimens was calculated from their weights in conditioned state before and after fungal attack. This method may have caused some inaccuracies in $\Delta m$ (due to sorption hysteresis and differential water sorption by healthy and rotten woods), but it was preferred to the method evaluating specimens in absolute dry state when colour change could be manifested at drying temperature of $103\ ^{\circ}\mathrm{C}$.

### 2.3. Colour Analyses of Woods

The colour analyses were performed for each specimen before and after fungal attack on the sanded top surfaces (point 2.1) in the same four places (Figure 1). The colour measurements were performed with the Color Reader CR-10 (Konica Minolta, Japan), having a CIE 10° standard observer, CIE standard illuminate D65, sensor head with a diameter of 8 mm (i.e., the measuring area was 50 mm$^2$), and a detector with 6 silicon photocells.

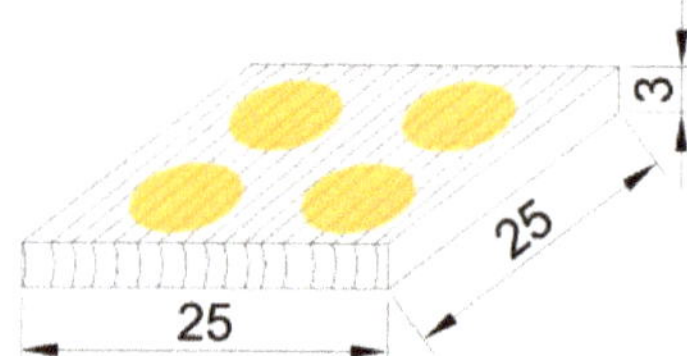


**Figure 1.** The same four places on the top surface of specimen in which colour measurements were performed in its original and fungal-attacked state.

The colourimetric parameters of each specimen were analysed according to the CIE 1976 $L^*a^*b^*$ colour system. A larger value of $L^*$, $a^*$, or $b^*$ means a lighter, redder, or yellower colour, respectively.

Based on the $L^*$, $a^*$, and $b^*$ colour coordinates, following the colour saturation—chromacity $C^*_{ab}$ and the hue tone angle $h_{ab}$ were calculated according to the [42] by Equations (1) and (2):

$$C^*_{ab} = \sqrt{a^{*2} + b^{*2}}, \tag{1}$$

$$h_{ab} = \tan^{-1}(b^*/a^*), \tag{2}$$

From the relative colour changes $\Delta L^*$, $\Delta a^*$, and $\Delta b^*$, namely differences between colour coordinates of the fungal-attacked and the original wood specimens, the total colour difference $\Delta E^*_{ab}$ was calculated by Equation (3) [42]:

$$\Delta E^*_{ab} = \sqrt{\Delta L^{*2} + \Delta a^{*2} + \Delta b^{*2}}, \tag{3}$$

Selected colourimetric parameters of tropical woods determined in the original state ($a^*$, $b^*$, $C^*_{ab}$, $h_{ab}$), as well as in the fungal-attacked state ($a^*_F$, $b^*_F$, $C^*_{ab\,F}$, $h_{ab\,F}$), were finally analysed in relation to their lightness ($L^*$ or $L^*_F$) by linear correlations using Equation (4):

$$Colourimetric\ parameter = A + B{\cdot}L^*, \tag{4}$$

Equation (4) was used as well as for searching relations between the $\Delta E^{*}_{ab}$ * and the relative colour changes $\Delta L^{*}$, $\Delta a^{*}$, $\Delta b^{*}$, and the weight loss $\Delta m$, respectively.

Conversion of average value of the colourimetric coordinates $L^{*}$, $a^{*}$ and $b^{*}$ of the fungal-attacked and the original wood specimens was generated their colour (Pantone) swatch.

The *t*-test statistically analysed the colour changes of individual tropical woods due to the brown-rot fungus *C. puteana*.

## 3. Results and Discussion

### 3.1. Colour of Original Tropical Woods

Table 2 documents colour characteristics of 21 tropical woods in the original state—a visualisation (colour and surface structure), and the colourimetric parameters $L^{*}$, $a^{*}$, $b^{*}$, $C^{*}_{ab}$ and $h_{ab}$.

**Table 2.** Visualisations and colorimetric parameters of 21 tropical woods.

| Wood Species | Visual | Lightness $L^{*}$ | Redness $+a^{*}$ | Yellowness $+b^{*}$ | Colour Saturation $C^{*}_{ab}$ | Hue Tone Angle $h_{ab}$ |
|---|---|---|---|---|---|---|
| Ipé | | 42.59 (0.79) | 8.87 (0.55) | 16.46 (1.01) | 18.70 (2.59) | 61.66 (1.22) |
| Okoumé | | 75.36 (1.13) | 5.61 (0.61) | 16.89 (0.57) | 17.81 (0.65) | 71.66 (1.69) |
| Tineo | | 47.69 (4.98) | 14.56 (1.81) | 16.23 (0.99) | 21.83 (1.77) | 48.26 (2.83) |
| Darkred meranti | | 62.03 (1.52) | 11.76 (0.81) | 16.35 (0.64) | 20.15 (0.87) | 54.30 (1.56) |
| Yellow balau | | 54.59 (0.92) | 13.22 (0.83) | 21.92 (0.88) | 25.60 (1.12) | 58.93 (1.02) |
| Macassar ebony | | 59.40 (0.80) | 13.86 (0.68) | 21.29 (0.81) | 25.41 (0.91) | 56.93 (1.23) |
| Doussié | | 60.57 (6.61) | 10.28 (1.89) | 25.42 (2.09) | 27.50 (1.81) | 67.93 (4.80) |
| Cerejeira | | 49.93 (3.34) | 10.85 (1.52) | 18.19 (1.34) | 21.22 (1.59) | 59.24 (3.46) |
| Bubinga | | 46.81 (1.23) | 17.50 (0.89) | 16.98 (1.17) | 24.40 (1.20) | 44.11 (2.00) |
| Ovengol | | 53.56 (4.65) | 8.08 (1.07) | 19.63 (4.38) | 21.27 (4.33) | 67.36 (2.69) |
| Merbau | | 46.92 (1.57) | 15.43 (0.53) | 18.97 (1.12) | 24.46 (1.09) | 50.83 (1.37) |
| Santos rosewood | | 50.79 (3.39) | 14.68 (1.31) | 22.49 (3.13) | 26.88 (3.23) | 56.63 (2.44) |
| Zebrano | | 55.85 (4.91) | 10.09 (1.06) | 19.18 (1.49) | 21.68 (1.63) | 62.24 (2.18) |
| Wengé | | 34.88 (1.68) | 7.89 (0.64) | 10.33 (1.30) | 13.02 (1.24) | 52.42 (3.38) |

**Table 2.** *Cont.*

| Wood Species | Visual | Lightness $L^*$ | Redness $+a^*$ | Yellowness $+b^*$ | Colour Saturation $C^*_{ab}$ | Hue Tone Angle $h_{ab}$ |
|---|---|---|---|---|---|---|
| Padouk | | 43.78 (2.05) | 31.42 (1.61) | 26.72 (2.02) | 41.26 (2.32) | 40.34 (1.57) |
| Sapelli | | 53.04 (1.69) | 13.88 (0.88) | 18.34 (0.72) | 23.01 (0.94) | 52.90 (1.62) |
| Iroko | | 63.52 (6.78) | 8.49 (0.88) | 26.20 (5.69) | 27.57 (5.59) | 71.70 (2.54) |
| Karri | | 50.72 (2.07) | 17.36 (1.13) | 23.84 (1.59) | 29.50 (1.83) | 53.92 (1.32) |
| Blue gum | | 59.25 (3.37) | 8.05 (0.74) | 17.24 (1.23) | 19.03 (1.27) | 64.95 (1.98) |
| Maçaranduba | | 45.06 (2.16) | 17.61 (2.39) | 14.34 (0.99) | 22.75 (2.23) | 39.45 (3.79) |
| Makoré | | 49.85 (2.24) | 12.08 (0.88) | 18.95 (1.10) | 22.49 (1.15) | 57.47 (2.10) |

Notes: Mean values are from 16 measurements. Standard deviations are in italic and parentheses.

The studied tropical woods differed mainly in the lightness $L^*$, which ranged between 34.88 ("very dark" wengé), 43.78 ("dark" padouk), 63.58 ("light" iroko) and 75.36 ("very light" okoumé). Achieved results are in an accordance with similar works disserting specific colour and texture characteristics of selected tropical woods [7,40,43].

All 21 tropical woods had the colour parameters $a^*$ and $b^*$ in a positive sphere of distribution. The redness index ($+a^*$) ranged from 5.61 (okoumé) to 31.42 (padouk), and the yellowness index ($+b^*$) from 10.33 (wengé) to 26.72 (padouk). This result suggests that woods coloured a brighter red or yellow, such as padouk, may belong also to darker species. When the values of $a^*$ and $b^*$ were for 21 tropical wood species evaluated in comparison to their lightness $L^*$, different tendencies of linear correlation were found. Value $a^*$ had a negative correlation against the $L^*$ (Figure 2A; $R^2 = 0.18$), while the $b^*$ had a positive correlation against the $L^*$ (Figure 2B; $R^2 = 0.10$). Nishino et al. [43] determined similar dependences between $a^*$ or $b^*$ and $L^*$ for 97 wood species from French Guiana.

The colour saturation $C^*_{ab}$ indicates the distance from the chromatic point ($a^* = 0$, $b^* = 0$) on the CIE 1976 $L^*a^*b^*$ colour space. For tropical woods the values of colour saturation ranged from 13.02 (wengé—characterized by a little intensive red-yellow shade) to 41.26 (padouk—characterized by a strong intensive red-yellow shade). Graphical analysis of the $C^*_{ab}$ against the $L^*$ for 21 tropical woods is present in Figure 2C. The $C^*_{ab}$ exhibited only a minimal decrease at higher values of lightness $L^*$ with zero coefficient of determination ($R^2 = 0.00$), i.e., was not found significant correlation.

The hue tone angle $h_{ab}$ ranges for wood between $0°$ and $90°$ (the first quadrant), where $0°$ represents the red colour and $90°$ represents the yellow colour. The yellow shade prevailed to the red shade for a vast majority of tested tropical woods, typically for iroko and okoumé. On the contrary, for padouk, having the most red and yellow striking shades, a slightly more dominant was the red shade. A reasonable positive linear correlation between the hue tone angle $h_{ab}$ and the lightness $L^*$ of individual tropical woods is presented in Figure 2D ($R^2 = 0.46$).

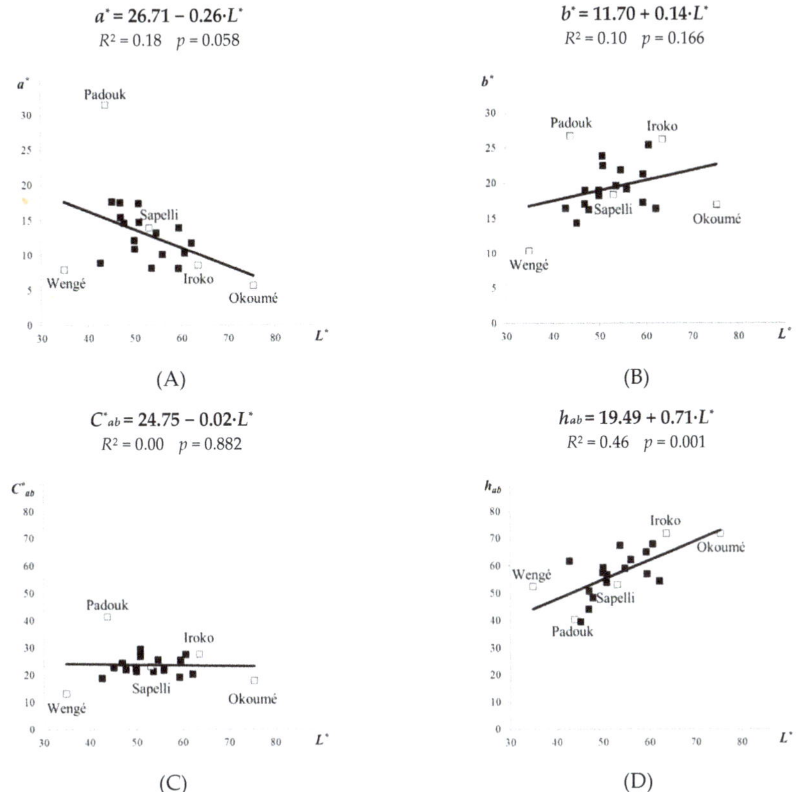



**Figure 2.** Linear correlations between the lightness $L^*$ and the colour coordinates $a^*$, $b^*$ (**A, B**), the colour saturation $C^*_{ab}$ (**C**), and the hue tone angle $h_{ab}$ (**D**)—for 21 original tropical wood species.

This result suggests that tropical woods with a dominant yellow shade are usually lighter. Németh [44] also found a linear correlation between the lightness $L^*$ and the hue angle $h_{ab}$ examining the colour co-ordinates of different temperate wood species.

### 3.2. Colour of Tropical Woods Exposed to the Fungus Coniophora puteana

Woods attacked by the brown-rot fungi gradually acquire deeper shades of brown of yellow [15,22,27,37,45]. However, specific colour changes during action of the brown-rot fungi can also occur, depending on the wood species [37,46], the history of wood ageing before fungal attack [15], the degree and uniformity of decay [47], or the specific enzymes, Fenton and other low-molecular degradation systems, and pigments produced by fungal mycelia [45].

Weight losses of 21 tropical woods attacked for 6 weeks by the brown-rot fungus *C. puteana* ranged from 0.08% to 6.48% (Table 3). The differences of the lightness $\Delta L^*$ and other colourimetric parameters $\Delta a^*$, $\Delta b^*$, $\Delta C^*_{ab}$, $\Delta h_{ab}$ and $\Delta E^*_{ab}$, determined as a difference between the fungal-attacked and the original tropical wood, are documented in Table 3. Visualization, together with the pantone swatches, of the top surfaces of selected tropical woods before and after their exposure to *C. puteana* is shown in Figure 3.

**Okoumé**     **Wengé**     **Padouk**     **Sapelli**     **Iroko**

**Figure 3.** Visualization and pantone swatches of the top surfaces of selected tropical woods before and after exposure to *C. puteana*.

After exposition to the brown-rot fungus *C. puteana*, the top surfaces of the darkest tropical woods wengé and ipé (Table 2) developed significantly lighter shades with the lightness increase $\Delta L^*$ +10.46 and +2.47 (Table 3). This "unexpected" result can be explained by a washout or deterioration of dark extractives present in darker wood species during the mycological test performed in humid environment in Petri dishes. On the contrary, the most noticeable darkening was observed in the top surfaces of the lightest and medium light tropical woods sapelli, okoumé, iroko and ovengol with the $\Delta L^*$ from −21.54 to −11.08 (Table 3). This "expected" result can be explained by degradation of white cellulose and hemicelluloses in presence of hydrolases and Fenton agent produced by *C. puteana* [45]. Previous studies have also reported a darkening of lighter European woods (beech and pine) due to decay processes [15,16].

A pronounced shade of the red due to *C. puteana* obtained only wengé, with $\Delta a^*$ +4.98. A lighter shade of red obtained okoumé, zebrano, iroko, and blue gum, with $\Delta a^*$ from +0.31 to +2.53 (Table 3). Conversely, the other tropical woods developed a greener shade, with $\Delta a^*$ from −0.22 to −7.76, the most markedly sapelli, padouk and macaranduba, with $\Delta a^* \geq −5.68$ (Table 3). The significant greening of padouk may be justified by its intensive red shade in the original state (Table 2).

The majority of tropical woods attacked by *C. puteana* obtained a more yellow shade, typically wengé, okoumé and blue gum, with $\Delta b^*$ from +4.42 to +10.05. However, three tropical woods—padouk, sapelli, and yellow balau—showed a significant tendency to become bluer, with $\Delta b^*$ from −3.08 to −8.74 (Table 3).

After fungal attack, a significant positive change in the colour saturation $\Delta C^*_{ab}$ had wengé +11.16, okoumé +5.93, and blue gum +5.15 (Table 3). It is an interesting knowledge, because the original wengé was the darkest species and okoumé the lightest one (Table 2). On the contrary, an evident negative change in $\Delta C^*_{ab}$ had sapelli, padouk, and yellow balau, with $\Delta C^*_{ab}$ from −4.04 to −10.84 (Table 3). However in a summary, the colour saturation of the tested original or fungal-attacked tropical woods had no significance to their lightness (Figure 2C, Figure 4C).

The positive differences in the hue tone angle $\Delta h_{ab}$ indicate that the wood surfaces changed due to the brown-rot fungus more towards yellowish as to reddish. The largest positive change in $\Delta h_{ab}$ was observed for maçaranduba, bubinga, dark red meranti, and sapelli in range from +13.46 to +7.21 (Table 3). A significantly negative $\Delta h_{ab}$, connected with more evident redness as yellowing, obtained padouk −5.49, and iroko −4.30. Statistically insignificant redness occurred for blue gum, ovengol, and zebrano, with $\Delta h_{ab}$ from −0.99 to −0.09 (Table 3).

**Table 3.** Weight loss and relative change of colourimetric parameters of 21 tropical woods caused by *C. puteana*.

| Wood species | $\Delta m$ | $\Delta L^*$ | | $\Delta a^*$ | | $\Delta b^*$ | | $\Delta C^*_{ab}$ | | $\Delta h_{ab}$ | | $\Delta E^*_{ab}$ |
|---|---|---|---|---|---|---|---|---|---|---|---|---|
| Ipé | 0.20<br>(0.03) | 2.47<br>(1.32) | a | −0.59<br>(0.50) | c | 0.01<br>(0.31) | d | −0.26<br>(0.29) | d | 1.68<br>(1.56) | b | 2.63<br>(1.26) |
| Okoumé | 6.48<br>(2.98) | −15.74<br>(1.55) | a | 1.20<br>(0.23) | b | 5.80<br>(0.50) | a | 5.93<br>(0.49) | a | 1.70<br>(0.51) | d | 16.84<br>(1.34) |
| Tineo | 5.38<br>(3.71) | −3.59<br>(1.63) | d | −1.88<br>(1.12) | b | 1.41<br>(0.43) | c | −0.11<br>(0.92) | d | 6.03<br>(2.10) | a | 4.44<br>(1.68) |
| Dark red meranti | 0.22<br>(0.09) | −8.44<br>(0.96) | a | −2.29<br>(0.27) | a | 3.28<br>(1.04) | a | 1.71<br>(0.94) | a | 9.90<br>(1.70) | a | 9.43<br>(0.70) |
| Yellow balau | 0.40<br>(0.08) | −8.61<br>(1.06) | a | −2.77<br>(1.00) | a | −3.08<br>(0.95) | a | −4.04<br>(1.17) | a | 2.04<br>(1.77) | b | 9.59<br>(1.47) |
| Macassar ebony | 0.52<br>(0.30) | −3.48<br>(1.87) | a | −1.88<br>(0.31) | a | −1.76<br>(0.84) | a | −2.49<br>(0.86) | a | 1.57<br>(0.49) | a | 4.73<br>(0.81) |
| Doussié | 1.63<br>(1.51) | −2.51<br>(1.68) | d | −0.91<br>(0.76) | d | −0.76<br>(0.66) | d | −1.23<br>(0.84) | d | 1.36<br>(1.07) | d | 3.05<br>(1.60) |
| Cerejeira | 0.30<br>(0.15) | −3.51<br>(2.26) | b | −2.89<br>(1.57) | b | −1.86<br>(1.28) | b | −2.89<br>(1.67) | b | 5.73<br>(2.12) | c | 4.98<br>(2.93) |
| Bubinga | 0.96<br>(0.13) | −3.81<br>(1.24) | a | −3.87<br>(0.85) | a | 1.89<br>(0.47) | a | −1.11<br>(0.90) | c | 10.29<br>(1.10) | a | 5.96<br>(0.15) |
| Ovengol | 0.61<br>(1.02) | −11.08<br>(1.28) | a | −0.46<br>(0.48) | d | −1.81<br>(0.84) | a | −1.83<br>(0.87) | b | −0.84<br>(1.10) | d | 11.28<br>(1.24) |
| Merbau | 0.48<br>(0.32) | −4.29<br>(1.71) | a | −0.22<br>(1.71) | d | −1.01<br>(1.25) | d | −0.88<br>(1.95) | d | −1.16<br>(1.99) | d | 4.87<br>(1.78) |
| Santos rosewood | 1.28<br>(0.31) | −0.83<br>(0.66) | d | −1.49<br>(1.67) | c | −1.48<br>(2.55) | d | −2.49<br>(0.86) | d | 1.57<br>(0.49) | d | 3.36<br>(1.89) |
| Zebrano | 1.65<br>(0.62) | −9.14<br>(1.01) | a | 0.82<br>(1.04) | d | 1.26<br>(1.18) | d | 1.52<br>(1.46) | d | −0.09<br>(1.69) | d | 9.39<br>(1.08) |
| Wengé | 0.67<br>(0.35) | 10.46<br>(2.17) | a | 4.98<br>(0.72) | a | 10.05<br>(1.87) | a | 11.16<br>(1.50) | a | 4.69<br>(2.60) | b | 15.54<br>(1.55) |
| Padouk | 0.95<br>(1.69) | −6.18<br>(1.32) | a | −5.91<br>(0.96) | a | −8.74<br>(1.93) | a | −9.93<br>(1.46) | a | −5.49<br>(2.70) | a | 12.41<br>(1.41) |
| Sapelli | 3.12<br>(1.15) | −21.54<br>(0.92) | a | −7.76<br>(1.26) | a | −7.87<br>(1.52) | a | −10.84<br>(1.99) | a | 7.21<br>(1.11) | a | 24.26<br>(1.52) |
| Iroko | 0.20<br>(0.97) | −15.08<br>(3.76) | a | 1.07<br>(0.57) | c | −2.89<br>(1.41) | c | −2.37<br>(1.49) | d | −4.30<br>(0.71) | a | 15.44<br>(3.89) |
| Karri | 0.69<br>(0.47) | −3.83<br>(1.37) | a | 0.31<br>(1.23) | d | 0.89<br>(1.41) | d | 0.91<br>(1.81) | d | 0.54<br>(0.89) | d | 4.43<br>(1.17) |
| Blue gum | 0.38<br>(0.21) | −6.27<br>(2.54) | a | 2.53<br>(1.46) | a | 4.42<br>(0.88) | a | 5.15<br>(1.52) | a | −0.99<br>(1.35) | d | 8.42<br>(1.91) |
| Maçaranduba | 0.08<br>(0.05) | −1.99<br>(1.21) | b | −5.68<br>(0.40) | a | 1.66<br>(0.40) | b | −2.81<br>(0.38) | a | 13.46<br>(1.08) | a | 6.35<br>(0.64) |
| Makoré | 1.78<br>(1.84) | −9.43<br>(1.69) | a | −2.44<br>(1.37) | a | −2.45<br>(1.65) | b | −3.33<br>(2.09) | b | 2.35<br>(1.28) | c | 10.19<br>(2.10) |

Note: Mean values of colourimetric parameters are from 16 measurements, and of weight losses from 4 specimens. Standard deviations are in italics and parentheses. The t-test analysed the colour changes in relation to the original wood at the 99.9% significance level (a), the 99% significance level (b), the 95% significance level (c), and without an evident significant difference at $p \geq 0.05$ (d).

The linear correlations between the colourimetric parameters $a^*_F$, $C^*_{ab\,F}$ or $h_{ab\,F}$ versus the lightness $L^*_F$ for the 21 fungal-attacked tropical woods (Figure 4A,C,D) remained almost the same as were determined for the 21 original tropical woods (Figure 2A,C,D).

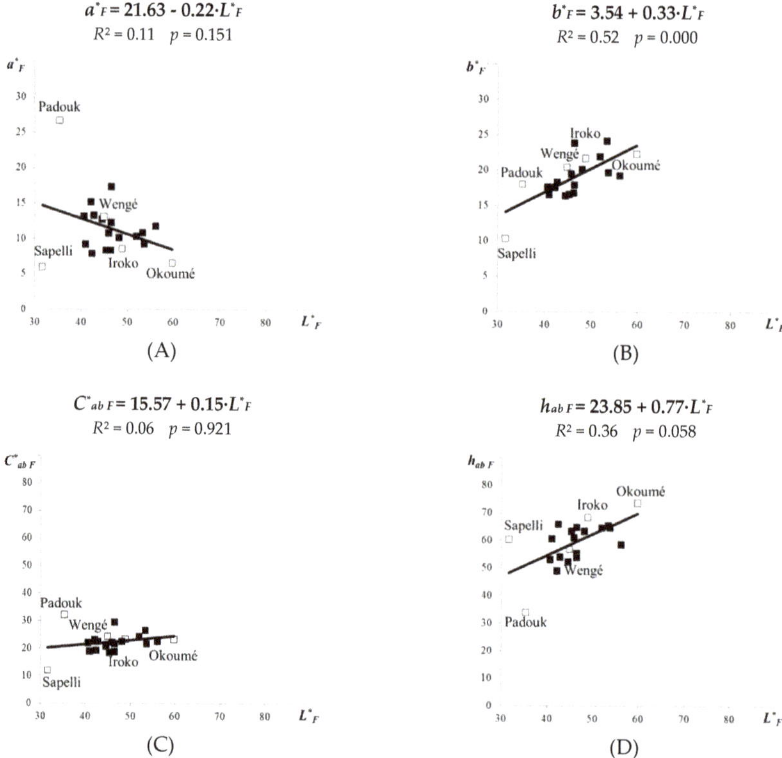

Figure 4. Linear correlations between the lightness $L^*$ and the colour coordinates $a^*$, $b^*$ (**A,B**), the colour saturation $C^*_{ab}$ (**C**), and the hue tone angle $h_{ab}$ (**D**)—for 21 tropical wood species attacked by the brown-rot fungus *C. puteana*.

In the fungal-attacked tropical woods, again no evident relationships were found between the lightness and redness ($R^2 = 0.11$), or the lightness and colour saturation ($R^2 = 0.06$). Indirectly, it can be stated that fungal attack did not have an apparent effect on these relationships. However, due to *C. puteana* the relationship between lightness and hue tone angle evidently decreased ($R^2 = 0.36$ for $h_{ab\ F}$—Figure 4D; while previously $R^2 = 0.46$ for $h_{ab}$—Figure 2D).

Conversely, the $b^*_F$ coordinate, which indicates yellowing, grew more evidently with $L^*_F$ for the 21 fungal-attacked tropical woods. It is evident from comparing the slope trend B and the coefficient of determination R$^2$ in Figure 2B ($b^* = 11.70 + 0.14 \cdot L^*$; B = 0.14; $R^2 = 0.10$) with the same parameters in Figure 4B ($b^*_F = 3.54 + 0.33 \cdot L^*_F$; B = 0.33; $R^2 = 0.52$). As the slope trend B increased more evidently only for the yellow colour coordinate $b^*$, from 0.14 to 0.33, there indirectly was confirmed that at brown rot some lighter tropical woods (such as okoumé, or dark red meranti) obtain a more yellow shade.

The total colour differences $\Delta E^*_{ab}$ of the fungal-attacked tropical woods were significantly similar to their lightness changes $\Delta L^*$ (Figure 5A), however, values of $\Delta E^*_{ab}$ were greater as $\Delta L^*$ in an accordance with Equation (3) (Table 3). Usually, the lightest and medium light tropical woods okoumé, sapelli, iroko and ovengol had the highest $\Delta E^*_{ab}$ values from 11.28 to 24.26. This result is in an accordance with other works dealing with the durability and colour changes of tropical woods due to biological deterioration [48,49]. However, it was simultaneously observed that after attack by *C. puteana* the top surfaces of wengé and padouk (darker tropical woods—Table 2) also had high values of $\Delta E^*_{ab}$ 15.80 and 12.41 (Table 3, Figure 5A). This result can be explained by washout of dark extractives presented in darker wood species during the mycological test, similarly mentioned for the lightness changes. On the contrary, effects of the $\Delta a^*$ and $\Delta b^*$ values of all tropical species on the $\Delta E^*_{ab}$ value were not significant (Figure 5B,C).

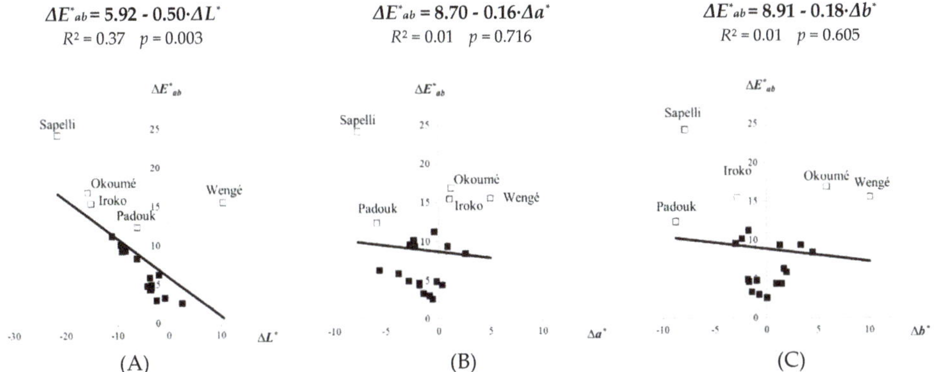


**Figure 5.** Linear correlations between the total colour difference $\Delta E^*_{ab}$ and changes of other colour parameter $\Delta L^*$ (**A**), $\Delta a^*$ (**B**), $\Delta b^*$ (**C**)—for 21 tropical woods attacked with the brown-rot fungus *C. puteana*.

Several studies have shown that there are some relationships between the colour parameters of wood and its decay resistance [15,16,46–50]. The relationship between the colour and weight loss of decayed wood is based on the type and amount of wood extractives, which have effect on the colour, and on its decay resistance. Such relationships could be encouraging but not always sufficient for predicting decay resistance of tested woods. Specifically, from our experiment it is evident that for 21 tropical woods no dependency was found between the total colour difference $\Delta E^*_{ab}$ and the decay resistance determined as weight loss $\Delta m$ (Figure 6; $R^2 = 0.10$).

Generally, the intensity and specificity of colour changes in woods attacked by brown rot—when browning can be connected with yellowing or bluing and also with reddening or greening—are influenced not only by the enzymatic and pigment specification of the individual brown-rot fungus, but also by the wood species (e.g., presence of specific extracts), the degree of its decay, and the environmental factors. Therefore, some of tested tropical woods at exposition to *C. puteana* had a more yellowish shade while other ones a more reddish shade, when all the colour changes depended probably on their molecular structure and natural durability. In future experiments, we would like to analyse these factors in more detail.

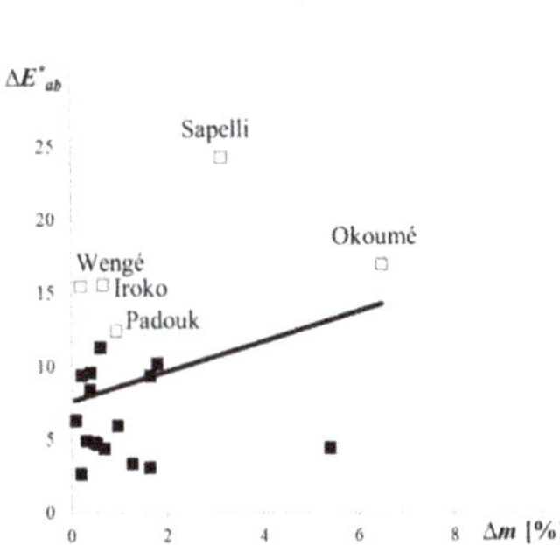


**Figure 6.** Linear correlations between the total colour differences $\Delta E^*_{ab}$ and the weight losses $\Delta m$ of 21 tropical woods exposed 6 weeks to *C. puteana*.

Colour changes of tropical woods exposed in interiors of buildings are unwanted. Brown-rot fungi decrease their strength and can worsen their colour and aesthetic. On the contrary, some benefits of rotting can be interesting for claddings and other decorative products, but in this situation the intentional decay has to be performed before they are installed. By processing spalted wood artists can create decorative material, such as fine art mosaics, furniture, and dishes, which have been used for centuries [51] and also hold an interesting niche modern market in decorative veneers or other decorative materials in North America and Europe [17].

## 4. Conclusions

Several tropical woods have positive surface characteristics, such as interesting texture and colour, and high natural durability, which is an essential requisite for wooden constructions exposed to the environment with a high risk of biodeterioration by fungi and insects. For architects, changes in their original colour at biodeterioration processes are important. The colour analyses of 21 tropical woods before and after intentional attack by the brown rot-fungus *C. puteana* led to the following conclusions:

- For the original tropical woods, the $a^*$ coordinate (redding) declined with increase of the lightness $L^*$, while the $b^*$ (yellowing) and $h^*_{ab}$ (hue ton angle) coordinates grew with the lightness. No significance was found between $C^*_{ab}$ (colour saturation) and $L^*$.

- For the fungal-attacked tropical woods, the linear correlations between the colour coordinates $a^*$, $C^*_{ab}$ or $h_{ab}$ and the lightness $L^*$ remained almost the same as for the original tropical woods, and only the $b^*$ coordinate grew more evidently in relation to $L^*$.

- The total discoloration $\Delta E^*_{ab}$ values were highest for the top surfaces of the lighter species (sapelli, okoumé, iroko) and the darkest species (wengé), when the $\Delta E^*_{ab}$ were justified by the marked change of the lightness $\Delta L^*$.

- Significant changes in the lightness and colouration of the fungal-attacked tropical woods indicated these colour changes could be caused not only by the biodegradation of polysaccharides, but also by biodegradation or leaching of some extractives during laboratory mycological tests.

**Author Contributions:** For research articles with several authors, a short paragraph specifying their individual contributions must be provided. The following statements should be used "conceptualization, Z.V. and L.R.; methodology, Z.V. and L.R.; software, Z.V. and L.R.; validation, Z.V. and L.R.; formal analysis, Z.V. and L.R.; investigation, Z.V. and L.R.; resources, Z.V. and L.R.; data curation, Z.V. and L.R.; writing—original draft preparation, Z.V. and L.R.; writing—review and editing, Z.V. and L.R.; visualization, Z.V.; supervision, L.R..; project administration, L.R.

**Acknowledgments:** This work was supported by the Slovak Research and Development Agency under the contract No. APVV-17-0583, and the VEGA project 1/0729/18.

**Conflicts of Interest:** The authors declare no conflicts of interest.

## References

1. Janin, G.; Gonçalez, J.C.; Ananías, R.A.; Charrier, B.; Silva, G.F.D.; Dilem, A. Aesthetics appreciation of wood colour and patterns by colorimetry. Part 1. Colorimetry theory for the CIE Lab system. *Maderas Cienc. Tecnol.* **2001**, *3*, 14.

2. Tolvaj, L.; Persze, L.; Lang, E. Correlation between hue angle and lightness of wood species grown in Hungary. *Wood Res.* **2013**, *58*, 141–145.

3. Slabejová, G.; Šmidriaková, M.; Fekiač, J. Gloss of transparent coating on beech wood surface. *Acta Fac. Xylologiae Zvolen* **2016**, *58*, 37–44.

4. Hon, D.N.-S.; Minemura, N. Color and discoloration. In *Wood and Cellulosic Chemistry*, 2nd ed.; Hon, D.N.-S., Shiraishi, N., Eds.; CRC Press: New York, USA, 2000; pp. 385–442.

5. Babiak, M.; Kubovský, I.; Mamoňová, M. Color space of the selected domestic species. In *Interaction of Wood with Various Forms of Energy*, 1st ed.; Kurjatko, S., Kúdela, J., Eds.; Technical University in Zvolen: Zvolen, Slovakia, 2004; pp. 113–117.

6. Katuščák, S.; Kucera, J. CIE orthogonal and cylindrical color parameters and the color sequences of the temperate wood species. *Wood Res.* **2000**, *45*, 9–21.

7. da Silva, R.A.F.; Setter, C.; Mazette, S.S.; de Melo, R.R.; Stangerlin, D.M. Colorimetry of wood from thirty tropical species. *Ciênc. Madeira* **2017**, *8*, 36–41.

8. Meints, T.; Teischinger, A.; Stingl, R.; Hansmann, C. Wood colour of central European wood species: CIE Lab characterisation and colour intensification. *Eur. J. Wood Wood Prod.* **2017**, *75*, 499–509. [CrossRef]

9. Klement, I.; Vilkovská, T. Color characteristics of red false heartwood and mature wood of beech (*Fagus sylvatica* L.) determining by different colour saturation city coordinates. *Sustainability* **2019**, *11*, 690. [CrossRef]

10. Mosedale, J.R.; Charrier, B.; Janin, G. Genetic control of wood colour, density and heartwood ellagitannin content of European oak (*Quercus petraea* and *Quercus robur*). *Forestry* **1996**, *69*, 111–124. [CrossRef]

11. Phelps, J.E.; McGinnes, E.A.; Garret, H.E.; Cox, G.S. Growth quality evaluation of black walnut wood. II. Color analyses of veneer produced on different sites. *Wood Fiber Sci.* **1982**, *15*, 177–185.

12. Derkyi, N.S.A.; Bailleres, H.; Chaix, G.; Thevenon, M.F. Colour variation in teak (*Tectona grandis*) wood from plantations across the ecological zones of Ghana. *Ghana J. For.* **2009**, *25*, 40–48. [CrossRef]

13. Kržišnik, D.; Lesar, B.; Thaler, N.; Humar, M. Influence of natural and artificial weathering on the colour change of different wood and wood-based materials. *Forests* **2018**, *9*, 488. [CrossRef]

14. Reinprecht, L.; Mamoňová, M.; Pánek, M.; Kačík, F. The impact of natural and artificial weathering on the visual, colour and structural changes of seven tropical woods. *Eur. J. Wood Wood Prod.* **2018**, *76*, 175–190. [CrossRef]

15. Reinprecht, L.; Hulla, M. Colour changes in beech wood modified with essential oils due to fungal and ageing-fungal attacks with *Coniophora puteana*. *Drewno* **2015**, *58*, 37–48.

16. Vidholdová, Z.; Slabejová, G.; Polomský, J. Colour changes of Scots pine wood due to action of the white-rot fungus *Trametes versicolor*. In *Protecting Trees and Wood*, 1st ed.; Hlaváč, P., Vidholdová, Z., Eds.; Technical University in Zvolen: Zvolen, Slovakia, 2016; pp. 61–66.

17. Van Court, R.C.; Robinson, S.C. Stimulating Production of Pigment-Type Secondary Metabolites from Soft Rotting Wood Decay Fungi ("Spalting" Fungi). In *Advances in Biochemical Engineering/Biotechnology*, 1st ed.; Springer Nature: Basel, Switzerland, 2019; p. 16.

18. Rayner, A.D.; Boddy, L. *Fungal Decomposition of Wood. Its Biology and Ecology*; Antony Rowe Ltd.: Chippenham, UK, 1997; p. 587.

19. Pandey, K.K.; Pitman, A.J. FTIR studies of the changes in wood chemistry following decay by brown-rot and white-rot fungi. *Int. Biodeter. Biodegr.* **2003**, *52*, 151–160. [CrossRef]

20. Zabel, R.A.; Morrell, J.J. *Wood Microbiology: Decay and Its Prevention*; Academic Press: London, UK, 2012; p. 476.

21. Nascimento, M.S.; Santana, A.L.B.D.; Maranhão, C.A.; Oliveira, L.S.; Bieber, L. Phenolic extractives and natural resistance of wood. In *Biodegradation-Life of Science*; Chamy, R., Rosenkranz, F., Eds.; InTech: Rijeka, Croatia, 2013; pp. 349–370.

22. Sablík, P.; Giagli, K.; Pařil, P.; Baar, J.; Rademacher, P. Impact of extractive chemical compounds from durable wood species on fungal decay after impregnation of nondurable wood species. *Eur. J. Wood Wood Prod.* **2016**, *74*, 231–236. [CrossRef]

23. Schmidt, O. Indoor wood-decay basidiomycetes: Damage, causal fungi, physiology, identification and characterization, prevention and control. *Mycol. Prog.* **2007**, *6*, 261–279. [CrossRef]

24. Frankl, J. Wood-damaging fungi in truss structures of baroque churches. *J. Perform. Constr. Facil.* **2015**, *29*, 04014138. [CrossRef]

25. Hyde, K.D.; Al-Hatmi, A.M.S.; Andersen, B.; Boekhout, T.; Buzina, W.; Dawson, T.L.; Eastwood, D.C.; Jones, E.B.G.; Hoog, S.; Kang, Y.; et al. The world's ten most feared fungi. *Fungal Divers.* **2018**, *93*, 161–194. [CrossRef]

26. Gabriel, J.; Švec, K. Occurrence of indoor wood decay basidiomycetes in Europe. *Fungal Biol. Rev.* **2017**, *31*, 212–217. [CrossRef]

27. Irbe, I.; Andersone, I.; Andersons, B.; Noldt, G.; Dizhbite, T.; Kurnosova, N.; Nuopponen, M.; Stewart, D. Characterisation of the initial degradation stage of Scots pine (*Pinus sylvestris* L.) sapwood after attack by brow-rot fungus *Coniophora puteana*. *Biodegradation* **2011**, *22*, 719–728. [CrossRef]

28. Robinson, S.C.; Laks, P.E. Wood species and culture age affect zone line production of *Xylaria polymorpha*. *Open Mycol. J.* **2010**, *4*, 18–21. [CrossRef]

29. Robinson, S.C.; Tudor, D.; Cooper, P.A. Wood preference of spalting fungi in urban hardwood species. *Int. Biodeter. Biodegr.* **2011**, *65*, 1145–1149. [CrossRef]

30. Robinson, S.C.; Tudor, D.; Cooper, P.A. Feasibility of using red pigment producing fungi to stain wood for decorative applications. *Can. J. For. Res.* **2011**, *41*, 1722–1728. [CrossRef]

31. Robinson, S.C. Developing fungal pigments for "painting" vascular plants. *Appl. Microbiol. Biotechnol.* **2012**, *93*, 1389–1394. [CrossRef]

32. Robinson, S.C.; Tudor, D.; Cooper, P.A. Utilizing pigment-producing fungi to add commercial value to American beech (*Fagus grandifolia*). *Appl. Microbiol. Biotechnol.* **2012**, *93*, 1041–1048. [CrossRef]

33. Beck, H.G.; Freitas, S.; Weber, G.; Robinson, S.C.; Morrell, J.J. Resistance of fungal derived pigments to ultraviolet light exposure. In *International Research Group in Wood Protection*; IRG/WP: St. George, UT, USA, 2014.

34. Vega Gutierrez, S.; Robinson, S.C. Microscopic analysis of pigments extracted from spalting fungi. *J. Fungi* **2017**, *3*, 15. [CrossRef]

35. Blanchette, R.A. Screening wood decayed by white rot fungi for preferential lignin degradation. *Appl. Environ. Microbiol.* **1984**, *48*, 647–653.

36. Stalpers, J.A.; Vlug, I. Confistulina, the anamorphs of *Fistulina hepatica*. *Can. J. Bot.* **1983**, *61*, 1660–1666. [CrossRef]

37. Hillis, W.E. *Heartwood and Tree Exudates*; Springer: Berlin/Heidelberg, Germany, 1987; p. 267.

38. Coulson, J. *Wood in Construction—How to Avoid Costly Mistakes*; John Wiley & Sons Ltd.: Chichester, UK, 2012; p. 208.

39. EN 350. *Durability of Wood and Wood-Based Products. Testing and Classification of the Durability to Biological Agents of Wood and Wood-Based Materials*; European Committee for Standardization: Brussels, Belgium, 2016.

40. Wagenführ, R. *Holzatlas*; Fachbuchverlag Leipzig, Carl Hanser Verlag: Munchen, Germany, 2007; p. 816.

41. URL 1. The IUCN Red List of Threatened Species. Version 2017-2. Available online: www.iucnredlist.org (accessed on 16 November 2017).

42. CIE. *Colorimetry—Part 4: CIE 1976 L*a*b Colour Space*; CIE DS 014-4.3/E:2007; CIE Central Bureau: Vienna, Austria, 2007.

43. Nishino, Y.; Janin, G.; Chanson, B.; Détienne, P.; Gril, J.; Thibaut, B. Colorimetry of wood specimens from French Guiana. *J. Wood Sci.* **1998**, *44*, 3–8. [CrossRef]

44. Németh, K. The colour of wood in CIE Lab system. *Az Erdészeti és Faipari Egyetem Tudományos Közleményei* **1982**, *2*, 125–135.

45. Eriksson, K.-E.L.; Blanchett, R.A.; Ander, P. *Microbial and Enzymatic Degradation of Wood and Wood Components*; Springer Series in Wood Science: Berlin/Heidelberg, Germany; New York, NY, USA, 1990; p. 407.

46. Gierlinger, N.; Jacques, D.; Grabner, M.; Wimmer, R.; Schwanninger, M.; Rozenberg, P.; Pâques, L.E. Colour of larch heartwood and relationships to extractives and brown-rot decay resistance. *Trees* **2004**, *18*, 102–108. [CrossRef]

47. Kokutse, A.D.; Stokes, A.; Baillères, H.; Kokou, K.; Baudasse, C. Decay resistance of Togolese teak (*Tectona grandis* Lf) heartwood and relationship with colour. *Trees* **2006**, *20*, 219–223. [CrossRef]

48. Costa, M.D.A.; Costa, A.F.D.; Pastore, T.C.M.; Braga, J.W.B.; Gonçalez, J.C. Characterization of wood decay by rot fungi using colorimetry and infrared spectroscopy. *Ciência Florestal* **2011**, *21*, 567–577.

49. Amusant, N.; Fournier, M.; Beauchene, J. Colour and decay resistance and its relationships in *Eperua grandiflora*. *J. Ann. For. Sci.* **2008**, *65*, 1–6. [CrossRef]

50. Stangerlin, D.M.; Costa, A.F.D.; Gonçalez, J.C.; Pastore, T.C.M.; Garlet, A. Monitoring of biodeterioration of three Amazonian wood species by the colorimetry technique. *Acta Amazon.* **2013**, *43*, 429–438. [CrossRef]

51. Blanchette, R.A.; Wilmering, A.M.; Baumeister, M. The use of green-stained wood caused by the fungus *Chlorociboria* in intarsia masterpieces from the 15th century. *Holzforschung* **1992**, *46*, 225–232. [CrossRef]

 *forests*

*Article*

# Analysis of Economic Feasibility of Ash and Maple Lamella Production for Glued Laminated Timber

Philipp Schlotzhauer [1],*, Andriy Kovryga [2], Lukas Emmerich [1], Susanne Bollmus [1], Jan-Willem Van de Kuilen [2,3] and Holger Militz [1]

[1] Dept. Wood Biology and Wood Products, Faculty of Forest Sciences and Forest Ecology, Georg-August-University of Göttingen, 37077 Lower Saxony, Germany
[2] Holzforschung München, Technical University of Munich, 80797 Bavaria, Germany
[3] Faculty of Civil Engineering and Geosciences, Delft University of Technology, 2628 CN Delft, The Netherlands
* Correspondence: philipp.schlotzhauer@uni-goettingen.de; Tel.: +49-551/39-33562

Received: 28 May 2019; Accepted: 24 June 2019; Published: 26 June 2019

**Abstract:** *Background and Objectives:* In the near future, in Europe a raised availability of hardwoods is expected. One possible sales market is the building sector, where medium dense European hardwoods could be used as load bearing elements. For the hardwood species beech, oak, and sweet chestnut technical building approvals already allow the production of hardwood glulam. For the species maple and ash this is not possible yet. This paper aims to evaluate the economic feasibility of glulam production from low dimension ash and maple timber from thinnings. Therefore, round wood qualities and the resulting lumber qualities are assessed and final as well as intermediate yields are calculated. *Materials and Methods:* 81 maple logs and 79 ash logs cut from trees from thinning operations in mixed (beech) forest stands were visually graded, cant sawn, and turned into strength-graded glulam lamellas. The volume yield of each production step was calculated. *Results:* The highest volume yield losses occur during milling of round wood (around 50%) and "presorting and planning" the dried lumber (56%–60%). Strength grading is another key process in the production process. When grading according to DIN 4074-5 (2008), another 40%–50% volume loss is reported, while combined visual and machine grading only produces 7%–15% rejects. *Conclusions:* Yield raise potentials were identified especially in the production steps milling, presorting and planning and strength grading.

**Keywords:** volume yield; European hardwoods; low quality round wood; strength grading; glulam

---

## 1. Introduction

The share of hardwoods in the wood stock of Central European forests is steadily increasing [1]. The higher availability of hardwoods requires the development of new markets and new value chains for an overall increase in use. A possible, large sales market is the application in load-bearing structures.

Medium dense hardwoods have preferable mechanical properties compared to softwood. The higher tensile strength of hardwoods leads to either smaller member dimensions or higher load carrying capacities. The high bending strength for hardwood glulam (up to 48 MPa) has been reported by Blaß et al. [2] and Frühwald et al. [3] for beech glulam and by Van de Kuilen and Torno [4] for ash glulam. In recent years, a number of technical approvals for hardwood glulam have been issued:

Beech glulam [5],
VIGAM oak glulam [6],
Schiller oak glulam [7], and
SIEROLAM glulam of chestnut [8].

Despite the attractive mechanical properties, the use of hardwoods in structural applications remains minor. According to Frühwald et al. [3] and Mack [9], more than 90% of the glulam products in Europe are made of softwood (mainly spruce). The survey by Ohnesorge et al. [10] on glulam producers in Germany, Switzerland, and Austria revealed that in the year 2005 out of 900,000 m$^3$ of glued rod-shaped solid wood products only 1% contained hardwood.

A number of technological reasons as well as historical and silvicultural reasons has led to the fact that mainly softwood is used in wood construction. The use of softwood has been favored over decades because the physical properties are quite predictable and differences between the different softwood species are small. Furthermore, softwood is characterized by long, straight logs with low degrees of taper, homogeneous assortments, and few knots that are usually evenly distributed [11]. There are several further technological constraints for the use of hardwoods in structural applications, such as lack of knowledge of the long-term behavior of hardwood gluing, or the less number of certified grading machines compared to softwood, non-harmonized standardization and production processes not optimized for hardwood species.

One major aspect for the broader use of hardwoods in construction (especially glulam) is the economic feasibility of the production. For hardwoods, at present, no calculated data from a production facility is available. Torno et al. [12] estimated the production cost of ash lamellas to be three times higher as of spruce lamellas. Thus, besides the higher load-bearing capacity of hardwood glulam, the cost-efficient use of the resource hardwood is required, in order to reduce this cost difference. This includes both the optimization of the production process and of the resources used. Processing cheap, particularly small diameter hardwood logs, which are usually used for energy recovery in Europe [12], is one of the frequently discussed issues. Exploiting small diameter hardwoods for material utilization, e.g., sawing, is an important issue in Northern America as well [13].

It is the aim of this paper to contribute to the overall goal of an effective use of the available hardwood resources by minimizing the waste of each production step (of glulam lamellas) separately and for the entire production. The use of small diameter logs from thinnings as a poor-quality resource is the focus of this yield analysis. In the current study, the yield analysis from log sections to planed and graded glulam lamellas is performed using state of the art processing technology. Moreover, the achieved yields are linked to the mechanical properties relevant for glulam lamellas and measured for the investigated samples. Doing so, the economic feasibility of lamella production out of small diameter logs of the rare hardwood species maple and ash can be estimated. The single production steps and technologies of the production of glulam from low-dimension maple and ash logs are analyzed and described.

## 2. Conversion Efficiency of Hardwoods

In literature, different terms exist to measure the conversion efficiency. In Northern America, the recovery rates with measures like lumber overrun, lumber recovery factor (LRF) and cubic lumber recovery (CLR) are used. In Europe, the term yield is most commonly used. All these definitions have in common that they calculate the volume ratio between the output sawn product and the input logs. The term yield goes even beyond that and can be determined for each production step separately. It can include final, as well as intermediate, products. This allows revealing and analyzing the weakest points of the production process. The use of waste material as side product or for energy production can also be considered. A higher lumber volume does not necessarily lead to higher lumber value. That is why it is important to distinguish between lumber volume recovery and lumber value recovery. For sawmill owners or managers, the latter is decision relevant [14].

Due to the low production volumes of hardwood glulam, yield values are known to only a very small extent. Studies on European hardwoods analyzing the yield from log to planed (dry-dressed) lumber are rare. Torno et al. [12] performed an extensive study on the production of beech lamellas and Van de Kuilen and Torno [15] on beech and ash lamellas. For lamellas sorted according to the German visual grading standard DIN 4074-5 [16], volume yield values as high as 26% for beech and

27.7% for ash were attained. When sorting the lamellas according to the more stringent sorting rules of the German technical approval Z 9.1 679 [5], for the production of glulam the total yield starting at round wood (middle diameter classes 2b–6) ended at only 22% for beech and 26.9% for ash. In this case, however, higher mechanical properties are presumed. As shown by Torno et al. [12], the cutting pattern and the sawing technology affect the final yield. For graded beech lamellas those can drop to 10% or rise to 26%. The highest yield was attained with the grade sawing method, where a vertical bandsaw headrig cuts "around the log" until only a heart plank is left. In these studies, in addition to the cutting pattern and the sawing technology, the quality and the diameter of the round wood had a major influence on the final yield. Frühwald et al. [3] estimate the total yield of the production of high-quality beech glulam from good to medium round wood qualities (B and C) to be around 28.5%.

The reported final yields for hardwood lamellas are below the ones for softwoods. Final yields of the latter range from 24.5%–38.5% [17,18]. Even higher yield values of 40% are stated by Torno et al. [12] for a modern spruce profiling unit. Frühwald et al. [3] mention that the final yield depends greatly on the size (production volume) of the glulam producing company. Only looking at the production of spruce glulam from dried sawn lumber, big producers are able to attain yields between 69% and 75%, while little glulam producers only reach yields of 53%.

Studies like the ones presented by Torno et al. [12] and Van de Kuilen and Torno [15] on the yield from logs to planed and strength graded hardwood lamellas are scarce. A few studies describe the yields of only individual production steps. Their results are summarized below.

## 2.1. Sawing/Milling

According to Steele [19], the following factors influence the lumber recovery in sawing (milling):

- Log diameter, length, taper, and quality
- Kerf width
- Sawing variation, rough green-lumber size, and size of dry-dressed lumber
- Product mix
- Decision making by sawmill personnel
- Condition and maintenance of mill equipment
- Sawing method

In the study of Lin et al. [14] in small US hardwood sawmills the factors log grade, diameter, sweep, length, species and sawmill specifications had a significant influence on the lumber volume recovery. It is also stressed that interactions between different factors can have a significant influence on the lumber volume recovery. Further influencing factors like board edging and trimming are also introduced. Richards et al. [20] simulate the volume and value yield of sawing hardwood lumber depending on the above mentioned factors. In their simulation the volume yield of live sawing is always higher than that of any four-sided sawing pattern (quadrant, cant, and decision), when sawing the same size logs. When sawing small logs with large core defects the value yield, though, is higher when applying a four-sided sawing pattern. The authors also emphasize the importance of the rotational position on the carriage for the first cut.

Ehlebracht [17] compares volume yield values of four German sawmills for the sawing of square-edged sawn lumber (rough green) from low dimension beech logs. The highest yield value of 57% is attained by a gang saw headrig utilizing the cant sawing method [20]. The lowest yield value of 36% is produced by a circular saw headrig, which produces a comparatively wide kerf. These values are consistent with the values reported by Emhardt and Pfingstag [21] and Fronius [22] that, when combining their findings, present values that range from 42%–47% for the production of square-edged sawn lumber from low dimension beech logs (middle diameter classes 2b and 3a). The lower yield values of Ehlebracht [17] are comparable to the 35% yield reported by Fischer [23] for the production of parquet friezes and pallet boards from low dimension oak logs. For five small US hardwood mills, Lin et al. [14] report cubic recovery percentages (CRP) of 53.2% for red oak (*Quercus rubra* L.) and 57.5%

for yellow polar (*Liriodendron tulipifera* L.). The CRP expresses the volume of rough green lumber as percentage of cubic log scale volume and is therefore comparable to the yield of the production step "sawing" analyzed by Ehlebracht [17]. The mean small-end diameter (SED) of the input logs in the study of Lin et al. [14] was 33 cm, i.e., also low dimension logs were sawn. All five sawmills used the grade sawing method—two with circular saw headrigs and three with bandsaw headrigs. The simulations of Richards et al. [20] for US hardwood mills result in volume yield values, which range from 54%–76%. The high values, though, are only attainable, when live sawing large logs. According to Fronius [22], a further yield drop of 15%–20% (relative to the original round wood volume) is to be expected when square edging live sawn lumber.

### 2.2. Drying

Drying losses arise from volumetric shrinkage and the quality of the sawn lumber after drying. For hardwoods such as oak, improper drying results in staining, checking, splitting, and warp, which leads to a reduced sawn wood value [24,25]. Therefore, proper drying schedules are of high importance.

Generally, the higher the specific gravity of the wood is the higher is also the volumetric shrinkage [26]. It varies within a species and even for lumber from the same log. The volumetric shrinkage during technical drying of rough green lumber to a moisture content of 12% ranges from 14%–21% for beech, from 12.8%–13.6% for ash and from 11.5%–11.8% for maple [17,27]. Spruce shrinkage losses are around 12% [27]. The volumetric shrinkage in the production of hardwood lamellas for glulam lies between 11% and 17% for beech and at 9.8% for ash [15].

### 2.3. Planing

Planing losses depend on the chosen oversize, the final product and the drying quality (i.e., warping and bowing). The resulting losses present a combination of planing away the oversize and sorting out (presorting) boards with intensive bowing. For example, when trimming the lamellas to shorter lengths, the oversize can be reduced and thus the planing losses are also reduced. In similar studies to the presented one [12,15], planing and presorting losses (due to bowing) for the production of hardwood glulam lamellas vary from 18%–46%—a relatively wide range.

### 2.4. Grading

Grading is an important step within the production, as the quality of sawn wood is assessed in terms of appearance (i.e., cladding, furniture) or mechanical properties predicted. As a consequence, a discrete value is assigned to a lumber specimen. Both the quality of the produced lumber in terms of achieved mechanical properties and the yield are of interest. For grading, the yield is the share of dry-dressed lumber (dried, jointed, and planed), which is assigned to a certain quality class and not rejected.

Data on hardwood grading yield in general, and on strength grading in particular, is scarce, since hardwoods are rarely strength graded. Generally, the yield losses depend on the grading method (machine vs. visual grading), wood quality, growth region, cross-section, and sawing pattern selected. For European hardwoods, the effect the single mentioned factors have on the grading yield, are known to only a small extent. If lamellas are sawn pith free, the grading losses are lower compared to other sawing patterns. This is because the pith is a general rejection criterion for visually graded hardwood lumber after the German visual grading standard for structural timber DIN 4074-5 [16]. Thus, Glos and Torno [28] report for 324 ash boards and 459 maple boards graded according to DIN 4074-5 rules for joists rejection rates of as high as 21% and 37% due to pith and extreme grain deviation. It should be mentioned, though, that for that study the visual assessment of the boards is only being made for that part of each specimen, which is selected as free testing length. In Torno et al. [12] the loss values for beech lamellas range from 37%–62%, if graded visually in accordance with the German visual grading rules for structural lumber DIN 4074-5 [16]. If lamellas are graded in accordance with the German technical building approval for beech glulam Z 9.1 679 [5] the rejection rate increases to 47% and 69%.

## 3. Test material

The round wood used for this investigation came from thinnings in mixed forest stands (mixed beech forests) of the state forestry offices Leinefelde and Heiligenstadt (Central Germany). The wood was harvested in the winter of 2014/2015 with harvester technology. Until the milling in June 2015, the round wood sections (logs) with a length between 3.20 and 3.40 m remained on the log yard of the department sawmill. According to the transport invoice 14.89 m$^3$ (79 logs) of ash (*Fraxinus excelsior* L.) and 16.25 m$^3$ (81 logs) of maple (80 logs of *Acer platanoides* L. and 1 log of *Acer pseudoplatanus* L.) were delivered (with bark). For the yield analysis, round wood sections (logs) with the following characteristics were ordered:

- Round wood quality C or worse (according to the Framework Agreement on Raw Timber Trade in Germany-RVR [29]);
- Length ≥ 3.20 m; and
- Round wood diameter classes 2–3

## 4. Production Steps and Determination of Characteristics

### 4.1. Round Wood Sections (Logs)

On the log yard the round wood sections were trimmed uniformly to a length of 3.15 m in order to be able to determine the heartwood coloring (i.e., brown heart) on both ends. At the top (small) end of each trunk a slice of 1–2 cm thickness was cut off. The final cut was performed at the bottom of each trunk (large end) to a length of 3.15 m. Thus, total log volumes were reduced to 14.3 m$^3$ for ash and 15.8 m$^3$ for maple. For each round wood section the minimum and the maximum diameter was determined in the middle of every 25 cm section. The last section only had a length of 15 cm. Using the mean diameter for each 25 cm section and the one 15 cm section ($d_{Mn}$), the section volumes were calculated with Huber's formula. The single section´s volumes were then added up resulting in Equation (1):

$$V_{Sec.} = \left( \sum_{1}^{12} \frac{\pi}{4} \times 0.25 \text{ m} \times d_{Mn}{}^2 \right) + \frac{\pi}{4} \times 0.15 \text{ m} \times d_{Mn}{}^2 \left[ \text{m}^3 \right] \tag{1}$$

The logs were sorted into diameter classes according to their small-end (top-end) diameter (SED) and into quality classes according to the specifications of the RVR [29] and DIN 1316-3 [30]. Both standards allow the assignment to classes from A (highest quality) to D (lowest quality). The quality-determining characteristics of the round wood sections were determined and recorded in accordance with Annex VIII (Measurement of the characteristics) of the RVR [29]. The characteristics shrinkage cracks, insect holes, tree cancer and the so-called moon ring (light discoloration in heartwood) were not recorded and thus were not part of sorting.

The RVR [29] offers no separate quality grading for maple and ash logs. Thus, depending on the particular characteristic, the oak grading rules (e.g., for knots, star shake, twigs, etc.) or those for beech (only for width of brown heart and heart shake) were used.

### 4.2. Sawing/Milling

The logs were milled with a mobile horizontal bandsaw headrig (Montana ME 90 2.0 from SERRA, Rimsting, Germany) with a kerf width of 2.45 mm. The cant sawing patterns used are shown in Figure 1.

The sawing patterns and the distribution of board dimensions were chosen for each log separately, mainly depending on the small-end log diameter ($d_z$ or SED). Thus, the maximum yield could be attained. The pattern A was used most. If side boards were produced (colored boards in pattern C), they were edged to square edged lumber on a circular saw. For maple, five different lumber dimensions were sawn, for ash three (see Table 1).

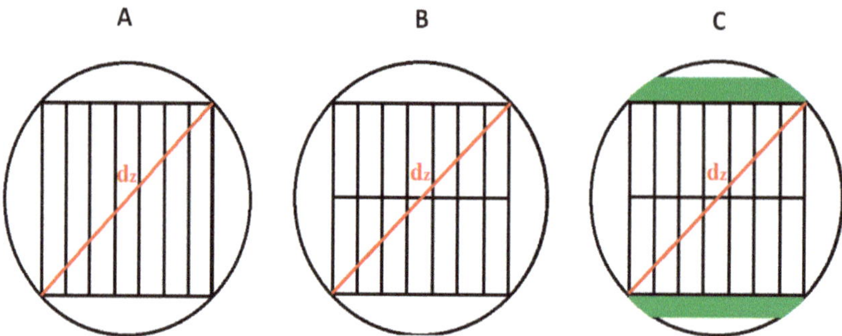

**Figure 1.** Cant sawing patterns of the milling.

**Table 1.** Nominal dimensions and quantities (*n*) of sawn green lumber and the resulting planed lamellas (dry-dressed lumber).

| Sawn Green Lumber | Planed Lamellas | Maple | Ash |
|---|---|---|---|
| Width × height (mm) × (mm) | Width × height (mm) × (mm) | *n* | *n* |
| 115 × 35 | 100 × 25 | 88 | - |
| 145 × 35 | 125 × 25 | 132 | - |
| 145 × 40 | 125 × 30 | 85 | 121 |
| 115 × 45 | 100 × 35 | 92 | 104 |
| 145 × 45 | 125 × 35 | 94 | 162 |

Only the main product glulam lamella was produced for this study. No side products, like trimming or baseboards, etc., were produced. The side products would raise the final yield. The final product—planed glulam lamellas—were subjected to destructive tensile testing after visual and machine strength grading (see sub sample "TH II" in Kovryga et al. [31]).

*4.3. Drying*

The technical drying took place in the in-house conventional dryer (HB Drying Systems, Almelo, The Netherlands). The drying parameters were chosen in order to ensure gentle drying of the boards. The drying process took 21 days. To determine the volumetric shrinkage, the dry lumber volume (at 12% moisture content) is subtracted from the sawn lumber (rough green) volume. For this purpose, for each dimension and wood species six lamellas were selected randomly. On these lamellas, the lengths (in mm) were determined with a tape measure on the rough green and the dry lumber. Lumber dimensions (in mm) were measured at intervals of 25 cm—starting and ending at the board ends.

*4.4. Presorting and Planing*

The dried boards were jointed and planed to glulam lamellas (dry-dressed lumber) with the nominal dimensions presented in Table 1. After the planing process, each lamella that could not attain the nominal dimension (cross-section) on the full length was sorted out (due to a combination of bowing and too little oversize). The volume of the remaining glulam lamellas was calculated by determining their lengths with a tape measure and using the nominal lamella dimensions.

*4.5. Strength Grading of Planed Boards*

4.5.1. Visual Strength Grading

To assess the quality of hardwood lamellas, different grading methods were used. First, each of the lamellas was visually classified according to the German visual strength grading standard DIN

4074-5 [16] over the entire length. The standard uses ten visual criteria to assign hardwood boards to visual strength grading classes. In the current study, the knottiness, presence of pith, bark inclusion, wane, and fiber deviation (grain angle) were considered.

All relevant grading criteria were measured as defined in DIN 4074-5 [16]. To assess the knottiness—one of the major parameters of strength grading—the criteria single knot (SK) and knot cluster (KC) were used. Single knot or *DIN Einzelast Brett (DEB)* relates the size of the single knot to the lamella width. For grading, the ratio (knot) with the highest value is indicative. Knot cluster (KC) or *DIN Astansammlung Brett (DAB)* is a multiple knot criterion, which considers all knots appearing in a (moving) window of 150 mm. Therefore, the spread of all knots over the 150 mm window is related to the width of the board. The edge knot criterion (E) or *Schmalseitenast* is an optional criterion for boards and represents the penetration depth of the knots appearing on the edge side only. A low value of these visual grading criteria stands for either rare occurrence or small size of the strength reducing knots and vice versa.

The only adjustment made concerns the measurement of the fiber deviation (grain angle). Fiber deviation is defined as an angle between the fibers and loading direction over a certain length and is measured in percent. The grain angle has a significant impact on strength [32]. Most grading standards indicate that the fiber deviation can be measured on drying checks or by the scribing method on the wood surface. Both methods are reported to have limited use for medium-dense hardwoods [33,34]. In the present study, the visible fiber deviation was detected on drying checks and, additionally, the surface was assessed qualitatively for fiber deviations exceeding the limits of DIN 4074-5 [16]. The specimens exceeding the limits are rejected.

Hardwood boards are assigned to the visual grades LS13 (highest quality), LS10 (medium quality) and LS7 (lowest quality) based on the boundary values listed in Table 2. To assign a lamella to a visual grade, all boundary values are to be met. Otherwise, the specimen is assigned to the next lower grade or rejected.

**Table 2.** Boundary values for grading of hardwood lamellas to visual grades (LS7 to LS13) after DIN 4074-5 [16].

|  | LS13 | LS10 | LS7 |
| --- | --- | --- | --- |
| **DEB (SK)** | 0.2 | 0.333 | 0.5 |
| **DAB (KC)** | 0.333 | 0.5 | 0.666 |
| **Edge knot (E)** | -* | -* | -* |
| **Pith** | no | no | no |
| **Fibre deviation** | 7% | 12% | 16% |

* No requirements set.

Additionally, to estimate the effect of the grading parameters pith and DAB on the yield, two grading combinations—one without any requirements on pith and one without any requirements on pith and DAB—are applied to the lamellas.

### 4.5.2. Combined Visual and Machine Strength Grading

Additionally, the boards were graded using a combined visual and machine grading approach. The procedure was suggested by Frese and Blaß [35] and is used for beech glulam produced after the German technical building approval Z-9.1-679 [5]. This grading approach combines visual grading parameters (i.e., SK and KC) with the dynamic Modulus of Elasticity ($MOE_{dyn}$), a parameter used in most state of the art grading machines for softwoods. The $MOE_{dyn}$ was determined using the "eigenfrequency" method (laboratory and grading machine ViSCAN by MiCROTEC, Bressanone/Brixen, Italy). In case of ViSCAN, the natural frequency ($f$) from longitudinal oscillation was combined with the density ($\varrho$) measured by an X-ray source, and the length ($l$) of the measured specimen (Equation (2)).

In the laboratory, the density was determined using the gravimetric method. Both measurements provide comparable results in terms of $R^2$ value (0.972).

$$MOE_{dyn} = 4 \times l^2 \times \rho \times f^2 \times 10^6 \qquad (2)$$

The combined approach uses separate boundary values for visual grading parameters (i.e., SK, KC) and $MOE_{dyn}$. The boundaries presented by Frese and Blaß [35] are fitted to beech lamellas. For the present study, the combined grading is optimized for ash and maple and presented by the paper Kovryga et al. [31]. Table 3 shows the combination of boundary values selected for the current study. As example, for maple the "Solution B" and for ash the "Solution C" proposed for combined grading by Kovryga et al. [31] is selected. The presented combination allows grading to three different grades plus reject group. The highest grade shows characteristic tensile strength values (above 38 N/mm$^2$) fitting the tensile strength of finger jointed lamellas stated by Van de Kuilen and Torno [4].

**Table 3.** Optimized grading rules for combined visual and machine strength grading of ash and maple (according to Kovryga et al. [31]; maple: "Solution B", ash: "Solution C").

| | Grade | Boundary Values | | | | | Resulting Tensile-Classes |
|---|---|---|---|---|---|---|---|
| | | DEB (SK) (–) | DAB (KC) (–) | Edge knot (E) (–) | Pith (–) | $MOE_{dyn}$ (kN/mm$^2$) | |
| Maple | 1 | 0.1 | 0.1 | -* | Allowed | 13.9 | DT38 |
| | 2 | 0.2 | 0.5 | -* | Allowed | 12.2 | DT25 |
| | 3 | 0.3 | 0.6 | -* | Allowed | 10.9 | T15 |
| | Reject | | | | | | |
| Ash | 1 | 0.2 | 0.2 | -* | Allowed | 16.5 | DT38 |
| | 2 | 0.3 | 0.3 | -* | Allowed | 15.5 | DT34 |
| | 3 | 0.4 | 0.4 | -* | Allowed | 11.6 | DT22 |
| | Reject | | | | | | |

* No requirements set.

### 4.6. Yield Calculation

For the determination of the total yield, the yields of each single production step are added up. The yield of each production step is calculated by dividing the output product volume by the input volume. How volumes of each intermediate product are calculated and what assumptions are made for these calculations is described above for each production step separately.

## 5. Results and Discussion

### 5.1. Grading of Logs

Table 4 shows the sorting of the maple and ash logs into diameter and quality classes. Following the descriptions of Van de Kuilen and Torno [15], the diameter sorting was carried out by considering the small-end diameter (SED) inside bark. The supplied round wood sections mainly cover the diameter classes from 2a to 3b, with individual sections with diameters below 20 cm and over 40 cm. For maple and ash, the bark shows a mean thickness of 0.5 cm. Maple shows a higher number of logs graded to the higher quality classes (B and C) compared to ash.

**Table 4.** Number of logs per species sorted after small-end diameter (inside bark) class and quality class according to RVR [29].

| Diameter Class | 2a | 2b | 3b | 4 | 1b | 2a | 2b | 3a | 3b | 1b | 2a | 2b | 3a | 3b |
|---|---|---|---|---|---|---|---|---|---|---|---|---|---|---|
| Quality Class | B | | | | C | | | | | D | | | | |
| Quantity — Maple | | 3 | 2 | 1 | 4 | 18 | 11 | 9 | 3 | 5 | 14 | 5 | 2 | 3 |
| Ash | 1 | | | | | 8 | 16 | 5 | 1 | 2 | 23 | 19 | 4 | |

Tables 5 and 6 show the results of the log quality sorting according to RVR [29] and DIN 1316 3 [30] in detail.

**Table 5.** Yields in % for quality sorting of logs according to RVR [29] separated after sorting criteria (log characteristics).

| Log Characteristics | Maple | | | | Ash | | | |
|---|---|---|---|---|---|---|---|---|
| | A | B | C | D | A | B | C | D |
| Callused knot (bump) | 23 | 1 | 76 | | 46 | | 54 | |
| Healthy knot | 63 | 29 | 9 | | 89 | 10 | 1 | |
| Decayed knot | 63 | 31 | 5 | 1 | 89 | 6 | 5 | |
| Twigs | 76 | 24 | | | 100 | | | |
| Bump on group of broken of twigs | 95 | 1 | 4 | | 100 | | | |
| Star shake/check | 60 | 29 | 11 | | 4 | 14 | 80 | 3 |
| Heart shake/check | 33 | 61 | 5 | 1 | 81 | 15 | 1 | 3 |
| Frost crack | 98 | | 3 | | 99 | | 1 | |
| Ring shake | 98 | 3 | | | 99 | | 1 | |
| Bow (Sweep and crook) | 48 | 14 | 6 | 33 | 38 | 1 | 5 | 56 |
| Spiral (twisted) grain | 98 | | 3 | | 100 | | | |
| Rot | 99 | | | 1 | 97 | | | 3 |
| Log length | 99 | | 1 | | 96 | | 4 | |
| Width of brown heart | 86 | 14 | | | 25 | 46 | 29 | |
| Final quality class of logs | | 8 | 56 | 36 | | 1 | 38 | 61 |

**Table 6.** Yields in % for quality sorting of logs according to DIN 1316-3 [30] separated after sorting criteria (log characteristics).

| Log characteristics | Maple | | | | Ash | | | |
|---|---|---|---|---|---|---|---|---|
| | A | B | C | D | A | B | C | D |
| Length | 99 | | 1 | | 96 | | 4 | |
| Mid-diameter | 14 | 11 | 70 | 5 | | 4 | 9 * | |
| Callused knot (bump) | 25 | | 34 | 41 | 47 | | 42 | 11 |
| Healthy knot | 95 | | 5 | | 91 | 3 | 6 | |
| Decayed knot | 90 | | 8 | 3 | 96 | | 4 | |
| Eccentricity of pith | 88 | 13 | | | 80 | 20 | | |
| Star shake/check | 60 | | 5 | 35 | 4 | | | 96 |
| Heart shake/check | 40 | 14 | 46 | | 87 | 4 | 9 | |
| Brown heart | 66 | | 34 | | 38 | 25 | 37 | |
| Bow (Sweep and crook) | 61 | 5 | 1 | 33 | 39 | 5 | | 56 |
| Rot | 99 | | | 1 | 97 | | | 3 |
| **Final quality class of logs** | **1** | **1** | **26** | **71** | | | **4** | **96** |

* No requirements set.

Tables 5 and 6 present the final assignment of the round wood sections into the quality classes (the last row of both tables) based on the individual class assignment for each sorting criterion. Each single criterion's influence on the grading can be seen as well as the total distribution of quality classes per species. For example, according to DIN 1316-3 [30], 71% of the maple logs are graded into the lowest quality class D (see Table 6). The final percentage value is a result of all wood characteristics combined. It can be seen that for maple the grading into the D class is mainly due to the characteristics callused knot, star shake, and bow. When sorting according to the RVR [29] specifications, mainly log bowing is decisive for sorting into class D (see Table 5). Especially in the second lowest grade C, it is observable that the two different quality sorting schemes weigh the different characteristics differently, i.e., have different characteristic's boundary values for the same class. While grading into RVR class C of maple is mainly due to callused knots (76%), DIN 1316-3 [30] sorting into class C is due to a number of characteristics (mid diameter, callused knots, heart shake, and brown heart). Both grading schemes sort the majority of the studied logs into the classes C and D.

In general, the two sorting guidelines for round wood use different lists of characteristics. For example, Table 6 shows that in the case of sorting according to DIN 1316-3 [30] the criterion mid-diameter leads to a classification into quality class C for 70% of the maple logs and for 96% of the ash logs. Compared to that, the diameter of the logs is not relevant, when sorting according to RVR [29]. The possible advantage of the absence of log size criteria is that the actual visible log quality can be assessed and used to qualify the logs for the production in addition to the diameter.

Looking at Tables 5 and 6, it also becomes obvious that—under the same storage conditions—ash logs tend to form more severe end cracks (star and heart shake) than maple. This cracking results in a serious deterioration of quality and leads to a reduced sawn lumber yield (mainly value yield). Thus, it is recommended to saw (mill) ash logs shortly after logging or adapt storage (e.g., water storage) to ensure the best possible lumber quality and highest yield. Short storage times' respectively adjusted storage conditions are also advised for maple logs, since fungal discoloration starting from the log ends presents problems [27]. For an end use as construction material, though, these discolorations may be of low significance, since they do not affect the elasto-mechanic properties of the lumber.

### 5.2. Yields from Logs to Unsorted Glulam Lamellas

Table 7 summarizes the volume losses and the resulting yields for each production step. It can be seen that the major production losses arise from sawing the logs and presorting the dried boards. Both species do not differ considerably.

**Table 7.** Yield for each production step from logs to planed lamellas (unsorted).

| Product | Production Step (PS) | Maple | | | | Ash | | | |
|---|---|---|---|---|---|---|---|---|---|
| | | Yield | | Waste/Loss | | Yield | | Waste/Loss | |
| | | in m$^3$ | in % | in m$^3$ | in % PS | in m$^3$ | in % | in m$^3$ | in % PS |
| Logs | | 15.8 | | | | 14.3 | | | |
| | Milling/sawing | | | 7.6 | 48.2 | | | 7.1 | 49.5 |
| Boards (green) | | 8.2 | 51.8 | | | 7.2 | 50.5 | | |
| | Drying | | | 0.7 | 8.7 | | | 0.8 | 10.7 |
| Boards (dry) | | 7.5 | 47.3 | | | 6.5 | 45.1 | | |
| | Presorting & planing | | | 4.2 | 56.3 | | | 3.8 | 59.6 |
| Planed lamellas | | 3.3 | 20.9 | | | 2.6 | 18.2 | | |

### 5.2.1. Sawing/Milling

The mean volume yield of sawing the 81 maple logs by the cant sawing method to square-edged lumber is 51.8%. The mean volume yield of sawing the 79 ash logs is with 50.5% slightly lower. The log diameter strongly influences the volume yield of this production step. The effect the mid-diameter has on the sawing yield of this study can be observed in Figure 2.

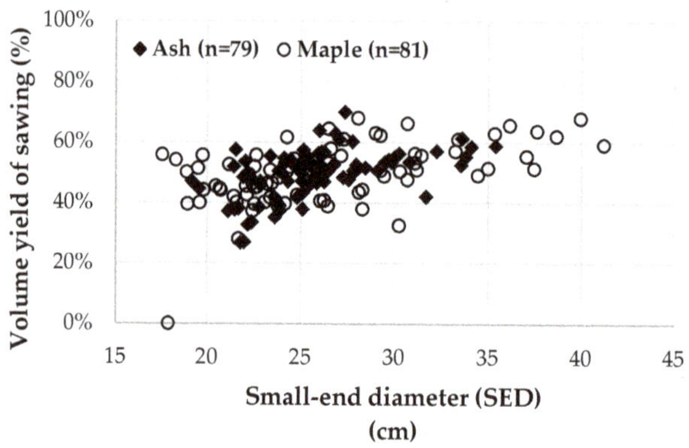


**Figure 2.** Volume yield of sawing ash and maple depending on the small-end log diameter (inside bark).

The higher the log diameter gets, the higher the yield gets. The variation in sawing yield values drops with increasing diameters, as the influence of the log grade (quality class) on the yield decreases. Compared to the log diameter, the quality class only has a minor influence. Wade et al. [36] analyzed data from 35 US hardwood mills and also concluded that a positive linear relationship between log diameter and sawing yield (in their case LRF) exists. In the simulation of Richards et al. [20] hardwood sawing yields of sawing low dimension logs (SED = 25 cm) by the cant sawing method start at 56.1%, while from logs with large diameters (SED = 71 cm) up to 67.2% rough green lumber can be produced. In Ehlebracht [17] only one of four hardwood mills attained a sawing yield of 57%, when cant sawing low-dimensional hardwood logs. Two mills achieved yields like this study (50% and 51%), while a mill with a circular saw headrig only reached 36% volume yield. All presented studies show that it is economically advantageous to sort out logs with diameters below a certain value. The boundary value for the diameter has to be determined for each production site and product separately. The results of Wiedenbeck et al. [13] give rise to the assumption that this boundary value also depends on the wood species sawn.

Lin et al. [14] prove that the log grade (quality class) has an effect on the hardwood volume recovery. In this study, this is only observed in the lower log diameter classes. The individual characteristics eccentricity of pith, ovality and taper show no significant influence on the yield of the first production step. The two latter characteristics are not part of the RVR [29] and DIN 1316-3 [30] sorting standards. Nonetheless, their influence on the yield during production is examined. For ash, the degree of bowing (in one direction) has no influence on the yield of milling. For maple, increased bowing (in one direction) leads to a decreased yield of milling. Multiple bows in one log (in one or more directions) decrease the yield of milling significantly. Comparing logs with one bow in only one direction with logs with multiple bows, the yield is reduced from 52.7% to 43.7% for maple and from 51.7% to 46.1% for ash. The same relationship—but less pronounced—can be found in so-called butt-cuts. In these first logs of trees taken above the stump, the milling process removes a high volume of wood from the large end of the log.

## 5.2.2. Drying

Drying of the green lumber was carried out for all dimensions and species with the same slow drying program, in order to avoid damages due to inadequate (i.e., too fast) drying. For maple, the volumetric shrinkage lies between 8.0% and 8.9% (average 8.7%), while for ash it lies between 9.6% and 11% (average 10.7%). For both species, these values lie in the lower range of the above-mentioned literature values. In some cases, the boards started warping (bowing, crooking, cupping, twisting, etc.) immediately after or even during the milling due to inherent tension in the trunks (eccentric pith, reaction wood, around big knots, etc.). Nonetheless, these boards were stacked and underwent drying.

## 5.2.3. Presorting and Planing

Before planing the dried boards, they were pre-sorted. Boards with extreme bowing were sorted out. If the infeed and outfeed rollers of the planer were able to press down the bow, resulting in fully planed board surfaces, the lamellas were not sorted out. Nevertheless, the volume loss of this production step is 56.3% for maple and 59.6% for ash. The resulting total yields of planed boards (unsorted glulam lamellas) are, thus, 20.9% for maple and 18.2% for ash. If the presorting was excluded from this calculation, i.e., if the bows were cut out (resulting in shorter lamella lengths) and thus all boards could be planed to the nominal dimensions, total yield values of 33.4% (maple) and 33.2% (ash) could be obtained. For future investigations, it is planned to evaluate the influence round wood quality has on presorting and planning losses. Especially for low-dimension logs of poor quality the question arises, how much of the resulting twisting and bowing in the dried lumber is due to the drying process and how much is already present in the rough green lumber.

### 5.3. Strength Grading of Glulam Lamellas (Planed Boards)

#### 5.3.1. Grading Results

As explained in Section 4.5, the planed boards were graded visually according to the German visual strength grading standard DIN 4074-5 [16]. Furthermore, the result of two adjusted grading schemes were compared—when the criterion "pith" is excluded from visual strength grading according to DIN 4074-5 [16] and when only single knots (DEB) are evaluated according to DIN 4074-5 [16]. Additionally, the lamellas were graded following the combined visual and machine grading proposed by Kovryga et al. [31] and presented in Table 3.

Figure 3 shows the grading results for ash and maple, respectively. The second box of each diagram gives the results of visual grading according to DIN 4074-5 [16]. For both ash and maple, only few boards are sorted into the classes LS7 and LS10. The majority is either sorted into the highest quality class LS13 or rejected. When excluding the criterion pith from DIN 4074-5 [16] sorting (see third box), no ash lamellas and four maple lamellas were rejected. The majority of the lamellas is graded into LS13 (ash: 195; maple: 238). Only applying the DIN 4074-5 [16] boundary values for the criterion DEB (single knot) gives almost identical sorting results. The combined grading proposed and optimized by Kovryga et al. [31] for the here studied lamellas result in a relatively even distribution of lamellas over the three grades. For ash 6.8% and for maple 15.7% of the lamellas are rejected.

For grading according to DIN 4074-5 [16], a high effect of the pith criterion on the grade class assignment can be stated. Grading with pith as rejection criterion results in a reject rate of 48% for the ash boards and 38% for the maple boards. If the pith criterion is excluded from grading, none of the ash boards and only 1% of the maple boards are rejected. Similar results are reported by Torno et al. [12], who detected pith in 26% and 30% of the graded beech boards. Here the sawing pattern was similar to this study, but logs with larger diameters were sawn. Van de Kuilen and Torno [15] calculated for their study the ratio of pith containing board volume to initial round wood volume (inside bark) to be 0.2% for ash and 0.9% for beech. In this study, this ratio is 9.1% for ash and 8.0% for maple. This much higher appearance of pith can be explained by the fact that lower dimension logs were sawn and the overall log quality was poorer. Furthermore, the study of Van de Kuilen and Torno [15] used a special sawing pattern ("sawing around the log" or "grade sawing") designed to produce boards without pith. Generally, it can be concluded that the sawing pattern and the low log dimensions chosen for this study resulted in a high amount of pith boards, which have to be sorted out, when sorting according to DIN 4074-5 [16]. Pith is also the main downgrading criterion in the grading of ash and maple lamellas studied by Glos and Torno [28]. The rest of the boards of this study show good quality for both species, resulting in a high proportion in LS13 grading.

One explanation for the higher amount of pith containing boards in the ash compared to the maple collective can be the fact that in ash trees the pith is typically "wandering", which is due to crooked growth in early years [17,37]. Other reasons can be more severe bowing of the ash logs or littler log dimensions. Figure 4 proves that the small-end diameters are not severely different for the 81 maple and 79 ash log sections.

The bowing of the raw material was according to RVR [29] specifications only measured for log sections that had one bow over the entire log length. This criterion shows now difference between the species ash and maple as well (see Figure 4). Checking the number of logs with compound bowing (bowing into two or more directions) reveals a different picture, though. While only 55% of the maple log sections are characterized by compound bowing, 77% of the ash log sections have compound bows. This could be an explanation for the higher amount of pith containing boards in the ash collective. Since the collected data does not contain information on the degree of compound bowing, one cannot distinguish between the influence of the "wandering pith" and the log section bowing.

To finalize the discussion of the effect of the grading parameters on the yield, the effect of the knot cluster criterion (DAB) is observed. Comparing both visual grading options—for DIN 4074-5 [16] "without considering pith" and "only DEB (without considering pith and DAB)"—little to no changes can be observed (see Figure 3). The added value (information) of DAB for grading is illustrated in Figure 5, which plots the maximum DEB against maximum DAB values for all ash and maple boards. The paired values (boards) on the bisector show those boards, where the maximum DEB is bigger than or equal to any found DAB. For all other boards a DAB greater then the DEB is reported. The grey area indicates those boards, for which the criterion DAB leads to a sorting class downgrading, when sorting according to for DIN 4074-5 [16]. This is the case for only twelve maple boards (3.7%) and three ash boards (1.4%). Therefore, the criterion knot cluster (DAB) is not decisive for downgrading into a lower sorting class, if graded after DIN 4074-5 [16].

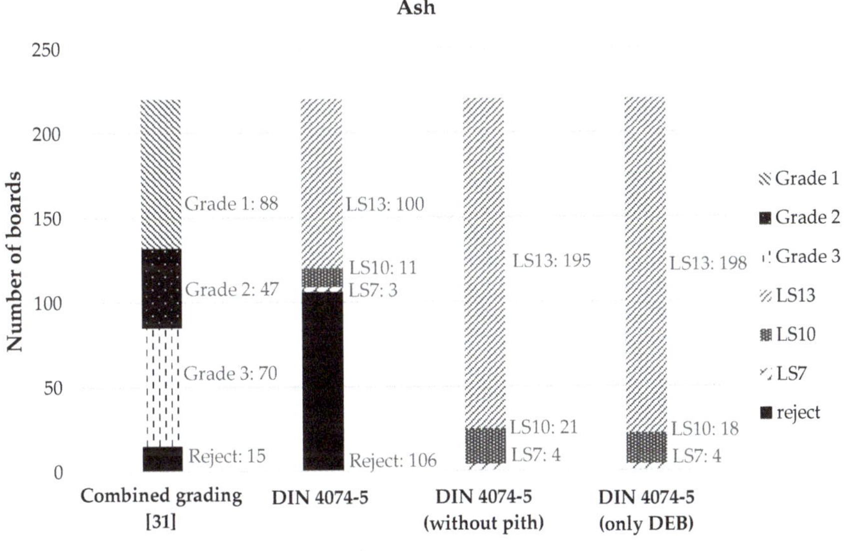


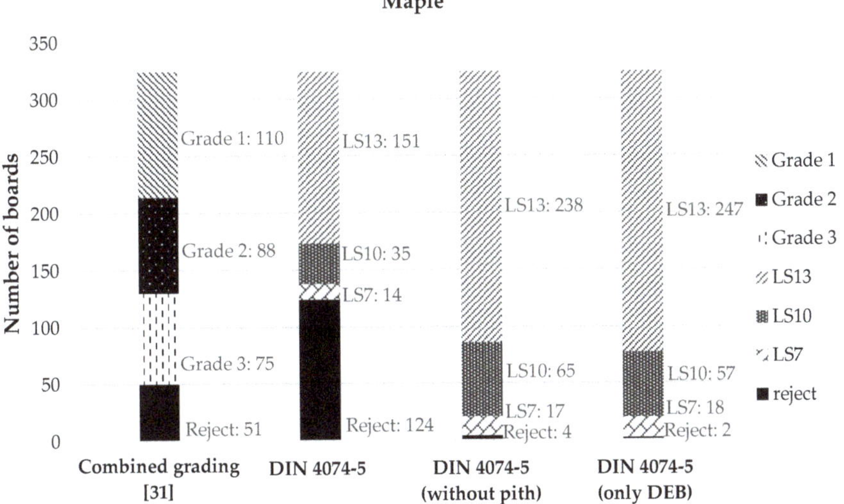


**Figure 3.** Yields for the combined (left bar) and visual grading (three right bars) of the planed ash and maple boards.

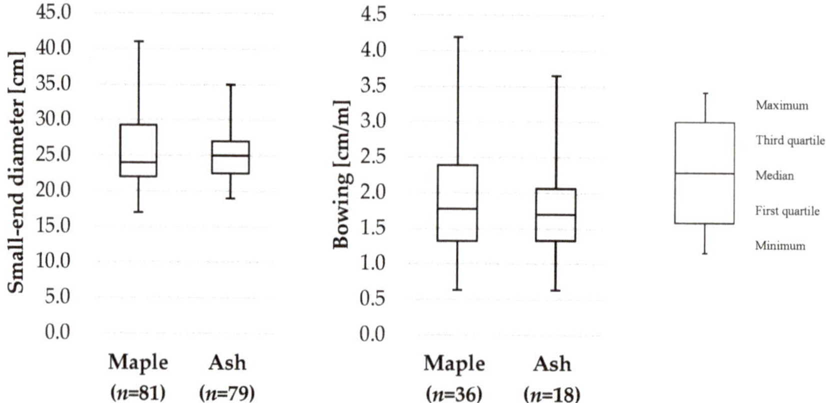



**Figure 4.** Boxplots for log section small-end diameter and bowing (only in one direction according to RVR instructions) separated after species (*n* = number of log sections).

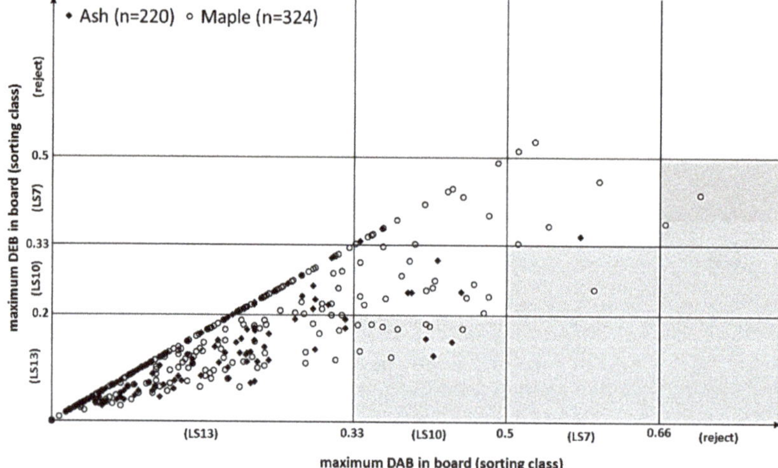



**Figure 5.** Maximum knot ratio of single knot (DEB) and knot cluster (DAB) for ash and maple boards. For all boards in grey area the criterion knot cluster (DAB) leads to downgrading into lower sorting class.

This confirms the findings made by other authors for the hardwood species beech. Frese and Riedler [38] postulate that for flat sawn beech lamellas (with lying annual rings) the sorting criterion DAB is not decisive for downgrading. Glos and Lederer [33] state that out of 219 beech boards only for one board the criterion DAB is sorting class determining. Blaß et al. [2] find similar results for a set of 350 beech boards (for 1.4% DAB decisive) and another set of 1888 beech boards (for 0.4% DAB decisive).

When applying stricter boundary values for the DAB than stated by the DIN 4074-5 [16], the DAB's influence on the grade rises. In the combined grading proposed by Kovryga et al. [31], the boundary values for DEB and DAB for ash were set to be identical—i.e., the strictest DAB setting possible. Thus, 13.6% of the ash lamellas of this study are downgraded due to the criterion DAB (cluster knot). For maple grading according to the settings proposed by Kovryga et al. [31] only 6.5% of the studied lamellas are downgraded due to the criterion DAB (cluster knot). This is because the proposed DAB boundary values for maple are not as strict as for ash. Regarding knot clusters (DAB), Figure 5 suggests that the investigated maple wood contains proportionally more knot clusters than the ash wood. Further analysis reveals only a difference of 6%, though. A total of 33% of the ash boards and 39%

of the maple boards contain knot clusters (greater than the max. DEB). Figure 6 shows that these maximum knot clusters are bigger in the maple collective than in the ash collective. The same holds for single knots. This leads to a higher proportion of LS7 and LS10 boards in the maple collective compared to the ash group (see Figure 3).

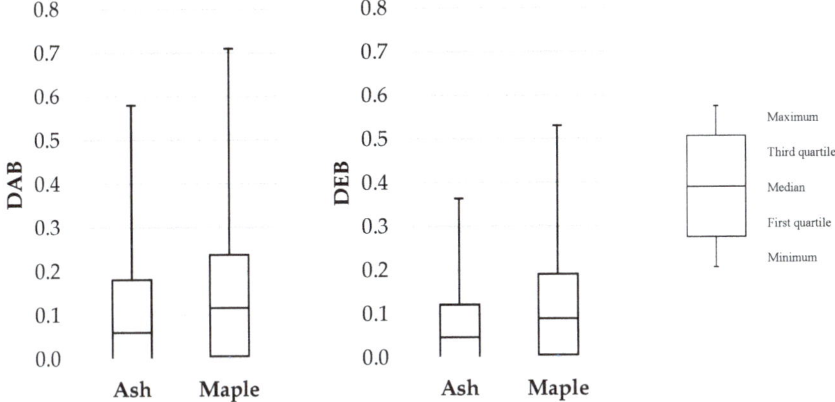


**Figure 6.** Boxplots separated after sorting criteria (maximum DEB and DAB in board) and species.

In general, special care must be taken when comparing grading results of different publications. The research material can be extremely diverse (i.e., species, origin, quality, sawing pattern, etc.), but also data acquisition for grading can be different. For example, Glos and Torno [28] grade the evaluated lumber only after the sorting criteria occurring within the tension test length, while for this study the entire board length is evaluated. Furthermore, the sorting criterion "grain angle" is a source of confusion, since its visual determination on unbroken boards is problematic [33,34].

### 5.3.2. Yields of Graded Lumber

The four sorting schemes presented in Figure 3 lead to different rates of so-called "rejects", i.e., boards that have to be sorted out. Table 8 lists the relative and absolute losses for the production step "grading" for each grading scheme and the resulting overall yields (referring to the round wood volume).

**Table 8.** Volume yields for the production step grading (from planed board to graded lamella) for four different grading schemes.

| Product | Options for Production Step Grading | Maple | | | | Ash | | | |
|---|---|---|---|---|---|---|---|---|---|
| | | Yield | | Waste/Loss | | Yield | | Waste/Loss | |
| | | in m$^3$ | in % | in m$^3$ | in % PS | in m$^3$ | in % | in m$^3$ | in % PS |
| Boards planed (unsorted lamellas) | | 3.3 | 20.9 | | | 2.6 | 18.2 | | |
| | Grading I (Combined grading according to Kovryga et al. [31]) | | | 0.5 | 15.7 | | | 0.2 | 6.8 |
| Combined grading lamellas | | 2.8 | 17.8 | | | 2.4 | 17.0 | | |
| | Grading II (4074-5) | | | 1.3 | 39.0 | | | 1.3 | 49.8 |
| 4074-5 lamellas | | 2.0 | 12.7 | | | 1.3 | 9.1 | | |
| | Grading III (4074-5 without pith) | | | 0.04 | 1.3 | | | 0 | 0 |
| 4074-5 lamellas without pith | | 3.3 | 20.6 | | | 2.6 | 18.2 | | |
| | Grading IV (4074-5, only DEB) | | | 0.02 | 0.7 | | | 0 | 0 |
| 4074-5 lamellas (only DEB) | | 3.3 | 20.8 | | | 2.6 | 18.2 | | |

When grading the lamellas according to DIN 4074-5 [16], for ash yield values lie around 9%, for maple around 13%. When excluding the sorting criterion "pith", total yields of ash are doubled (18.2%), those of maple rise to 20.6%. The difference between grading scheme III and IV is very little to none. This is due to the fact that the DAB (KC) has very little influence on the strength grading according to DIN 4074-5 [16].

As Table 8 shows, excluding the sorting criterion pith from the sorting scheme, raises the final yield considerably. Since board tension strength is the key influencing factor on glulam bending strength, tension testing of glulam lamellas has to show the effect the pith has on the board tension strength and stiffness (see [31]). If this influence is neglectable, the yield of grading can be raised extremely. It is important to state, though, that this does not hold equally for other strength properties. Glos and Torno [28], for example, prove for ash and maple that pith has a significant influence on the bending strength of square-edged lumber. They also stress the fact that the appearance of pith is often accompanied by bows, twists, and cracks. Similar results are presented by Glos and Lederer [33] for beech and oak square-edged lumber. Hübner [39] proves the pith's significant influence on the tension strength perpendicular to grain of ash glulam.

Further research has to work towards a hardwood strength grading system that is based on the mechanical properties of the resulting glulam. Kovryga et al. [31] proposes different optimized grading schemes for ash and maple glulam lamellas. For this study, one optimized combined grading solution from Kovryga et al. [31] was chosen for each species (see Table 3) to show an example of resulting yield. The chosen grading scheme distinguishes between three grades resulting in three board tensile strength classes based on destructive tension testing. For ash, the lowest class is DT22 with a characteristic tensile strength higher than 22 N/mm$^2$. For hardwoods, Kovryga et al. [40] proposes no tensile strength class lower than DT18. For lower mechanical properties softwood T-classes can be used. Therefore, for maple the lowest class is T15 (softwood tensile strength class) with a characteristic tensile strength not lower than 15 N/mm$^2$ (see Table 3). In this study, the proposed strength grading results in 15.7% rejects for maple and 6.8% rejects for ash. The resulting yields of 17.8% and 17.0% are considerably higher compared to grading according to DIN 4074-5 [16].

The economic feasibility of the production of hardwood glulam is strongly influenced by the final yield of glulam lamellas. Torno et al. [12] calculated that the production of beech glulam lamellas costs at least three times as much as that of spruce lamellas, calculating with beech round wood prices of €53.50–€80.00 per cubic meter. Since final yield figures of this study and Torno et al. [12] lie in a similar range, these costs can also be assumed for the lamellas of this study. This makes raising the yield inevitable, if a competitive hardwood product shall be produced.

When evaluating the competitiveness of a product, not only the production cost, but also the added value should be considered. Following the proposed combined grading of Kovryga et al. [31], for this study strength classes with a characteristic tensile strength as high as 38 N/mm$^2$ can be produced. With ash lamellas of this characteristic strength, glulam with bending strength values of as high as 48 N/mm$^2$ can be achieved [4]. Via "upgrading", i.e., cutting out large knots, the characteristic tensile strength of ash lamellas can be raised up to 54 N/mm$^2$ [31]. Using the combined grading approach for beech lamellas, Erhardt et al. [41] report tensile strength values of as high as 50 N/mm$^2$. This raised strength allows the production of more slender structures, which means material savings but also more construction possibilities for the architect and engineer. The listed benefits would yield obviously in higher reward for the producer. Although the present market situation has not led to a wide spread use of hardwood glulam, future changes in spruce availability, round wood prices (especially hardwood) and wood processing technology (etc.) might make the production lucrative.

## 6. Conclusions

For this study, the volume yields of the production of glulam lamellas from low quality and low dimension ash and maple log sections are investigated. For this purpose, 16.25 m$^3$ of maple (81 log sections) and 14.89 m$^3$ of ash (79 log sections) were harvested from natural forest stands (mixed beech forests) in central Germany and were turned into dry-dressed lumber (unsorted lamellas) with state of the art technologies. The resulting board volumes amount for only 20.9% (maple) and 18.2% (ash) of the original log volumes. The most waste is produced in the production step "presorting and planing" (maple: 56%; ash: 60%), since here a high percentage of the boards has to be sorted out due to bowing. By trimming these boards to shorter lengths, the waste of this production step could be reduced considerably. In addition, the sawing (milling) of the boards produced in both cases around 50% waste, which is in line with the above-mentioned literature values for sawing low-quality hardwoods. Nonetheless, with an adjusted sawing technology, this waste can be reduced (e.g., through shorter log sections and optimized machine combinations). It is also advisable to define a minimum input log diameter, since the lower the log diameter is, the lower the volume yield of milling becomes. Another approach to a raised final volume and value yield is the diversification of final products. Thus, as an example, glulam lamellas could be produced as a low-quality co-product from the production of high quality lumber for furniture production.

Strength grading of lamellas lowers final volume yields even further. When sorting the lamellas according to DIN 4074-5 [16], final volume yields of 12.7% for maple and 9.1% for ash are attained. One way of raising the final volume and also value yield could be the adjustment of the sorting (grading) scheme. For example, by excluding the criterion "pith" from sorting, final yield values of 20.6% (maple) and 18.2% (ash) can be achieved. Generally, it is advisable to combine visual and machine sorting to an assortment and species adjusted combined grading, which is optimized after the criteria "desired tensile strength and stiffness" but also "yield". The paper Kovryga et al. [31] is attempting this. Resulting total yields, when applying the selected optimized combined grading of Kovryga et al. [31] to this study´s lumber, lie between 17% and 18%. This yield is considerably lower than that obtained for softwood glulam lamellas. Factors like the higher attainable tensile strength, if compared to 30 N/mm$^2$ possible for softwoods [42], and the appealing appearance of hardwood glulam may make up for the yield disadvantages. In general, the economic feasibility of hardwood glulam is influenced by a serious of factors, which have to be analyzed in detail for each final product and production plant separately.

**Author Contributions:** Conceptualization, methodology, investigation, formal analysis, visualization, and writing—original draft preparation: P.S. and A.K.; data curation and investigation: L.E.; funding acquisition, project administration, supervision, and writing—review and editing: S.B.; resources, supervision, and writing—review, and editing: J.W.V.d.K. and H.M.

**Funding:** This study was funded by the German Federal Ministry of Food and Agriculture (grant number 22024211).

**Acknowledgments:** We thank Antje Gellerich for giving helpful advice during data acquisition.

**Conflicts of Interest:** The authors declare no conflict of interest.

## References

1. Sauter, U. WP 1: Hardwood Resources in Europe—Standing Stock and Resource Forecasts. Presented at Workshop "European Hardwoods for the Building Sector", Garmisch-Patenkirchen, Germany, 2016.
2. Blaß, H.J.; Denzler, J.; Frese, M.; Glos, P.; Linsenmann, P. *Biegefestigkeit Von Brettschichtholz Aus Buche [Bending Strength of Beech Glulam]*; Karlsruher Berichte zum Ingenieurholzbau, Band 1 Karlsruher Berichte zum Ingenieurholzbau; Universitätsverlag Karlsruhe: Karlsruhe, Germany, 2005.
3. Frühwald, A.; Ressel, J.B.; Bernasconi, A.; Becker, P.; Pitzner, B.; Wonnemann, R.; Mantau, U.; Sörgel, C.; Thoroe, C.; Dieter, M.; et al. *Hochwertiges Brettschichtholz Aus Buchenholz [High Quality Glulam Made of Beech Wood]*; Final Report; University of Hamburg: Hamburg, Germany, 2003.
4. Van de Kuilen, J.W.; Torno, S. *Materialkennwerte von Eschenholz Für Den Einsatz in Brettschichtholz [Material Properties of Ash Wood for the Use in Glulam]*; Final Report; Holzforschung München, Technische Universität München: Munich, Germany, 2014.
5. DIBt. BS-Holz aus Buche und BS-Holz Buche-Hybridträger [Glulam and hybrid glulam made of beech]; German technical building approval Z-9.1-679; holder of approval: Studiengemeinschaft Holzleimbau e.V., Germany; issued by Deutsches Institut für Bautechnik (DIBt), Germany; valid until 27.10.2019. 2014.

6.  OiB. VIGAM—Glued laminated timber of oak; ETA-13/0642 European Technical Approval; holder of approval: Elaborados y Fabricados Gámiz, S.A., Spain; issued by Austrian Institute of Construction Engineering (OiB), Austria; valid until 27.06.2018. 2013.

7.  DIBt. Holz Schiller Eiche-Pfosten-Riegel-Brettschichtholz [Timber Schiller oak post and beam glulam]; German technical building approval Z-9.1-821; holder of approval: Holz Schiller GmbH, Germany; issued by Deutsches Institut für Bautechnik (DIBt), Germany; valid until 02.03.2018. 2013.

8.  OiB. Sierolam—Glued laminated timber of chestnut; ETA-13/0646 European Technical approval; holder of approval: Siero Lam, S.A., Spain; issued by Austrian Institute of Construction Engineering (OiB), Austria; valid until 27.06.2018. 2013.

9.  Mack, H. The European Market for Glulam—Results of a Survey within the European Glued Timber Industry. In Proceedings of the "Wiener Leimholz Symposium 2006", Vienna, Austria, 23–24 March 2006; pp. 35–50.

10. Ohnesorge, D.; Hennig, M.; Becker, G. Bedeutung von Laubholz bei der Brettschichtholzherstellung [Significance of hardwoods in glulam production]. *Holztechnologie* **2009**, *6*, 47–49.

11. Welling, J. Differences between hard—and softwoods and their influence on processing and use. In Proceedings of the "Gülzower Fachgespräche Stoffliche Nutzung von Laubholz", Würzburg, Germany, 6–7 September 2012.

12. Torno, S.; Knorz, M.; Van de Kuilen, J.W. Supply of beech lamellas for the production of glued laminated timber. In Proceedings of the 4th International Scientific Conference on Hardwood Processing, Florence, Italy, 7–9 October 2013; pp. 210–217.

13. Wiedenbeck, J.; Scholl, M.S.; Blankenhorn, P.R.; Ray, C.D. Lumber volume and value recovery from small—diameter black cherry, sugar maple, and red oak logs. *BioResources* **2017**, *12*, 853–870. [CrossRef]

14. Lin, W.; Wang, J.; Wu, J.; DeVallance, D. Log Sawing Practices and Lumber Recovery of Small Hardwood Sawmills in West Virginia. *For. Prod. J.* **2011**, *61*, 216–224. [CrossRef]

15. Van de Kuilen, J.W.; Torno, S. *Untersuchungen Zur Bereitstellung von Lamellen Aus Buchen- Und Eschenholz Für Die Produktion von Brettschichtholz [Studies on the Supply of Lamellas for the Production of Glulam from Beech and Ash Wood]*; Final Report X37; Holzforschung München, Technische Universität München: Munich, Germany, 2014.

16. DIN 4074-5. *Strength Grading of Wood—Part 5: Sawn Hardwood*; German Institute for Standardization (DIN): Berlin, Germany, 2008.

17. Ehlebracht, V. Untersuchung Zur Verbesserten Wertschöpfung Bei Der Schnittholzerzeugung Aus Schwachem Buchenstammholz (Fagus Sylvatica L.) [Investigation on improved Added Value in the Production of Sawn Lumber from Low Dimension Beech Timber]. Dissertation, Georg-August-Universität Göttingen, Göttingen, Germany, 2000.

18. Eickers, A. Verschnittuntersuchungen Bei Der Verarbeitung von Brettschichtholzlamellen [Yield Loss in the Production of Glulam Lamellas]. Diploma Thesis, Fachhochschule Hildesheim/Holzminden, Hildesheim, Germany, 1997.

19. Steele, P.H. *Factors Determining Lumber Recovery in Sawmillin*; General Technical Report FPL-39; U.S. Department of Agriculture, Forest Service, Forest Products Laboratory: Madison, WI, USA, 1984.

20. Richards, D.B.; Adkins, W.K.; Hallock, H.; Bulgrin, E.H. *Lumber Values from Computerized Simulation of Hardwood Log Sawing*; Research Paper FPL-356; U.S. Department of Agriculture, Forest Service, Forest Products Laboratory: Madison, WI, USA, 1980.

21. Emhardt, M.; Pfingstag, S. *Schnittholzqualität und Ausbeute von Schwachem Buchenstammholz [Sawn Lumber Quality and Yield Produced from Low Dimension Beech Logs]*; Final Report; FVA Baden-Würtemberg: Freiburg, Germany, 1993.

22. Fronius, K. *Arbeiten Und Anlagen Im Sägewerk—Spaner, Kreissäge, Bandsäge [Work and Machines in a Sawmill—Chipper Canter, Circular Saw, Band Saw]*, 2nd ed.; DRW-Verlag Weinbrenner: Leinfelden-Echterdingen, Germany, 1989.

23. Fischer, H. *Neue Verwendungsmöglichkeiten von Laubschwachholz [New Possible Uses for Low Dimensions Hardwood Timber]*; Internal Report; Forstliche Versuchsanstalt Rheinland-Pfalz: Trippstadt, Germany, 1996.

24. Lamb, F.M.; Wengert, E.M. Techniques and procedures for the quality drying of oak lumber. In Proceedings of the 41st Meeting of the Western Dry Kiln Association, Corvallis, OR, USA, 1990.

25. Denig, J.; Wengert, E.M.; Simpson, W.T. *Drying Hardwood Lumber*; General technical Report FPL-GTR-118; U.S. Department of Agriculture, Forest Service, Forest Products Laboratory: Madison, WI, USA, 2000.

26. Kollmann, F.; Côté, W.A., Jr. *Principles of Wood Science and Technology, I. Solid Wood*; Springer: Berlin, Heidelberg, 1968.

27. Wagenführ, R. *Holzatlas*; Fachbuchverlag im Carl Hanser Verlag: Leipzig, Germeny, 2007.

28. Glos, P.; Torno, S. *Aufnahme Der Einheimischen Holzarten Ahorn, Esche und Pappel in Die europäische Norm EN 1912: "Bauholz – Festigkeitsklassen— Zuordnung von Visuellen Sortierklassen und Holzarten" [Inclusion of tha Native Wood Species Maple, Ash and Poplar in the European Standard EN 1912: "Structural Timber—Strength Classes—Assignment of Visual Grades and Species"]*; Final Report 06517; Holzforschung München, Technische Universität München: Munich, Germeny, 2008.

29. RVR. *Rahmenvereinbarung Für Den Rohholzhandel in Deutschland (RVR) [Framework Agreement on Raw Timber Trade in Germany]*; Deutscher Forstwirtschaftsrat & Deutscher Holzwirtschaftsrat: Berlin, Germany, 2015.

30. DIN 1316-3. *Hardwood Round Timber—Qualitative Classification—Part 3: Ash and Maples and Sycamore*; German Institute for Standardization (DIN): Berlin, Germany, 1997.

31. Kovryga, A.; Schlotzhauer, P.; Stapel, P.; Militz, H.; Van de Kuilen, J.W. Visual and machine strength grading of European ash and maple for glulam application. *Holzforschung* **2019**, *0*. [CrossRef]

32. Hankinson, R.L. *Investigation of Crushing Strength of Spruce at Varying Angles of Grain*; Air Service Information Circular No. 259; U.S. Air Service: Washington, DC, USA, 1921.

33. Glos, P.; Lederer, B. *Sortierung von Buchen-Und Eichenschnittholz Nach Der Tragfähigkeit Und Bestimmung Der Zugehörigen Festigkeits-Und Steifigkeitskennwerte [Strength Grading of Beech and Oak Lumber and Determination of Characteristic Strength and Stiffness Values]*; Final Report; Institut für Holzforschung, Technische Universität München: Munich, Germany, 2000.

34. Frühwald, A.; Schickhofer, G. Strength grading of hardwoods. In Proceedings of the 14th International Symposium on Nondestructive Testing of Wood, Eberswalde, Germany, 2–4 May 2005.

35. Frese, M.; Blaß, H.J. Beech glulam strength classes. In Proceedings of the Meeting 38 of the International Council of Research and Innovation in Building and Construction, Working Commission W18—Timber Structures, Karlsruhe, Germany, 2005.

36. Wade, M.W.; Bullard, S.H.; Steele, P.H.; Araman, P.A. Estimating Hardwood Sawmill Conversion Efficiency Based on Sawing Machine and Log characteristics. *For. Prod. J.* **1992**, *42*, 21–26.

37. Torno, S.; Van de Kuilen, J.W. Esche für tragende Verwendungen—Festigkeitseigenschaften visuell sortierten Eschenholzes [Ash for load bearing use—Strength properties of visually graded ash lumber]. *LWF Aktuell* **2010**, *77*, 18–19.

38. Frese, M.; Riedler, T. Untersuchung von Buchenschnittholz (Fagus sylvatica L.) hinsichtlich der Eignung für Brettschichtholz [Examination of suitability of beech lumber for glulam]. *Eur. J. Wood Wood Prod.* **2010**, *68*, 445–453. [CrossRef]

39. Hübner, U. Mechanische Kenngrößen von Buchen-, Eschen- Und Robinienholz Für Lastabtragende Bauteile [Mechanical parameters of Beech, Ash and Locust Lumber for Load Bearing Members]. Ph.D. Thesis, Technische Universität Graz, Graz, Austria, 2013.

40. Kovryga, A.; Stapel, P.; Van De Kuilen, J.W. Tensile strength classes for hardwoods. In Proceedings of the 49th INTER & CIB Meeting, Graz, Austria, 2016.

41. Erhardt, T.; Fink, G.; Steiger, R.; Frangi, A. Strength grading of European beech lamellas for the production of GLT & CLT. In Proceedings of the 49th INTER & CIB Meeting, Graz, Austria, 2016.

42. EN 338. *Structural Timber–Strength Classes*; European Committee for Standardization: Brussels, Belgium, 2016.

 *forests*

*Article*

# Construction of Wood-Based Lamella for Increased Load on Seating Furniture

Nadežda Langová, Roman Réh, Rastislav Igaz, Ľuboš Krišťák *, Miloš Hitka and Pavol Joščák

Faculty of Wood Sciences and Technology, Technical University in Zvolen, 960 01 Zvolen, Slovakia;
langova@tuzvo.sk (N.L.); reh@tuzvo.sk (R.R.); igaz@tuzvo.sk (R.I.); hitka@tuzvo.sk (M.H.); joscak@tuzvo.sk (P.J.)
* Correspondence: kristak@tuzvo; Tel.: +421-4552-06836

Received: 9 May 2019; Accepted: 18 June 2019; Published: 25 June 2019

**Abstract:** The research on population shows that the count of overweight people has been constantly growing. Therefore, designing and modifying utility items, e.g., furniture should be brought into focus. Indeed, furniture function and safety is associated with the weight of a user. Current processes and standards dealing with the design of seating furniture do not meet the requirements of overweight users. The research is aimed at designing flexible chairs consisting of lamellae using the finite element method (FEM). Three types of glued lamellae based on wood with different number of layers and thickness were made and subsequently, their mechanical properties were tested. Values for modulus of elasticity and modulus of rupture were used to determine stress and deformation applying the FEM method for modelling flexible chairs. In this research, the methodology for evaluating the ultimate state of flexible chairs used to analyse deformation and stability was defined. The analysis confirms that several designed constructions meet the requirements of actual standards (valid for the weight of a user up to 110 kg) but fail to meet the requirements for weight gain of a population.

**Keywords:** glued lamella; flexible chair; weight of a user; ultimate state

## 1. Introduction

Requirements of the construction design of furniture for sitting arise from the needs to ensure that healthy sitting provides physical, mental and social comfort for users. Promoting correct posture with high quality lumbar support (total surface pressure is reduced as much as possible) and the ability to change positions while sitting are two ways to make users feel comfortable over long periods of usage. Correct sitting positions may prevent permanent spinal deformity or lower quality of life physiology, such as breathing, digestion, etc. [1]. Several requirements must be taken into consideration while designing seating furniture, but two of them are considered essential: Various measurements of the human body (especially height), and different weight and human body shapes must be taken into account.

Determining the appropriate single weight for all users is a difficult process as weight gain has recently been reported all over the world. In many countries, population weight gain is seen as part of the global obesity epidemic [2–8]. Data from 591 local and 369 national research studies were used by the author [9]. Another study based on 450 national studies determined the trends in weight gain from 1990 to 2020 [10]. The data mentioned in both research studies, as well as in many others, have showed overall weight gain in recent decades [11–13]. Regional and national studies in European countries (Figure 1) show that the situation is very similar all over Europe [14–17]. In 2002, the 95th, 98th and 99th percentiles for the body weight of men in the US were 114.6 kg, 131.61 kg and 141.17 kg [18].

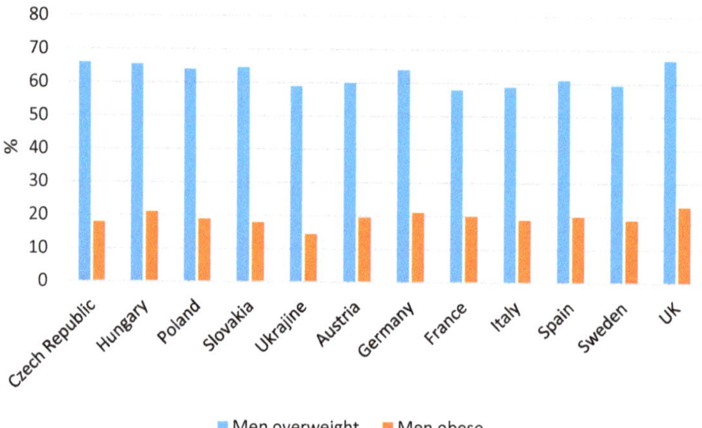



**Figure 1.** Overweight and obese men in European countries in 2016 [19].

Similar increase in average weight of users has been observed in the Slovak population as well. In 2017, the weight of more than 5% Slovak men was 110 kg (Figure 2). Moreover, the weight of 11% of these men was more than 130 kg. Based on BMI data in Slovakia, in 2017, 400,000 men in Slovakia suffered from obesity and 90,000 men suffered from severe obesity [19–23].

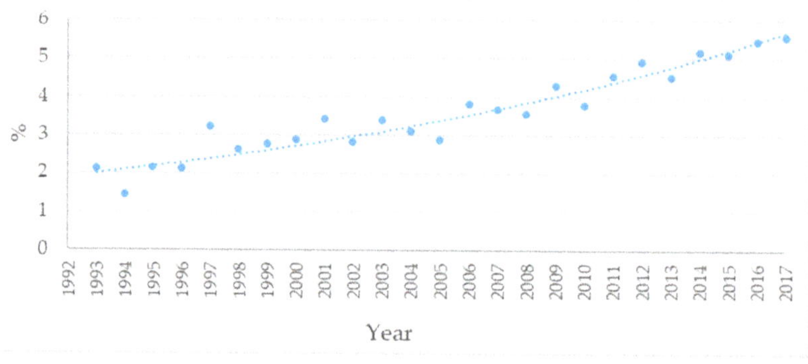



**Figure 2.** Percentage of Slovak men with weight more than 110 kg.

Various industrial sectors, such as automation, aviation, furniture manufacturing, footwear, and clothing industries have been affected by the current trend in human dimensions, especially steady weight gain and an increase in human height over the last few years [24–26]. In the case of furniture, these trends have been applied in some countries in the world recently, e.g., in the US, the standard BIFMA X6.1 (2012), as a new safety and performance standard for educational seating was accepted by the National Standards Institute (ANSI). Three sizes of school chairs were defined in the standard: A (seat height of less than 352 mm, user weight of 35 kg—it corresponds with the 95th percentile for boys aged 6), B (352 mm to 425 mm, 75 kg—it corresponds with the 95th percentile for girls aged 12), and C (more than 425 mm, 115 kg—it corresponds with 95% for adult male population) [27,28]. The standard resulted from long-term research that aimed at the importance of designing appropriate classroom furniture for schoolchildren [29–33]. Furniture for users with weight from 253 lb (115 kg) up to 400 lb (181 kg), which corresponds with the 99.5th percentile for men in the US, is specified in another accepted standard BIFMA X5.11 (2015) [34]. Similar standards were also accepted in Australia, e.g., the standard AFRDI 142 (2012) focused on four categories of users of "heavy duty" office chairs: 135 kg for a single shift (8 h), 135 kg for multiple shifts, 160 kg for single/multiple shifts [35]. Another Australian standard AFRDI 151 (2014) deals with chairs for home, designed for users with weight more than 100 kg (four options—135kg, 160 kg, 185 kg and chairs for bariatric patients with the weight more than 300 kg) [36]. In Europe, there is the standard BS 5459-2 focused on static and dynamic load of office pedestal chairs for persons with weight up to 225 kg [37]. This standard was designed by the company Satra, furniture testing facility in the UK (ISO 17025:2017) [38,39].

There are not many research studies dealing with furniture dimensions and construction in connection with overweight population or persons with disabilities [40]. References [41–44] suggest using anthropometric measurements in the process of designing furniture. The research of the authors [45,46] is focused on static analysis and testing the chairs in connection with the weight of users. In Slovakia, the effect of body weight on the size of chair joints was analysed in the study [46]. At the same time, the effect of a secular trend on functional dimensions of furniture was studied [47]. Czech authors [48] dealt with the use of anthropometric data in connection with seating and bed furniture as well. The authors [49] discussed the use of wider beds by healthcare providers in the case of patients with weight of more than 159 kg. The use of specific bed size for users with weight more than 147 kg or BMI score greater than 55 is suggested in another study [50]. Oversized beds for patients with BMI greater than 45 are recommended in the study [51].

Native wood and wood composites, besides plastics, and metals, are the most used materials in the manufacturing process of seating furniture. Fixed and flexible seating arrangement can be recognised in terms of constructing and joining structural elements of seating furniture. Stiffness required, especially in the case of dining chairs, is a typical feature of fixed seating arrangement made out of solid timber [52]. Flexible seating arrangement is especially used in manufacturing chairs designed for relaxation or as office chairs [53]. Wood is modified or wood-based composite materials are made of it in order to increase wood flexibility (as well as wood strength). Laminated furniture panels—lamellae and plywood—are widely used in furniture manufacturing. Properties of lamellae and plywood used in furniture projects depend on many various factors, such as moisture content of veneers, temperature, pressure and pressing time [54–58]. Adhesive properties, its viscosity, thickness of adhesive layer, quality of adhesive application, mechanical properties of veneers, treatment quality or removal of small elements from the surface (saw dust) are other factors affecting the bending strength [59–63]. Due to high bending strength of lamellae during dynamic loading, laminated wood is preferred in furniture manufacturing, especially chairs and beds [64].

At present, there are two directions in the research into chair anatomy. The first direction is focused on experimental testing of furniture construction. Experimental measurements and calculations are focused mainly on the weakest point—the joint—during static and dynamic loading and on the effect of tenon size on the ratio of dynamic to static loading rate [65–72]. The second direction deals with furniture design and construction using numerical and analytical methods. The finite element method (FEM) used to estimate or determine the load capacity of individual joint dimensions is the most often used method [73–78].

FEM allows manufacturers to optimise the shape and size of chairs. The developed models establish procedures to perform virtual testing on laminated bamboo chair to reduce product design and testing time [79]. This virtual testing results in design improvement and development of the laminated bamboo chair. The research study [80] is focused on classification of chairs according to their performance. Three hundred and fifteen chairs were tested and following the test results, acceptable light, medium and heavy design loads were determined for wood chair performance. Moreover, these values are in compliance with the allowable design loads.

Current European Standards associated with seating furniture (EN 1728:2012, EN 12520:2015, EN 1022:2018) are based on users with body weights of up to 110 kg [81–83]. Based on results of weight gain all around the world, the aim of the research is to determine the effect of the human weight on the load-carrying capacity and the dimensions of flexible chair consisting of lamellae. Mentioned data are required to a large extent by chair manufacturers.

The aim of this paper was to analyse the effect of laminated furniture panels with various thicknesses on the function of chair frame construction. Suggested minimum lamella thickness meeting the requirements of chairs for users with weight up to 110 kg and 150 kg resulted from the conducted research. For the ultimate load-carrying capacity and ability to use lamellae in flexible chairs, three thicknesses of lamellae were studied. Other thicknesses of lamellae, required to ensure overweight users feel safe, were tested. The methodology for evaluating the ultimate limit state of flexible chairs used based on ergonomics and chair safety can be considered for further research; normal and shear strength must be evaluated as well.

## 2. Materials and Methods

Three types of lamellae with various numbers of layers and total thickness were examined in the research on mechanical properties. Individual types of lamellae consisted of 9, 11 or 13 veneer layers created the final thickness of lamellae of 11.0 mm (type A), 13.5 mm (type B) or 16.0 mm (type C).

The lamellae were made of veneers of European beech wood (*Fagus sylvatica* L.) without defects by rotary peeling process using a 4-foot lathe (Královopolská strojírna, Brno, Czech Republic) at the Technical University in Zvolen, Slovakia from plasticized round wood. Beech wood is the most used wood species in furniture manufacturing in Slovakia. Its mechanical properties make it ideal for veneer production. The average thickness of veneers after drying to the moisture content of 6 ± 1% was 1.23 mm. Direction of wood grain in all veneers in lamella set was the same. PVAC dispersion Rakoll E WB 0301 (H. B. Fuller, Minnesota, USA) was used for gluing. The viscosity of the adhesive mixture was 5.500 mPa·s and pH value was 3.5 at the time of gluing. Adhesive was applied to the second veneer on both sides using a glue spreader with two rollers and an adhesive layer formed was 220 g·m$^{-2}$.

Veneer set pressing was carried out in a hydraulic press using a press mold to form the final lamella shape. Forasmuch as the molds were under stress, the pressure during pressing process was 0.8 N·mm$^{-2}$, at a temperature of 20 °C for 30 min. The total dimension of pressed semi-finished products was as follows: length of 600 mm and width of 280 mm. The angle between the two adjacent lamella sides was 103° with radius of curvature of 80 mm. After stabilizing (120 h in standard climatic conditions), the semi-finished product was cut into final lamellae with width of 50 mm. Subsequently, individual lamellae were smoothed with 80-, 120- and 150-grit sandpaper to improve final surface quality.

Afterwards, test specimens were formed from lamellae in order to determine mechanical properties. Thirty test specimens of each type (A, B and C) with dimension of 250 × 50 mm were formed from the straight part of the lamellae. From the mold lamella part, 30 mold test specimens for each type (A, B and C) were formed. Subsequently, test specimens were air-conditioned at a temperature of 20 ± 2 °C and air humidity of 65 ± 5%. The moisture content (EMC) of specimens after air-conditioning was 12 ± 1%.

Flat common specimens were tested using the standard methodology of the three-point bend test according to the standard EN 310: 1993. Mold unconventional specimens were tested by modified methodology created for the needs of this research. Mechanical testing of mold specimens was carried out using the modified three-point bend test. The load was spread evenly and the specimen was broken after 60 ± 30 s.

Wood is a material whose properties possess orthogonal anisotropy, i.e., its physical and mechanical properties differ in three principal planes [84]. Three symmetry planes are differentiated in wood: cross-section perpendicular to the grain direction, longitudinal-radial and longitudinal-tangential, which are parallel with the wood grain direction and at the same time are mutually perpendicular. Due to its structural organization, lamella can be considered to be an anisotropy material in the plane perpendicular to the grain direction. The mechanical properties of lamella in both planes perpendicular to the grain direction are almost identical. Therefore, wood-based lamellae can be defined as transverse-axial anisotropic material. In the presented research, lamellae were formed as an orthogonal anisotropic material. Anisotropy must be taken into account in the modeling with the finite element method. The physical and mechanical properties of lamellae used in the modelling are summarized in Table 1.

**Table 1.** Material constants for laminated beech lamellae, y direction is along the grain [85].

| Young's Modulus (MPa) | | | Poisson's Ratio (-) | | | Shear Modulus (MPa) | | |
|---|---|---|---|---|---|---|---|---|
| $E_x$ | $E_y$ | $E_z$ | $\mu_{xy}$ | $\mu_{yz}$ | $\mu_{xz}$ | $G_{xy}$ | $G_{yz}$ | $G_{xz}$ |
| 1130.0 | 16670.0 | 630.0 | 0.044 | 0.33 | 0.027 | 1200.0 | 190.0 | 930.0 |

In the research on seating construction, a chair consisting of two frames was created. Base chair frame consisted of two U-shaped profiles were joined with transverse rails. The frame of seat and back was flexible and joined with transverse elements. Glued joints were used for chair construction because in comparison to other mechanical joining components, their stiffness was higher and they transferred the load better. Anthropometric measurements of users were taken into account for dimensions, construction and shape of the chair. Basic dimensions of designed chair are mentioned in Figure 3. Lamella dimensions and shape used in the project corresponded with those made and tested experimentally. While creating a chair model, three types of tested lamellae were used one after another (type A, B and C).

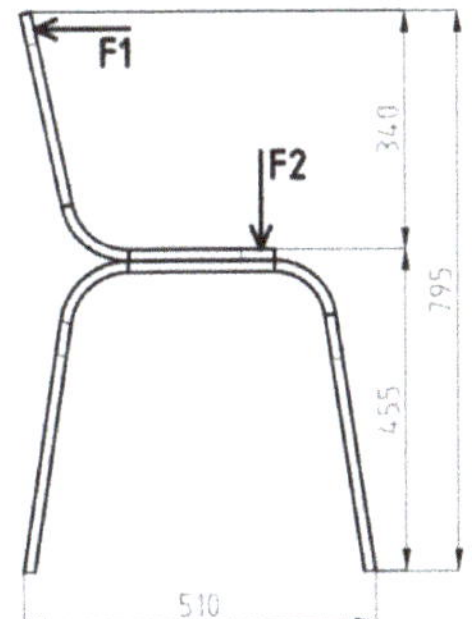


**Figure 3.** Static loading of the chair according to the standard EN 1728:2012 + dimensions of the designed chair.

The methodology for testing the chairs, especially loading, was based on the standard EN 1728: 2012. The users' weight of 110 kg was regarded as maximum weight in the standard, while horizontal force acting on the back was $F_1 = 450$ N and the vertical force acting on the seat was $F_2 = 1300$ N. In the case of users with weight more than 110 kg, acting forces were not defined. However, statistics dealing with the weight of adult population showed the fact that designing the furniture for users with weight of up to 110 kg did not meet the actual requirements. Therefore, the forces resulting from the load caused by the overweight users had to be defined. 150 kg was the maximum user's weight set and the forces were determined using multiple linear regression. Acting forces of $F_1 = 613$ N and $F_2 = 1775$ N and user's weight of 150 kg were used in the process of creating a chair model. Direction and the point at which the force was applied are defined in Figure 3.

The loading analysis of the tested chair was conducted using the program ANSYS. In the software environment, a 3D volume model taking into consideration the orthotropic properties of wood-based lamellae was created. A coordinate system used was as follows: Y-axis was in the grain direction, X-axis was perpendicular to the grain in the radial direction and Z-axis was perpendicular to the grain in tangential direction. When mold lamellae were created—base, seat and chair back—the properties of lamellae were changed in relation to the lamella shape. Chair base lamella was created from three parts. Properties in individual planes were changed in relation to grain orientation. In the mold parts of lamella, the values of loading perpendicular to the grain were defined. Material constants are defined in Table 1. Every element had to be assigned to a particular material. 3D element Solid 95 with 20 nodes was an element type. Boundary conditions were according to the standard EN 1728:2012. Supports of the back legs of a chair were regarded as fixed (fixed supports) in order to ensure that the loading was evenly transferred to the construction. Displacement supports were used in the front legs of the chair, movement in the *y-axis* direction was available. All joints in chair construction were considered fixed (bonded).

In terms of dimensioning the structural elements, limit state design requires the construction to meet two principal criteria: the ultimate limit state (ULS) used to evaluate the strength of construction, i.e., design strength and the serviceability limit state (SLS) used to evaluate the construction deformation.

The serviceability limit state, i.e., maximum deformation of flexible chair frame is defined neither in scientific journals nor in standards. It can result from an ergonomic chair design, suggested dimensions and seat-to-back angle. The angle recommended for designing a relaxed chair ranges from 103° to 110°. When 110° was the maximum value of an angle that could not be exceeded during loading, then the maximum displacement of a chair back was 40 mm backwards (Figure 4). This value of displacement was considered the maximum value for evaluating the serviceability limit state.

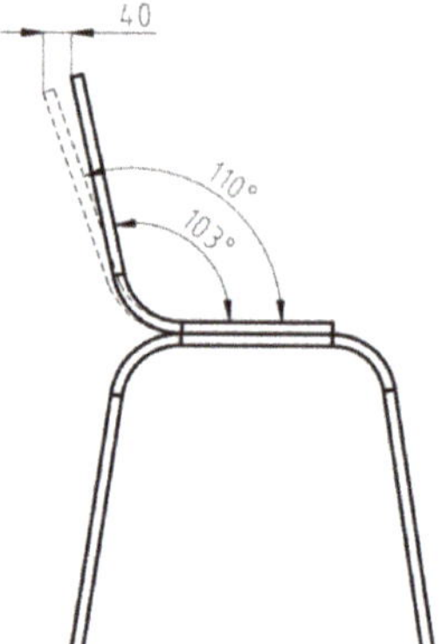


**Figure 4.** Deformation of the flexible chair.

When determining the serviceability limit state for flexible chairs, maximum limit displacement of the chair back in the highest point could be $u_{max} = 40$ mm. It resulted from the suggested seat-to-back angle of 110° (Figure 4). Reliability of the designed displacement $u_d$ (determined by the FEM calculation) is:

$$u_d \leq u_{max} \tag{1}$$

However, in terms of safety, a chair with mentioned limit displacement of back must be safe and stable, i.e., backward overturning must not occur (chair must not tip over). Calculation of stability is mentioned in the standard EN 1022:2018. Loading is shown in Figure 5. Considering the flexibility, the studied lamella chair was a chair with variable geometry. According to the mentioned standard, the chair was considered stable when it does not tip after applying a load of $m = 110$ kg. When the seat-to-back angle was 110°, the centre of gravity of the load could not be positioned behind the tipping point of a chair, i.e., the point when the back leg is in contact with the floor. The position of the centre of gravity of the load can be defined using the graphical method (Figure 5).

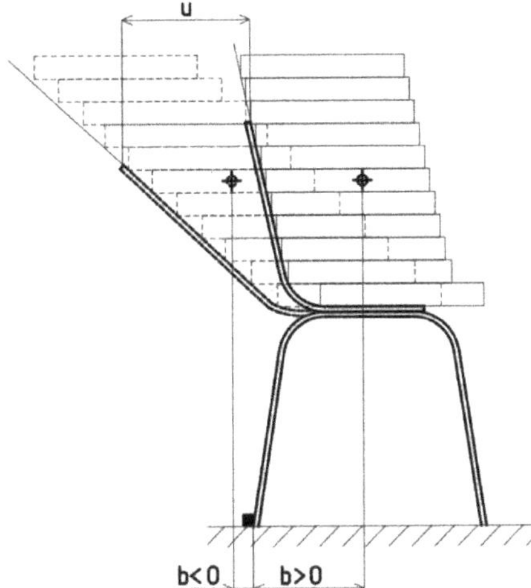


**Figure 5.** Defining the maximum possible deformation of the flexible chair.

The serviceability limit state boundary conditions resulting from the lamella stiffness determined experimentally was defined specifically in the chair construction. Following the results of the FEM analysis, normal and shear stresses were determined. When structural elements were dimensionised, normal and shear stresses were considered to be a design stress. In the process of dimensioning the chair components according to the serviceability limit state, requirements for reliable molding had to be met:

$$\sigma_{0,d} \leq f_{b,0,d} \tag{2}$$

where: $\sigma_{0,d}$—design stress in the beech lamella mold (MPa),

$f_{b,0,d}$—design strength in the beech lamella mold (MPa).

The value of characteristic strength had to be determined in order to calculate the design strength of lamella. Mean value of the bending strength ($\overline{\sigma}$) of tested lamellae achieved experimentally at a temperature of $t$ = 20 °C and $\varphi$ = 65% was an essential condition to determine the characteristic strength. Characteristic bending strength is a value corresponding with $\alpha$ quantile of assumed statistical division of evaluated strength. When $\alpha$ = 5%, the formula is:

$$f_{b,0,k} = \overline{\sigma} \cdot (1 - t_{95} \cdot \vartheta_x) \tag{3}$$

where: $f_{b,0,k}$—characteristic bending strength of glued lamella (MPa),

$\overline{\sigma}$—mean value of bending strength (MPa),

$t_{95}$—quantile of Student's t-distribution (one-side test), when $t_{95}$ = 1.64,

$\vartheta_x$—coefficient of variation (absolute value) (MPa).

When the characteristic bending strength is known, design strength $f_{b,0,d}$ is determined using the formula:

$$f_{b,0,d} = k_{mod} \cdot \frac{f_{b,0,k}}{\gamma_M} \tag{4}$$

where: $f_{b,0,k}$—characteristic strength of beech lamella in mold (MPa),

$\gamma_M$—partial safety factor (-), for wood-based materials $\gamma_M = 1.3$,

$k_{mod}$—modification coefficient (-) (takes into account the effect of loading time and moisture content on the characteristic strength of material) for the action/load with the shortest design situation $k_{mod} = 1.10$.

## 3. Results and Discussion

Values of bending strength were determined experimentally using the specimens made of lamellae described in methodology. Bending strength was defined individually for flat and mold parts of lamellae. Mean values of bending strength, characteristic values, as well as design values for flat and mold lamella parts determined experimentally are summarised in the following tables (Tables 2 and 3). The values in Table 3 highlighted in bold ($f_{b,0,d}$) were used for evaluation of the ultimate limit state.

**Table 2.** Calculated values of flat lamella.

| Type of Lamella | Mean Value $\sigma$ (MPa) | Coefficient of Variation $\vartheta$ (%) | Characteristic Bending Strength $f_{b,0,k}$ (MPa) | Design Bending Strength $f_{b,0,d}$ (MPa) |
|---|---|---|---|---|
| A | 111.85 | 4.96 | 102.76 | 86.86 |
| B | 104.64 | 6.57 | 93.37 | 79.01 |
| C | 93.80 | 6.11 | 84.41 | 71.42 |

**Table 3.** Calculated values of mold lamella.

| Type of Lamella | Mean Value $\sigma$ (MPa) | Coefficient of Variation $\vartheta$ (%) | Characteristic Bending Strength $f_{b,0,k}$ (MPa) | Design Bending Strength $f_{b,0,d}$ (MPa) |
|---|---|---|---|---|
| A | 123.85 | 4.55 | 114.61 | **96.98** |
| B | 98.13 | 3.59 | 92.35 | **78.15** |
| C | 89.48 | 5.97 | 80.70 | **68.28** |

*3.1. Ultimate Limit State Assessment*

With dependence on the type of chair construction, the joint between the side rail and back leg or the seat-back joint is the most stressed joint [86–88]. This fact was confirmed in the process of lamella chair construction with the stress concentrated especially in the mold of seat frame. In terms of anisotropy, lamella mold is stressed in a direction perpendicular to the grain. Due to the direction of chair loading and according to the theory of simple bending, the inner mold part is affected by the compression parallel to the grain direction; on the other hand, outer mold part is affected by tension parallel to the grain direction. Bending strength of wood perpendicular to the grain direction is greater than the compression strength parallel to the grain direction and lower than the tensile strength parallel to the grain direction. Therefore, when evaluating the ultimate limit state, design bending strength of lamella $f_{b,0,d}$ (gathered experimentally) is compared to maximum normal stress (in tension $\sigma_{t,0,d}$ or on compression $\sigma_{c,0,d}$) gathered using FEM. Design stress determined by FEM cannot exceed the value of design bending strength of lamella resulting from specimen testing in order to meet the conditions associated with the ultimate limit state. Due to the fact that the most significant effect of stresses is in lamella mold, values determined in mold lamellae were used for comparison. Maximum values of normal stress achieved using the FEM for chairs made of lamellae (type A, B and C) and for loading of 110 kg and 150 kg are mentioned in Table 4. FEM visual outputs of stresses are shown in Figure 6. Values highlighted in red colour are not suitable in terms of ultimate limit state.

**Table 4.** Maximum values of normal stress for the pitch seat-to-back angle of 103°.

| Type of Lamella | Loading of 110 kg | | Loading of 150 kg | |
| --- | --- | --- | --- | --- |
| | Design Stress FEM (MPa) | | Design Stress FEM (MPa) | |
| | In Tension $\sigma_{t,0,d}$ | In Compression $\sigma_{c,0,d}$ | In Tension $\sigma_{t,0,d}$ | In Compression $\sigma_{c,0,d}$ |
| A | 85.90 | 122.55 | 117.06 | 167.38 |
| B | 69.04 | 104.52 | 82.55 | 142.77 |
| C | 47.18 | 65.14 | 64.25 | 88.98 |

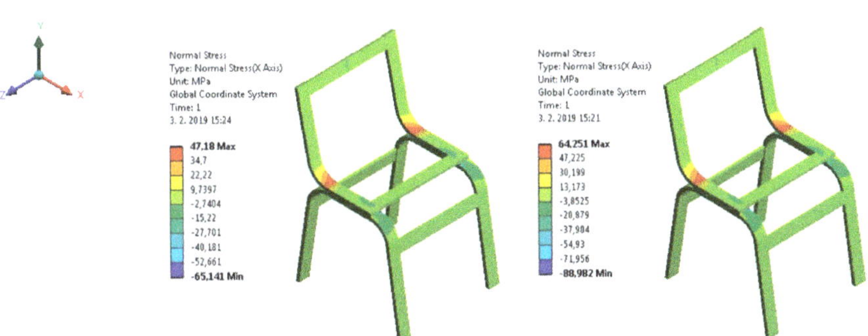

**Figure 6.** FEM visual outputs and stress concentration in tension and compression parts of lumbar curve of chair. Left for 110 kg tension (max. 47.18 MPa) and compression (max. 65.14 MPa) design stress and right for 150 kg tension (max. 64.25 MPa) and compression (max. 88.98 MPa).

Following the analysis of data gathered by comparing design values of bending strength determined experimentally and the values of design bending strength resulting from the use of FEM, the fact that ultimate limit state conditions were met can be stated. The values in Table 4 show that lamella with thickness of 11 mm (type A) met the requirements for use for a user with weight of 110 kg, only in the case of tensile stress. The value of compression stress was exceeded by 25.57 MPa. When the customer's weight was 150 kg, design tensile stress was exceeded by 20.08 MPa and design compression stress by 70.40 MPa. Following the results, the fact that this type of lamella cannot be used in chair construction for overweight users can be stated.

Lamella with thickness of 13.5 mm (type B) met the requirements of the ultimate limit state when the user's weight was 110 kg. In the case of a user with weight of 150 kg, design values of tensile stress were exceeded by 4.40 MPa and design compression stress by 64.62 MPa. Therefore, the lamella cannot be used when the chair is loaded with 150 kg.

The thickness of the last tested lamella was 16 mm (type C). It met the ultimate limit state conditions when the weight of a user is 110 kg. However, in the case of weight of 150 kg, it only met conditions in terms of tensile stress. Design value of compression stress was exceeded by 20.70 MPa. It means that lamella C cannot be used for chair construction for a user with weight of 150 kg as well.

### 3.2. Serviceability Limit State Assessment

Maximum values of the displacement of the upper edge of the seat $u$ (mm) with loading of 110 kg and 150 kg and corresponding values of the distance of the centre of gravity of the load from the tipping point $b$ (mm) in the direction of $x$-axis are mentioned in Table 5. FEM visual outputs to analyze the displacement of the back are shown in Figure 7. The values of displacement highlighted in red color are not satisfactory in terms of the serviceability limit state.

**Table 5.** Maximum values of the backward displacement of the back *u* (mm) and values of the distance of the centre of gravity of the load from the tipping *b* (mm) in the direction of x-axis.

| Type of Lamella. | Loading of 110 kg | | Loading of 150 kg | |
|---|---|---|---|---|
| | *u* (mm) | *b* (mm) | *u* (mm) | *b* (mm) |
| A | 289.15 | −63.5 | 343.15 | −140.1 |
| B | 189.13 | +33.7 | 256.16 | −54.6 |
| C | 96.72 | +157.4 | 128.34 | +74.2 |

Note: In case the back is not loaded, the distance between the centre of gravity of the load and the back leg is b = +237.2 mm in the direction of *x*-axis.

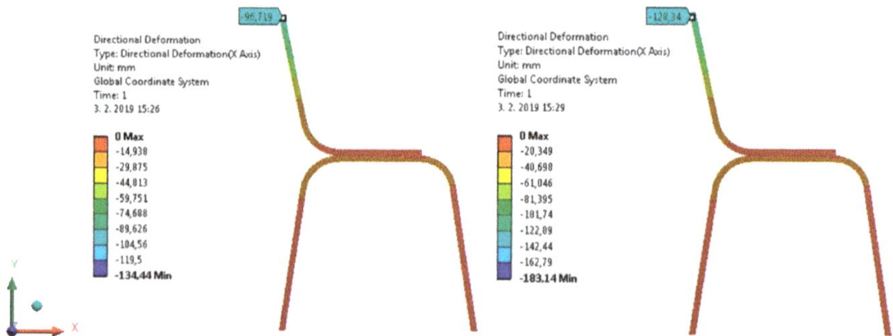


**Figure 7.** FEM visual outputs of deformation and displacement of the back while the load is applied. Left figure for 110 kg and right figure for 150 kg.

Analyzing the data summarized in Table 5, the serviceability limit state conditions can be evaluated. According to the requirements, seat-to-back angle must not exceed 110° (meeting the conditions results from the displacement of the upper edge of the seat). At the same time, the position of the centre of gravity must not be behind the tipping point and b value must not be negative in the direction of the *x*-axis. Tipping point is defined in the position of the back edge of the back leg.

Following the values determined by FEM for the lamella-type A, it is clear that the lamella did not meet defined conditions in the case of the loading of neither 110 kg nor 150 kg. In both loadings, allowable value of the displacement of the upper edge of the chair back was exceeded, and the value describing the position of the centre of gravity was negative in the direction of *x*-axis. Support provided by this lamella in the chair back was not adequate. Therefore, there was a danger of tipping over.

The lamella-type B did not meet the requirements for allowable back deformation for user weight of 110 kg and 150 kg. In terms of the position of the centre of gravity, the requirement is met only in case of loading of 110 kg. When user weight is 150 kg, there is a danger of tipping over because the centre of gravity was positioned behind the back leg of the chair.

The lamella-type C met the requirements for the position of the centre of gravity in the case of both weights of users. In spite of these findings, its use was not accepted due to the deformation of the chair back. Its value exceeded the allowable value for user weight of 110 kg or 150 kg.

The mentioned findings associated with meeting the requirements of the ultimate limit state as well as the serviceability limit state and the use of lamellae implied that no lamella type can be used in any tested cases of chair construction. Albeit the lamella-type C met the requirement for the ultimate limit state for the user with weight of 110 kg, the requirements for the serviceability limit state were not met.

### 3.3. Lamella Construction Meeting the Requirements of Ultimate States

Following the mentioned findings, the fact that lamella used in given chair construction should consist of a higher number of layers, thus, with greater thickness can be stated. Therefore, the group of specimens of lamella (type D) with 17 layers with total thickness of 21 mm was formed. Following the testing, design value of bending strength $\bar{\sigma} = 35.83$ MPa with the coefficient of variation of $\vartheta = 5.3\%$ was determined. Consequently, FEM analysis was carried out to determine the values of design stresses and deformation of the chair back. Calculated values are summarized in Table 6. FEM visual outputs of the stresses and displacement of the chair back are shown in Figure 8.

**Table 6.** Values of design compression and tensile stresses and values of the backward displacement of the chair back $u$ (mm) in the case of the lamella type D when loading is 150 kg.

| Type of Lamella | Design Stress-FEM (MPa) | | Displacement of the Chair Back |
|---|---|---|---|
| | in Tension | in Compression | $u$ (mm) |
| D | 28.16 | 26.78 | 37.41 |

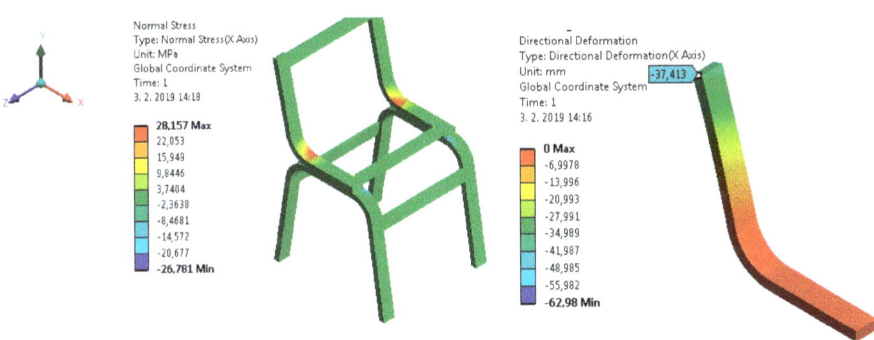

**Figure 8.** FEM visual outputs of stresses and displacement of the chair back with thickness of 21 mm when loading is 150 kg. Left: Values of design compression (23,781 MPa) and tensile stresses (28,157 MPa), right: value of the backward displacement of the chair back (37,413 mm).

Comparing the values of design bending strength and design values of stresses and values of the displacement of the upper edge of chair back achieved by the FEM analysis, it is clear that the lamella-type D with thickness of 21 mm would meet the requirements for both ultimate states in the case of loading of 150 kg. Due to the fact that in the research only a small sample size of this lamella type was made, testing lamella type D will offer an excellent opportunity for further research focused on dynamic loading.

Comparing the results to other studies dealing with chair modeling using FEM is quite difficult because of the evaluation stress according to von Mises mentioned by most authors. Wood is a material whose properties possess orthogonal anisotropy with nonlinear performance in elastic and plastic deformation. According to our opinion, two mentioned facts are key factors not allowing researchers to evaluate the stresses in wood-based material according to von Mises. The von Mises stress criterion is weighing the different oriented stresses to one "mixed" stress, which is not suitable to be compared to a scalar failure value for wood [89].

## 4. Conclusions

Various areas of economy, including furniture design and construction, have been affected by weight gain trends across populations in the last years. In Slovakia, the average weight of the population has increased by almost 10 kg over the last 25 years. A similar trend is observed globally. Almost 6% of the Slovak population is men with weight more than 110 kg. Therefore, the current standards must be re-evaluated. Valid legislation dealing with furniture design takes into account users' weight of 110 kg. However, according to anthropometric studies, 150 kg is the weight of users that should be taken into account in the future.

- The research presented was focused on the assessment of two ultimate states of flexible chair construction. Minimum thickness requirements for lamellae needed for chairs for users with weight up to 110 kg and 150 kg resulted from the research.

- Following the mechanical properties of laminated veneer lamellae and the assessment of ultimate limit state and serviceability limit state, as well as the use of lamella in flexible chairs, four thicknesses of lamellae were examined.

- Requirements for the strength of structural elements were evaluated by the ultimate limit state and allowable deformation of chair construction and the position of the centre of gravity during the loading were evaluated by the serviceability limit state. Following the results of the research, the fact that three types of tested lamellae (thickness 11 mm, 13.5 mm and 16 mm) did not meet the requirements of the both ultimate states. Lamella with thickness of 21 mm met the requirements for both ultimate states in the case of loading of 110 kg and 150 kg.

- The methodology to evaluate the serviceability limit state of flexible chairs based on ergonomy and chair safety can be considered as another contribution of the research.

Weight gain is a global problem affecting the industrial goods sector. In the case of research, the cooperation of professionals in anthropology, ergonomics, construction, design and health is needed, in order to modify the size and function of furniture. Designing wooden furniture should be connected with a sustainable strategy of economy aimed at efficient use of local renewable resources. Only a complex approach can contribute to meeting the goals of sustainability of the furniture industry leading to sustainability of standards and timeliness.

**Author Contributions:** Conceptualization, N.L., M.H., P.J. and R.I.; Data curation, N.L., R.R. and M.H.; Visualization, R.R., Ľ.K., R.I.; Writing-original draft, N.L., R.R., Ľ.K. and R.I.

**Acknowledgments:** This research was supported by the Slovak Research and Development Agency under Contract No. APVV-16-0297, Contract No. APVV-14-0506, Contract No. APVV-17-0583, VEGA 1/0717/19, VEGA 1/0556/19 and KEGA 012TU Z-4/2017.

**Conflicts of Interest:** The authors declare no conflict of interest.

## References

1. Joščák, P.; Dudas, J.; Gáborík, J.; Gaff, M.; Langová, N.; Navrátil, V.; Slabejová, G. *Wood and Wood-Based Furniture Constructions*; Technická univerzita vo Zvolene: Zvolen, Slovakia, 2014.

2. Wang, Y.; Beydoun, M.A. The obesity epidemic in the United States—Gender, age, socioeconomic, racial/ethnic, and geographic characteristics: A systematic review and meta-regression analysis. *Epidemiol. Rev.* **2007**, *29*, 6–28. [CrossRef] [PubMed]

3. Stevens, G.A.; Singh, G.M.; Lu, Y.; Danaei, G.; Lin, J.K.; Finucane, M.M.; Bahalim, A.N.; McIntire, R.K.; Gutierrez, H.R.; Cowman, M.; et al. National, regional, and global trends in adult overweight and obesity prevalences. *Popul. Health Metr.* **2012**, *10*, 22. [CrossRef] [PubMed]

4. Tremmel, M.; Gerdtham, U.G.; Nilsson, P.M.; Saha, S. Economic Burden of Obesity: A Systematic Literature Review. *Int. J. Environ. Res. Public Health* **2017**, *14*, 435. [CrossRef] [PubMed]

5. González-Rodríguez, L.G.; Perea Sánchez, J.M.; Aranceta-Bartrina, J.; Gil, Á.; González-Gross, M.; Serra-Majem, L.; Varela-Moreiras, G.; Ortega, R.M. Intake and Dietary Food Sources of Fibre in Spain: Differences with Regard to the Prevalence of Excess Body Weight and Abdominal Obesity in Adults of the ANIBES Study. *Nutrients* **2017**, *9*, 326. [CrossRef] [PubMed]

6. Burke, N.L.; Shomaker, L.B.; Brady, S.; Reynolds, J.C.; Young, J.F.; Wilfley, D.E.; Sbrocco, T.; Stephens, M.; Olsen, C.H.; Yanovski, J.A.; et al. Impact of Age and Race on Outcomes of a Program to Prevent Excess Weight Gain and Disordered Eating in Adolescent Girls. *Nutrients* **2017**, *9*, 947. [CrossRef] [PubMed]

7. Chan, R.S.; Woo, J. Prevention of Overweight and Obesity: How Effective is the Current Public Health Approach. *Int. J. Environ. Res. Public Health* **2010**, *7*, 765–783. [CrossRef]

8. Peltzer, K.; Pengpid, S.; Samuels, T.A.; Özcan, N.K.; Mantilla, C.; Rahamefy, O.H.; Wong, M.L.; Gasparishvili, A. Prevalence of Overweight/Obesity and Its Associated Factors among University Students from 22 Countries. *Int. J. Environ. Res. Public Health* **2014**, *11*, 7425–7441. [CrossRef]

9.  Finucane, M.M.; Stevens, G.A.; Cowan, M.J.; Danaei, G.; Lin, J.K.; Paciorek, C.J.; Singh, G.M.; Gutierrey, H.R.; Lu, Y.; Bahalim, A.N. Global burden of metabolic risk factors of chronic diseases collaborating group body mass index. *Lancet* **2011**, *377*, 557–567. [CrossRef]

10. De Onis, M.; Blössner, M.; Borghi, E. Global prevalence and trends of overweight and obesity among preschool children. *Am. J. Clin. Nutr.* **2010**, *92*, 1257–1264. [CrossRef]

11. Rockholm, B.; Baker, J.L.; Sorensen, T.I.A. The levelling off of the obesity epidemic since the year 1999- A review of evidence and perspectives. *Obes. Rev.* **2010**, *11*, 835–846. [CrossRef]

12. Odunitan-Wayas, F.; Okop, K.; Dover, R.; Alaba, O.; Micklesfield, L.; Puoane, T.; Uys, M.; Tsolekile, L.; Levitt, N.; Battersby, J.; et al. Food Purchasing Characteristics and Perceptions of Neighborhood Food Environment of South Africans Living in Low-, Middle- and High-Socioeconomic Neighborhoods. *Sustainability* **2018**, *10*, 4801. [CrossRef]

13. Manzano, S.; Doblaré, M.; Hamdy Doweidar, M. Altered Mechano-Electrochemical Behavior of Articular Cartilage in Populations with Obesity. *Appl. Sci.* **2016**, *6*, 186. [CrossRef]

14. Ng, M.; Fleming, T.; Robinson, M.; Tomson, B.; Graetz, N.; Margono, C.; Mullany, E.C.; Biryukov, S.; Abbafati, C.; Abera, S.F.; et al. Global, regional, and national prevalence of overweight and obesity in children and adults during 1980-2013: A systematic analysis for the Global Burden of Disease Study 2013. *Lancet* **2014**, *384*, 766–781. [CrossRef]

15. Di Cesare, M.; Bentham, J.; Stevens, G.A.; Zhou, B.; Danaei, G.; Lu, Y.; Bixby, H.; Cowan, M.J.; Riley, L.M.R.; Hajifathalian, K.; et al. Trends in adult body-mass index in 200 countries from 1975 to 2014: A pooled analysis of 1698 population-based measurement studies with 19.2 million participants. *Lancet* **2016**, *387*, 1377–1396.

16. Hitka, M.; Hajduková, A. Antropometrická optimalizácia rozmerov lôžkového nábytku. *Acta Facultatis Xylologiae Zvolen* **2013**, *55*, 101–109.

17. Čuta, M.; Kukla, L.; Novák, L. Modelling the development of body height (length) in children using parental height data. *Československá Pediatrie* **2010**, *65*, 159–166.

18. Harrison, C.R.; Robinette, K.M. *CAESAR > Summarz Statistics for the Adult Population (Ages 18–65) of the United States of America*; United States Air Force Research Laboratorz: Wright-Patterson AFB, OH, USA, 2002.

19. Réh, R.; Krišťák, Ľ.; Hitka, M.; Langová, N.; Joščák, P.; Čambál, M. Analysis to Improve the Strength of Beds Due to the Excess Weight of Users in Slovakia. *Sustainability* **2019**, *11*, 624. [CrossRef]

20. Ministry of Health of the Slovak Republic. Report on Health Status in Slovakia. Health 2016. Available online: http://health.gov.sk (accessed on 31 September 2018).

21. Public Health Authority of the Slovak Republic. Annual Report on the Activities of the Public Health Office for 2017. Available online: http://uvzsr.sk/docs/vs/vyrocna_sprava_2017.pdf (accessed on 31 September 2018).

22. Statistical Office of the Slovak Republic. View of Health Status of the Slovak Population and Its Determinants (Results of EHIS 2016). Available online: http://slovak.statistics.sk (accessed on 31 September 2018).

23. World Health Organization Regional Office for Europe. The Health Systems in Transition (HiT). Available online: http://euro.whoint/en/countries/slovakia (accessed on 31 September 2018).

24. Porta, J.; Saco-Ledo, G.; Cabanas, M.D. The ergonomics of airplane seats: The problem with economy class. *Int. J. Ind. Ergon.* **2019**, *69*, 90–95. [CrossRef]

25. Quigley, C.; Southall, D.; Freer, M.; Moody, A.; Porter, M. Anthropometric Study to Update Minimum Aircraft Seating Standards. Available online: https://dspace.lboro.ac.uk/2134/701 (accessed on 18 January 2019).

26. Roggla, G.; Moser, B.; Roggla, M. Seat space on airlines. *Lancet* **1999**, *353*, 1532. [CrossRef]

27. BIFMA X6.1. *Educational Seating. American National Standard for Office Furnishings (Revised in 2018)*; BIFMA: Grand Rapids, MI, USA, 2018.

28. Bellingar, T.A.; Benden, M.E. New ANSI/BIFMA standard for testing of educational seating. *Ergon. Des. Q. Hum. Factors Appl.* **2015**, *23*, 23–27. [CrossRef]

29. Parcells, C.; Stommel, M.; Hubbard, R.P. Mischmatch of classroom furniture and student body dimensions: Empirical findings and health implications. *J. Adolesc. Health* **1999**, *24*, 265–273. [CrossRef]

30. Panagiotopoulou, G.; Christoulas, K.; Papanckolaou, A.; Mandroukas, K. Classroom furniture dimensions and anthropometric measures in primary school. *Appl. Ergon.* **2004**, *35*, 121–128. [CrossRef] [PubMed]

31. Mohamed-Thariq, M.G.; Munasinghe, H.P.; Abeysekara, J.D. Designing chairs with mounted desktop for university students Ergonomics and comfort. *Int. J. Ind. Ergon.* **2010**, *40*, 8–18. [CrossRef]

32. Oyewole, S.A.; Haight, J.M.; Freivalds, A. The ergonomic design of classroom furniture/computer work station for first graders in the elementary school. *Int. J. Ind. Ergon.* **2010**, *40*, 437–447. [CrossRef]

33. Agha, S.R.; Alnahhal, M.J. Neural network and multiple linear regression to predict school children dimensions for ergonomic school furniture design. *Appl. Ergon.* **2012**, *43*, 979–984. [CrossRef] [PubMed]

34. BIFMA X5.11. *Large Occupant Office Chair Standard. American National Standard for Office Furnishings*; BIFMA: Grand Rapids, MI, USA, 2015.

35. AFRDI 142. *Certification for Heavy Duty Office Chairs*; Australian Furnishings Research and Development Institute: Launceston, Australia, 2012.

36. AFRDI 151. *Rated Load Standard for Fixed Height Chairs*; Australian Furnishings Research and Development Institute: Launceston, Australia, 2014.

37. BS 5459-2:2000 and A2:2008. *Specification for Performance Requirements and Tests for Office Furniture*; British Standards Institutions: London, UK, 2000.

38. ISO 17025. *General Requirements for the Competence of Testing and Calibration Laboratories*; International Organization for Standardization: Geneva, Switzerland, 2017.

39. SATRA. Satra Technology—International Product Research, Testing and Supply Chain Quality. Available online: http://satra.com (accessed on 31 September 2018).

40. Bonenberg, A.; Branowski, B.; Kurcyewski, P.; Lewandowska, A.; Sydor, M.; Torzynski, D.; Zablocki, M. Designing for human use: Examples of kitchen interiors for person with disability and elderly people. *Hum. Factors Ergon. Manuf.* **2019**, *29*, 177–186. [CrossRef]

41. Mehrparvar, A.H.; Mirmohammadi, S.J.; Hafezi, R.; Mostaghaci, M.; Davari, M.H. Static anthropometric dimensions in a population of Iranian high school students: Considering ethnic differences. *Hum. Factors J. Hum. Factors Ergon. Soc.* **2015**, *57*, 447–460. [CrossRef]

42. Qutubuddin, S.M.; Hebbal, S.S.; Kumar, C.S. Anthropometric consideration for designing students desks in engineering colleges. *Int. J. Curr. Eng. Technol.* **2013**, *3*, 1179–1185.

43. Shin, D.; Kim, J.Y.; Hallbeck, M.S.; Haight, J.M.; Jung, M.C. Ergonomic hand tool and desk and chair development process. *Int. J. Occup. Saf. Ergon.* **2008**, *14*, 247–252. [CrossRef]

44. Masson, A.E.; Hignett, S.; Gyi, D.E. Anthropometric Study to Understand Body Size and Shape for Plus Size People at Work. *Procedia Manuf.* **2015**, *3*, 5647–5654. [CrossRef]

45. Smardzewski, J. *Furniture Design*; Springer International Publishing: Cham, Switzerland, 2015.

46. Hitka, M.; Joščák, P.; Langová, N.; Krišťák, L.; Blašková, S. Load-Carrying Capacity and the Size of Chair Joints D etermined for Users with a Higher Body Weight. *Bioresources* **2018**, *13*, 6428–6443.

47. Hitka, M.; Sedmák, R.; Joščák, P.; Ližbetinová, L. Positive Secular Trend in Slovak Population Urges on Updates of Functional Dimensions of Furniture. *Sustainability* **2018**, *10*, 3474. [CrossRef]

48. Dvouletá, K.; Káňová, D. Utilization of anthropometry in the sphere of sitting and bed furniture. *Acta Univ. Agric. Silvic. Mendel. Brun.* **2014**, *62*, 81–90. [CrossRef]

49. Muir, M.; Archer-Heese, G. Essentials of a bariatric patient handling program. *OJIN Online J. Issues Nurs.* **2009**, *14*, 5.

50. Gourash, W.; Rogula, T.; Schauer, P.R. *Essential Bariatric Equipment: Making Your Facility More Accommodating to Bariatric Surgical Patients*; Springer: New York, NY, USA, 2007.

51. Wiggermann, N.; Smith, K.; Kumpar, D. What bed size does a patient need? The relationship between body mass index and space required to turn in bed. *Nurs. Res.* **2017**, *66*, 483–489. [CrossRef] [PubMed]

52. Ko, Y.C.; Lo, C.H.; Chen, C.C. Influence of Personality Traits on Consumer Preferences: The Case of Office Chair Selection by Attractiveness. *Sustainability* **2018**, *10*, 4183. [CrossRef]

53. Nüesch, C.; Kreppke, J.N.; Mündermann, A.; Donath, L. Effects of a Dynamic Chair on Chair Seat Motion and Trunk Muscle Activity during Office Tasks and Task Transitions. *Int. J. Environ. Res. Public Health* **2018**, *15*, 2723. [CrossRef]

54. Borůvka, V.; Dudík, R.; Zeidler, A.; Holeček, T. Influence of Site Conditions and Quality of Birch Wood on Its Properties and Utilization after Heat Treatment. Part I—Elastic and Strength Properties, Relationship to Water and Dimensional Stability. *Forests* **2019**, *10*, 189. [CrossRef]

55. Wei, P.; Rao, X.; Yang, J.; Guo, Y.; Chen, H.; Zhang, Y.; Chen, S.; Deng, X.; Wang, Z. Hot pressing of wood-based composites: A review. *For. Prod. J.* **2016**, *66*, 419–427. [CrossRef]

56. Derikvand, M.; Kotlarewski, N.; Lee, M.; Jiao, H.; Nolan, G. Characterisation of Physical and Mechanical Properties of Unthinned and Unpruned Plantation-Grown Eucalyptus nitens H. Deane & Maiden Lumber. *Forests* **2019**, *10*, 194.

57. Šubic, B.; Fajdiga, G.; Lopatič, J. Bending Stiffness, Load-Bearing Capacity and Flexural Rigidity of Slender Hybrid Wood-Based Beams. *Forests* **2018**, *9*, 703. [CrossRef]

58. Réh, R.; Igaz, R.; Krišťák, Ľ.; Ružiak, I.; Gajtanska, M.; Božíková, M.; Kučerka, M. Functionality of beech bark in adhesive mixtures used in plywood and its effect on the stability associated with material systems. *Materials* **2019**, *12*, 1298. [CrossRef] [PubMed]

59. Morales, G.A. Potential of Gmelina arborea for solid wood products. *New For.* **2004**, *28*, 331–337. [CrossRef]

60. Sikora, A.; Svoboda, T.; Záborský, V.; Gaffová, Z. Effect of selected factors on the bending deflection at the limit of proportionality and at the modulus of rupture in laminated veneer lumber. *Forests* **2019**, *10*, 401. [CrossRef]

61. Zhang, X.; Zhu, Y.; Yu, Y.; Song, J. Improve Performance of Soy Flour-Based Adhesive with a Lignin-Based Resin. *Polymers* **2017**, *9*, 261. [CrossRef] [PubMed]

62. Gejdoš, M.; Tončíková, Z.; Němec, M.; Chovan, M.; Gergeľ, T. Balcony cultivator: New biomimicry design approach in the sustainable device. *Futures* **2018**, *98*, 32–40. [CrossRef]

63. Bekhta, P.; Hiziroglu, S.; Potapova, O.; Sedliačik, J. Shear Strength of Exterior Plywood Panels Pressed at Low Temperature. *Materials* **2009**, *2*, 876–882. [CrossRef]

64. Erdil, Y.Z.; Zhang, J.; Eckelman, C.A. Withdrawal and bending strength of dowel-nuts in plywood and oriented strandboard. *For. Prod. J.* **2003**, *53*, 54–57.

65. Prekrat, S.; Smardzewski, J. Effect of gluline shape on strength of mortise and tenon joint. *Drv. Ind.* **2010**, *61*, 223–228.

66. Kilic, H.; Kasal, A.; Kuskun, T.; Acar, M.; Erdil, Y.Z. Effect of Tenon Size on Static Front to Back Loading Performance of Wooden Chairs in Comparison with Acceptable Design Loads. *Bioresources* **2018**, *13*, 256–271.

67. Kuskun, T.; Kasal, A.; Haviarova, E.; Kilic, H.; Uysal, M.; Erdil, Y.Z. Relationship between static and cyclic front to back load capacity of wooden chairs, and evaluation of the strength values according to acceptable design values. *Wood Fiber Sci.* **2018**, *50*, 1–9. [CrossRef]

68. Oktaee, J.; Ebrahimi, G.; Layeghi, M.; Ghofrani, M.; Eckelman, C.A. Bending moment capacity of simple and haunched mortise and tenon furniture joints under tension and compression loads. *Turk. J. Agric. For.* **2014**, *38*, 291–297. [CrossRef]

69. Likos, E.; Haviarova, E.; Eckelmna, C.A.; Erdil, Y.Z.; Ozcifci, A. Technical note: Static versus cyclic load capacity of side chairs constructed with mortise and tenon joints. *Wood Fiber Sci.* **2013**, *45*, 223–227.

70. Gaff, M.; Gašparík, M.; Boruvka, V.; Haviarová, E. Stress simulation in layered wood-based materials under mechanical loading. *Mater. Des.* **2015**, *87*, 1065–1071. [CrossRef]

71. Grič, M.; Joščák, P.; Tarvainen, I.; Ryonankoski, H.; Lagaňa, R.; Langová, N.; Andor, T. Mechanical properties of furniture self-locking frame joints. *Bioresources* **2017**, *12*, 5525–5538. [CrossRef]

72. Branowski, B.; Zablocki, M.; Sydor, M. Experimental analysis of new furniture joints. *Bioresources* **2018**, *13*, 370–382. [CrossRef]

73. Hajdarević, S.; Busuladžič, I. Stiffness Analysis of Wood Chair Frame. *Procedia Eng.* **2015**, *100*, 746–755. [CrossRef]

74. Staneva, N.; Genchev, Y.; Hristodorova, D. Approach to designing an upolstered furniture frame by the finite element method. *Acta Fac. Xylologiae* **2018**, *60*, 61–70.

75. Hu, W.G.; Guan, H.Y. Research on withdrawal strength of mortise and tenon joint by numerical and analytic methods. *Wood Res.* **2017**, *62*, 575–586.

76. Derikvand, M.; Ebrahimi, G. Finite element analysis of stress and strain distributions in mortise and loose tenon furniture joints. *J. For. Res.* **2014**, *25*, 677–681. [CrossRef]

77. Hu, W.G. Study of finite element analysis of node in solid wood structure furniture based on ANSYS. *Furnit. Inter. Des.* **2015**, *46*, 65–67.

78. Kasal, A.; Eckelman, C.A.; Haviarova, E.; Yalcin, I. Bending moment capacities of L-shaped mortise and tenon joints under compression and tension loadings. *Bioresources* **2016**, *1*, 6836–6853. [CrossRef]

79. Diler, H.; Efe, H.; Erdil, Y.Z.; Kuskun, T.; Kasal, A. Determination of allowable design loads for wood chairs. In Proceedings of the XXVIII International Conference Research for Furniture Industry, Poznan, Poland, 21–22 September 2017.

80. Laemlaksakul, V. Innovative Design of Laminated Bamboo Furniture Using Finite Element Method. *Int. J. Math. Comput. Simul.* **2008**, *3*, 274–284.

81. EN 1728. *Furniture. Seating. Test Methods for the Determination of Strength and Durability*; BSI: London, UK, 2012.

82. EN 12520. *Furniture—Strength, Durability and Safety—Requirements for Domestic Seating*; BSI: London, UK, 2015.

83. EN 1022. *Domestic Furniture. Seating. Determination of Stability*; BSI: London, UK, 2018.
84. Sydor, M.; Wieloch, G. Construction properties of wood taken into consideration in engenering practice. *Drewno* **2009**, *52*, 63–73.
85. Požgaj, A.; Chovanec, D.; Kurjatko, S.; Babiak, M. *Štruktúra a Vlastnosti Dreva*; Príroda a.s.: Bratislava, Slovakia, 1993.
86. Sedlecký, M.; Kvietková, M.S.; Kminiak, R. Medium-density fiberboard (MDF) and Edge-glued Panels (EGP) after edge milling—Surface roughness after machining with different parameters. *Bioresources* **2018**, *13*, 2005–2021. [CrossRef]
87. Němec, M.; Kminiak, R.; Danihelová, A.; Gergel, T.; Ondrejka, V. Vibrations and workpiece surface quality at changing feed speed of CNC machine. *Akustika* **2017**, *28*, 117–124.
88. Ke, Q.; Zhang, F.; Zhang, Y. Optimisation design of pine backrest chair based on Taguchi method. *Int. Wood Prod. J.* **2017**, *8*, 18–25. [CrossRef]
89. Szalai, J. Festigkeitstheorien von anisotropen Stoffen mit sprodem bruchverhalten. *Acta Silv. Lign. Hung.* **2008**, *4*, 61–79.

*Article*

# Preferences for Urban Building Materials: Does Building Culture Background Matter? †

Olav Høibø [1,*], Eric Hansen [2], Erlend Nybakk [3] and Marius Nygaard [4]

[1] Faculty of Environmental Sciences and Natural Resource Management, Norwegian University of Life Sciences (NMBU), P.O. Box 5003, NO-1432 Ås, Norway
[2] Department of Wood Science & Engineering, Oregon State University, 119 Richardson Hall, Corvallis, OR 97331, USA; eric.hansen@oregonstate.edu
[3] Department of Marketing, Economics and Innovation, Kristiania University College, P.O. 1190 Sentrum, 0107 Oslo, Norway; Erlend.Nybakk@kristiania.no
[4] Department of Architecture, Oslo School of Architecture and Design, P.O. Box 1633, Vika, 0119 Oslo, Norway; Marius.Nygaard@aho.no
* Correspondence: olav.hoibo@nmbu.no; Tel.: +47-67-231-743
† Some of the result in this article was presented by Olav Høibø [1,*], Eric Hansen [2], Erlend Nybakk [3] and Marius Nygaard [4] in a poster at SWST 2015 International Convention, Jackson Lake Lodge, Grand Teton National Park, Wyoming, USA; 2015-06-07–2015-06-12.

Received: 24 July 2018; Accepted: 14 August 2018; Published: 17 August 2018

**Abstract:** A fast-growing global population, increasing urbanization, and an increasing flow of people with different building cultural backgrounds bring material use in the housing sector into focus. The aim of this study is to identify material preferences in the building environment in cities and to determine if the building cultural background impacts those preferences. The data in this study consisted of responses from two groups of dwellers in Norway, including immigrants from countries where wood is an uncommon building material and native Norwegians from a building culture where wood is common. We found that the most preferred materials were often the same as the most common materials currently used in city buildings. Only small differences were found between the two groups of dwellers that were studied. Most differences were related to concerns about material choice in general and where individuals wanted to live. Respondents who preferred city living preferred commonly used city materials, such as concrete and steel. For cladding materials, stone/bricks were the most preferred. However, stained or painted wood was one of the most preferred, even though it is not commonly used in city buildings.

**Keywords:** marketing; material preference; urban housing; immigrants; building culture background; building material

---

## 1. Introduction

A fast-growing global population [1] and a focus on sustainable development and climate change bring the housing sector and materials used for housing into focus. United Nations estimates place the global population at approximately 9.6 billion by 2050 [2]. Currently, the global demand for new housing is approximately five million units per year [3]. Given the state of housing stock and the mentioned growth in the population, a significant increase of housing units is needed by 2050. A growing proportion of the global population will reside in urban areas, where housing density is a factor in sustainable development [4].

In addition to the fast-growing global population and increasing urbanization, immigrant flow is accelerating due to differences in income, social networks, and various state policies, thus leading to an overall growing number of immigrant cities [5]. In Western Europe, an unprecedented number of newcomers have arrived during the last two decades. When considering cities with more than 100,000 immigrants, North American and Western European cities are key immigrant destinations [6].

Impending climate change means that the carbon footprint has gained importance as a key metric in the assessment of the environmental impacts of buildings. Embodied energy and emissions of materials are vital parts of this picture. In the future, embodied energy and choice of material will be even more important since energy consumption from operational use will decrease and building material consumption will increase [7]. Therefore, timber-framed buildings, which are found to have lower global warming potential than concrete and steel structures [8,9], might play an important role with regard to the reduction of environmental consequences of city buildings. However, wood is not a common modern city building material, and might therefore be a material less preferred by consumers. Further, residents from countries where wood is hardly used in any buildings might have lower preferences for wood than people coming from countries where wood is more common.

In Norway, developments close to city centers are mainly buildings of four to eight stories. These building types are easily constructed with wood-based products [10]. New building codes and more sprinkler systems further facilitate timber use.

There is a growing body of consumer preference studies on building materials [11,12]. However, little research has been done on material preferences in the context of the urban built environment and changes in demographics resulting from immigration and movement.

As city officials, urban planners, architects, and construction companies plan for future housing, it is imperative that they understand the housing [1] and material preferences of city dwellers, especially in light of the changing demographics of regions resulting from immigration and movement to urban locales. Additionally, in Norway, the population is urbanizing. In a recent forecast, it was suggested that the Oslo region will receive up to 310,000 new inhabitants by 2020, thus adding to its current population, and an additional 600,000 in the period from 2020 to 2040. Housing these new arrivals will significantly impact the Oslo region [13]. The newcomers will partly come from Norway, where wooden houses are common, but newcomers will also come from countries where wood is hardly used. Accordingly, this study seeks to identify the differences between consumers with different building material backgrounds with regard to their preferences for materials in structural, interior, and exterior urban housing applications.

In the remainder of this article, we first provide a background regarding material preferences and the context of housing related to an urbanizing population that includes a significant proportion of immigrants. Next, we provide a background leading to research questions regarding the material preference differences between residents that have immigrated to Norway from countries where wood is hardly used in any buildings and native Norwegians. We use this as an example that may be considered in other global settings as cities plan their future housing expansion. We then discuss the methods used in the study, provide a description of the results, discuss those results, and provide specific policy and business implications.

*Background*

The materials used in buildings are a function of the availability and suitability of materials, as well as various cultural norms and traditions. For instance, in regions with termites, wood is less frequently used, and brick and stone buildings are more common. In some cultures or countries, wood-based housing is seen as inferior [12] and can even be considered a material associated with low social status, while in other countries, the traditions for using wood are strong. In Norway, a long tradition of using wood is illustrated with more than 800-year-old wooden buildings, and today, approximately 78% of the dwellings in Norway are one- and two-story wood structures [14].

Earlier studies have found relationships between tradition and material preferences [15] and between personal tradition and residential choice from a life style perspective [16], and have also revealed that choices are related to familiarity [1]. Extensive research has investigated the relationship between preferences and social expectations and the idea that the exterior of a house conveys meaning about the owner to others [17,18]. Individuals may also use the house exterior to define their identity [19]. Hauge and Kolstad [20] suggest that there may be differences between genders or among ethnicities and cultural backgrounds with regard to what the interior and exterior of a house says about the owner. Accordingly, we might expect that people coming from different regions with different material traditions have different preferences. On the other hand, since preferences are also most likely related to where the material is used and modern building traditions in cities around the world tend to be similar, less differences between people from different parts of the world with regard to what they expect and prefer regarding materials used in multistory city housing may be expected.

In addition to studying differences between people with backgrounds from regions with different building material traditions, our study also includes analyses of the stated preferences for how and where to live. Since the exterior of a house might convey meaning about the owner to others [17,18] and people might use the house to define their identity [19], individuals who prefer city living might, to a greater extent, identify themselves with and be more positive regarding the buildings made of materials that are common in cities compared to buildings made of materials that are more common outside cities.

## 2. Materials and Methods

The work described below is partly based on the same survey as that used by Høibø et al. [21]. Here, we emphasize how material preferences are related to the respondent's origin and where and how they want to live, while Høibø et al. [21] focused on material preferences related to attitudes regarding durability and solidity, how environmentally friendly the material is, knowledge about wood, and experience with remodeling.

We collected responses through an online survey from individuals in immigrant families coming from countries where wood is not commonly used in houses (hereafter referred to as immigrants) and native Norwegians. Native Norwegians in this study are defined as those born in Norway with both parents from Norway. Immigrants are those with both parents born outside Norway (the individual respondent could be born inside or outside Norway).

### 2.1. Sampling

A total of 1751 persons were asked to participate in the study. However, the collection of data stopped when six hundred and sixty two people had completed the questionnaire. The respondents were part of the TNS Gallup As (today Kantar TNS AS) recruited probability panel, certified according to ISO 9001, ISO 20252, and ISO 26362:2009. The recruitment of the Gallup panel is mainly done through telephone listings and their sampling matrix design weights for biases based on how easy different groups of people are to reach. Panel members do not know the nature of the study before they access the electronic questionnaire. Demographic data about the respondents from the TNS Gallup AS database were added to our data set.

TNS Gallup AS did not have a large enough panel of immigrants specific to Oslo and several surrounding communities (Oslo region). Therefore, an additional set of respondents outside the Oslo region was targeted, in addition to the survey that Høibø et al. [21] used. Of the 662 responses received, 532 responses fit our definition of native Norwegians and immigrants, thus resulting in 437 native Norwegian responses and 95 immigrant responses. Of the 95 immigrant respondents, 67 were born in countries other than Norway. Thirty-nine immigrant respondents reside outside the Oslo region. Since there were two groups of immigrants, one residing in the same counties as the native Norwegian respondents, and a group residing in other counties, the total material is not completely random. Adjustments with a dummy variable were therefore made in some analyses. If there was no significant effect of this grouping, all respondents were considered to represent the same region. Most of the immigrants (71) had a background from Asia, Africa, or South America, while the rest came from Poland, which represents a large immigrant group in Norway.

Figure 1 shows distributions of the types of houses, in terms of structural material, that the immigrants and the native Norwegians had mainly lived in until they were 16 years old. The respondents that did not know how they lived in this period are not included in the figure. Table 1 provides additional information about the respondents.

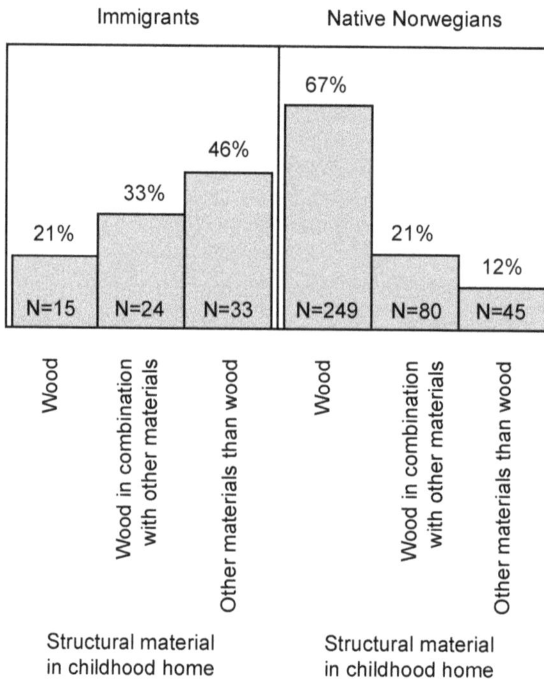


**Figure 1.** Distributions in N (number) and % of what kind of houses with respect to the structural material that the immigrants and the native Norwegians respectively had lived in the most until they were 16 years old.

**Table 1.** Statistics for the respondents.

| Description of the Respondents | N (Number) |
| --- | --- |
| Region of origin immigrants/native Norwegians | 95/437 |
| Age (mean = 45.4, Standard deviation = 16.6) | |
| Gender female/male | 277/255 |
| Currently living in city | 343 |
| Currently living in large town | 87 |
| Currently living in small town | 87 |
| Currently living rural | 13 |
| Currently living in apartment, 3-story building or more | 271 |
| Currently living in row house | 125 |
| Currently living in detached house | 104 |
| Currently living in other type of housing | 30 |

*2.2. Description of Variables and Questions*

Even though it may have been more difficult for respondents, we did not want them to react to visual images, but rather to provide their more basic material preferences. To help mitigate the issue of lack of knowledge, we provided an "I don't know" category for many questions. The questionnaire was tested on a small sample of respondents. The feedback was positive, and we made no major changes to the questionnaire.

All variables used in the statistical tests and models are shown in Table 2. For importance, knowledge, and preferences, a nine-point scale was used. For example, the scale ranged from "not important" to "very important" or from "do not like" to "like very much" [21]. Because we collected data via a questionnaire, all measures of preferences were stated preferences, as is commonly recommended [1].

**Table 2.** Variable definitions and abbreviation list.

| Variables | Abbreviation | N Levels |
|---|---|---|
| Type of material for structural use | MStr | 3 types<br>9 point scale |
| Type of material used on indoor walls and ceilings | MInd | 5 types<br>9 point scale |
| Type of material for outdoor cladding | MCla | 4 types<br>9 point scale |
| Preference for living in city, population more than 100,000 | PrCity | 9 point scale |
| Preference for living in rural areas | PrRur | 9 point scale |
| Respondents' region of origin: immigrants and native Norwegians | ResOr | 2 levels |
| Importance of the structural materials used | ImpStru | 9 point scale |
| Importance of the materials used for indoor walls and ceilings | ImpInd | 9 point scale |
| Importance of the materials used for outdoor cladding | ImpCla | 9 point scale |
| Effect of different sampling between immigrants and inside and outside the Oslo region | EfSamp | 2 levels |

Three main questions about material preferences were included in the questionnaire [21]. One question was about the materials used in the structural part of the building. Answers were given individually for concrete, steel, and wood [21]. The next question was about the materials used for cladding. Answers were given individually for untreated wood cladding, painted or stained wood cladding, metal sheeting, and stone/bricks [21]. The last question was about the materials used for inside walls and ceilings. Untreated wood; lacquered, stained, or painted wood; paint or wallpaper on gypsum boards; paint or wallpaper on wood-based boards; and paint or wallpaper on concrete were the options [21]. Individual questions about the importance of the material used for structural purposes, outdoor cladding, and indoor walls and ceilings were also included [21].

Other questions included the following:

In what setting would the respondent prefer to live?

> In a city (population more than 100,000).
> In a large town (population between 10,000 and 100,000).
> In a small town (population less than 10,000), or
> In a rural area.

In what type of housing would they prefer to live?

> In a detached house.
> In a row house, or
> In an apartment in an apartment block with three stories or more.

Importance of closeness to stores, schools and other services.
Importance of closeness to family, friends, and acquaintances.
Importance of low price.
Relationship to and knowledge about buildings and the construction industry.

*2.3. Analysis*

The statistical software JMP version 10.0 from the SAS Institute Inc. (Cary, NC, USA) [22] was used in the data analyses. Where appropriate, contrasts were tested with F-tests. However, some of the data exhibited heteroscedasticity and nonnormality, and so we chose to use a logistic regression and chi square tests in most analyses. For the comparison of groups, we used chi square tests and, when necessary, we merged cells to maintain greater than 80% of cells with five or more responses. The responses of "do not know" were not included in the analyses. The logistic regression calculated the probabilities for each level of the response and gave nine probabilities, depending on the values of the independent variables. To do this, eight fitting lines were calculated (when a nine-point scale was used) (see figure caption Figure 3).

A small effect of the difference in sampling between immigrants inside and outside the Oslo region was found for the analyses on indoor wall and ceiling material preferences and where they want to live, thus requiring a correction via a dummy variable. If nothing else was said, variables were rejected if the probability of type I error was smaller than 0.05.

## 3. Results

*3.1. Material Preferences*

Native Norwegians had somewhat higher mean preferences for concrete and steel structural materials than immigrants (Figure 2a). However, the differences were small. For wood as a structural material, it was the opposite, but the difference was minor (Figure 2a). The differences in preferences between wood and concrete and between wood and steel, respectively, were not significantly different between immigrants and native Norwegians.

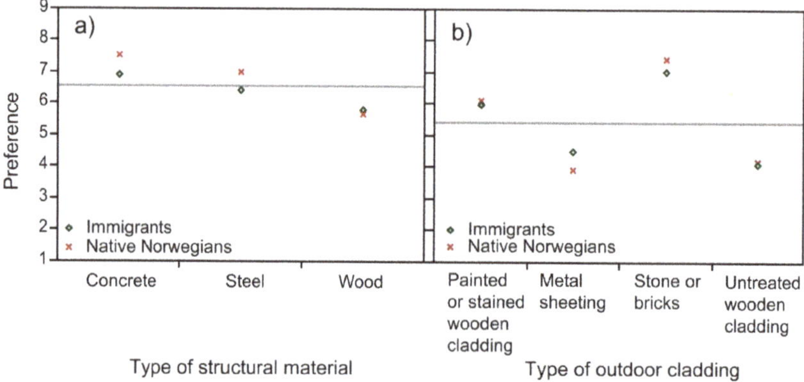


**Figure 2.** Mean preferences for different structural materials (**a**) and cladding materials (**b**).

For outdoor cladding, the only significant difference between immigrants and native Norwegians was the native Norwegians' somewhat higher preference for stone/brick cladding ($p = 0.038$, chi-square test). However, when testing the difference with an F-test, the $p$-value was only 0.088. Immigrants had a somewhat higher preference for metal sheeting, but this difference was not statistically significant (Figure 2b). For painted or stained wooden cladding and untreated wooden cladding, native Norwegians and immigrants had almost the same preferences (Figure 2b). Overall, there were significant differences in the preferences for the different cladding materials ($p < 0.0001$, chi-square test).

No significant differences were found between native Norwegians and immigrants for the five indoor materials. Overall, there were significant differences in indoor material preferences ($p < 0.0001$, chi-square test).

### 3.2. Location and House Type Preference

No significant differences were found between immigrants in the Oslo region and native Norwegians with respect to detached and row houses. However, across all respondents, a significant difference in the preferences between types of house was found ($p < 0.0001$, chi-square test). Detached houses were the most preferred, while row houses and apartments in multistory buildings had almost the same preference for respondents from the Oslo region. For immigrants outside the Oslo region, apartments in multistory buildings were less preferred than other types of housing.

Closeness to stores, schools, and other services was significantly more important for immigrants from the Oslo region than for non-Oslo region immigrants and native Norwegians ($p = 0.023$, chi-square test). Closeness to family was also significantly more important for immigrants from the Oslo region than for native Norwegians ($p = 0.029$, chi-square test). For immigrants outside the Oslo region, the importance for living close to family was less important than it was for immigrants in the Oslo region. The effect of the dummy variable was almost significant at the 5% level. When excluding this effect, no significant effect of respondents' region of origin was found. No significant difference between immigrants and native Norwegians was found for the importance of closeness to friends and acquaintances.

Finally, low price was significantly more important for immigrants, regardless of where they currently reside, than for native Norwegians ($p < 0.0001$, chi-square test).

### 3.3. Multiple Models

Model 1 (Table 3) includes the variables that were found to be important for structural material preferences. Concrete was the most preferred material (largest probability for 9 and 8 preferences, left column plots, Figure 3).

**Table 3.** Statistics for the multiple logistic regressions.

| | Model 1 Structural Materials | Model 2 Outdoor Cladding | Model 3 Indoor Walls & Ceilings |
|---|---|---|---|
| **Summary statistics for the different models** | | | |
| Entropy $R^2$ (Coef. of determin.)/Gen $R^2$ (Coef. of determin.) | 0.049/0.18 | 0.079/0.29 | 0.024/0.098 |
| N | 1165 | 1725 | 2325 |
| **$p$-values for the independent variables in the different regressions** | | | |
| MStr | <0.0001 | | |
| ResOr | 0.73 | 0.35 | 0.31 |
| PrCity | <0.0001 | 0.0011 | 0.0025 |
| MStr × PrCity | 0.021 | | |
| ImpStru | <0.0001 | | |
| ResOr × ImpStru | 0.0006 | | |
| PrRur | 0.0001 | 0.0010 | |
| PrCity × PrRur | 0.0002 | | |
| MInd | | | <0.0001 |
| MInd × PrCity | | | 0.0033 |
| ImpInd | | | <0.0001 |
| ResOr × ImpInd | | | 0.0036 |
| MInd × ImpInd | | | 0.0002 |
| MCl | | <0.0001 | |
| MCl × PrCity | | 0.0841 | |
| ImpCla | | <0.0001 | |
| ResOr × ImpCla | | 0.0038 | |
| MCl × ImpCla | | 0.014 | |
| EfSamp | | | 0.048 |

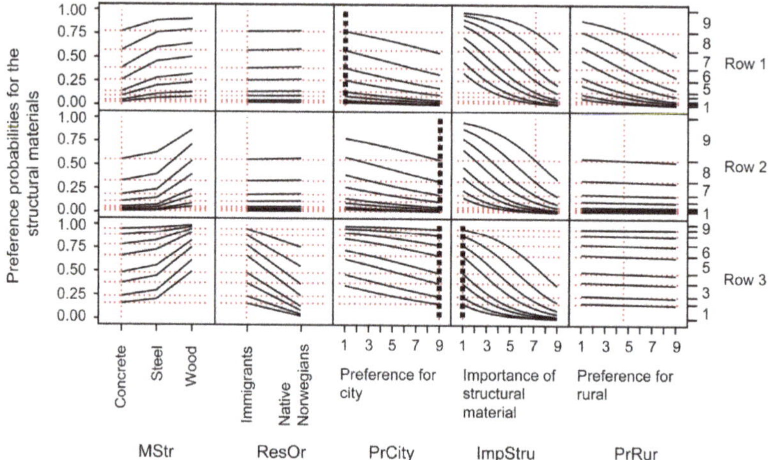



**Figure 3.** Profile plots showing the preferences for different structural materials, depending on the values of various independent variables in Model 1. The profile plots show the preferences for different structural materials, depending on the values of various independent variables in Model 1 (Table 3). Three rows of plots are included to show how the preferences for the different structural materials change with changes in the independent variable values. The thick vertical dashed lines indicate where the researcher set the value of the independent variables. The distances between the horizontal lines in the first column of plots show the probability for the different preference values. The probability for preference 9 is the distance between the upper most line and 1.00. The distance between lines 7 and 8 shows the probability for preference value 8. The probability for the lowest preference value is between 0.00 and the lowest line. For example, in row 2, column 1, approximately 40% of the respondents rated their preference for concrete as the highest value of 9, given by the independent variables setting values, as shown by the vertical dashed lines. Although some of the data in Figure 2 is categorical, the lines between categories are provided only for the ease of the visual interpretation of changes in level from one category to the next. This figure text is partly the same as the figure text in Figure 1 in the article of Høibø et al. [21].

Model 1 includes a significant interaction effect between the type of structural material and preference for living in a city. A higher preference for living in a city corresponds with an increasing preference for steel and concrete (changes are larger for steel than concrete) rather than wood (Figure 3, row 1 and row 2). Preference for living in a city was the only variable with a significant interaction effect on the 5% level with the type of structural material.

The other interaction effects are mainly related to the level of structural material preferences. Nevertheless, for respondents saying that the structural material type in general is of little importance, the probability for the highest preference decreases the most for concrete, somewhat less for steel, and the least for wood, compared to that of the respondents who reported that structural material was important for them (Figure 3, row 2 and row 3). However, the probability for the lowest preference increases the most for wood with the respondents saying that the structural material type in general is of little importance, compared to that of the respondents who stated that structural material was important (Figure 3, row 2 and row 3). Model 1 also includes a significant interaction effect between the respondent's region of origin and the importance of the structural material. For the respondents saying that the type of structural material was of little importance, the immigrants responded with lower preferences than the native Norwegians (Figure 3, row 3). A significant interaction effect between preferences for living in cities and preferences for living in rural areas was also included. This effect only affected the levels of preferences across the different structural materials.

An interaction effect between the type of structural material and the respondent's region of origin was also tested, but it was not significant. This result means that the differences in preferences between the different structural materials were not significantly different between immigrants and native Norwegians.

Model 2 (Table 3) includes variables that were important for the preferences for the different materials used for outdoor cladding. Stone/bricks were the most preferred cladding material, followed by painted or stained wood. Metal sheeting and untreated wood were the least preferred (Figure 4, row 1, column 1). Model 2 includes an interaction effect between the type of material for outdoor cladding and preference for living in a city, even though it was only significant at the 8.4% level. The interaction effect resulted in higher preferences for stone/bricks than for the other materials when preferences for living in a city were high (Figure 4, row 1 and row 2). A significant interaction effect between the type of material for outdoor cladding and the importance of the material used for outdoor cladding was also included in Model 2. The interaction effect resulted in fewer differences in preferences between the different types of claddings when the importance of the material used for outdoor cladding was small (Figure 3, row 2 and row 3). Model 2 also includes a significant interaction effect between the respondent's region of origin and the importance of the materials used for outdoor cladding. For respondents saying that the material used for outdoor cladding was not important, the native Norwegians responded with higher material preferences in general than immigrants (Figure 4, row 3). Higher preferences for rural living adjusted the level of the fitted lines, thus resulting in a general increase in preferences across the different cladding materials (Figure 4, the last column of plots).

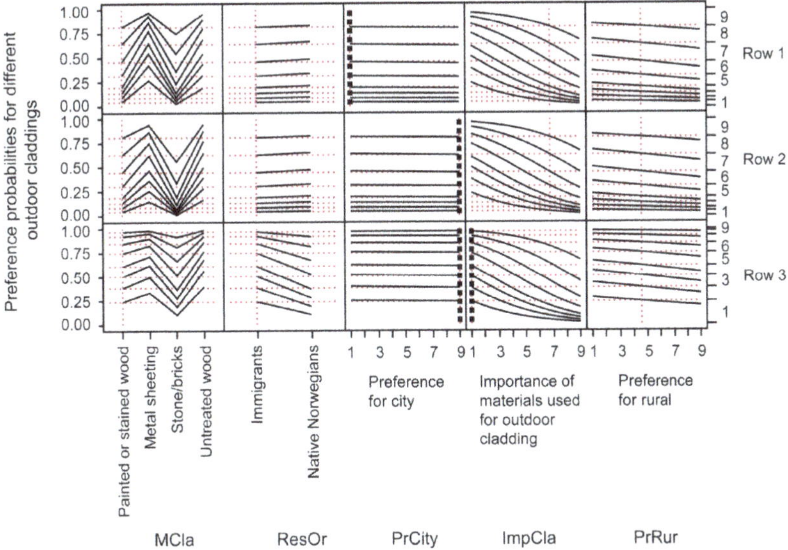


**Figure 4.** Profile plots showing the preferences for different outdoor cladding materials, which depend on the values of the different independent variables in Model 2. Three rows of plots are included in the figure to show how the preferences for the different outdoor cladding materials change with differing settings of the independent variables. This figure text is partly the same as that in Figure 4 in the article of Høibø et al. [21].

Model 3 includes the variables found to be important for the preferences for different materials used on indoor walls and ceilings (Table 3). Model 3 shows that lacquered, stained or painted wood, and paint or wallpaper on different boards were the most preferred (Figure 4 first row). Paint or wallpaper on concrete together with untreated wood were the least preferred indoor materials for respondents that did not prefer to live in a city (Figure 5, row 1). When respondents preferred to live in a city, the preference for untreated wood was the lowest (Figure 5, row 2). Respondents who preferred to live in a city and also said that the material used on indoor walls and ceilings was of low importance, preferred paint or wallpaper on concrete the most (Figure 5, row 3). For these respondents, lacquered, stained, or painted wood together with untreated wood were the least preferred materials (higher probability for the lower preferences).

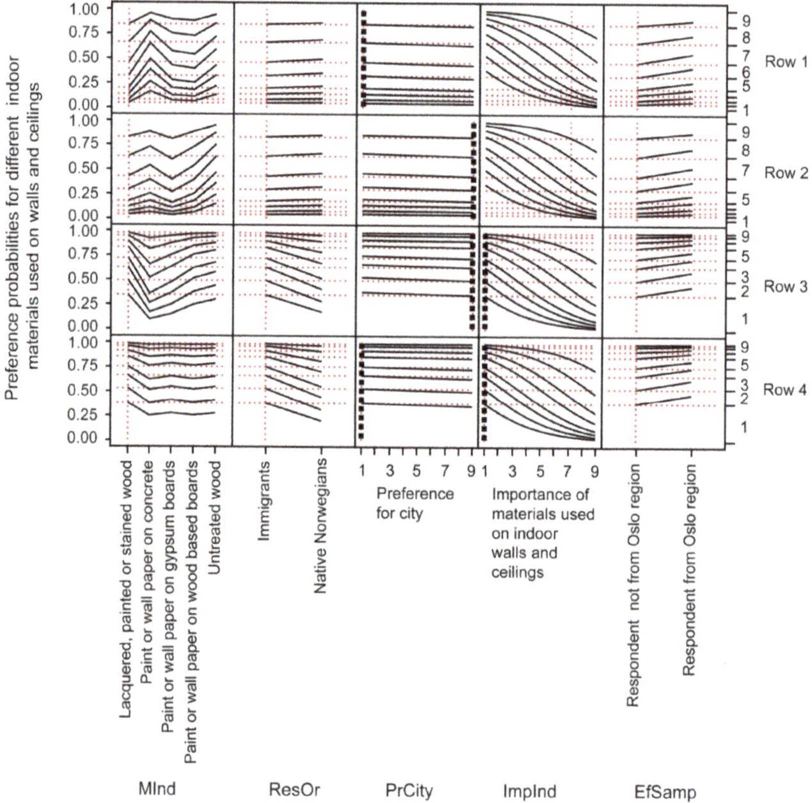


**Figure 5.** Profile plots showing the preferences for different indoor materials, which depend on the values of the different independent variables in Model 3. The four rows of plots are included in the figure to show how the preferences for the different indoor materials change with changes in independent variables. This figure text is partly the same as that in Figure 3 in the article of Høibø et al. [21].

Respondents with low preferences for living in large cities who also said that the material used on indoor walls and ceilings was of low importance had relatively equal scores for the different materials (Figure 5, row 4). When the materials used on indoor walls and ceilings was of low importance, the native Norwegians gave a higher score for all materials than the immigrants. Nevertheless, the relative difference between the different materials did not vary much between native Norwegians and immigrants.

## 4. Discussion

Craig et al. [15] found relationships between material traditions and material preferences, Vasanen [1] found that choices are related to familiarity, and Hauge and Kolstad [20] suggested that there might be differences between genders or among ethnicities and cultural backgrounds with respect to what the interior and exterior of a house says about the owner. Accordingly, we could expect to find differences between native Norwegians and immigrants regarding material preferences since the two groups have different building cultural backgrounds and different experiences with regard to building environments. However, only small differences were found between the two groups. Even with the extensive use of wood in one- and two-story houses in Norway, which represents approximately 78% of housing [14], the structural material preferences of Norwegians did not differ from those of immigrants. For both groups, concrete was the most preferred structural material, followed by steel. Wood was the least preferred. Since native Norwegians also had a significantly higher preference for stone/brick cladding than immigrants, and the immigrant group in this study have a building culture background from areas where wood is less frequently used and stone is more frequently used, the strong Norwegian wood tradition outside cities appears to play almost no positive role for wood used in a city context.

However, the overall high preferences for concrete and low preferences for wood fit well with the material tradition [15] in cities. Concrete is the dominant structural material in large buildings in Norway [23], while wood structures in multistory city buildings are uncommon. It therefore seems that preferences are mostly related to city building material traditions. However, according to the hypothesis of the material traditions in cities, we would not have expected painted or stained wooden cladding to fare that well in the evaluation. The visual presence of wood in landmark buildings and a general focus on wood as a natural and renewable resource may have influenced general attitudes. The diverse profiles and variety of colors associated with timber architecture may have had a positive impact on the acceptance of wooden cladding also in an urban setting.

Nevertheless, the most preferred cladding material was stone/bricks, which is also a common city cladding material. This finding is in accordance with those of McManus and Baxter [24], Craig et al. [15], and National Association of Home Builders (NAHB) [11], who found bricks to be the most preferred outdoor cladding material. The high preference for stone/bricks, particularly for native Norwegians, might be related to attitudes connected to the upmarket status, since stone and brick claddings are more expensive than the other cladding materials. It might also be related to the high focus in Norway on the maintenance and durability of claddings [21] and that stone bricks are regarded to be the most durable [15], which also fits with the finding of increasing differences in preferences among cladding materials for respondents that are more concerned about outdoor cladding (Figure 4, row 2 and row 3).

Our findings show that there are other factors that have a stronger influence on respondents' material preferences in city buildings than if they come from a country where wood is an uncommon building material, or from a country where wood is extensively used. We found some significant differences between the two groups of dwellers studied. However, most of the differences with regard to material preferences were related to different attitudes, such as preferences regarding where the respondents wanted to live. Since increased preferences for city living increased the preferences for materials that are common in cities, individuals who prefer city living to a greater extent identify themselves with buildings made of materials that are common in cities rather than buildings made of materials that are more common outside cities. This finding is in accordance with those of Nasar and Kang [17] and Sadalla and Sheets [18], who say that the exterior of a house might convey a meaning about the owner to others, and Després [19], who found that people might use the house to define identity. Higher preferences for city living combined with higher preferences for common city building structural materials (Figure 3, row 1 and row 2) therefore fits well with the expected relationship between tradition and material preferences in a city context, personal tradition, and residential choice from a life style perspective [16], and other choices related to familiarity [1]. This finding is also in accordance with our findings on residential choice, where we found that immigrants and native Norwegians inside the Oslo region prefer apartments in multistory buildings more than immigrants outside the Oslo region, who preferred apartments in multistory buildings the least.

Within the category of respondents who prefer to live in a city and said that the material that is used indoors is not important, the highest preferred indoor material was paint or wall paper on concrete, while for the respondents who said that indoor materials was important, paint or wallpaper on concrete was one of the least preferred materials. Our study therefore shows that there were different preferences for using concrete as a structural material and concrete used as an indoor wall and ceiling material, but that this difference depended to a large extent both on the respondent's attitudes regarding the importance of indoor material use and preferences for living in a city (Figure 5, row 2 and row 3 and Figure 3). The experience of concrete as a "cold" absorber of body heat radiation might be the reason why respondents who both prefer city living and said that indoor materials are important state low preferences for indoor concrete surfaces. Mechanical resistance and noise related to the boring of holes and simple interior modifications may also play a role. Our findings show that for concrete structures, inner surfaces other than paint or wallpaper on concrete should be considered.

It is logical that the same materials see different preferences for structural use and indoor surface use because each fulfills different needs. For visual surfaces, both visual and tactile properties are important. Brandt and Shook [25] found that consumers' quality attributes for forest products are usually visual or tactile. Consumer preferences for wood are found to depend on harmony, activity, and social status [26]. Harmony is related to homogeneity [27], while a positive relationship between visual homogeneity and preferences is found for decking materials [28]. For structural materials, physical properties such as strength properties, fire safety, and sound insulation are more important.

The correspondence of preferences may be attributed to the common features of all urbanizing regions and may also relate to the fundamental role of buildings as the stable framework for social life. In Norway, the functional and technological standards of buildings are homogenous and highly regulated on governmental and municipal levels. Rental housing constitutes a very small part of the housing market. Varieties of individual and shared ownership dominate, also among immigrants. The typical, cooperative housing associations in Oslo require participation in decisions regarding maintenance and investments, which may be a strong integrating factor that may influence attitudes towards the design and materiality of buildings.

## 5. Conclusions

Our findings provide a few primary insights into consumer preferences for city-based housing in Norway. First, with respect to material preferences in city housing, there are only minor differences between Norwegian natives who represent countries where wood is extensively used in houses outside cities and immigrants coming from a country where wood is hardly used at all. Differences that do exist between respondents are more related to where an individual would prefer to live. Second, despite the longstanding tradition of wood use in single-family houses in Norway, other materials that have traditionally been used in city buildings are more preferred. The preferences seem, therefore, to be more related to material traditions and to the context in which the materials are used. Individuals who prefer city living seem to a greater extent to identify themselves with buildings made of materials that are common in cities rather than with buildings made of materials that are more common outside cities. Since the material tradition and the context seem to be important factors for consumers, consumer information is important when a material is introduced in a new context. Landmark wooden city buildings for housing are a useful tool for developers to introduce, teach, and make wood more familiar in a city context.

## 6. Limitations

As our goal was not the generalization of a population, but rather to compare two groups of Norwegians, we did not attempt to obtain a pure random sample. With respect to immigrant participants, our sample size was smaller than ideal. This made the result less robust and hindered our ability to validate the models. Additionally, the variation within the immigrant group was large since they come from different continents and countries. This large variation may have decreased the probability of finding significant effects.

**Author Contributions:** O.H. contributed to the following parts: project administration, research design and data collection. He also conducted the statistical analyses and lead the writing. E.H. contributed to the following parts: project administration, research design, discussion around the statistical analyses and writing. E.N. contributed to the following parts: project administration, research design, discussion around the statistical analyses and writing. M.N. was the project leader and contributed to the following parts: project administration, research design, and writing.

**Funding:** The majority of funding for this project comes from The Research Council of Norway (project: Increased Use of Wood in Urban Areas, project number 225345).

**Acknowledgments:** The authors would like to thank the various parties responsible for making this research possible. We also thank Eva Fosby Livgard at TNS Gallup AS (today Kantar TNS AS) for giving us feedback on the research design. We also thank the Fulbright Foundation who partly funded a stay at Oregon State University for Olav Høibø, where most of this research was done.

**Conflicts of Interest:** The authors declare no conflict of interest. The founding sponsors had no role in the design of the study, in the collection, analyses, or interpretation of the data, in the writing of the manuscript, and in the decision to publish the results.

## References


1. Vasanen, A. Beyond stated and revealed preferences: The relationship between residential preferences and housing choices in the urban region of Turku, Finland. *J. Hous. Built Environ.* **2012**, *27*, 301–315. [CrossRef]
2. UN Department of Economic and Social Affairs Population Division. Available online: http://www.un.org/en/development/desa/population/publications/pdf/trends/WPP2012_Wallchart.pdf (accessed on 19 October 2014).
3. UN Habitat. Available online: http://mirror.unhabitat.org/content.asp?cid=5809&catid=206&typeid=6 (accessed on 19 October 2014).
4. Dunse, N.; Thanos, S.; Bramley, G. Planning policy, housing density and consumer preferences. *J. Prop. Res.* **2013**, *30*, 221–238. [CrossRef]
5. Price, M.; Benton-Short, L. Counting Immigrants in Cities across the Globe. Migration Policy Institute. Available online: http://www.migrationpolicy.org/article/counting-immigrants-cities-across-globe (accessed on 14 December 2007).
6. Price and Benton-Short. Available online: http://www.migrationpolicy.org/article/counting-immigrants-cities-across-globe (accessed on 13 November 2014).
7. Hernandez, P.; Kenny, P. Development of a methodology for life cycle building energy ratings. *Energy Policy* **2011**, *39*, 3779–3788. [CrossRef]
8. Robertson, A.B.; Lam, F.C.; Cole, R.J. A comparative cradle-to-gate life cycle assessment of mid-rise office building construction alternatives: Laminated timber or reinforced concrete. *Buildings* **2012**, *2*, 245–270. [CrossRef]
9. Ritter, M.A.; Skog, K.; Bergman, R. *Science Supporting the Economic and Environmental Benefits of Using Wood and Wood Products in Green Building Construction*; General Technical Report FPL–GTR–206; United States Department of Agriculture, Forest Service, Forest Products Laboratory: Washington, DC, USA, 2011; p. 9.
10. Mahapatra, K.; Gustavsson, L. Multi-storey timber buildings: Breaking industry path dependency. *Build. Res. Inf.* **2008**, *36*, 638–648. [CrossRef]
11. NAHB. Available online: http://www.prnewswire.com/news-releases/new-nahb-study-shows-national-consumers-prefer-brick-197850191.html (accessed on 28 November 2014).
12. Davies, I.; Walker, B.; Pendlebury, J. *Timber Cladding in Scotland*; ARCA Publications: Edinburgh, UK, 2002.
13. Tandberg, E.; Morstad, P. *Uttalelse Til Nasjonal Transportplan 2014–2023*; Byrådsavdeling for Miljø og Samferdse: Kirkegaten, Lillehammer, 2012; p. 7.
14. Statistics Norway. Available online: http://www.ssb.no/boligstat (accessed on 5 November 2014).

15. Craig, A.; Abbott, L.; Laing, R.; Edge, M. Assessing the Acceptability of Alternative Cladding Materials in Housing: Theoretical and Methodological Challenges. Available online: http://www.researchgate.net/publication/27250534_assessing_the_acceptability_of_alternative_cladding_materials_inhousing_theoretical_and_methodological_challenges (accessed on 16 December 2002).

16. Ærø, T. Residential choice from a lifestyle perspective. *Hous. Theory Soc.* **2006**, *23*, 109–130. [CrossRef]

17. Nasar, J.L.; Kang, J. House style preference and meanings across taste cultures. *Landsc. Urban Plan.* **1999**, *44*, 33–42. [CrossRef]

18. Sadalla, E.K.; Sheets, V.L. Symbolism in building materials. *Environ. Behav.* **1993**, *25*, 155–180. [CrossRef]

19. Desprès, C. The meaning of home: Literature review and directions for further research and theoretical development. *J. Archit. Plan. Res.* **1991**, *8*, 96–115.

20. Hauge, Å.L.; Kolstad, A. Dwelling as an expression of identity. A comparative study among residents in high-priced and low-priced neighbourhoods in Norway. *Hous. Theory Soc.* **2007**, *24*, 272–292. [CrossRef]

21. Høibø, O.; Hansen, E.; Nybakk, E. Building material preferences with a focus on wood in urban housing: Durability and environmental impacts. *Can. J. For. Res.* **2015**, *45*, 1617–1627. [CrossRef]

22. JMP, Version 10.0.0. Statistical Discovery, SAS Institute Inc.: Cary, NC, USA, 2012.

23. Store Norske Leksikon. Available online: https://snl.no/h\T1\oyhus (accessed on 25 November 2014).

24. McManus, B.R.; Baxter, D.O. Revealed preferences for building materials: A survey of low and moderate income households. *Hous. Soc.* **1981**, *8*, 45–51. [CrossRef]

25. Brandt, J.P.; Shook, S.R. Attribute elicitation: Implications in the research context 1. *Wood Fiber Sci.* **2007**, *37*, 127–146.

26. Broman, N.O. Means to Measure the Aesthetic Properties of Wood. Ph.D. Thesis, Luleå University of Technology, Luleå, Sweden, 2000.

27. Nyrud, A.Q.; Roos, A.; Rødbotten, M. Product attributes affecting consumer preference for residential deck materials. *Can. J. For. Res.* **2008**, *38*, 1385–1396. [CrossRef]

28. Høibø, O.; Nyrud, A.Q. Consumer perception of wood surfaces: The relationship between stated preferences and visual homogeneity. *J. Wood Sci.* **2010**, *56*, 276–283. [CrossRef]

*Article*

# Investigation of Bamboo Grid Packing Properties Used in Cooling Tower


**Li-Sheng Chen [1,2], Ben-Hua Fei [1,2], Xin-Xin Ma [1,2], Ji-Ping Lu [3] and Chang-Hua Fang [1,2,*]**

1   Department of Biomaterials, International Center for Bamboo and Rattan, Beijing 100102, China; chenlisheng@icbr.ac.cn (L.-S.C.); feibenhua@icbr.ac.cn (B.-H.F.); maxx@icbr.ac.cn (X.-X.M.)
2   SFA and Beijing Co-built Key Laboratory of Bamboo and Rattan Science & Technology, State Forestry Administration, Beijing 100102, China
3   Hengda Bamboo Filler Limited Company, Yixing 214200, China; hdlwz@163.com
*   Correspondence: cfang@icbr.ac.cn; Tel.: +86-010-84789786



Received: 1 November 2018; Accepted: 5 December 2018; Published: 7 December 2018



**Abstract:** Due to its advantages of good heat-resistance, environmental-friendliness, and low cost, bamboo grid packing (BGP) has become a promising new type of cooling packing. It is being increasingly used in Chinese industrial cooling towers to replace cooling packings made of polyvinyl chloride, cement, and glass fiber reinforced plastic. However, mechanical properties and fungal resistance are a concern for all bamboo applications. In this study, the modulus of rupture (MOR), modulus of elasticity (MOE), density, crystallinity, and environment scanning electron microscope (ESEM) properties were compared between fresh BGPs and those that had been in service for nine years in the cooling towers. The results showed that the MOR, MOE, density, crystallinity, and the crystal size of the used BGPs decreased to some extent, but still met the requirements for normal use in a cooling tower. The ESEM observation showed that the used BGPs were not infected by fungi. The decrease in mechanical properties could be caused by the decrease of density, crystallinity, and the decomposition of the chemical components of bamboo, but not by fungal infection.



**Keywords:** bamboo grid packing; cooling packing; cooling tower; mechanical properties; fungi; bamboo


---

## 1. Introduction

Hyperbolic cooling towers are widely installed at power plants, steel mills, petroleum refineries, and petrochemical plants due to their high capacity for heat rejection and energy saving. Compared to package-type cooling towers, hyperbolic cooling towers are generally much larger in size and require much more cooling packing. Cooling packing is the core component of cooling towers, and is responsible for 60%–70% of the heat dissipation in the cooling tower [1–4]. The type of packing material has an important role, as it provides a large surface area for evaporative heat and mass transfer from hot water to the ambient air and increases the contact time between both the two [5]. Different materials such as polyvinyl chloride (PVC), cement, and glass fiber reinforced plastic have been used as cooling packing. Currently the most popular cooling packing is made of PVC, with a market share exceeding 70% in China [6]. However, the PVC packing industry is facing major challenges, such as diminishing availability of petrochemical resources, increases in their prices, and the residues of PVC in the environment beyond its functional life [7]. In China, there are many power plants, steel mills, petroleum refineries, and petrochemical plants, from which the discarded cooling packing could cause severe environmental pollution. Furthermore, PVC packing has a short service life and poor anti-fouling properties. Thus, researchers and entrepreneurs have been seeking environmentally-friendly and longer-serving alternatives to PVC.

Recently, packing material made of bamboo has been put to use in several hyperbolic cooling towers in China [8]. Bamboo, as one of the fastest-growing and most versatile plants, grows widely across tropical and temperate zones with wet climate. It is a raw material with great economic importance that has been used since ancient times. Accounting for around 1% of the world's total forest area [9], there is a total area of 31.5 million hectares of bamboo, 60% of which are concentrated in rapidly developing countries, such as China, India, and Brazil [10]. Bamboo is also an invasive plant in some parts of the world [11], the expansion of which tends to reduce the biological diversity in local environment, affect the physico-chemical properties and microbial composition of soil, weaken the ecosystem function, and change the forest landscape [12]. However, due to the combination of advantages of fast growth, short life cycle, high mechanical strength, and low energy consumption [13], bamboo also has an outstanding natural potential for use as packing material, which may help control the expansion of bamboo forests, reduce greenhouse gas, and provide carbon sequestration. Compared to PVC packing, bamboo grid packing (BGP) also has certain advantages in temperature adaptability, anti-fouling properties [6,8], as well as a good cooling performance [14–17]. However, durability has always been a concern for any application of bamboo materials. The compromising of mechanical properties caused by continuous exposure to hot water flow in the cooling tower can be problematic. In addition, bamboo is vulnerable to fungal infection because of its richness in nutrients. Fungal infection could be fatal as it decreases the mechanical strength of bamboo and subsequently shortens the BGP's service life. The lack of research in this aspect hinders the development of BGP for industrial application.

To fill this gap, the mechanical properties, such as the modulus of rupture (MOR) and modulus of elasticity (MOE), of BGPs that had been used for nine years in cooling towers were investigated and compared to the properties of unused control samples. Density and crystallinity were also investigated, and samples were observed under environment scanning electron microscope (ESEM) to gain a better understanding of the changes in mechanical properties, as well as BGP's fungal resistance.

## 2. Materials and Methods

### 2.1. Materials

Raw materials were obtained from Moso bamboo (*Phyllostachys edulis* (Carrière) J.Houz) grown in Shaowu, Fujiian Province, China. Bamboo culms were cut into strips (1200 mm in longitudinal direction and 40 mm in tangential direction). Three holes with a diameter of 10 mm were made on the strips. Round bamboo sticks were inserted into the holes to connect the bamboo strips, as shown in Figure 1. The dimension of one piece of BGP was 1200 mm × 600 mm × 40 mm, and the spacing between the bamboo strips was 50 mm. BGP units were stacked to a height of 1.5 m in a hyperbolic cooling tower (Figure 1). Control samples were collected from the fresh BGP units.

Nine-year-old BGP units were collected from two hyperbolic cooling towers located respectively in Fujian and Shandong Province. The BGP collected from Fujian Province (FJBGP) was used in a hyperbolic cooling tower of a thermal power plant. The one collected from Shandong Province (SDBGP) was used in a hyperbolic cooling tower of a steel mill. In both cooling towers, the inlet water temperature was 45 to 50 °C, and the water mass flux was around 6500 kg/(h*m$^2$). Prior to the experiment, all specimens were conditioned at 21 ± 2 °C, with relative humidity of 65 ± 3%, to reach the equilibrium moisture content (EMC).

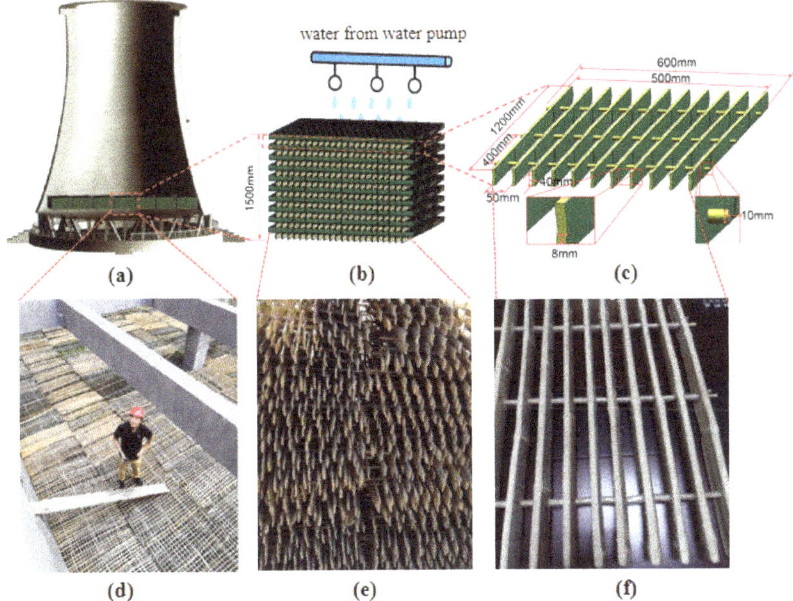



**Figure 1.** (**a**,**d**) Application of bamboo grid packing (BGP) in hyperbolic cooling towers; (**b**,**e**) stack of BGP units; (**c**,**f**) BGP assembled with bamboo strips.

### 2.2. Mechanical Properties and Density Test

Three-point static bending tests were performed according to GB/T 15780-1995 to obtain both the MOR and the MOE [18]. Since the load on the BGP is mainly in the tangential direction in the actual working environment of a cooling tower, the bending tests of bamboo strips were also performed in the tangential direction. The dimension of test specimens was 160 mm (longitudinal) × 10 mm (tangential) × t (thickness of the bamboo culm wall). Tests were loaded along the tangential direction at a rate of 6 mm/min. The moisture content was measured after the bending tests. The specimen density was measured with the drainage method according to GB/T 1933-2009 [19]. The number of specimens in the control group, the FJBGP, and the SDBGP for the determination of mechanical properties and density testing were ten, twenty, and sixteen, respectively.

### 2.3. Crystallinity Test

Eighty mesh bamboo powder was used as the experimental material. Bamboo powder was processed by a mill, which comprised a means of randomly mixing and distributing the small fiber lengths in various directions at surface pressures far below those that would induce crystallite fracture [20]. X-ray diffraction (XRD) measurements were performed to assess the crystalline properties of air-dried bamboo cell walls using an X-ray diffractometer (AV300, Panalytical Co., Amsterdam, The Netherlands) at a wavelength of 0.154 nm. The incident X-ray radiation was measured as the characteristic Cu X-ray passing through a nickel filter with a power of 40 kV and 40 mA. The XRD spectrum of every specimen was recorded in the angles ($2\theta$) of 5–50°. Three replicates were tested in this section. The cellulose crystallinity was calculated by the following Segal method [21]:

$$C_{rI} = \frac{I_{002} - I_{am}}{I_{002}} \times 100\% \tag{1}$$

where $C_{rI}$ represents the crystallinity of cellulose (%), $I_{002}$ represents the reflection intensity of (002) plane diffraction, and $I_{am}$ represents the intensity at the minimum near 18° of $2\theta$ angle.

The Scherrer equation, in X-ray diffraction and crystallography, relates the size of sub-micrometer particles, or crystallites, in a solid to the broadening of a peak in a diffraction pattern [22]. This study used this method to determine the size of crystal particles in cellulose. The Scherrer equation was used to calculate the mean size of ordered (crystalline) domains and can be written as [23]:

$$D = \frac{K\lambda}{\beta \cos\theta} \qquad (2)$$

where $D$ represents the mean size of the crystal region (nm), $K$ represents a dimensionless shape factor (0.9), $\lambda$ represents the X-ray wavelength, $\beta$ represents the line broadening at half the maximum intensity, and $\theta$ represents the scattering angle.

*2.4. Microstructure Observation*

An environmental scanning electron microscope (ESEM, XL30 FEG, FEI Co., Hillsboro, OR, USA) was used to observe the microstructure of bamboo strips. Cubic specimens (5 mm × 5 mm × 5 mm) were carefully prepared with razor blades and microtome to obtain a neat surface. Then, the surfaces of specimens were coated with elemental gold film (8–10 nm) and observed under the ESEM.

*2.5. Statistical Analysis*

Multiple comparisons were first subjected to an analysis of variance (ANOVA) using SPSS 19.0 (IBM SPSS Corporation, Chicago, IL, USA), and significant differences between average values of control and used BGP specimens were determined using Duncan's test at 0.05 significance level.

## 3. Results and Discussion

*3.1. Mechanical Properties and Density*

The mechanical properties of the BGP strips are presented in Table 1. The MOR of FJBGP and SDBGP were 106.16 MPa and 107.91 MPa, respectively, and the MOE of FJBGP and SDBGP were 8869.66 MPa and 8986.50 MPa. The difference in mechanical properties between FJBGP and SDBGP was not statistically significant. However, the MOE and MOR of both FJBGP and SDBGP were all significantly lower than those of the control samples. These results demonstrated that the hygrothermal conditions in the cooling towers had a negative effect on the BGP's mechanical properties. This might be related to the degradation of bamboo components by the water flow, which decreased its density and crystallinity.

**Table 1.** The average MOR, MOE, and density of used BGPs and control samples.

| Materials | MOR (MPa) | MOE (MPa) | Density (g/cm$^3$) | Retention of Properties (%) | | |
|---|---|---|---|---|---|---|
| | | | | MOR | MOE | Density |
| Control | 143.15 ± 16.68a | 10,250.83 ± 1091.67a | 0.7165 ± 0.4452a | 100 | 100 | 100 |
| FJBGP | 106.16 ± 19.14b | 8869.66 ± 1737.26b | 0.6596 ± 0.5839b | 74.16 | 86.53 | 92.06 |
| SDBGP | 107.91 ± 26b | 8986.50 ± 2010.31b | 0.6531 ± 0.8499b | 75.38 | 87.67 | 91.15 |

Note: Standard deviation are presented after ±. The different letters in the same column indicate a significant difference at the 0.05 level.

Compared to the control samples, the MOR retention of FJBGP and SDBGP were 74.16% and 75.38%, respectively, and the MOE retention of FJBGP and SDBGP were 86.53% and 87.67%. The retention of MOE exceeded that of MOR, which could be attributed to the fact that the hemicellulose and cellulose are more susceptible to degradation than lignin in hygrothermal condition [24,25]. Bamboo is mainly composed of cellulose, hemicellulose, and lignin. Cellulose in the cell walls of bamboo acts as a framework that provides both elasticity and strength, while lignin acts as a hard and solid substance that contributes to hardness and rigidity [26]. Hemicellulose acts as a matrix material that ensures the toughness, hardness, and strength of bamboo [24]. The thermal stability of lignin is better than that of cellulose and hemicellulose [26].

According to the "Technical specifications for bamboo filler of fossil fuel plants cooling tower" [27], the normal requirements for MOR and MOE are 100 MPa and 8500 MPa, respectively. Although the mechanical properties of the used BGP decreased, they still met the requirements despite nine years of use.

The average densities and standard deviations are shown in Table 1. The difference in density between FJBGP and SDBGP was not statistically significant. The densities of FJBGP and SDBGPT were significantly lower than that of control samples at the 0.05 level. The density retention of FJBGP and SDBGP were 92.06% and 91.15%, respectively, compared to control samples. The decrease of density could be explained by the degradation of bamboo cell wall components and the loss of water-soluble starch in the cell lumina (see Section 3.3) caused by the circulating hot water. However, the effect of starch loss on the density decrease could be neglected because the percentage of starch content in Moso bamboo is only around 0.1% [28]. The decrease of density could also explain the decrease of mechanical properties of the used BGP, since there was a significant correlation between density and mechanical properties, as presented in Figure 2, which was consistent with previous research [29]. However, the values of used BGPs were lower than those of control samples with similar densities for MORs (see the dotted circle of Figure 2), but not for MOEs. The reason could be the degradation and hydrolysis of hemicellulose and cellulose far exceeded that of lignin, as explained above. In addition, the decrease in cellulose crystallinity may also be an important factor (see Section 3.2).

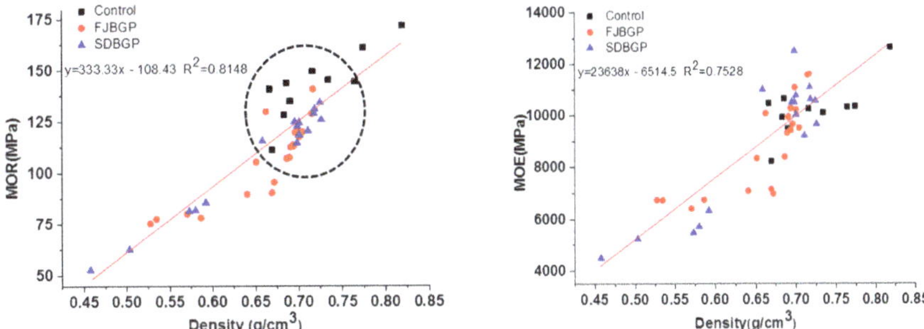


**Figure 2.** Relationships between MOR/MOE and density. Details of the dotted circles are explained in the text. FJBGP: bamboo grid packing collected from Fujian; SDBGP: bamboo grid packing collected from Shandong.

### 3.2. Crystallinity

Cellulose consists of amorphous and crystalline regions. Crystalline cellulose is tightly packed and hard to degrade, while the amorphous region easily decomposes under high temperature [30]. Cellulose crystallinity refers to the percentage of crystalline cellulose in the total cellulose and reflects the degree of crystallization that occurs during cellulose accumulation. The degree of orientation and relative content of the crystalline region have a significant effect on the fracture strength, toughness, and elastic modulus of bamboo [24]. When the cellulose is subjected to external forces, the molecular chains will be aligned along the direction of the external force to produce the preferential orientation, and the interaction between molecules will be greatly enhanced.

The XRD patterns of the three groups of samples are shown in Figure 3. It is clear that there was no change in the position of the cellulose diffraction peak. However, after BGP had been used in the cooling tower for nine years, the intensity of cellulose diffraction peak (002) significantly dropped. The crystallinity and crystal size of samples are as shown in Table 2. Compared to the control samples, the crystallinity retention of FJBGP and SDBGP were respectively 87.76% and 89.77%, and the crystal size retention of FJBGP and SDBGP were 85.21% and 86.88%. Hot water in the cooling tower could degrade the amorphous matrix of the used BGPs, which may increase the $C_{rI}$. However, in this study, the crystallinity of FHBGP and SDBGP decreased. This could be explained by the following reasons: The hot water cleaved acetyl groups from hemicellulose side chains to yield acetic and uronic acid [31], which catalyzed the hydrolysis of cellulose. Acid degraded not only the amorphous region of cellulose, but also the crystalline region [32], which is consistent with the decrease in crystal size (Table 2). The proportion of the crystalline region in the cell walls was low. When acid cleaved the molecular chain in the crystalline region, part of the crystalline region would be turned into an amorphous region, resulting in lower crystallinity and smaller crystal size of SDBGP and FJBGP.

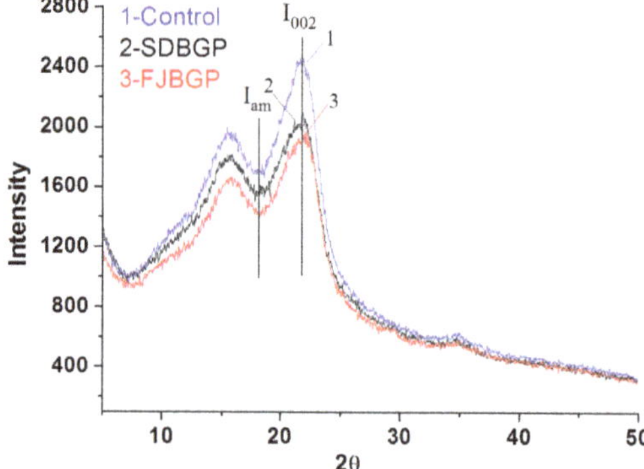


**Figure 3.** **X-ray diffraction** (XRD) patterns of the three sample groups.

**Table 2.** Crystallinity parameters of the three sample groups.

| Sample | 2θ | $C_{rI}$ | $D$ (nm) | Retention of Properties (%) | |
|---|---|---|---|---|---|
| | | | | $C_{rI}$ | $D$ |
| Control | 21.88 | 0.3276 | 2.82 | 100 | 100 |
| FJBGP | 21.76 | 0.2875 | 2.4 | 87.76 | 85.21 |
| SDBGP | 21.83 | 0.2941 | 2.45 | 89.77 | 86.88 |

The decrease in crystallinity led to low fracture strength and elastic modulus of the BGP, which can partly account for the decrease of MOR and MOE of used BGPs (Table 1).

### 3.3. Microstructure Characteristics

Bamboo is susceptible to insect and fungal infestation, which can modify the microstructure and reduce the mechanical properties of bamboo. According to the classification of durability, bamboo belongs to the third grade (non-durable grade). In general, when untreated bamboo is used in the outdoor environment, its service life does not exceed seven years [33]. The poor durability of bamboo is due to its richness in nutrients, which provide a food source for insects and rot fungi. After being infected by rot fungi, the microstructure of the bamboo changes, i.e., holes eroded by rot fungi appear in cell walls and mycelium in the cell cavities.

As shown in Figure 4a,b, the parenchyma cells of the control specimens contained many starch granules—several cells were virtually full of it. Furthermore, the starch granules were big. The inner walls of the cells were smooth. After being subjected to a rotting test for three weeks and 15 weeks, respectively, starch granules in the parenchyma cells disappeared and had been completely digested by rot fungi (Figure 4c,d) [33]. There were a large number of mycelia in the cell cavities, and the inner walls of the cells have formed big holes due to erosion by rot fungi. Noticeably, the following phenomena were observed in this study of both SDBGP and FJBGP: In the parenchyma cells of used BGPs, large-size starch granules disappeared (Figure 4e,f), but a few small-sized granules were still present. The inner walls of all cells remained smooth, without mycelium. The holes observed in the inner walls of cells were not the result of erosion by rot fungi, but were pits. These phenomena indicated that the BGPs were not attacked by fungi after being used for nine years, which could be explained by the following reasons: Firstly, the surfaces of the bamboo strips were covered with a 0.1–0.5 mm thick water film due to the continuous presence of water flow in the cooling tower [34]. Bamboo decay fungi are mostly aerobic fungi. The water film prevented them from obtaining oxygen. Secondly, the water temperature in the cooling towers generally exceeded 40 °C, which is higher than the suitable temperature for fungal growth (3~38 °C) [33]. Moreover, the starch granules in the bamboo were partially dissolved by the circulating hot water, resulting in the decreased amount and size.

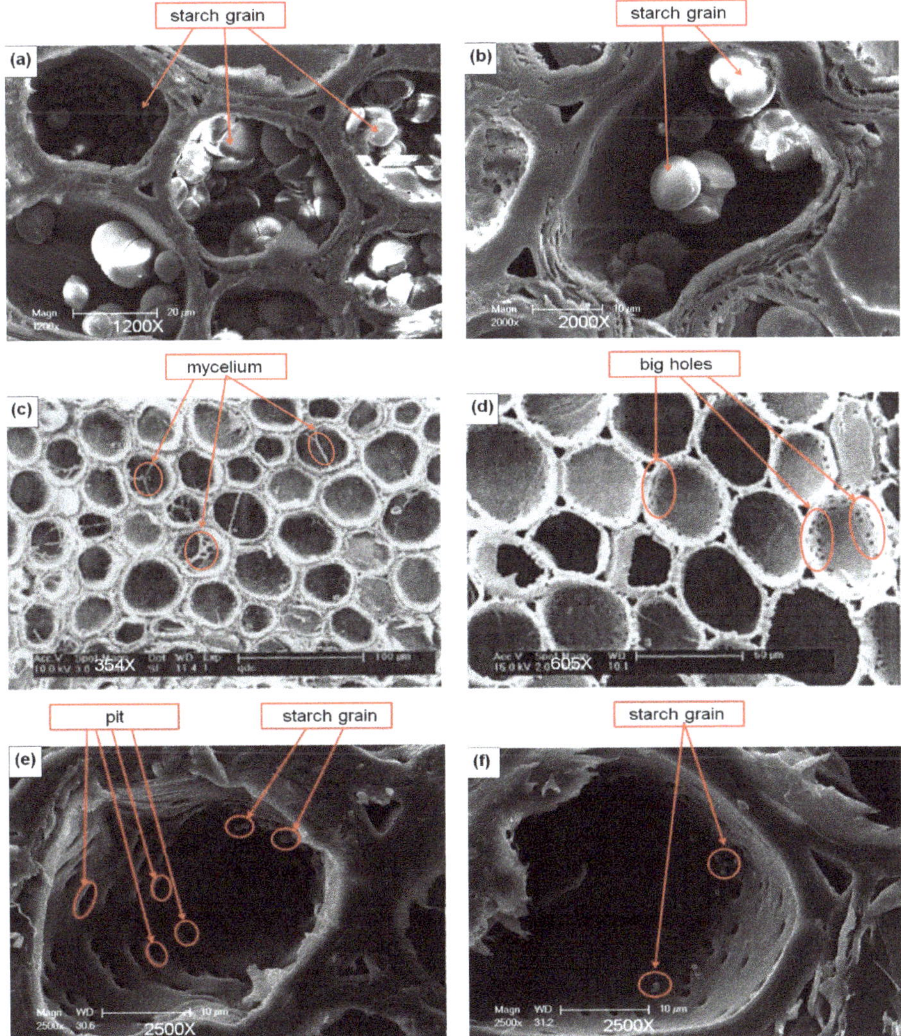


**Figure 4.** Environment scanning electron microscope (ESEM) images of bamboo cells. (**a**,**b**) Control samples; (**c**,**d**) specimens after three and 15 weeks of decay, respectively (from Qin [33]); (**e**,**f**) specimens of the used BGPs.

Combining all the findings and analyses, it can be concluded that the decreases in the mechanical properties of used BGPs were caused by decreases in density and crystallinity instead of fungal infection, despite that, for most bamboo applications, reduced durability caused by fungal damage is one of the major problems.

## 4. Conclusions

The properties of BGP that had been in use for nine years in two industrial cooling towers were investigated and compared to those of unused control samples. It was found that the MOR and MOE retentions were around 75% and 87%, respectively, which still met the normal use requirements for cooling towers. The density decreased with retention more than 91%. The crystallinity and crystal size retentions were about 89% and 86%, respectively. Bamboo is vulnerable to fungal infection due to its rich content of nutrients. However, a small amount of small-size starch granules was still present in the cell lumina of used BGP, and bamboo cell walls were also free from mycelia, which indicated that the used BGPs were not infected by fungi. No significant differences in the properties were found for the BGPs collected from the two cooling towers, which had similar conditions.

This study provides primary data for BGP used in industrial cooling towers. Further evaluation of BGP's service life is essential, as it may help promote the application of BGP as one of many uses of the abundant biomass resource of bamboo.

**Author Contributions:** L.-S.C., C.-H.F., and B.-H.F. conceived and designed the experiments; L.-S.C. performed the experiments; X.-X.M. analyzed the data; J.-P.L. contributed reagents/materials/analysis tools; L.-S.C. wrote the paper, and C.-H.F. revised it.

**Funding:** This research was funded by the Foundation of International Centre for Bamboo and Rattan (grant number: 1632016015) and the Forestry Patent Industrialization Guidance Project (grant number: Forestry Patent 2017-2).

**Acknowledgments:** The authors acknowledge the financial supports of the Foundation of International Centre for Bamboo and Rattan (No. 1632016015) and the Forestry Patent Industrialization Guidance Project (Forestry Patent 2017-2).

**Conflicts of Interest:** The authors declare no conflict of interest.

## References

1. Zhao, Z.G. A New Thermo-characteristic Formula for the Packing of Cooling Tower and It's use. *J. Hydrodyn.* **1996**, *11*, 599–605.
2. Chen, J.B.; Wang, Y.H.; Wang, J.; Zhang, P.; De-Xing, L.I. Experimental study on performance of cooling tower packing. *Ind. Water Wastewater* **2010**, *41*, 65–68.
3. Goshayshi, H.R.; Missenden, J.F. The investigation of cooling tower packing in various arrangements. *Appl. Therm. Eng.* **2000**, *20*, 69–80. [CrossRef]
4. Chen, J.B.; Shi, Y.H.; Wang, Y.H.; Hou, H.L.; Li, D.X. Experimental Study of Cooling Tower Packing and Deduce the New Formula of the NTU. *Adv. Mater. Res.* **2012**, *383–390*, 6134–6138. [CrossRef]
5. Lemouari, M.; Boumaza, M. Experimental investigation of the performance characteristics of a counterflow wet cooling tower. *Energy* **2010**, *36*, 5815–5823. [CrossRef]
6. Ma, X.X.; Lu, J.P.; Qin, D.C.; Fei, B.H. Application of bamboo in the circulating water cooling system. *J. For. Eng.* **2016**, *1*, 33–37.
7. Fazita, M.R.N.; Jayaraman, K.; Bhattacharyya, D.; Haafiz, M.K.M.; Saurabh, C.K.; Hussin, M.H.; Abdul, K.H.P.S. Green Composites Made of Bamboo Fabric and Poly (Lactic) Acid for Packaging Applications—A Review. *Materials* **2016**, *9*, 435. [CrossRef] [PubMed]
8. Chen, Y.L.; Shi, Y.F.; Xie, D.X. Performance Comparison between Bamboo Grid Packing and PVC Film Packing and its Applications. *Power Stn. Aux. Equip.* **2016**, *37*, 37–41.
9. Lobovikov, M.; Paudel, S.; Piazza, M.; Ren, H.; Wu, J. World bamboo resources. A thematic study prepared in the framework of the Global Forest Resources Assessment 2005. *Non-Wood For. Prod.* **2007**. [CrossRef]
10. Chirici, G.; Corona, P.; Portoghesi, L. Global forest resources assessment. *Ital. J. For. Mt. Environ.* **2013**, *4*, 269–273.

11. Lima, R.A.F.; Rother, D.C.; Muler, A.E.; Lepsch, I.F.; Rodrigues, R.R. Bamboo overabundance alters forest structure and dynamics in the Atlantic Forest hotspot. *Boil. Conserv.* **2012**, *147*, 32–39. [CrossRef]

12. Yang, Q.P.; Yang, G.Y.; Song, Q.N.; Shi, J.M.; Ou, Y.M.; Qi, H.Y.; Fang, X.M. Ecological studies on bamboo expansion: Process, consequence and mechanism. *Chin. J. Plant Ecol.* **2015**, *39*, 110–124.

13. Fang, C.H.; Jiang, Z.H.; Sun, Z.J.; Liu, H.R.; Zhang, X.B.; Zhang, R.; Fei, B.H. An overview on bamboo culm flattening. *Constr. Build. Mater.* **2018**, *171*, 65–74. [CrossRef]

14. Xi, M.J.; Zhang, R.P.; Wang, X.G.; Li, N.; Lu, S.J.; Yu, Z.X. Analysis of Characteristic of Bamboo Grating Filler in Cooling Tower. *Hebei Electr. Power* **2000**, *2*, 42–43.

15. Chu, X.M.; Wu, Y.C. The Cooling Effect is Significant after Cooling Tower Replaces the Bamboo Grid Packing. *Power Stn. Aux. Equip.* **1996**, *2*, 22–23.

16. Wang, Z.Q.; Liang, Q.; Chen, X. Transformation & economic analysis of cooling tower trickling filler in 300 MW unit. *Ningxia Electr. Power* **2005**, *1*, 42–46.

17. Fei, B.H.; Ma, X.X.; Qin, D.C.; Lu, J.P. New Materials for Industrial Cooling Tower: Bamboo Packing. *World Bamboo Rattan* **2016**, *14*, 31–35.

18. *GB/T 15780–1995 Testing Methods for Physical and Mechanical Properties of Bamboos*; General Administration of Quality Supervision, Inspection and Quarantine of the People's Republic of China/Standardization Administration of the People's Republic of China: Beijing, China, 1995.

19. *GB/T 1933-2009 Method for Determination of the Density of Wood*; General Administration of Quality Supervision, Inspection and Quarantine of the People's Republic of China/Standardization Administration of the People's Republic of China: Beijing, China, 2009.

20. Aeo, B.S.; Chaabouni, Y.; Msahli, S.; Sakli, F. Morphological and crystalline characterization of NaOH and NaOCl treated *Agave americana* L. fiber. *Ind. Crops Prod.* **2012**, *36*, 257.

21. Li, J. *Wood Spectroscopy*; Science Press: Beijing, China, 2003.

22. Patterson, A.L. The Scherrer Formula for X-ray Particle Size Determination. *Phys Rev.* **1939**, *56*, 978–982. [CrossRef]

23. Chen, M.; Wang, C.; Fei, B.; Ma, X.; Zhang, B.; Zhang, S.; Huang, A. Biological Degradation of Chinese Fir with Trametes Versicolor (L.) Lloyd. *Materials* **2017**, *10*, 834. [CrossRef]

24. Liu, Y.X.; Zhao, G.J. *Wood Resources Materials Science*; China Forestry Publishing House: Beijing, China, 2004.

25. Tjeerdsma, B.F.; Militz, H. Chemical changes in hydrothermal treated wood: FTIR analysis of combined hydrothermal and dry heat-treated wood. *Holz Roh-Werkst* **2005**, *63*, 102–111. [CrossRef]

26. Qin, L. *Effect of Thermo-Treatment on Physical, Mechanical Properties and Durability of Reconstituted Bamboo Lumber*; Chinese Academy of Forestry: Beijing, China, 2010.

27. *DL/T 1361-2014 Technical Specifications for Bamboo Filler of Fossil Fuel Plants Cooling Tower*; National Energy Administration of the People's Republic of China: Beijing, China, 2014.

28. Fan, F.Y. *Seasonal Variation of Starch Content in Dendrocalamus giganteus and Phyllostachys edulis and Its Relationship to Insect Damages*; Southwest Forestry University: Kunming, China, 2010.

29. Xu, Y.M. *Wood Science*; China Forestry Publishing House: Beijing, China, 2006.

30. Bhuiyan, M.T.R.; Hirai, N.; Sobue, N. Changes of crystallinity in wood cellulose by heat treatment under dried and moist conditions. *J. Wood Sci.* **2000**, *46*, 431–436. [CrossRef]

31. Sattler, C.; Labbé, N.; Harper, D.; Elder, T.; Rials, T. Effects of Hot Water Extraction on Physical and Chemical Characteristics of Oriented Strand Board (OSB) Wood Flakes. *CLEAN-Soil Air Water* **2010**, *36*, 674–681. [CrossRef]

32. Li, X.J.; Liu, Y.; Gao, J.M.; Wu, Y.Q.; Yi, S.L.; Wu, Z.P. Characteristics of FTIR and XRD for wood with high-temperature heating treatment. *J. Beijing For. Univ.* **2009**, *31*, 104–107.

33. Qin, D.C. *A Fundamental Sudy on the Application of CuAz Preservatives for Bamboo*; Chinese Academy of Forestry: Beijing, China, 2004.

34. Northwest Electric Power Design Institute. *Power Engineering Water Design Manual*; China Electric Power Press: Beijing, China, 2005.

*Article*

# Effects of Hygrothermal Environment in Cooling Towers on the Chemical Composition of Bamboo Grid Packing

**Li-Sheng Chen [1,2], Ben-Hua Fei [1,2], Xin-Xin Ma [1,2], Ji-Ping Lu [3] and Chang-Hua Fang [1,2,*]**

[1] Department of Biomaterials, International Center for Bamboo and Rattan, Beijing 100102, China; chenlisheng@icbr.ac.cn (L.-S.C.); feibenhua@icbr.ac.cn (B.-H.F.); maxx@icbr.ac.cn (X.-X.M.)
[2] SFA and Beijing Co-Built Key Laboratory of Bamboo and Rattan Science & Technology, State Forestry Administration, Beijing 100102, China
[3] Hengda Bamboo Filler Limited Company, Yixing 214200, China; hdlwz@163.com
* Correspondence: cfang@icbr.ac.cn; Tel.: +86-010-84789842

Received: 28 February 2019; Accepted: 16 March 2019; Published: 19 March 2019

**Abstract:** Bamboo grid packing (BGP) is a new kind of cooling packing, used in some Chinese hyperbolic cooling towers, which has excellent potential to complement or replace cooling packing made of polyvinyl chloride (PVC), cement, and glass fiber-reinforced plastic. For bamboo applications, mechanical properties and service life are matters of concern; this is strongly associated with bamboo's chemical composition and mass loss. To better understand the mechanics of mechanical property deterioration and service life reduction, this study investigated the effects of hygrothermal environments in cooling towers on the chemical and elemental composition, mass loss, Fourier-transform infrared (FTIR) spectrum, and color changes of BGP. The results showed that BGP that had been in service for nine years in cooling towers exhibited major decreases in content of hemicellulose and benzene-ethanol extractives, as well as a significant increases in the content of $\alpha$-cellulose and lignin. Exposure to the hygrothermal environment led to a decrease of oxygen content and around 8% mass loss, as well as an increase in carbon content compared to control samples. The hot water flow in cooling towers not only hydrolyzed hemicellulose, but also degraded some functional groups in cellulose and lignin. The lightness ($L^*$) and chromaticity ($a^*$ and $b^*$) parameters of the used BGP all decreased, except for the $a^*$ value of the outer skin. The total color change ($\Delta E^*$) of the inner skin of used BGP exceeded that of the outer skin.

**Keywords:** bamboo grid packing; cooling packing; cooling tower; chemical composition; elemental composition; FTIR; color

## 1. Introduction

A hyperbolic cooling tower is a device wherein hot water from the system is cooled by the ambient air with the assistance of cooling packing [1], and it is widely used in industries due to its high capacity for heat rejection and energy saving [2]. As the core component of a cooling tower, good cooling packing not only increases effective contact between air and water, which promotes heat and mass transfer, but also provides less resistance to the movement of air to reduce pressure drop [3]. In order to improve heat dissipation efficiency and reduce cost, many different materials have been used as cooling packing, such as polyvinyl chloride (PVC), cement, and glass fiber-reinforced plastic. PVC packing with smooth- and cross-ribbing is the most popular kind due to its outstanding cooling performance and lightness in weight. It is used in 96% of cooling towers [4]. However, PVC packing also has lots of disadvantages, such as short service life, poor anti-fouling properties, environmental burdens, etc. Furthermore, the price of PVC is on the rise. Therefore, many attempts have been made to seek an environmentally-friendly, low cost, and longer-serving alternative to PVC.

Owing to its fast growth speed, short rotation, great mechanical strength, and low energy consumption [5], bamboo, an abundant and sustainable plant resource, has been used as an innovative material for cooling packing. The product, known as bamboo grid packing (BGP), has been used in some hyperbolic cooling towers in China in recent years [6]. BGP has showed good temperature adaptability and anti-fouling properties, and also costs less than its PVC counterpart, which makes it a promising substitute for PVC packing. Mechanical properties and service life are a concern for all bamboo applications; this is strongly associated with its chemical composition and mass loss. The mechanical properties of bamboo were studied in our previous report [2]. However, it is still largely unknown how the hygrothermal environment in cooling towers affects the chemical composition of BGP. The lack of research in this aspect hinders the development of BGP for wider industrial application.

To better understand the reasons of mechanical property deterioration and service life reduction, the effects of hygrothermal environments in cooling towers on the chemical and elemental composition of BGP were investigated. In order to prevent damage to BGP's chemical composition and structure, Fourier-transform infrared (FTIR) spectroscopy was applied in this study. The presence of compounds or functional groups can be determined according to the number, shape, position, and intensity of the spectral bands, thus revealing the structure of the compound and its variation [7,8]. The color changes of BGP were also investigated, as they suggested variation in the chemical composition and structure. This study not only offers a new perspective for understand the changing patterns of the chemical composition, chemical structure, and color of BGP used in cooling towers, but may also contribute to the exploration of new approaches to alleviate hygrothermal aging.

## 2. Materials and Methods

### 2.1. Materials

The materials used in this study were identical to those in the previous study [2]. Raw materials were obtained from Moso bamboo (*Phyllostachys edulis* (Carr.) J. Houz), aged for four years, grown in Shaowu, Fujiian Province, China. The bamboo culms were sawn into segments 1200 mm long (Figure 1a) and then split longitudinally into several strips (1200 mm in longitudinal direction and 40 mm in tangential direction) (Figure 1b). Three holes with a diameter of 10 mm were made on the strips with a distance about 400 mm between each hole (Figure 1c). As shown in Figure 1e, round bamboo sticks 600 mm in length were inserted into the holes to connect the bamboo strips. The spacing between the strips was 50 mm. The dimensions of one piece of BGP were around 1200 mm × 600 mm × 40 mm. BGP units were stacked to a height of around 1.5 m in hyperbolic cooling towers [2]. BGP that has been in use for nine years was collected from two hyperbolic cooling towers located in Fujian and Shandong Provinces, respectively. The BGP collected from Fujian Province (FJBGP) and Shandong Province (SDBGP) were used in hyperbolic cooling towers of a thermal power plant and a steel mill, respectively. The bamboo materials used to fabricate the BGPs in these two towers were obtained from the same bamboo species and the same place as mentioned above. In both cooling towers, the inlet water temperature was around 45 to 50 °C, and the water mass flux was around 6500 kg/(h·m$^2$). Control samples were collected from unused BGP units. Prior to testing, all specimens were kept in a conditioned room at 21 ± 2 °C and 65 ± 3% relative humidity until their weights stabilized.

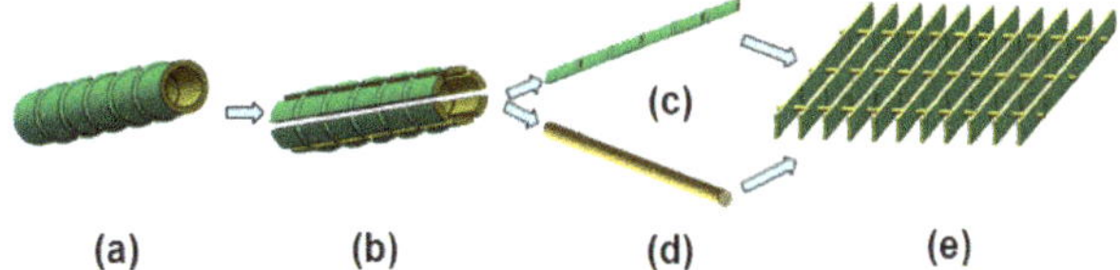



**Figure 1.** Preparation process of bamboo grid packing (BGP): (**a**) Raw bamboo; (**b**) radial splitting; (**c**) bamboo strip; (**d**) round bamboo stick; (**e**) bamboo grid packing.

*2.2. Chemical Composition Measurements*

Chemical composition of the used BGP and control samples were both measured. Specimen from each experiment was ground to pass a 40-mesh screen. The following contents in the ground meal was measured in accordance with relative standards: Acid-insoluble lignin using GB/T 2677.8-1994 [9], holocellulose content using GB/T 2677.10-1995 [10], $\alpha$-cellulose content using GB/T 744-2004 [11], and benzene-ethanol extractives using GB/T 2677.6-1994 [12]. The content of hemicellulose is the difference between the content of holocellulose and that of $\alpha$-cellulose.

*2.3. Elemental Composition and Mass Loss*

The used BGP and control samples were ground to fine powder and passed through different sieves to obtain a particle size between 0.2 and 0.5 mm. The powder was then conditioned at 103 °C for 24 h and stored in closed bottles before analysis. Elemental analyses were performed using a Thermofinnigam Flash EA1112 micro-analyzer (Sundy Co., Hunan, China). The content of carbon, oxygen, and hydrogen were measured. At least three replicates were tested for each sample. Mass loss due to chemical degradation caused by the hygrothermal environment in cooling towers was calculated according to the following equation:

$$ML(\%) = \frac{m_0 - m_1}{m_0} \times 100\% \tag{1}$$

where $m_0$ is the initial anhydrous mass of the sample before being put to use in cooling towers, and $m_1$ is the anhydrous mass of the same sample after nine years of service.

*2.4. FTIR Spectroscopy Analysis*

Before being ground, the used BGP was gently brushed to remove impurities from the surface. Then all specimens for the FTIR analysis were prepared by being ground in a mill with a 100-mesh screen (FW100, TAISITE Co., Tianjin, China). The FTIR spectra of the samples were obtained in a Nexus 670 spectrometer (Nicolet, WI, USA) within the range of 500–4000 cm$^{-1}$, with a resolution of 4 cm$^{-1}$ and 64 scans. The KBr pellet, consisting of KBr and randomly-selected bamboo powder with a weight ratio of 100:1, was prepared prior to the measurement. The FTIR analysis for each sample was performed in quintuplicate. For evaluation of the spectra, only the area between wavenumbers 800 and 1800 cm$^{-1}$ will be discussed; this includes the most important values of lignocellulose materials [13].

*2.5. Color Measurements*

The changes in color of BGP surface due to hygrothermal treatment were measured using a Technibrite Brightmeter micro S-5 colorimeter (Technidyne Corporation, New Albany, IN, USA) with an aperture size of 1 cm$^2$. CIELAB (CIEL*a*b*) is a color space that describes all the colors visible to the human eye and it is created to serve as a device-independent model used as a reference [14]. In the CIELAB system, the $L^*$ axis represents the lightness ($L^*$ varies from 100 (for white) to zero (for black)), $a^*$ and $b^*$ describe the chromatic coordinates on the green–red and blue–yellow axis, respectively ($+a^*$ is for red, $-a^*$ for green, $+b^*$ for yellow, $-b^*$ for blue) [15]. $\Delta E^*$ was calculated using the following equation:

$$\Delta E^* = \sqrt{(\Delta L^*)^2 + (\Delta a^*)^2 + (\Delta b^*)^2} \qquad (2)$$

where $\Delta L^*$, $\Delta a^*$, and $\Delta b^*$ are the difference of initial (control samples) and final (used BGP) values of $L^*$, $a^*$, and $b^*$, respectively.

### 2.6. Statistical Analysis

The effects of hygrothermal environment on the chemical composition and color changes in comparison with control samples were analyzed by analysis of variance (ANOVA) using SPSS 19.0 (IBM SPSS Corporation, Chicago, IL, USA). The significance ($p < 0.05$) between average values of control samples and used BGP specimens was compared using Duncan's test. Different letters given along with the average values of tested parameters indicated a significant difference by Duncan's test.

## 3. Results and Discussion

### 3.1. Chemical Composition Analysis

The structure of bamboo is stable because the natural tissues of the cell walls are composed of three bio-based chemicals—cellulose, hemicellulose, and lignin. Cellulose provides structural support in the cell walls. Hemicellulose and lignin are the binding and packing materials for the skeleton structure. The three components are interwoven into thin layers that together form bamboo cell walls [16]. In addition to lignocellulosic structures, bamboo also contains a variety of low molecular weight organic compounds known as extractives—such as terpenes, resins, fatty acids, waxes, and phenols—which can be extracted using solvents [17,18]. The temperature of inlet water in both cooling towers was around 45 to 50 °C, and the effects of hot water flow on the chemical composition of the used BGP are presented in Table 1.

**Table 1.** Analysis of main chemical composition of the used BGP and control samples.

| Sample | Holocellulose (%) | $\alpha$-Cellulose (%) | Hemicellulose (%) | Lignin (%) | Benzene-Ethanol Extractives (%) |
|---|---|---|---|---|---|
| Control | 64.08a (1.25) | 41.03a (1.94) | 23.05c (0.09) | 24.37a (2.23) | 5.77c (0.86) |
| FJBGP | 66.27ab (2.01) | 45.43b (4.08) | 20.84b (2.75) | 27.68b (2.44) | 1.62b (1.33) |
| SDBGP | 67.81b (2.35) | 48.35b (2.24) | 19.46a (2.65) | 26.63b (4.06) | 1.15a (1.23) |

Note: FJBGP: bamboo grid packing collected from Fujian; SDBGP: bamboo grid packing collected from Shandong. Values in parentheses are coefficient of variation. The different letters in the same column indicate a significant difference at the 0.05 level.

The hemicellulose content of FJBGP and SDBGP were 20.84% and 19.46%, respectively. Both data were significantly lower than those of control samples, which indicates that hot water flow in the cooling towers partially degraded hemicellulose. Hemicellulose is a heterogeneous low molecular weight polysaccharide composed of acetyl groups, aldonic acid groups, and different glycosyl groups, most of which are thermally labile, especially the acetyl groups. Furthermore, because of its branched structure and amorphous tissues, hemicellulose is considerably more susceptible to thermal degradation than other chemical components [19].

The benzene-ethanol extractive contents of FJBGP and SDBGP were 1.62% and 1.15%, respectively, which were also significantly lower than those of control samples. The changes in benzene-ethanol extractive content of the used BGP were not consistent with the results of heat-treated bamboo and wood [20,21]. Changes in extraction content are related to the degradation of hemicellulose and cellulose. Volatiles, extractives, and water are products that result from hemicellulose and cellulose degradation [22]; the phenolic compounds, which can be very soluble in benzene-ethanol mixture, can also be formed in depolymerization reactions [23]. The decrease in the benzene-ethanol extractive content of the used BGP may be attributed to the dissolution of extractives in hot water flow and the degradation of hemicellulose.

The α-cellulose contents of FJBGP and SDBGP were 45.43% and 48.35%, respectively, and the lignin contents were 27.68% and 26.63%. These numbers were all significantly higher than those of control samples, which indicates that the degradation of α-cellulose and lignin by hot water flow was less significant than that of hemicellulose. α-Cellulose is a homogeneous polysaccharide consisting of the same type of glucosyl groups. It is also a straight-chain structural macromolecule without branched chain, therefore its thermal stability is relatively good. Lignin is a complex phenolic polymer synthesized from three alcohol monomers (namely *p*-coumaryl alcohol, coniferyl alcohol, and sinapyl alcohol) [24]. It has large molecular weight, few lyophilic groups, and is insoluble in water and general solvents. Lignin is the most stable compound for which thermal degradation is observed above 100 °C, although high temperatures can result in cleaved bonds within the lignin [25]. Because cellulose and lignin have fewer lyophilic groups and more stable chemical structures, they are more difficult to hydrolyze by hot water flow than hemicellulose. The increase of α-cellulose and lignin contents in SDBGP and FJBDP may largely be the result of the hydrolysis of benzene-ethanol extractive content and the degradation of hemicellulose.

The effects of hygrothermal environments on the chemical composition of BGP may be less conspicuous in the short term because the highest water temperature is usually below 65 °C. The actual extent of change in the chemical composition of used BGP was due to years of accumulation.

Our previous experiment studied the mechanical properties of used BGP that had been in service for nine years in cooling towers [2]; the reduction in the mechanical properties of SDBGP and FJBGP was mainly caused by the depolymerization reactions of polymers, which led to the reduction in hemicellulose, the most thermochemically sensitive component of bamboo. The degradation of hemicellulose is primarily responsible for initial strength loss [26,27].

### 3.2. Elemental Composition and Mass Loss

Elemental composition and mass loss of the samples are presented in Table 2. Results indicate that the hygrothermal environment led to a drop in oxygen content and an increase of carbon content of used BGP compared to control samples, which was in line with the results of heat-treated wood [21,28,29]. The decrease of the O/C ratio of used BGP manifested severe dehydration due to higher degradation of amorphous polysaccharides and/or higher amounts of carbonaceous materials within the bamboo structure, as reported for heat-treated wood [30–32]. The degradation of polysaccharides involves depolymerization by transglycosylation and dehydration reactions. As a result, the production of anhydro monosaccharides that could participate in subsequent reactions led to the formation of carbonaceous materials [17,33].

**Table 2.** Elemental composition and mass loss of used BGP and control samples.

| Sample | Carbon (%) | Oxygen (%) | Hydrogen (%) | O/C [a] | Mass Loss [b] (%) |
|--------|-----------|-----------|--------------|---------|-------------------|
| Control | 48.68 | 45.11 | 6.14 | 0.695 | - |
| FJBGP | 48.95 | 44.95 | 6.06 | 0.689 | 7.94 |
| SDBGP | 49.15 | 44.73 | 6.06 | 0.683 | 8.85 |

Note: [a] = atomic ratio; [b] = mass loss due to hygrothermal treatment.

Elaieb et al. reported that the O/C ratio was negatively correlated with mass loss and inferred that the O/C ratio is a good indicator for mass loss of wood after thermo-degradation [34]. Moreover, the O/C ratio is related to the content of hemicellulose [28]. The O/C ratio and hemicellulose contents of FJBGP were 0.689 and 20.84%, respectively, exceeding those of SDBGP, while the mass loss of FJBGP was 7.94%, lower than that of SDBGP. These indicate that hygrothermal degradation of SDBGP was more significant than of FJBGP.

### 3.3. FTIR Analysis

To investigate the changes of chemical structure that took place in used BGP, FTIR spectroscopy was applied in this study. Spectra of specimens are shown in Figure 2. Bamboo bears basic structural similarities to wood because of the similar chemical constituents. Therefore, characterization and assignment of IR peaks in bamboo was done by reference to wood, as previously described by Wang and Ren [35,36]. The assignments of the peaks to structural components were as follows: 1730 cm$^{-1}$ for unconjugated C=O in xylan (hemicellulose), 1603 cm$^{-1}$ and 1510 cm$^{-1}$ for aromatic skeleton in lignin, 1460 cm$^{-1}$ for CH$^3$ deformation in lignin and CH$^2$ bending in xylan, 1425 cm$^{-1}$ for CH$^2$ scissor vibration in cellulose, 1370 cm$^{-1}$ for C-H deformation in cellulose and hemicellulose, 1330 cm$^{-1}$ for C-H vibration in cellulose and C$_1$-O vibration in syringyl derivatives, 1240 cm$^{-1}$ for syringyl ring and C-O stretch in lignin and xylan, 1160 cm$^{-1}$ for C-O-C vibration in hemicellulose and cellulose, 1122 cm$^{-1}$ for aromatic skeletal and C-O stretch in lignin and cellulose, 1048 cm$^{-1}$ for C-O stretch in cellulose and hemicellulose, 897 cm$^{-1}$ for C-H deformation in cellulose [37–39].

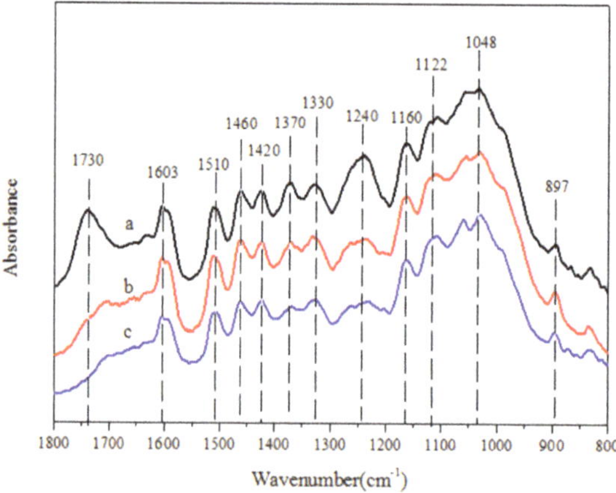


**Figure 2.** Fourier-transform infrared (FTIR) spectra of used BGP and control samples. (**a**) Control samples; (**b**) BGP collected from Fujian Province (FJBGP); (**c**) BGP collected from Shandong Province (SDBGP).

Hygrothermal treatment seemed to have a significant impact on the chemical structure of used BGP (Figure 2). Peaks at 1730, 1370, 1240, and 1048 cm$^{-1}$ for used BGP decreased compared to control samples. The intensity of the unconjugated C=O peak at 1730 cm$^{-1}$, often used as an important baseline for the degradation of hemicellulose [25], experienced a dramatic decrease, indicating that hemicellulose was degraded most severely. The degradation of unconjugated C=O was caused by deacetylation that occurred during hygrothermal treatment, which was a result of the cleavage of acetyl groups linked as ester groups to the hemicellulose [40]. The hygrothermal treatment caused a marked reduction in the 1370 cm$^{-1}$ peak in used BGP due to damage of C-H bonds in hemicellulose and cellulose. The peak intensity of 1240 cm$^{-1}$ of used BGP was lower than that of control samples owing to the degradation of syringyl ring and C-O bonds in lignin and xylan. The decrease in intensities of absorption bands at 1048 cm$^{-1}$ indicates the loss of C-O in cellulose and hemicellulose. The difference in FTIR spectra between FJBGP and SDBGP was rather slight, except for the absorption bands at 1048 cm$^{-1}$. The intensities of absorption bands at 1048 cm$^{-1}$ in SDBGP was lower than that of FJBGP, and the absorption peak at 1048 cm$^{-1}$ in SDBGP was split into two small peaks, which showed the degradation of C-O bonds in SDBGP was more noticeable than that of FJBGP. Although the content of cellulose and lignin increased, the C-H and C-O bonds in cellulose and syringyl ring in lignin were both partially degraded.

The intensity of the remaining bands for used BGP were almost the same with control samples, which indicates that functional groups corresponding to these peaks were relatively unaffected by hot water flow.

### 3.4. Color Changes

The color of lignocellulosic material is related to its chemical composition. Since the main chemical composition and structure of bamboo is similar to wood, the theory and method of analyzing color changes of wood can be applied to analyze bamboo. The effects of hygrothermal environment on color changes of the outer and inner skin of all samples are shown in Table 3. Control samples were light-colored. $L^*$ values of the outer and inner skin were 66.27 and 74.03, respectively. The surface color grew significantly darker after nine years of use. In terms of the outer and inner skin, the $L^*$ value of FJBGP decreased from 66.27 to 40.95 and from 74.03 to 36.30, respectively, and that of SDBGP decreased from 66.27 to 50.85 and from 74.03 to 41.95. Changes in color reflected alteration in the chemical composition and structure of BGP during hygrothermal treatment. Some studies reported that that a decrease in $L^*$ caused by a loss of lightness was possibly due to a degradation of hemicellulose and an increase in lignin content [41,42], as well as the degradation of oxygen-containing groups such as carboxyl groups and acetyl groups [19]. The darkening of used BGP may mainly be attributable to the degradation of acetyl groups in hemicellulose and oxidation of lignin.

**Table 3.** Color changes of used BGP and control samples.

| Samples | | CIELAB | | | |
|---|---|---|---|---|---|
| | | $L^*$ | $a^*$ | $b^*$ | $\Delta E^*$ |
| Outer skin | Control | 66.27c (5.08) | 1.70a (3.23) | 14.34c (12.48) | — |
| | FJBGP | 40.95a (4.91) | 3.40b (9.12) | 11.69a (4.19) | 25.51 |
| | SDBGP | 50.85b (5.68) | 4.44c (7.21) | 13.61b (3.38) | 15.68 |
| Inner skin | Control | 74.03c (2.34) | 5.16c (8.91) | 19.91c (4.82) | — |
| | FJBGP | 36.30a (3.5) | 2.87a (7.77) | 9.99a (8.51) | 39.08 |
| | SDBGP | 41.95b (12.35) | 3.84b (6.14) | 13.32b (9.67) | 32.78 |

Note: Values in parentheses are coefficient of variation. The different letters in the same column indicate a significant difference at the 0.05 level.

The $a^*$ value of FJBGP and SDBGP outer skin increased from 1.7 to 3.4 and from 1.7 to 4.44, respectively. However, the changes in the inner skin demonstrated an opposite trend, decreasing from 5.16 to 2.87 (FJBGP) and from 5.16 to 3.84 (SDBGP), respectively. The original brightly-colored lignocellulosic material turned dark and reddish under the hygrothermal environment, as a result of the volatilization of a large amount of colored extracts. However, the color indices of lignocellulosic material (near neutral color) only increase slightly due to the "darkening" effect caused by the rapid oxidation of the chemical composition [43]. Naturally, the original color of bamboo outer skin is green, while the inner skin is mostly red-brownish. After most of the extractives and chromophores in used BGP were hydrolyzed by hot water flow and some chemical composition were oxidized, the color of the outer and inner skin of used BGP become somewhat alike. In other words, the outer skin reddened while the inner skin grew comparatively green.

For the $b^*$ value, FJBGP decreased from 14.34 to 11.69 and from 19.91 to 9.99 for the outer and inner skin, respectively, and SDBGP dropped from 14.34 to 13.61 and from 19.91 to 13.32. The lower $b^*$ values of used BGP indicates that it became bluer compared to control samples. This specific change was induced by the modification of hemicellulose following hygrothermal treatment [15]. The total color difference ($\Delta E^*$) of used BGP's inner skin was greater compared to the outer skin because the inner skin underwent a greater change in terms of lightness. The $\Delta E^*$ was caused by the volatilization of color extracts and the oxidation, degradation and polymerization of the chemical composition, which also had a significant correlation with the content of holocellulose, $\alpha$-cellulose, and acid insoluble lignin [44].

## 4. Conclusions

The results of this study confirmed that the hygrothermal environment in cooling towers influenced the chemical and elemental composition, mass loss, chemical structure, and color changes of BGP. Statistically significant differences were observed in the content of hemicellulose, benzene-ethanol extractives, $\alpha$-cellulose, and lignin between used BGP and control samples. Specifically, there was a decrease of hemicellulose and benzene-ethanol extractive content in used BGP and an increase in $\alpha$-cellulose and lignin content. The O/C ratio of used BGP decreased in general, but it was higher in FJBGP than in SDBGP with less mass loss. The FTIR spectra showed the unconjugated C=O, C-H, and C-O bonds in hemicellulose; C-H and C-O bonds in cellulose; and the syringyl ring in lignin were partially degraded. However, the peak intensities and positions of the rest of the functional groups remained mostly unchanged. Color parameters such as lightness ($L^*$) and chromatic coordinates $a^*$ and $b^*$ of used BGP surface all decreased due to the hygrothermal environment's influence, with the exception of the $a^*$ value of the outer skin. The total color difference ($\Delta E^*$) of used BGP's inner skin was more substantial than that of the outer skin.

This study provides primary data for BGP used in industrial cooling towers. Still more work needs to be done in order to promote the application of BGP, such as optimizing its design, evaluating its service life, devising appropriate methods to alleviate the effects of hygrothermal aging, etc.

**Author Contributions:** L.-S.C., X.-X.M., and B.-H.F. conceived and designed the experiments; L.-S.C. performed the experiments; C.-H.F. analyzed the data; J.-P.L. contributed reagents/materials/analysis tools; L.-S.C. wrote the paper, and C.-H.F. revised it.

**Funding:** This research was funded by Foundation of International Centre for Bamboo and Rattan (grant numbers: 1632016015) and the Forestry Patent Industrialization Guidance Project (grant numbers: Forestry Patent 2017-2).

**Acknowledgments:** The authors acknowledge the financial supports of the Foundation of International Centre for Bamboo and Rattan (No. 1632019002) and the Forestry Patent Industrialization Guidance Project (Forestry Patent 2017-2).

**Conflicts of Interest:** The authors declare no conflict of interest.

## References

1. Seetharamu, K.N.; Swaroop, S. The effect of size on the performance of a fluidized bed cooling tower. *Wärme-und Stoffübertragung* **1991**, *26*, 17–21. [CrossRef]
2. Chen, L.S.; Fei, B.H.; Ma, X.X.; Lu, J.P.; Fang, C.H. Investigation of Bamboo Grid Packing Properties Used in Cooling Tower. *Forests* **2018**, *9*, 762. [CrossRef]
3. Lemouari, M.; Boumaza, M. Experimental investigation of the performance characteristics of a counterflow wet cooling tower. *Int. J. Therm. Sci.* **2010**, *49*, 2049–2056. [CrossRef]
4. Goshayshi, H.R.; Missenden, J.F. The investigation of cooling tower packing in various arrangements. *Appl. Therm. Eng.* **2000**, *20*, 69–80. [CrossRef]
5. Fang, C.H.; Jiang, Z.H.; Sun, Z.J.; Liu, H.R.; Zhang, X.B.; Zhang, R.; Fei, B.H. An overview on bamboo culm flattening. *Constr. Build. Mater.* **2018**, *171*, 65–74. [CrossRef]
6. Chen, Y.L.; Shi, Y.F.; Xie, D.X. Performance Comparison between Bamboo Grid Packing and PVC Film Packing and its Applications. *Power Stn. Aux. Equip.* **2016**, *37*, 37–41.
7. Slahor, J.J.; Hassler, C.C.; Degroot, R.C.; Gardner, D.J. Preservative Treatment Evaluation of Red Maple and Yellow-Poplar with ACQ-B. *For. Prod. J.* **1997**, *47*, 50–54.
8. Deng, Q.P.; Li, D.G.; Zhang, J.P. FTIR Analysis on Changes of Chemical Structure and Compostions of Waterlogged Archaeological Wood. *J. Northwest For. Univ.* **2008**, *23*, 149–153.

9. *GB/T 2677.8-1994 Fibrous Raw Material-Determination of Acid-insoluble Lignin*; General Administration of Quality Supervision, Inspection and Quarantine of the People's Republic of China; Standardization Administration of the People's Republic of China: Beijing, China, 1994.

10. *GB/T 2677.10-1995 Fibrous Raw Material-Determination of Holocellulose*; General Administration of Quality Supervision, Inspection and Quarantine of the People's Republic of China; Standardization Administration of the People's Republic of China: Beijing, China, 1995.

11. *GB/T 744-2004 Pulps-Determination of Alkali Resistance*; General Administration of Quality Supervision, Inspection and Quarantine of the People's Republic of China; Standardization Administration of the People's Republic of China: Beijing, China, 2004.

12. *GB/T 2677.6-1994 Fibrous Raw Material-Determination of Solvent Extractives*; General Administration of Quality Supervision, Inspection and Quarantine of the People's Republic of China; Standardization Administration of the People's Republic of China: Beijing, China, 1994.

13. Gelbrich, J.; Mai, C.; Militz, H. Evaluation of bacterial wood degradation by Fourier Transform Infrared (FTIR) measurements. *J. Cult. Herit.* **2012**, *13*, S135–S138. [CrossRef]

14. Schimleck, L.R.; Espey, C.; Mora, C.R.; Evans, R.; Taylor, A.; Muniz, G. Characterization of the wood quality of pernambuco (Caesalpinia echinata Lam) by measurements of density, extractives content, microfibril angle, stiffness, color, and NIR spectroscopy. *Holzforschung* **2009**, *63*, 457–463. [CrossRef]

15. Pandey, K.K. Study of the effect of photo-irradiation on the surface chemistry of wood. *Polym. Degrad. Stab.* **2005**, *90*, 9–20. [CrossRef]

16. Xu, Y.M. *Wood Science*; China Forestry Publishing House: Beijing, China, 2006.

17. Rowell, R.M. *Handbook of Wood Chemistry & Wood Composites*; CRC Press: Boca Raton, FL, USA, 2005.

18. Ma, Q.Z. *The Research on Utilization Approaches of High-Grade Resource Recovering of Bamboo Resources*; Central South University of Forestry and Technology: Changsha, China, 2011.

19. Windeisen, E.; Strobel, C.; Wegener, G. Chemical changes during the production of thermo-treated beech wood. *Wood Sci. Technol.* **2007**, *41*, 523–536. [CrossRef]

20. Meng, F.D.; Yu, Y.L.; Zhang, Y.M.; Yu, W.J.; Gao, J.M. Surface chemical composition analysis of heat-treated bamboo. *Appl. Surf. Sci.* **2016**, *371*, 383–390. [CrossRef]

21. Mohareb, A.; Sirmah, P.; Pétrissans, M.; Gérardin, P. Effect of heat treatment intensity on wood chemical composition and decay durability of Pinus patula. *Eur. J. Wood Wood Prod.* **2012**, *70*, 519–524. [CrossRef]

22. Hill, C.A.S. *Wood Modification: Chemical, Thermal and Other Processes*; John Wiley and Sons: Hoboken, NJ, USA, 2006.

23. Herrera, R.; Erdocia, X.; Llano-Ponte, R.; Labidi, J. Characterization of hydrothermally treated wood in relation to changes on its chemical composition and physical properties. *J. Anal. Appl. Pyrolysis* **2014**, *107*, 256–266. [CrossRef]

24. Yang, S.M.; Jiang, Z.H.; Ren, H.Q.; Fei, B.H.; Yao, W.B. Study status and development tendency of bamboo lignin. *Wood Process. Mach.* **2008**, *19*, 23–33.

25. Tjeerdsma, B.F.; Militz, H. Chemical changes in hydrothermal treated wood: FTIR analysis of combined hydrothermal and dry heat-treated wood. *Holz Roh-Werkst.* **2005**, *63*, 102–111. [CrossRef]

26. Levan, S.L.; Ross, R.J.; Winandy, J.E. *Effects of Fire Retardant Chemicals on the Bending Properties of Wood at Elevated Temperatures*; U.S. Department of Agriculture, Forest Service, Forest Products Laboratory: Madison, WI, USA, 1990.

27. Winandy, J.E. *Effects of Fire Retardant Treatments After 18 Months of Exposure at 150 °F (66 °C)*; Res. Note FPL-RN-0264; U.S. Department of Agriculture, Forest Service, Forest Products Laboratory: Madison, WI, USA, 1995.

28. Inari, G.N.; Pétrissans, M.; Pétrissans, A.; Gérardin, P. Elemental composition of wood as a potential marker to evaluate heat treatment intensity. *Polym. Degrad. Stab.* **2009**, *94*, 365–368. [CrossRef]

29. Candelier, K.; Dumarcay, S.; Petrissans, A.; Desharnais, L.; Gerardin, P. Comparison of chemical composition and decay durability of heat treated;wood cured under different inert atmospheres: Nitrogen or vacuum. *Polym. Degrad. Stab.* **2013**, *98*, 677–681. [CrossRef]

30. Alén, R.; Kotilainen, R.; Zaman, A. Thermochemical behavior of Norway spruce (Picea abies) at 180–225 °C. *Wood Sci. Technol.* **2002**, *36*, 163–171. [CrossRef]

31. Inari, G.N.N.; Petrissans, M.; Lambert, J.; Ehrhardt, J.J.; Gérardin, P. XPS characterization of wood chemical composition after heat-treatment. *Surf. Interface Anal.* **2010**, *38*, 1336–1342. [CrossRef]

32. Nguila, I.G.; Steeve, M.; Stéphane, D.; Mathieu, P.; Philippe, G. Evidence of char formation during wood heat treatment by mild pyrolysis. *Polym. Degrad. Stab.* **2007**, *92*, 997–1002.

33. Fabbri, D.; Chiavari, G.; Prati, S.; Vassura, I.; Vangelista, M. Gas chromatography/mass spectrometric characterisation of pyrolysis/silylation products of glucose and cellulose. *Rapid Commun. Mass Spectrom.* **2010**, *16*, 2349–2355. [CrossRef]

34. Elaieb, M.; Candelier, K.; Pétrissans, A.; Dumarçay, S.; Gérardin, P.; Pétrissans, M. Heat treatment of Tunisian soft wood species: Effect on the durability, chemical modifications and mechanical properties. *Maderas Cienc. Tecnol.* **2015**, *17*, 699–710. [CrossRef]

35. Wang, X.; Ren, H. Comparative study of the photo-discoloration of moso bamboo (Phyllostachys pubescens Mazel) and two wood species. *Appl. Surf. Sci.* **2008**, *254*, 7029–7034. [CrossRef]

36. Wang, X.Q.; Ren, H.Q. Surface deterioration of moso bamboo (Phyllostachys pubescens) induced by exposure to artificial sunlight. *J. Wood Sci.* **2009**, *55*, 47–52. [CrossRef]

37. Pandey, K.K.; Pitman, A.J. FTIR studies of the changes in wood chemistry following decay by brown-rot and white-rot fungi. *Int. Biodeterior. Biodegrad.* **2003**, *52*, 151–160. [CrossRef]

38. Sun, B.; Liu, J.; Liu, S.; Yang, Q. Application of FT-NIR-DR and FT-IR-ATR spectroscopy to estimate the chemical composition of bamboo (Neosinocalamus affinis Keng). *Holzforschung* **2011**, *65*, 689–696. [CrossRef]

39. Tomak, E.D.; Topaloglu, E.; Gumuskaya, E.; Yildiz, U.C.; Ay, N. An FT-IR study of the changes in chemical composition of bamboo degraded by brown-rot fungi. *Int. Biodeterior. Biodegrad.* **2013**, *85*, 131–138. [CrossRef]

40. Carrasco, F.; Roy, C. Kinetic study of dilute-acid prehydrolysis of xylan-containing biomass. *Wood Sci. Technol.* **1992**, *26*, 189–208. [CrossRef]

41. Shangguan, W.; Gong, Y.; Zhao, R.; Ren, H. Effects of heat treatment on the properties of bamboo scrimber. *J. Wood Sci.* **2016**, *62*, 383–391. [CrossRef]

42. Bekhta, P.; Proszyk, S.; Krystofia, T. Colour in short-term thermo-mechanically densified veneer of various wood species. *Eur. J. Wood Wood Prod.* **2014**, *72*, 785–797. [CrossRef]

43. Zhang, S.J.; Zhang, S.C.; Zhang, S.H. Effect of hot treatment on wood color of different treespecies. *For. Technol.* **1996**, *21*, 44–46.

44. Zhang, Y.M. *Study on the Effect of Color and Physical-Mechanical Properties for Heat-Treated Bamboo*; Chinese Academy of Forestry: Beijing, China, 2010.

MDPI

St. Alban-Anlage 66

4052 Basel

Switzerland

Tel. +41 61 683 77 34

Fax +41 61 302 89 18

www.mdpi.com

*Forests* Editorial Office

E-mail: forests@mdpi.com

www.mdpi.com/journal/forests